RANDOM TENSORS

Random Tensors

Razvan Gurau

CNRS, École Polytechnique, Université Paris-Saclay, France

OXFORD
UNIVERSITY PRESS

OXFORD
UNIVERSITY PRESS

Great Clarendon Street, Oxford, OX2 6DP,
United Kingdom

Oxford University Press is a department of the University of Oxford.
It furthers the University's objective of excellence in research, scholarship,
and education by publishing worldwide. Oxford is a registered trade mark of
Oxford University Press in the UK and in certain other countries

First Edition published in 2017

Published in the United States of America by Oxford University Press
198 Madison Avenue, New York, NY 10016, United States of America

British Library Cataloguing in Publication Data
Data available

Library of Congress Control Number: 2016942564

ISBN 978–0–19–878793–8

To Delia

Contents

Chapter 1

Introduction

Random tensor models generalize random matrix models and provide a framework for the study of random geometries in any dimension.

Random matrices [59, 88, 128] have been introduced by Wishart [165] for the statistical analysis of large samples and first used in physics by Wigner [164] for the study of the spectra of heavy nuclei. They are probability distributions for $N \times N$ random variables $\mathbb{M}_{ab}$, which are invariant under the conjugation of $\mathbb{M}$ by the unitary group.

The moments and partition function of a matrix model can be evaluated as sums over ribbon Feynman graphs, with weights fixed by the Feynman rules. The relation between random matrices and random geometries comes from the fact that these ribbon graphs are dual to surfaces. As the probability distribution of the surfaces (weights of the graphs) is fixed by the Feynman rules, random matrices yield a *canonical* theory of random two-dimensional surfaces.

Of course this evaluation in terms of Feynman graphs suffers from the well known flaw of perturbative expansions in quantum field theory: it diverges. However, the situation is much more subtle in matrix models than in usual quantum field theory. In his seminal work [108] 't Hooft showed that the size of the matrix N plays a distinctive role in this case. Indeed, a matrix model is endowed with a natural small parameter, $1/N$, which is absent in usual quantum field theory, and one can reorganize the perturbative expansion of a matrix model as a series in $1/N$. This series turns out to be indexed by the genus [108] of the graphs (or equivalently of their dual surfaces).

At leading order in $1/N$ the planar graphs [37, 53] dominate. Planar graphs can be arbitrarily large (i.e. can have arbitrarily many vertices) but can be explicitly enumerated [47, 50, 156] and form a summable family. This holds order by order in $1/N$ and constitutes the fundamental feature of matrix models: the $1/N$ expansion reorganizes the perturbative series into nontrivial but manageable packages of graphs of fixed genus.

The large N limit is governed by the behavior of the planar series which has a nonzero radius of convergence. When approaching the boundary of the domain of analyticity of some series, convergence is lost due to the contribution of the higher order terms. It follows that, tuning the coupling constants of a matrix model to some critical values, infinite planar graphs will dominate and the model will undergo a phase transition to a continuum theory of random, infinitely refined, surfaces [54, 113].

Matrix models offer a great deal more. It turns out that they encode a complete theory of infinitely refined random surfaces (i.e. two-dimensional quantum gravity) coupled to matter [6, 35, 36, 54, 59, 60, 72, 114, 115, 125]. Subsequently, the critical behavior of matter in random geometry can be mapped onto the one of matter in fixed geometry through the KPZ correspondence [116, 62, 64]. Subleading terms in the $1/N$ series are accessed in the double scaling limit [38, 63, 78] which corresponds to two-dimensional quantum gravity at finite Newton's constant. Random matrices are also crucial in noncommutative quantum field theory [83, 101], and in particular for the study of the well known Grosse Wulkenhaar model, which has been shown asymptotically safe at all orders [61] and solved in the planar sector [84, 85, 86, 87].

From a more mathematical perspective, the $1/N$ expansion of random matrices has been analyzed in detail [65, 88, 158], related to enumerative combinatorics [47, 50, 156] and led to the free probability theory [159, 160, 161]. The emergent infinitely refined random geometry, know in the mathematical literature as the Brownian map [120, 121, 122], has been rigorously studied. Surprisingly, although one deals with random surfaces, the Brownian map has Hausdorff dimension 4. Albeit not proven, it is a widely believed conjecture that the spectral dimension of the Brownian map is 2.

One can bypass the perturbative expansion in matrix models and use for instance the eigenvalue decomposition or the Schwinger Dyson equations [68, 67, 66, 126, 10, 89] to study them. However, the link between matrix models and random surfaces must be revisited in this case: the $1/N$ expansion in terms of surfaces seems at first sight a perturbative artifact. In order to obtain a theory of random surfaces one first expands in perturbations and subsequently reorganizes the perturbative series in powers of $1/N$. As the original perturbative expansion is not summable, what is the precise meaning of this repackaging? As recent results show [100], one can in some cases extend the link between random surfaces and matrix models beyond the perturbation series. Generalizing the results of [100] to more general models is an ongoing research program that has recently been fueled by the advent of tensor models.

A number of very interesting and difficult questions concerning matrix models are still open. For example, the phase transition to the continuum takes place when tuning the coupling constants to *negative* critical values, in a range of parameters where the models are unstable. This tuning is meaningful after restricting to the leading order planar series (which is absolutely convergent), but what, if any, is the meaning of this tuning to criticality beyond perturbation theory?

The steady development of matrix models over the past decades is one of the most impressive achievements of modern theoretical physics. This success inspired their generalization in the 1990s to random tensor models [3, 82, 153, 155, 154] designed to describe random geometries in higher dimensions. However, for twenty years random tensors essentially failed to match the success of random matrices because, for a long time, a $1/N$ expansion for tensors could not be found.

Generalizing random matrices, random tensors generate Feynman graphs that can be interpreted as topological spaces. However, the spaces generated in this way are quite nontrivial: one obtains not only all the manifolds, but also all the pseudo manifolds of a fixed dimension[1]. Several models [34, 133, 134, 9, 8], mainly under the guise of "group field theories" (see [135, 136] for more recent developments in this direction), which are tensor models decorated by extra data, have been proposed in the attempt to tackle this problem. Although different in some important respects (for instance the metric of the underlying space is encoded quite differently), these models do not bring any insight into the problem of the $1/N$ expansion.

Starting in 2009 [93, 91, 90], new results and techniques led to the discovery of the $1/N$ expansion for tensors [94, 102, 96]. They form the backbone of the modern theory of random tensors, which is the object of this book.

The $1/N$ expansion was first obtained for the colored models [93, 91, 90] and subsequently extended to all invariant tensor models [31]. The graphs of the new tensor models always encode genuine D-dimensional cellular complexes hence D-dimensional topological spaces. The $1/N$ series [94, 102, 96, 26, 52, 98] is indexed by a positive integer called the *degree*, which plays in higher dimensions the same role the genus played for matrix models (unlike the genus, however, the degree is *not* a topological invariant).

In the large N limit a specific family of graphs, called *melonic* [30, 106, 99], is selected. The melonic graphs triangulate the D-dimensional sphere in any dimension [94, 102, 96] and form a summable family. Like matrix models, tensor models undergo a phase transition to a theory of continuous random spaces [30, 31] when tuning to criticality, and exhibit a double scaling limit [107, 51, 33]. In stark contrast to matrix models, for $3 \leq D \leq 5$, in this limit only an exponentially bounded family of triangulations of the sphere contribute. For $D \geq 6$ the family selected in the double scaling limit is neither exponentially bounded nor restricted to the spherical topology. The $1/N$ expansion of tensor models exhibits very strong universality properties [99, 92, 97] and has been put on a firm mathematical footing [98] in the sense of constructive field theory [76] by means of the loop vertex expansion [141, 142, 124, 103].

The study of the ensuing theory of random geometries in higher dimensions has attracted increased attention [105, 16, 148, 43, 44, 32, 22, 95, 92, 97, 118, 27, 25, 29, 17, 14, 7, 13]. Breaking the unitary invariance of the quadratic part, tensor models become tensor field theories [18, 19, 12, 11, 150, 15, 151, 152, 46, 41, 45, 17, 12, 19, 14, 20], and a regime parallel to the large N limit of tensor models is reached via a genuine renormalization group flow, and not by sending the parameter N to infinity.

The scope of the present work is to present a self-contained, *ab initio* introduction to random tensors. Most of the results presented are therefore explained in full detail and the reader is supposed to have no prior familiarity with the field. As some of these results are quite technical, many of the proofs can be skipped at a first read.

Random tensors are the straightforward generalization of random matrices and ideas coming from tensor models allow one to obtain new results for random matrices [100], which are not accessible by other means. It should be stressed, however, that the

[1]A further complication comes from the fact that the initial models proposed only generate 2-complexes.

$D = 2$ case of matrices is very special. Indeed, random tensors behave by and large quite differently from random matrices. The origin of this resides in the fact that the melonic family, which dominates in tensor models, is very different from (and in fact much more restricted than) the planar family dominating matrix models.

This book is divided into two parts. In the first part, consisting of Chapters 2, 3, 4, 5 and 6, we introduce the general framework and the main results on random tensors.

Chapter 2 sets the stage of this book. In this chapter we introduce the basic notions of trace invariants built out of a tensor and its complex conjugate and we discuss the representation of these invariants as edge colored graphs. We subsequently introduce the invariant probability measures, the main object of interest for the rest of this book.

In Chapter 3 we study in detail the edge colored graphs. We show that each graph is a D-complex (whose two cells, called *faces*, will play an important role in tensor models) dual to a vertex colored triangulation. We then discuss open graphs and their boundary graphs. Finally, we introduce a positive integer associated to a connected edge colored graph, the *degree*, which is central in tensor models.

In Chapter 4 we perform the full classification of graphs at fixed degrees. It turns out that for any fixed degree one obtains an infinite but exponentially bounded family of graphs whose generating function can be written explicitly.

In Chapter 5 we detail the graphs of degree zero. These graphs, called *melonic*, play in higher dimension a role similar to the one played by the planar graphs in two dimensions. We study the geometry of the infinite family of melons and show that, when endowed with a certain metric, this family reproduces the behavior of branched polymers.

In Chapter 6 we prove that invariant tensor measures have a very strong universality property. Assuming a certain uniform bound in N on the cumulants of an invariant measure, the measure becomes Gaussian in the large N limit. We discuss at the end of the chapter under which conditions one can escape this universal behavior.

In the second part of this book, consisting of Chapters 7, 8, 9, 10 and 11 we present in detail specific examples of random tensors models.

Chapter 7 presents a quick review of some of the feature of matrix models, which will be generalized for tensor models in the subsequent chapters.

In Chapter 8 we consider the most general perturbed Gaussian measure such that the perturbation scales with N at the same rate as the Gaussian part. We show that, in the sense of *perturbation series* (i.e. term by term in the perturbative expansion), the partition function and expectations of such measures admit a $1/N$ expansion and that furthermore such measures are properly uniformly bounded. In particular this means that, in the sense of perturbation theory, the universality theorem applies to all these measures.

In Chapter 9 we study at length the Gaussian measure with a quartic perturbation. In this case we are able to establish analyticity results for the partition function and the cumulants in the *constructive sense*, that is for the full resummed functions and not term by term for their perturbative expansions. We show that the $1/N$ expansion and the proper uniform boundedness hold, and that, moreover, in the case of the quartic model, the perturbative expansion of Chapter 8, while divergent, is Borel summable.

In Chapter 10 we present the double scaling limit of the quartic tensor model. When tuning the coupling constant of the quartic model to some critical value, the terms in the $1/N$ series diverge. Crucially, the more a term is suppressed in $1/N$ the more it diverges at criticality. We are thus able to identify a regime in which the approach to criticality is tuned with the large N limit and an infinity of terms in the $1/N$ series contributes.

In Chapter 11 we analyze a quartic tensor model with one interaction for a tensor of arbitrary rank. We show that in this case the critical point corresponds to a phase transition in the tensor model associated to a breaking of the unitary symmetry. We discuss the model in the two phases and prove that, in a double scaling limit, the symmetric phase corresponds to a theory of infinitely refined random surfaces, while the broken phase corresponds to a theory of infinitely refined random nodal surfaces. At leading order in the double scaling limit planar surfaces dominate in the symmetric phase, and planar nodal surfaces dominate in the broken phase.

Finally, in Chapter 12 we discuss the interpretation of tensor models as generating Euclidean dynamical triangulations and draw the conclusions of this book.

Part I

Toolbox for random tensors

Chapter 2

Preliminaries

The basic building blocks of random tensor models are generic tensors, that is tensors having no symmetry properties under permutations of their indices.

In this first chapter we first recall some basic definitions and properties of generic tensors. We then study the invariants one can build out of a tensor and its complex conjugate (its dual) and introduce a complete set of invariants, called *trace invariants*, which are canonically represented as edge colored graphs. Finally, we briefly introduce the invariant probability measures encoding random tensors.

2.1 Tensors

Let us denote $\mathcal{H}$ a complex Hilbert space of dimension N, $\langle \cdot, \cdot \rangle$ the sesquilinear scalar product of $\mathcal{H}$ and $\mathcal{H}^\vee$ the dual of $\mathcal{H}$. The dual space $\mathcal{H}^\vee$ is identified with the complex conjugate $\bar{\mathcal{H}}$ via the conjugate linear isomorphism:

$$z \to z^\vee(\cdot) = \langle z, \cdot \rangle \ .$$

Let $\{\psi_a \,|\, a = 1, \dots N\}$ be an orthonormal basis in $\mathcal{H}$ and $\psi^a \equiv \psi_a^\vee(\cdot) = \langle \psi_a, \cdot \rangle$ its dual basis. The components (in the dual basis $\{\psi^a\}_{a=1\dots N}$) of the dual of a vector are the complex conjugate of the components (in the direct basis $\{\psi_a\}_{a=1\dots N}$) of that vector:

$$\Big(\sum_{a=1}^{N} z^a \psi_a \Big)^\vee (\cdot) = \sum_{a=1}^{N} \overline{z^a} \, \langle \psi_a, \cdot \rangle \equiv \sum_{a=1}^{N} (z^\vee)_a \psi^a \ .$$

A covariant tensor of rank D is a multi linear form on a tensor product of D distinct Hilbert spaces $\mathcal{H}_c$, $c = 1, \dots D$:

$$\mathbf{T} : \mathcal{H}_1 \otimes \cdots \otimes \mathcal{H}_D \to \mathbb{C} \ .$$

We denote N^c the dimension of $\mathcal{H}_c$. Choosing a basis $\{[\psi^c]_{a^c}\}_{a^c=1,\dots N^c}$ in each $\mathcal{H}_c$, we denote the components (entries) of the tensor in the tensor product basis by:

$$\mathbb{T}_{a^1\dots a^D} \equiv \mathbf{T}\big([\psi^1]_{a^1},\dots,[\psi^D]_{a^D}\big) ,$$
$$\mathbf{T} = \sum_{a^c=1\dots N^c,\ \forall c} \mathbb{T}_{a^1\dots a^D}\ [\psi^1]^{a^1} \otimes\dots\otimes [\psi^D]^{a^D} .$$

A generic tensor $\mathbf{T}$ has no symmetry properties under permutation of its arguments. In fact, as each one of the arguments belongs to a different Hilbert space, it might very well be that the dimensions N^c are all different. Consequently, the components $\mathbb{T}_{a^1\dots a^D}$ of such a tensor have no symmetry properties either. It follows that the indices $a^1,\dots a^D$ have a well defined position. We call the position of an index its *color*, and we denote $\mathcal{D}$ the set of colors $\{1,\dots D\}$.

A tensor can be seen as a linear map between tensor product spaces. There are in fact as many choices as there are subsets $\mathcal{C}\subset\mathcal{D}$. Let us denote:

$$a^{\mathcal{C}} \equiv (a^c, c\in\mathcal{C}) ,$$

the indices with colors in $\mathcal{C}$. The complementary indices are then naturally denoted $a^{\mathcal{D}\setminus\mathcal{C}} \equiv (a^c, c\in\mathcal{D}\setminus\mathcal{C})$. In keeping with this notation, we will often denote the D-uple of indices of the tensor, $(a^1,\dots a^D)$, by $a^{\mathcal{D}}$. From now, when indexing a sum, a^c is understood to run between 1 and N^c.

For any subset $\mathcal{C}$ the tensor can be seen as a multi linear map:

$$\mathbf{T} : \bigotimes_{c\in\mathcal{C}} \mathcal{H}_c \to \bigotimes_{c\in\mathcal{D}\setminus\mathcal{C}} \bar{\mathcal{H}}_c .$$

The action of this map on an element $z^{\mathcal{C}}$ in the tensor product vector space $\bigotimes_{c\in\mathcal{C}}\mathcal{H}_c$ is written in the tensor product basis as:

$$z^{\mathcal{C}} = \sum_{a^c,\ \forall c\in\mathcal{C}} z^{\{a_c\}_{c\in\mathcal{C}}} \bigotimes_{c\in\mathcal{C}} [\psi^c]_{a_c} ,$$
$$\mathbf{T}(z^{\mathcal{C}}) = \sum_{a^c, c\in\mathcal{D}} \mathbb{T}_{a^1\dots a^D}\ z^{\{a_c\}_{c\in\mathcal{C}}} \Big(\bigotimes_{c\in\mathcal{D}\setminus\mathcal{C}} [\psi^c]^{a_c}\Big) .$$

The matrix elements of this linear map (in the tensor product basis) are:

$$T_{a^{\mathcal{D}\setminus\mathcal{C}}a^{\mathcal{C}}} \equiv \mathbb{T}_{a^1\dots a^D} \ \text{ with } a^c \in a^{\mathcal{C}} \cup a^{\mathcal{D}\setminus\mathcal{C}} .$$

As we deal with complex inner product spaces, the dual tensor $\mathbf{T}^\vee$ is a contravariant tensor of rank D defined by:

$$\mathbf{T}^\vee\big([z^1]^\vee,\dots[z^D]^\vee\big) \equiv \overline{\mathbf{T}(z^1,\dots z^D)} .$$

Taking into account that:

$$\sum_{a^{\mathcal{D}}} \overline{\mathbb{T}_{a^1\dots a^D}}\ \overline{[z^1]^{a^1}}\dots\overline{[z^D]^{a^D}} = \sum_{a^{\mathcal{D}}} \overline{\mathbb{T}_{a^1\dots a^D}}\ ([z^1]^\vee)_{a^1}\dots[z^D]^\vee)_{a^D} ,$$

the dual tensor and its components write:

$$\mathbf{T}^{\vee} = \sum_{a^1,\dots a^D} \overline{\mathbb{T}_{a^1\dots a^D}}\ \psi_{a^1} \otimes \dots \otimes \psi_{a^D} ,$$
$$(\mathbf{T}^{\vee})^{a^1\dots a^D} = \overline{\mathbb{T}_{a^1\dots a^D}} .$$

It follows that the components (in the dual basis) of the dual tensor are the complex conjugate of the components (in the direct basis) of the tensor. The dual tensor can be seen as a conjugated linear map:

$$\mathbf{T}^{\vee} : \bigotimes_{c\in\mathcal{D}\setminus\mathcal{C}} \bar{H}_c \to \bigotimes_{c\in\mathcal{C}} H_c ,$$
$$(T^{\vee})^{a^{\mathcal{C}} a^{\mathcal{D}\setminus\mathcal{C}}} \equiv \overline{\mathbb{T}_{a^1\dots a^D}} \ \text{ with } a^c \in a^{\mathcal{C}} \cup a^{\mathcal{D}\setminus\mathcal{C}} ,$$

that is the adjoint of the map T.

From now on we denote the components of the dual tensor by $\bar{\mathbb{T}}_{\bar{a}_1\dots\bar{a}_D}$. We will write all the indices in subscript and we add a bar for the contravariant indices in order to distinguish them from the covariant ones. We also denote $\delta_{a^{\mathcal{C}}\bar{a}^{\mathcal{C}}}$ the product $\prod_{c\in\mathcal{C}} \delta_{a^c\bar{a}^c}$. Indices are always understood to be listed in increasing order of their colors.

Under an *independent* change of basis in each $\mathcal{H}_c$, $[\chi^c]_{a^c} = \sum_{b^c} U^{(c)}_{a^c b^c} [\psi^c]_{b^c}$, the components of the tensor and of its dual transform as:

$$\mathbb{T}'_{a^1\dots a^D} = \sum_{b^1\dots b^D} U^{(1)}_{a^1 b^1} \cdots U^{(D)}_{a^D b^D}\ \mathbb{T}_{b^1\dots b^D} ,$$
$$\bar{\mathbb{T}}'_{\bar{a}^1\dots\bar{a}^D} = \sum_{\bar{b}^1\dots\bar{b}^D} \bar{U}^{(1)}_{\bar{a}^1\bar{b}^1} \cdots \bar{U}^{(D)}_{\bar{a}^D\bar{b}^D}\ \bar{\mathbb{T}}_{\bar{b}^1\dots\bar{b}^D} , \tag{2.1}$$

that is they transform with the *external* tensor product of D fundamental (respectively anti fundamental) representations of the unitary groups $U(N^c)$. We emphasize that the unitary operators $U^{(1)}, \dots U^{(D)}$ above are all distinct, and in fact can act on spaces of different dimensions.

A tensor of rank $D \geq 3$ is a multidimensional generalization of a matrix. Notions like the determinant of a matrix generalize naturally to tensors. The hyperdeterminant [74] of a tensor $\mathbf{T}$ is a polynomial in the tensor entries, which is zero if and only if there exist D nontrivial complex vectors $0 \neq [z^c] \in \mathcal{H}_c$, $c \in \mathcal{D}$ such that:

$$[z^c] = \sum_{a^c=1}^{N^c} [z^c]_{a^c} [\psi^c]_{a^c} ,$$
$$\frac{\partial \mathbf{T}([z^1], \dots [z^D])}{\partial [z^c]_{a^c}} = 0 , \qquad \forall c, \forall a^c .$$

The hyperdeterminant is said to be of format $(N^1 - 1, \dots, N^D - 1)$ and it exists and is unique if and only if [74]:

$$\max_c \{N^c - 1\} \leq \sum_{c=1}^{D} (N^c - 1) - \max_c \{N^c - 1\} .$$

A format is called a boundary format in the case of equality, and the hyperdeterminant greatly simplifies in that case. The hyperdeterminant is an example of a polynomial in the tensor entries which is invariant [74] under a general change of basis as in Eq. (2.1).

From now on, for simplicity, we restrict to $N^c = N, \forall c$, but all the results we present in this book can be generalized to arbitrary formats.

2.2 Invariants

The hyperdeterminant is an example of an invariant polynomial. We are interested in understanding in more detail the structure of such invariant polynomials. The complete study of such invariance is a vast chapter in mathematics (see for instance [132]) and a detailed introduction is beyond the scope of this book. However, as we will be using invariants throughout this book we present here a very selective introduction to this topic. In particular, in this section we will identify a *complete set* of invariants, such that any invariant polynomial can be decomposed as a linear combination of invariants in the set. However, we will see that, although very convenient, the invariants in this set are not independent at finite N.

Let us consider an arbitrary invariant polynomial:

$$I(\mathbb{T},\bar{\mathbb{T}}) = \sum_{k,\bar{k}\geq 0}\sum_{a,\bar{a}} I_{\{a_j^{\mathcal{D}},\bar{a}_j^{\mathcal{D}}\}_{j=1,\dots k}} \Big(\prod_{j=1}^{k}\mathbb{T}_{a_j^{\mathcal{D}}}\Big)\Big(\prod_{j=1}^{\bar{k}}\bar{\mathbb{T}}_{\bar{a}_j^{\mathcal{D}}}\Big) ,$$

where the sum runs over all the as and all the $\bar{a}$s from 1 to N (and we recall that $a^{\mathcal{D}}$ denotes the D-uple of indices $(a^1,\dots a^D)$).

Being an invariant, $I(\mathbb{T},\bar{\mathbb{T}})$ does not change under the action of the unitary group, and in particular equals its average. Denoting $[dU]$ the Haar measure on the unitary group, we have:

$$I(\mathbb{T},\bar{\mathbb{T}}) = \sum_{k,\bar{k}\geq 0}\Bigg[\sum_{a,\bar{a},b,\bar{b}} I_{\{a_j^{\mathcal{D}},\bar{a}_j^{\mathcal{D}}\}_{j=1,\dots k}} \int_{U(N)^{\otimes D}}[dU^{(1)}]\dots[dU^{(D)}] \tag{2.2}$$
$$\times\Big(\prod_{j=1}^{k} U^{(1)}_{a_j^1 b_j^1}\cdots U^{(D)}_{a_j^D b_j^D}\mathbb{T}_{b_j^1\dots b_j^D}\Big)\Big(\prod_{j=1}^{\bar{k}}\bar{U}^{(1)}_{\bar{a}_j^1\bar{b}_j^1}\cdots\bar{U}^{(D)}_{\bar{a}_j^D\bar{b}_j^D}\bar{\mathbb{T}}_{\bar{b}_j^1\dots\bar{b}_j^D}\Big)\Bigg] .$$

The integrals over the unitary group can be explicitly evaluated in terms of the Weingarten functions [163, 48, 49]:

$$\int_{U(N)}[dU]\prod_{j=1}^{k}U_{a_jb_j}\prod_{j=1}^{\bar{k}}\bar{U}_{\bar{a}_j\bar{b}_j} = \delta_{k\bar{k}}\sum_{\sigma,\tau\in\mathfrak{S}(k)}\Big(\prod_{j=1}^{k}\delta_{a_j\bar{a}_{\sigma(j)}}\delta_{b_j\bar{b}_{\tau(j)}}\Big)\,\mathrm{Wg}(N,\sigma\tau^{-1}) , \tag{2.3}$$

where $\mathfrak{S}(k)$ denotes the set of permutations over k elements and $\mathrm{Wg}(N,\sigma\tau^{-1})$ is a Weingarten function. These functions are given explicitly as [48, 49]:

$$\mathrm{Wg}(N,\sigma\tau^{-1}) = \frac{1}{k!^2}\sum_{\lambda\vdash k}\frac{\chi^{\lambda}(1)^2\chi^{\lambda}(\sigma\tau^{-1})}{s_{\lambda,N}(1)} ,$$

where the sum runs over the integer partitions λ of k, χ^λ is the character of the symmetric group corresponding to λ and $s_{\lambda,N}(x)$ is the Schur polynomial, hence $s_{\lambda,N}(1)$ is the dimension of the irreducible representation of $U(N)$ associated to λ. The exact form of the Weingarten functions plays a secondary role in tensor models.

A permutation τ of k elements, $\tau \in \mathfrak{S}(k)$ is said to have *cycle structure*:

$$1^{C_1(\tau)} \dots q^{C_q(\tau)} \dots ,$$

or simply $C_1(\tau), \dots C_q(\tau), \dots$ if it has $C_q(\tau)$ cycles of length q. In particular $k = \sum_{q\geq 1} qC_q(\tau)$. The Weingarten functions $\mathrm{Wg}(N, \sigma\tau^{-1})$ depend only on the cycle structure of the permutation $\sigma\tau^{-1}$, and moreover [48, 49]:

$$\lim_{N\to\infty} N^{2k-C(\sigma\tau^{-1})}\, \mathrm{Wg}(N,\sigma\tau^{-1}) = \prod_{q\geq 1}\left[(-1)^{q-1}\frac{1}{q}\binom{2q-2}{q-1}\right]^{C_q(\sigma\tau^{-1})} ,$$
$$\mathrm{Wg}(N,(1)) = \frac{1}{N} , \tag{2.4}$$

where $C(\sigma\tau^{-1}) = \sum_{q\geq 1} C_q(\sigma\tau^{-1})$ denotes the total number of cycles of the permutation $\sigma\tau^{-1}$ and (1) is the unique permutation of one element in cycle notation. A self contained physicists derivation of these two properties of the Weingarten functions is give in Appendix A.

Substituting Eq. (2.3) into Eq. (2.2) and summing over the indices a and $\bar{a}$, we obtain that any invariant can be written as:

$$I(\mathbb{T},\bar{\mathbb{T}}) = \sum_{k\geq 0} \sum_{\tau^{(1)}\dots\tau^{(D)}\in\mathfrak{S}(k)} I_k(\tau^{(1)},\dots\tau^{(D)})$$
$$\times \sum_{b,\bar{b}}\Big(\prod_{j=1}^{k} \mathbb{T}_{b_j^{\mathcal{D}}}\bar{\mathbb{T}}_{\bar{b}_j^{\mathcal{D}}}\Big)\Big(\prod_{c=1}^{D}\prod_{j=1}^{k}\delta_{b_j^c \bar{b}^c_{\tau^{(c)}(j)}}\Big) , \tag{2.5}$$

for some coefficients $I_k(\tau^{(1)},\dots\tau^{(D)})$. The second line in Eq. (2.5) is itself an invariant, henceforth called a *trace invariant* [31, 99]. Trace invariants are built by contracting in all possible ways pairs of covariant and contravariant indices in a product of tensor entries. We use the shorthand notation $\tau^{\mathcal{D}}$ to designate D permutations $(\tau^{(1)}\dots\tau^{(D)})$ of k elements.

Remark 2.1. We have the following remark.

Invariance under conjugation. The trace invariants are invariant under a simultaneous left right multiplication of all the permutations $\tau^{(c)}$ by two fixed permutations $\bar{\pi}$ and π. In order to show this we consider the invariant:

$$\sum_{b,\bar{b}}\Big(\prod_{j=1}^{k} \mathbb{T}_{b_j^{\mathcal{D}}}\bar{\mathbb{T}}_{\bar{b}_j^{\mathcal{D}}}\Big)\Big(\prod_{c=1}^{D}\prod_{j=1}^{k}\delta_{b_j^c \bar{b}^c_{\tau^{(c)}(j)}}\Big) .$$

As all the indices are just dummy summation variables, let us first relabel $b_j^{\mathcal{D}}$ as $b_{\pi(j)}^{\mathcal{D}}$. As this relabeling does not concern the indices $\bar{b}$, the invariant is written:

$$\sum_{b,\bar{b}}\Big(\prod_{j=1}^{k} \mathbb{T}_{b_{\pi(j)}^{\mathcal{D}}}\bar{\mathbb{T}}_{\bar{b}_j^{\mathcal{D}}}\Big)\Big(\prod_{c=1}^{D}\prod_{j=1}^{k}\delta_{b^c_{\pi(j)} \bar{b}^c_{\tau^{(c)}(j)}}\Big)$$

$$= \sum_{b,\bar{b}} \Big(\prod_{j'=1}^{k} \mathbb{T}_{b_{j'}^{\mathcal{D}}} \Big) \Big(\prod_{j=1}^{k} \bar{\mathbb{T}}_{\bar{b}_j^{\mathcal{D}}} \Big) \Big(\prod_{c=1}^{D} \prod_{j'=1}^{k} \delta_{b_{j'}^c, \bar{b}^c_{\tau^{(c)}\pi^{-1}(j')}} \Big) ,$$

where $j' = \pi(j)$. Relabeling now the indices $\bar{b}_j^{\mathcal{D}}$ as $\bar{b}^{\mathcal{D}}_{\bar{\pi}(j)}$ we obtain:

$$\sum_{b,\bar{b}} \Big(\prod_{j'=1}^{k} \mathbb{T}_{b_{j'}^{\mathcal{D}}} \Big) \Big(\prod_{j=1}^{k} \bar{\mathbb{T}}_{\bar{b}^{\mathcal{D}}_{\bar{\pi}(j)}} \Big) \Big(\prod_{c=1}^{D} \prod_{j'=1}^{k} \delta_{b_{j'}^c, \bar{b}^c_{\bar{\pi}\tau^{(c)}\pi^{-1}(j')}} \Big) ,$$

and rearranging the $\bar{\mathbb{T}}$ terms we obtain the desired result.

This remark implies that in Eq. (2.5) one must symmetrize over these equivalent invariants and write $I(\mathbb{T}, \bar{\mathbb{T}})$ as the linear combination:

$$I(\mathbb{T}, \bar{\mathbb{T}}) = \sum_{k\geq 0} \sum_{\tau^{\mathcal{D}} \in [\mathfrak{S}(k)]^D} \frac{1}{k!k!} \, I_k^s(\tau^{\mathcal{D}}) \sum_{b,\bar{b}} \Big(\prod_{j=1}^{k} \mathbb{T}_{b_j^{\mathcal{D}}} \bar{\mathbb{T}}_{\bar{b}_j^{\mathcal{D}}} \Big) \Big(\prod_{c=1}^{D} \prod_{j=1}^{k} \delta_{b_j^c \bar{b}^c_{\tau^{(c)}(j)}} \Big) ,$$

$$I_k^s(\tau^{\mathcal{D}}) = \sum_{\pi, \bar{\pi} \in \mathfrak{S}(k)} \mathcal{I}_k(\bar{\pi}\tau^{(1)}\pi^{-1}, \dots \bar{\pi}\tau^{(D)}\pi^{-1}) \, . \tag{2.6}$$

The trace invariants are a complete set: by integration over the unitary group, any invariant is written as a sum over trace invariants. Before discussing whether these invariants form a basis (that is whether the decomposition is unique), we first introduce a convenient graphical representation.

2.2.1 Edge colored graphs

The trace invariants admit a very straightforward graphical representation as edge colored graphs.

Definition 2.1. *A* bipartite, closed, edge D-colored graph *is a graph $\mathcal{B} = (\mathcal{V}(\mathcal{B}), \mathcal{E}(\mathcal{B}))$ with vertex set $\mathcal{V}(\mathcal{B})$ and edge set $\mathcal{E}(\mathcal{B})$ such that:*

- *$\mathcal{V}(\mathcal{B})$ is bipartite, i.e. it is written as the disjoint union $\mathcal{V}(\mathcal{B}) = \mathcal{V}^w(\mathcal{B}) \cup \mathcal{V}^b(\mathcal{B})$, such that $\forall e \in \mathcal{E}(\mathcal{B})$, then $e = (v, \bar{v})$ with $v \in \mathcal{V}^w(\mathcal{B})$ and $\bar{v} \in \mathcal{V}^b(\mathcal{B})$. We call $v \in \mathcal{V}^w(\mathcal{B})$ the white vertices and $\bar{v} \in \mathcal{V}^b(\mathcal{B})$ the black vertices of $\mathcal{B}$.*

- *The edge set is partitioned into D disjoint subsets $\mathcal{E}(\mathcal{B}) = \bigcup_{c=1}^{D} \mathcal{E}^c(\mathcal{B})$, where $\mathcal{E}^c(\mathcal{B}) = \{e^c = (v, \bar{v})\}$ is the subset of edges with color c. We orient all the edges from the white vertex v to the black vertex $\bar{v}$.*

- *All vertices are D-valent with all edges incident to a vertex having distinct colors.*

We denote $k(\mathcal{B})$ the number of white vertices of the graph $\mathcal{B}$. As all the vertices are hooked to exactly one edge for each color, we have:

$$k(\mathcal{B}) = |\mathcal{V}^w(\mathcal{B})| = |\mathcal{E}^c(\mathcal{B})| \, , \; \forall c = |\mathcal{V}^b(\mathcal{B})| = \frac{1}{2} |\mathcal{V}(\mathcal{B})| \, .$$

We denote $C(\mathcal{B})$ the number of connected components of the graph $\mathcal{B}$, which we label $\mathcal{B}_{(\rho)}$. For the sake of brevity, whenever there is no risk of confusion, we will refer to bipartite, closed, edge D-colored graphs simply as D-colored graphs.

A trace invariant, which is specified by D permutations $\tau^{\mathcal{D}}$ of k elements, can be canonically represented as a D-colored graph (see Figure 2.1 for some examples of this construction) obtained as follows:

- we represent $\mathbb{T}_{b_j^1 \dots b_j^D}$ (respectively $\bar{\mathbb{T}}_{\bar{b}_j^1 \dots \bar{b}_j^D}$) by a white vertex, which we label v_j (respectively a black vertex which we label $\bar{v}_j$),

$$\mathcal{V}^w = \{v_j | j = 1, \dots k\} , \quad \mathcal{V}^b = \{\bar{v}_j | j = 1, \dots k\} .$$

- we connect the vertex v_j with the vertex $\bar{v}_{\tau^{(c)}(j)}$ by an edge of color c,

$$\forall c , \quad \mathcal{E}^c = \{(v_j, \bar{v}_{\tau^{(c)}(j)}) | j = 1 \dots k\} .$$

We denote this graph by $\text{Gr}(\tau^{\mathcal{D}})$:

$$\text{Gr}(\tau^{\mathcal{D}}) \equiv \left\{ \{v_j, \bar{v}_j | j = 1, \dots k\} , \bigcup_{c=1}^{D} \{(v_j, \bar{v}_{\tau^{(c)}(j)}) | j = 1 \dots k\} \right\} .$$

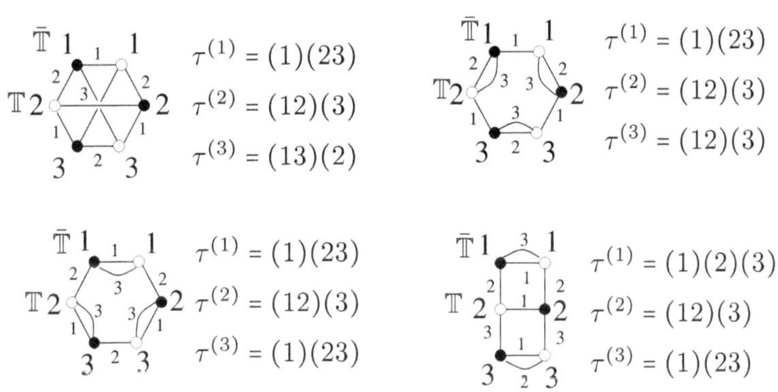

Figure 2.1: D-colored graphs representing trace invariants. The permutations $\tau^{(c)}$ are written in cycle notation, the capital labels designate the vertices and the lower-case labels designate the colors of the edges.

The edges inherit the color of the indices whose contraction they represent and always connect a white and a black vertex. Conversely, the trace invariant associated to a (bipartite, closed, edge) D-colored graph $\mathcal{B}$ is:

$$\text{Tr}_{\mathcal{B}}(\mathbb{T}, \bar{\mathbb{T}}) = \sum_{b, \bar{b}} \delta^{\mathcal{B}}_{b\bar{b}} \prod_{v_j, \bar{v}_j \in \mathcal{V}(\mathcal{B})} \mathbb{T}_{b_j^{\mathcal{D}}} \bar{\mathbb{T}}_{\bar{b}_j^{\mathcal{D}}} , \quad \delta^{\mathcal{B}}_{b\bar{b}} \equiv \prod_{c=1}^{D} \prod_{e^c = (v_j, \bar{v}_k) \in \mathcal{E}^c(\mathcal{B})} \delta_{b_j^c \bar{b}_k^c} .$$

We call the operator $\delta^{\mathcal{B}}_{b\bar{b}}$ the *trace invariant operator associated to* $\mathcal{B}$. The trace invariant associated to the D-colored graph $\mathcal{B}$ is called *connected* if $\mathcal{B}$ is connected.

Remark 2.2. Several remarks are in order.

Counting of graphs. There are $k!^D$ bipartite edge D-colored graphs with $2k$ labeled vertices, as many as there are distinct D-uples of permutations $\tau^{\mathcal{D}}$ of k elements.

The fundamental melon. There exists a unique D-colored graph with two vertices consisting in D edges connecting the two vertices. It is encoded in the permutations:

$$\tau^{(c)} = (1), \forall c\,.$$

We call it the *fundamental melon* and we denote it $\mathcal{B}^{(2)}$. The trace invariant operator associated to $\mathcal{B}^{(2)}$ is:

$$\delta^{\mathcal{B}^{(2)}}_{a\bar{a}} = \prod_{c=1}^{D} \delta_{a^c\bar{a}^c} \equiv \delta_{a^{\mathcal{D}}\bar{a}^{\mathcal{D}}}\,.$$

Cycles and connected components. The number of cycles of the permutation $\sigma\tau^{-1}$ is the number of connected components of the 2-colored graph associated with the permutations σ and τ. This explains why we denoted this number of cycles by $C(\sigma\tau^{-1})$, and the number of connected components of a graph $\mathcal{B}$ by $C(\mathcal{B})$.

The graphs $\mathcal{B}$ have labeled vertices.

Definition 2.2. *We say that two bipartite D-colored graphs $\mathcal{B}_1$ and $\mathcal{B}_2$ are* isomorphic *or* equivalent up to a vertex relabeling, *and we denote $\mathcal{B}_1 \sim \mathcal{B}_2$, if there exists a pair of bijections $f : \mathcal{V}^w(\mathcal{B}_1) \to \mathcal{V}^w(\mathcal{B}_2)$, $\bar{f} : \mathcal{V}^b(\mathcal{B}_1) \to \mathcal{V}^b(\mathcal{B}_2)$ such that:*

$$\forall c\,, \quad (v,\bar{v}) \in \mathcal{E}^c(\mathcal{B}_1) \Leftrightarrow (f(v),\bar{f}(\bar{v})) \in \mathcal{E}^c(\mathcal{B}_2)\,.$$

Two graphs encoded into the two D-uples of permutations $\tau_1^{\mathcal{D}}$ and $\tau_2^{\mathcal{D}}$ are isomorphic if and only if there exist two permutations π and $\bar{\pi}$ such that:

$$\tau_1^{(c)} = \bar{\pi}\tau_2^{(c)}\pi^{-1}, \forall c\,. \tag{2.7}$$

In particular, from Remark 2.1 we conclude that the trace invariants are invariant under a vertex relabeling:

$$\mathcal{B}_1 \sim \mathcal{B}_2 \Rightarrow \mathrm{Tr}_{\mathcal{B}_1}(\mathbb{T},\bar{\mathbb{T}}) = \mathrm{Tr}_{\mathcal{B}_2}(\mathbb{T},\bar{\mathbb{T}})\,.$$

An *unlabeled* D-colored graph is an equivalence class $[\mathcal{B}]$ of D-colored graphs up to vertex relabelings. The trace invariants are class functions depending only on unlabeled graphs.

We conclude that by averaging over the unitary group any invariant is written as a sum:

$$\boxed{\begin{aligned} I(\mathbb{T},\bar{\mathbb{T}}) &= \sum_{k\geq 0,\,\tau^{\mathcal{D}}\in[\mathfrak{S}(k)]^D} \frac{t_{[\mathrm{Gr}(\tau^{\mathcal{D}})]}}{k!k!} \mathrm{Tr}_{\mathrm{Gr}(\tau^{\mathcal{D}})}(\mathbb{T},\bar{\mathbb{T}}) \\ &= \sum_{\mathcal{B}} \frac{t_{[\mathcal{B}]}}{k(\mathcal{B})!k(\mathcal{B})!} \mathrm{Tr}_{\mathcal{B}}(\mathbb{T},\bar{\mathbb{T}})\,, \end{aligned}} \tag{2.8}$$

where $\mathcal{B}$ runs over all the D-colored graphs with labeled vertices. An invariant is called *connected* if it is a sum over connected graphs only:

$$\sum_{\substack{\mathcal{B} \\ \mathcal{B} \text{ connected}}} \frac{t_{[\mathcal{B}]}}{k(\mathcal{B})!k(\mathcal{B})!} \operatorname{Tr}_{\mathcal{B}}(\mathbb{T}, \bar{\mathbb{T}}) .$$

2.2.2 Matrices

The same discussion goes through for matrices, that is tensors with two indices $\mathbb{M}_{a^1a^2}$. The trace invariants for $D = 2$ are represented as 2-colored graphs. All the vertices of such a graph are 2-valent and the two edges incident to a vertex have distinct colors (1 and 2). The connected invariants are cycles with colors 1 and 2.

Example 2.1. Consider, for example, the couple of permutations (in cycle notation):

$$\tau^{(1)} = (1)(2)\dots(k) , \qquad \tau^{(2)} = (1,2,\dots k) .$$

The graph $\operatorname{Gr}(\tau^{(1)}, \tau^{(2)})$ is a 2-colored cycle with k white vertices. The trace invariant associated to $\operatorname{Gr}(\tau^{(1)}, \tau^{(2)})$ is:

$$\begin{aligned} \operatorname{Tr}_{\operatorname{Gr}(\tau^{(1)},\tau^{(2)})}(\mathbb{M}, \bar{\mathbb{M}}) &= \sum_{a,\bar{a}} \Big(\prod_{j=1}^{k} \delta_{a_j^1 \bar{a}^1_{\tau^{(1)}(j)}} \delta_{a_j^2 \bar{a}^2_{\tau^{(2)}(j)}} \Big) \Big(\prod_{j=1}^{k} \mathbb{M}_{a_j^1 a_j^2} \bar{\mathbb{M}}_{\bar{a}_j^1 \bar{a}_j^2} \Big) \\ &= \sum_{a^2, \bar{a}^2} \Big(\prod_{j=1}^{k} \delta_{a_j^2 \bar{a}^2_{\tau^{(2)}(j)}} \Big) \Big(\prod_{j=1}^{k} (\mathbb{M}^\dagger \mathbb{M})_{\bar{a}_j^2 a_j^2} \Big) = \operatorname{Tr}\big[(\mathbb{M}^\dagger \mathbb{M})^k \big] . \end{aligned} \tag{2.9}$$

In the case of matrices there exists another graphical representation of the connected trace invariants which is quite common in the literature. One represents the matrices $\mathbb{M}$ and $\mathbb{M}^\dagger$ as ribbon halfedges, and associates their indices to the sides of the ribbons (also called *strands*). An invariant is then represented as a ribbon vertex (for example, $\operatorname{Tr}\big[(\mathbb{M}^\dagger\mathbb{M})^2\big]$ is drawn in Figure 2.2). The mapping between the ribbon vertex repre-

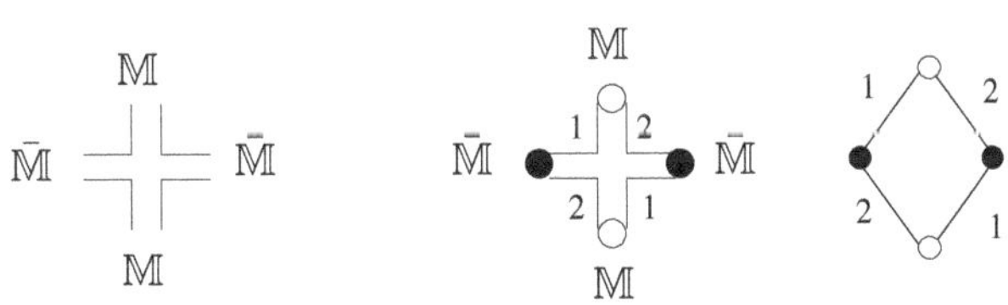

Figure 2.2: From ribbon vertices to colored graphs.

sentation and the 2-colored graph representation used in this paper is trivial: one adds a black or white vertex at the end of each ribbon halfedge, labels the strands by the position of the index of the matrix, and stretches the drawing to obtain the associated colored graph representation.

The general decomposition of invariants in terms of trace invariants becomes for $D = 2$:

$$I(\mathbb{M}, \bar{\mathbb{M}}) = \sum_{k \geq 0} \sum_{\tau^{(1)}, \tau^{(2)} \in \mathfrak{S}(k)} \frac{t_{[\operatorname{Gr}(\tau^{(1)},\tau^{(2)})]}}{k!k!} \operatorname{Tr}_{\operatorname{Gr}(\tau^{(1)},\tau^{(2)})}(\mathbb{M}, \bar{\mathbb{M}}) .$$

The determinant of a matrix is an invariant. It is given explicitly as a linear combination of trace invariants by the Cayley-Hamilton theorem. A similar relation must hold between the hyperdeterminant and the trace invariants in higher dimension, but to the best of our knowledge, is not yet explicitly known.

Remark 2.3. Several remarks are in order.

Factorization of the trace invariants. Observe that the trace invariant associated to the permutations $(\tau^{(1)}, \tau^{(2)})$ factorizes over the cycles of the permutation $[\tau^{(1)}]^{-1}\tau^{(2)}$:

$$\begin{aligned}
\mathrm{Tr}_{\mathrm{Gr}(\tau^{(1)},\tau^{(2)})}(\mathbb{M},\bar{\mathbb{M}}) &= \sum_{a,\bar{a}} \Big(\prod_{j=1}^{k} \delta_{a_j^1 \bar{a}^1_{\tau^{(1)}(j)}} \delta_{a_j^2 \bar{a}^2_{\tau^{(2)}(j)}}\Big)\Big(\prod_{j=1}^{k} \mathbb{M}_{a_j^1 a_j^2} \bar{\mathbb{M}}_{\bar{a}_j^1 \bar{a}_j^2}\Big) \\
&= \sum_{a,\bar{a}} \Big(\prod_{j=1}^{k} \delta_{a_j^1 \bar{a}_j^1} \delta_{a_j^2 \bar{a}^2_{[\tau^{(1)}]^{-1}\tau^{(2)}(j)}}\Big)\Big(\prod_{j=1}^{k} \mathbb{M}_{a_j^1 a_j^2} \bar{\mathbb{M}}_{\bar{a}_j^1 \bar{a}_j^2}\Big) \\
&= \sum_{a,\bar{a}} \Big(\delta_{a_j^2 \bar{a}^2_{[\tau^{(1)}]^{-1}\tau^{(2)}(j)}}\Big)\Big(\prod_{j=1}^{k} [\mathbb{M}^\dagger \mathbb{M}]_{\bar{a}_j^2 a_j^2}\Big) \\
&= \prod_{q\geq 1} \Big(\mathrm{Tr}\big[(\mathbb{M}^\dagger \mathbb{M})^q\big]\Big)^{C_q([\tau^{(1)}]^{-1}\tau^{(2)}]} .
\end{aligned}$$

Symmetrized invariants. Let us compute the symmetrized trace invariants:

$$\sum_{\mathcal{B}'\in[\mathcal{B}]} \frac{1}{k(\mathcal{B}')!k(\mathcal{B}')!} \mathrm{Tr}_{\mathcal{B}'}(\mathbb{M},\bar{\mathbb{M}}) ,$$

in the case of matrices. Using the previous remark, these invariants can be written as a sum over permutations:

$$\sum_{\substack{\tau^{(1)},\tau^{(2)}\in\mathfrak{S}(k(\mathcal{B}))\\ \mathrm{Gr}(\tau^{(1)},\tau^{(2)})\in[\mathcal{B}]}} \frac{1}{k(\mathcal{B})!k(\mathcal{B})!} \prod_{q\geq 1} \Big[\mathrm{Tr}\big[(\mathbb{M}^\dagger \mathbb{M})^q\big]\Big]^{C_q([\tau^{(1)}]^{-1}\tau^{(2)})} .$$

Observe that the cycles of the permutation $[\tau^{(1)}]^{-1}\tau^{(2)}$ are exactly the connected components of the graph $\mathrm{Gr}(\tau^{(1)}, \tau^{(2)})$.

In order to compute this sum we must first identify the equivalence class of the 2-colored graph $\mathcal{B}$. From Definition 2.2:

$$\begin{aligned}
&\mathrm{Gr}(\tau_1^{(1)}, \tau_1^{(2)}) \sim \mathrm{Gr}(\tau_2^{(1)}, \tau_2^{(2)}) \\
&\qquad \Leftrightarrow \exists\, \bar{\pi}, \pi \in \mathfrak{S}(k) \text{ such that } \tau_1^{(c)} = \bar{\pi}\tau_2^{(c)}\pi^{-1} , \quad c = 1,2 .
\end{aligned}$$

On the other hand:

$$\begin{cases} \tau_1^{(1)} = \bar{\pi}\tau_2^{(1)}\pi^{-1} \\ \tau_1^{(2)} = \bar{\pi}\tau_2^{(2)}\pi^{-1} \end{cases} \Leftrightarrow [\tau_1^{(1)}]^{-1}\tau_1^{(2)} = \pi[\tau_2^{(1)}]^{-1}\tau_2^{(2)}\pi^{-1} ,$$

where, once π is identified from the right hand side above at fixed τs, $\bar{\pi}$ is defined by the relation $\bar{\pi} \equiv \tau_1^{(1)}\pi[\tau_2^{(1)}]^{-1}$. It follows that the equivalence class $[\mathrm{Gr}(\tau^{(1)},\tau^{(2)})]$ is one to one to the equivalence class of $[\tau^{(1)}]^{-1}\tau^{(2)}$ under conjugation.

Recall that two permutations τ and τ' are equivalent (conjugated) if and only if they have the same cycle structure, $C_q(\tau) = C_q(\tau')$, $\forall q$, and that there are exactly:

$$\frac{[\sum_q qC_q(\tau)]!}{\prod_{q\geq 1} C_q(\tau)!\, q^{C_q(\tau)}} ,$$

permutations in the equivalence class of τ. Denoting $C_q(\mathcal{B})$ the number of connected components with $2q$ vertices of the graph $\mathcal{B}$, the symmetrized invariant computes to:

$$\begin{aligned}
&\sum_{\substack{\tau^{(1)},\tau^{(2)}\in\mathfrak{S}(k(\mathcal{B}))\\ C_q([\tau^{(1)}]^{-1}\tau^{(2)})=C_q(\mathcal{B})}} \frac{1}{k(\mathcal{B})!k(\mathcal{B})!}\prod_{q\geq 1}\Big[\operatorname{Tr}\big[(\mathbb{M}^\dagger\mathbb{M})^q\big]\Big]^{C_q(\mathcal{B})}\\
&= \sum_{\substack{\tau\in\mathfrak{S}(k)\\ C_q(\tau)=C_q(\mathcal{B})}} \frac{1}{k(\mathcal{B})!}\prod_{q\geq 1}\Big[\operatorname{Tr}\big[(\mathbb{M}^\dagger\mathbb{M})^q\big]\Big]^{C_q(\mathcal{B})}\\
&= \prod_{q\geq 1}\frac{1}{C_q(\mathcal{B})!}\left(\frac{\operatorname{Tr}[(\mathbb{M}^\dagger\mathbb{M})^q]}{q}\right)^{C_q(\mathcal{B})} .
\end{aligned}$$

Consequences of the factorization. The coefficients $t_{[\mathrm{Gr}(\tau^{(1)},\tau^{(2)})]}$ depend only on the equivalence class of $\mathrm{Gr}(\tau^{(1)},\tau^{(2)})$, that is they depend only on the cycle structure of the permutation $[\tau^{(1)}]^{-1}\tau^{(2)}$. Let us assume furthermore that the coefficients t factorize as:

$$t_{[\mathrm{Gr}(\tau^{(1)},\tau^{(2)})]} = \prod_{q\geq 1} t_q^{C_q([\tau^{(1)}]^{-1}\tau^{(2)})} .$$

An invariant whose coefficients factorize is written, then, as:

$$\begin{aligned}
Z(\mathbb{M},\bar{\mathbb{M}}) &= \sum_{k\geq 0}\sum_{\tau^{(1)},\tau^{(2)}\in\mathfrak{S}(k)}\frac{1}{k!k!}\prod_{q\geq 1}\Big[t_q\operatorname{Tr}\big[(\mathbb{M}^\dagger\mathbb{M})^q\big]\Big]^{C_q([\tau^{(1)}]^{-1}\tau^{(2)})}\\
&= \sum_{k\geq 0}\frac{1}{k!}\sum_{\tau\in\mathfrak{S}(k)}\prod_{q\geq 1}\Big[t_q\operatorname{Tr}\big[(\mathbb{M}^\dagger\mathbb{M})^q\big]\Big]^{C_q(\tau)}\\
&= \sum_{k\geq 0}\frac{1}{k!}\sum_{C_q\geq 0,\,\forall q}^{\sum_q qC_q=k}\frac{[\sum_q qC_q]!}{\prod_q C_q!\, q^{C_q}}\prod_{q\geq 1}\Big[t_q\operatorname{Tr}\big[(\mathbb{M}^\dagger\mathbb{M})^q\big]\Big]^{C_q}\\
&= \prod_{q\geq 1}\left[\sum_{C_q\geq 0}\frac{1}{C_q!}\left(\frac{t_q\operatorname{Tr}\big[(\mathbb{M}^\dagger\mathbb{M})^q\big]}{q}\right)^{C_q}\right] = e^{\sum_{q\geq 1}\frac{t_q}{q}\operatorname{Tr}\left[(\mathbb{M}^\dagger\mathbb{M})^q\right]} .
\end{aligned}$$

Starting from the same constants t_q one can on the other hand build the connected invariant:

$$W(\mathbb{M},\bar{\mathbb{M}}) = \sum_{k\geq 0} \sum_{\substack{\tau^{(1)},\tau^{(2)}\in\mathfrak{S}(k)\\ \mathrm{Gr}(\tau^{(1)},\tau^{(2)})\,\text{connected}}} \frac{t_{[\mathrm{Gr}(\tau^{(1)},\tau^{(2)})]}}{k!k!} \mathrm{Tr}_{\mathrm{Gr}(\tau^{(1)},\tau^{(2)})}(\mathbb{M},\bar{\mathbb{M}}) .$$

The graph $\mathrm{Gr}(\tau^{(1)},\tau^{(2)})$ is connected if and only if the permutation:

$$[\tau^{(1)}]^{-1}\tau^{(2)} ,$$

has only one cycle of length k, that is $C_q = 0, q < k$ and $C_k = 1$. Thus:

$$W(\mathbb{M},\bar{\mathbb{M}}) = \sum_{k\geq 1} \frac{1}{k!} \frac{k!}{C_k!\, k^{C_k}} \Big[t_k \,\mathrm{Tr}\big[(\mathbb{M}^\dagger \mathbb{M})^k\big]\Big]^{C_k} \bigg|_{C_k=1}$$
$$= \sum_{k\geq 1} \frac{1}{k} t_k \,\mathrm{Tr}\big[(\mathbb{M}^\dagger \mathbb{M})^k\big] ,$$

and $Z(\mathbb{M},\bar{\mathbb{M}})$ and $W(\mathbb{M},\bar{\mathbb{M}})$ are related by:

$$Z(\mathbb{M},\bar{\mathbb{M}}) = e^{W(\mathbb{M},\bar{\mathbb{M}})} . \tag{2.10}$$

2.3 Connected and disconnected trace invariants

A relation parallel to Eq. (2.10) holds also for $D \geq 3$. Let us consider the most general case of a (formal) series Z that is written as a sum over some combinatorial species [24] whose members are denoted G, with $n(G)$ labeled vertices of "amplitudes" $t_{[G]}$. We assume that the amplitudes $t_{[G]}$ do not depend on the labels (that is, as the notation suggests, they are class functions under the equivalence relation $G_1 \sim G_2$ if G_1 and G_2 are isomorphic):

$$Z = \sum_G \frac{1}{n(G)!}\, t_{[G]} .$$

The combinatorial species G can consist in graphs, combinatorial maps (see Definition 3.9), forests (i.e. undirected acyclic graphs), etc. For example, in the previous section we dealt with the species of permutations. Let us denote $G_{(\nu)}$ the connected members of the species (that is connected graphs, connected combinatorial maps, trees, etc.). The $G_{(\nu)}$s have labeled vertices also.

Every G is the union of its connected components $G = \cup_\rho G_{(\rho)}$. Some of these connected components are isomorphic, that is they are identical after a vertex relabeling. Grouping all these components together, every G is written uniquely as:

$$G = \bigcup_{\nu\geq 1} G_{(\nu)}^{q_\nu} , \quad n(G) = \sum_{\nu\geq 1} n(G_{(\nu)}) q_\nu ,$$

where it is understood that the q_ν copies of the same $G_{(\nu)}$ are labeled independently. The integer $q_{(\nu)}$ is the multiplicity of the connected component $G_{(\nu)}$, i.e. the number of times a relabeling of $G_{(\nu)}$ is encountered in the decomposition of G into connected components $G_{(\rho)}$.

Lemma 2.1. *If the amplitude $t_{[G]}$ factorizes over the connected components of G, then the logarithm of Z is the sum over the connected members of the species:*

$$\ln Z = \sum_{\nu\geq 1} \frac{1}{n(G_{(\nu)})!} t_{[G_{(\nu)}]} \ .$$

Proof. As the amplitude factorizes we have:

$$Z = \sum_{G} \frac{1}{n(G)!} t_{[G]} = \sum_{G_1^{q_1} \cup G_2^{q_2} \cup \dots} \frac{1}{n(G)!} \prod_{\nu \geq 1} t^{q_\nu}_{[G_{(\nu)}]}$$
$$= \sum_{q_1, q_2, \dots \geq 0} \frac{1}{\left(\sum_{\nu\geq 1} n(G_{(\nu)}) q_\nu\right)!} \prod_{\nu\geq 1} t^{q_\nu}_{[G_{(\nu)}]} \sum_{G = G_1^{q_1} \cup G_2^{q_2} \dots} 1 \ .$$

Observe that there are exactly:

$$\frac{\left(\sum_{\nu\geq 1} n(G_{(\nu)}) q_\nu\right)!}{\prod_{\nu\geq 1} q_\nu! \left[n(G_{(\nu)})!\right]^{q_\nu}} \ ,$$

permutations of the labels of G corresponding to the same decomposition into connected members $G_{(\nu)}$ with labeled vertices, therefore we obtain:

$$Z = \sum_{q_1\geq 0, q_2 \geq 0, \dots} \frac{1}{\prod_{\nu\geq 1} q_\nu!} \prod_{\nu \geq 1} \left(\frac{1}{n(G_{(\nu)})!} t_{[G_{(\nu)}]} \right)^{q_\nu} = e^{\sum_{\nu\geq 1} \frac{1}{n(G_{(\nu)})!} t_{[G_{(\nu)}]}} \ .$$

□

For bipartite graphs a similar relation holds, as expressed in the following corollary.

Corollary 2.1. *If Z is a formal series that is written as a sum over bipartite graphs $\mathcal{G}$ with $2k(\mathcal{G})$ labeled vertices:*

$$Z = \sum_{\mathcal{G}} \frac{1}{k(\mathcal{G})! k(\mathcal{G})!} t_{[\mathcal{G}]} \ ,$$

where the amplitude $t_{[G]}$ does not depend on the labels of the vertices and factorizes over the connected components of $\mathcal{G}$ then, denoting $\mathcal{G}_{(\nu)}$ the bipartite connected graphs, the logarithm of Z is:

$$\ln Z = \sum_{\nu\geq 1} \frac{1}{k(\mathcal{G}_{(\nu)})! k(\mathcal{G}_{(\nu)})!} t_{[\mathcal{G}_{(\nu)}]} \ .$$

Proof. The proof follows the same lines as the one of Lemma 2.1, but one needs to count that there are:

$$\frac{k(\mathcal{G})! k(\mathcal{G})!}{\prod_{\nu\geq 1} q_\nu! [n(\mathcal{G}_{(\nu)})! n(\mathcal{G}_{(\nu)})!]^{q_\nu}}$$

permutations of the black and white labels of $\mathcal{G}$ corresponding to the same decomposition into q_1 components $\mathcal{G}_1$, q_2 components $\mathcal{G}_2$, and so on.

□

The trace invariant associated with a graph $\mathcal{B} = \cup_\rho \mathcal{B}_{(\rho)}$ factorizes over its connected components $\mathcal{B}_{(\rho)}$. Indeed:

$$\mathrm{Tr}_{\mathcal{B}}(\mathbb{T},\bar{\mathbb{T}}) = \sum_{b,\bar{b}} \left(\prod_{c=1}^{D} \prod_{e^c=(v_j,\bar{v}_k)\in\mathcal{E}^c(\mathcal{B})} \delta_{b_j^c \bar{b}_k^c} \right) \prod_{v_j,\bar{v}_j\in\mathcal{V}(\mathcal{B})} \mathbb{T}_{b_j^D} \bar{\mathbb{T}}_{\bar{b}_j^D} = \prod_\rho \mathrm{Tr}_{\mathcal{B}_{(\rho)}}(\mathbb{T},\bar{\mathbb{T}}) ,$$

hence, denoting $\mathcal{B}_{(\rho)}$ the connected components of $\mathcal{B}$:

$$\boxed{\begin{aligned} &\sum_{\mathcal{B}} \frac{1}{k(\mathcal{B})!k(\mathcal{B})!} \left(\prod_\rho t_{[\mathcal{B}_{(\rho)}]} \right) \mathrm{Tr}_{\mathcal{B}}(\mathbb{T},\bar{\mathbb{T}}) \\ &\qquad = \exp\left\{ \sum_{\mathcal{B}\text{ connected}} \frac{1}{k(\mathcal{B})!k(\mathcal{B})!} t_{[\mathcal{B}]} \,\mathrm{Tr}_{\mathcal{B}}(\mathbb{T},\bar{\mathbb{T}}) \right\} . \end{aligned}} \tag{2.11}$$

2.4 Uniqueness of the decomposition on trace invariants

A natural question is whether the trace invariants form a basis in the space of invariants. The answer to this question is somewhat subtle. Let us first consider the case of matrices. The symmetrized trace invariants are, according to Remark 2.3:

$$\prod_{q\ge 1} \frac{1}{C_q(\mathcal{B})!} \left(\frac{\mathrm{Tr}[(\mathbb{M}^\dagger\mathbb{M})^q]}{q} \right)^{C_q(\mathcal{B})} .$$

While these invariants are a complete set, they *are not independent* at finite N. Indeed, any invariant polynomial is a symmetric polynomial in the principal values of $\mathbb{M}$ (i.e. the eigenvalues of $\mathbb{M}^\dagger\mathbb{M}$). As such, it admits a unique decomposition in products of power sum polynomials:

$$\mathrm{Tr}[(\mathbb{M}^\dagger\mathbb{M})^q] ,$$

but only the first N power sum polynomials are independent (all the other power sum polynomials can be decomposed on lower order power sum polynomials using the Cayley Hamilton theorem). This problem is washed out in the $N\to\infty$ limit where one needs all the single trace invariants in order to fix the spectral measure of $\mathbb{M}^\dagger\mathbb{M}$.

The fact that in the case of matrices, at finite N, the decomposition in Eq. (2.8) is not unique suggests that for $D \ge 3$ the decomposition will not be unique either. The crucial point is that the symmetrized trace invariants:

$$\sum_{\mathcal{B}'\in[\mathcal{B}]} \frac{1}{k(\mathcal{B}')!k(\mathcal{B}')!} \mathrm{Tr}_{\mathcal{B}'}(\mathbb{T},\bar{\mathbb{T}}) ,$$

reproduce themselves under averaging over the unitary group *only if N is larger than* $k(\mathcal{B})$.

Lemma 2.2 (Convolution inverse of the Weingarten functions). *For $N \geq k$ and any permutations $\tau, \pi \in \mathfrak{S}(k)$, one has:*

$$\sum_{\sigma \in \mathfrak{S}(k)} N^{C(\pi^{-1}\sigma)} \, \mathrm{Wg}(N, \sigma\tau^{-1}) = \begin{cases} 1 & \text{if } \pi = \tau \\ 0 & \text{otherwise} \end{cases} . \tag{2.12}$$

Proof. For any permutation π we have:

$$\begin{aligned}
\left(\prod_{j=1}^{k} \delta_{b_j \bar{b}_{\pi(j)}}\right) &= \sum_{a,\bar{a}} \left(\prod_{j=1}^{k} \delta_{a_j \bar{a}_{\pi(j)}}\right) \int_{U(N)} [dU] \left(\prod_{j=1}^{k} U_{a_j b_j}\right)\left(\prod_{j=1}^{k} \bar{U}_{\bar{a}_j \bar{b}_j}\right) \\
&= \sum_{a,\bar{a}} \left(\prod_{j=1}^{k} \delta_{a_j \bar{a}_{\pi(j)}}\right) \sum_{\sigma,\tau \in \mathfrak{S}(k)} \left(\prod_{j=1}^{k} \delta_{a_j \bar{a}_{\sigma(j)}} \delta_{b_j \bar{b}_{\tau(j)}}\right) \mathrm{Wg}(N, \sigma\tau^{-1}) \\
&= \sum_{\sigma,\tau \in \mathfrak{S}(k)} \left(\prod_{j=1}^{k} \delta_{b_j \bar{b}_{\tau(j)}}\right) N^{C(\pi^{-1}\sigma)} \mathrm{Wg}(N, \sigma\tau^{-1}) .
\end{aligned}$$

Let us apply this equality for $b_j = j$ and $\bar{b}_{\tau'(j)} = j$ for some fixed τ'. Observe that this is possible only if $N \geq k$. On the right hand side the term $\tau = \tau'$ (and only this term) survives. On the left hand side one obtains 0 unless $\tau' = \pi$, and in this last case one obtains 1.

□

For $N \geq k(\mathcal{B})$, a symmetrized trace invariant reproduces itself after averaging over the unitary group:

$$\begin{aligned}
&\sum_{\substack{\pi^{\mathcal{D}} \in [\mathfrak{S}(k(\mathcal{B}))]^D \\ \mathrm{Gr}(\pi^{\mathcal{D}}) \in [\mathcal{B}]}} \frac{1}{k(\mathcal{B})! k(\mathcal{B})!} \sum_{a,\bar{a}} \left(\prod_{c=1}^{D} \prod_{j=1}^{k(\mathcal{B})} \delta_{a_j \bar{a}_{\pi^{(c)}(j)}}\right) \left(\prod_{j=1}^{k(\mathcal{B})} \mathbb{T}_{a_j^{\mathcal{D}}} \bar{\mathbb{T}}_{\bar{a}_j^{\mathcal{D}}}\right) \\
&= \sum_{\substack{\pi^{\mathcal{D}} \in [\mathfrak{S}(k(\mathcal{B}))]^D \\ \mathrm{Gr}(\pi^{\mathcal{D}}) \in [\mathcal{B}]}} \sum_{\sigma^{\mathcal{D}}, \tau^{\mathcal{D}} \in [\mathfrak{S}(k(\mathcal{B}))]^D} \frac{1}{k(\mathcal{B})! k(\mathcal{B})!} \sum_{a,\bar{a},b,\bar{b}} \left(\prod_{c=1}^{D} \prod_{j=1}^{k(\mathcal{B})} \delta_{a_j \bar{a}_{\pi^{(c)}(j)}}\right) \\
&\quad \times \left[\prod_{c=1}^{D} \left(\prod_{j=1}^{k(\mathcal{B})} \delta_{a_j \bar{a}_{\sigma^{(c)}(j)}} \delta_{b_j \bar{b}_{\tau^{(c)}(j)}}\right) \mathrm{Wg}(N, \sigma^{(c)}[\tau^{(c)}]^{-1})\right] \left(\prod_{j=1}^{k} \mathbb{T}_{b_j^{\mathcal{D}}} \bar{\mathbb{T}}_{\bar{b}_j^{\mathcal{D}}}\right) \\
&= \sum_{\substack{\pi^{\mathcal{D}} \in [\mathfrak{S}(k(\mathcal{B}))]^D \\ \mathrm{Gr}(\pi^{\mathcal{D}}) \in [\mathcal{B}]}} \sum_{\sigma^{\mathcal{D}}, \tau^{\mathcal{D}} \in [\mathfrak{S}(k(\mathcal{B}))]^D} \frac{1}{k(\mathcal{B})! k(\mathcal{B})!} \\
&\quad \times \left[\prod_{c=1}^{D} N^{C([\pi^{c}]^{-1}\sigma^{(c)})} \mathrm{Wg}(N, \sigma^{(c)}[\tau^{(c)}]^{-1})\right] \\
&\quad \times \left(\sum_{b,\bar{b}} \prod_{c=1}^{D} \prod_{j=1}^{k(\mathcal{B})} \delta_{b_j \bar{b}_{\tau^{(c)}(j)}} \prod_{j=1}^{k} \mathbb{T}_{b_j^{\mathcal{D}}} \bar{\mathbb{T}}_{\bar{b}_j^{\mathcal{D}}}\right) \\
&= \sum_{\substack{\pi^{\mathcal{D}} \in [\mathfrak{S}(k(\mathcal{B}))]^D \\ \mathrm{Gr}(\pi^{\mathcal{D}}) \in [\mathcal{B}]}} \frac{1}{k(\mathcal{B})! k(\mathcal{B})!} \left(\sum_{b,\bar{b}} \prod_{c=1}^{D} \prod_{j=1}^{k(\mathcal{B})} \delta_{b_j \bar{b}_{\pi^{(c)}(j)}} \prod_{j=1}^{k} \mathbb{T}_{b_j^{\mathcal{D}}} \bar{\mathbb{T}}_{\bar{b}_j^{\mathcal{D}}}\right) ,
\end{aligned}$$

but *does not* reproduce itself for $N < k(\mathcal{B})$. A trace invariant with a degree larger than N is written, after averaging over the unitary group, as a linear combination of other trace invariants. This problem is washed away in the $N \to \infty$ limit.

2.5 Invariant probability measures

A general introduction to probability measures is provided in Appendix B. In this section we present the specific notions relevant for probability measures for random tensors.

Consider a collection of N^D complex random variables arranged in a D dimensional array $\mathbb{T}_{a^1 \dots a^D}$, with $a^c = 1, \dots N$. We denote a probability measure for the joint distribution of these variables by $\mu(\mathbb{T}, \bar{\mathbb{T}})$ (or simply by μ). In order for these N^D variables to constitute a random tensor, $\mu(\mathbb{T}, \bar{\mathbb{T}})$ must have some invariance properties.

The *moments* of $\mu(\mathbb{T}, \bar{\mathbb{T}})$:

$$\left\langle \mathbb{T}_{a_1^{\mathcal{D}}} \dots \mathbb{T}_{a_q^{\mathcal{D}}} \bar{\mathbb{T}}_{\bar{a}_1^{\mathcal{D}}} \dots \bar{\mathbb{T}}_{\bar{a}_k^{\mathcal{D}}} \right\rangle_{\mu} \equiv \int d\mu(\mathbb{T}, \bar{\mathbb{T}}) \quad \mathbb{T}_{a_1^{\mathcal{D}}} \dots \mathbb{T}_{a_q^{\mathcal{D}}} \bar{\mathbb{T}}_{\bar{a}_1^{\mathcal{D}}} \dots \bar{\mathbb{T}}_{\bar{a}_k^{\mathcal{D}}} ,$$

are the expectations of products of q variables $\mathbb{T}_{a_i^{\mathcal{D}}}, i = 1, \dots q$ and k complex conjugated variables $\bar{\mathbb{T}}_{\bar{a}_j^{\mathcal{D}}}, j = 1, \dots k$. We consider only even distributions, that is such that the moments are nontrivial only if the numbers of $\mathbb{T}$ and $\bar{\mathbb{T}}$ are equal, $q = k$.

We denote $\mathbf{P}$ a partition of the set $\mathcal{V} = \{a_1^{\mathcal{D}} \dots a_k^{\mathcal{D}}, \bar{a}_1^{\mathcal{D}} \dots \bar{a}_k^{\mathcal{D}}\}$ into disjoint bipartite subsets and we denote $|\mathbf{P}|$ the number of parts in $\mathbf{P}$. The parts of $\mathbf{P}$ are denoted $\mathcal{V}(\alpha)$ for $\alpha = 1, 2, \dots |\mathbf{P}|$, and the sets $\mathcal{V}(\alpha)$ are bipartite in the sense that:

$$|\mathcal{V}(\alpha) \cap \{a_1^{\mathcal{D}} \dots a_k^{\mathcal{D}}\}| = |\mathcal{V}(\alpha) \cap \{\bar{a}_1^{\mathcal{D}} \dots \bar{a}_k^{\mathcal{D}}\}| \equiv k(\alpha) , \quad \sum_{\alpha=1}^{|\mathbf{P}|} k(\alpha) = k .$$

The *cumulants* of μ are defined implicitly by the formula:

$$\left\langle \mathbb{T}_{a_1^{\mathcal{D}}} \bar{\mathbb{T}}_{\bar{a}_1^{\mathcal{D}}} \dots \mathbb{T}_{a_q^{\mathcal{D}}} \bar{\mathbb{T}}_{\bar{a}_q^{\mathcal{D}}} \right\rangle_{\mu} = \sum_{\mathbf{P}} \prod_{\alpha=1}^{|\mathbf{P}|} \kappa_{2k(\alpha)} \left[\mathbb{T}_{a_i^{\mathcal{D}}}, \bar{\mathbb{T}}_{\bar{a}_j^{\mathcal{D}}} | a_i^{\mathcal{D}}, \bar{a}_j^{\mathcal{D}} \in \mathcal{V}(\alpha) \right] ,$$

and can be computed in terms of the moments using the Möbius inversion formula. The generating function of the moments of the measure μ is:

$$Z(J, \bar{J}; \mu) = \int d\mu(\mathbb{T}, \bar{\mathbb{T}}) \quad e^{\sum \bar{\mathbb{T}}_{\bar{a}^{\mathcal{D}}} J_{\bar{a}^{\mathcal{D}}} + \sum J_{a^{\mathcal{D}}} \mathbb{T}_{a^{\mathcal{D}}}} ,$$

where the sums run over all the (multi) indices $a^{\mathcal{D}}$ and $\bar{a}^{\mathcal{D}}$. The cumulants of μ can be computed as derivatives of the logarithm of Z:

$$\kappa_{2k} \left(\mathbb{T}_{a_1^{\mathcal{D}}}, \bar{\mathbb{T}}_{\bar{a}_1^{\mathcal{D}}}, \dots \mathbb{T}_{a_k^{\mathcal{D}}}, \bar{\mathbb{T}}_{\bar{a}_k^{\mathcal{D}}} \right) = \frac{\partial^{(2k)}}{\partial J_{a_1^{\mathcal{D}}} \partial \bar{J}_{\bar{a}_1^{\mathcal{D}}} \dots \partial J_{a_k^{\mathcal{D}}} \partial \bar{J}_{\bar{a}_k^{\mathcal{D}}}} \ln Z(J, \bar{J}; \mu) \Big|_{\substack{J=0 \\ \bar{J}=0}} .$$

A random tensor model is a probability measure that is invariant under unitary transformations.

Definition 2.3. *We call:*

- trace invariant *a probability measure* μ *of* N^D *complex random variables* $\mathbb{T}_{a^{\mathcal{D}}}, \bar{\mathbb{T}}_{\bar{a}^{\mathcal{D}}}$ *whose* cumulants *are linear combinations of trace invariant operators:*

$$\kappa_{2k}[\mathbb{T}_{a_1^{\mathcal{D}}}, \bar{\mathbb{T}}_{\bar{a}_1^{\mathcal{D}}} \dots \mathbb{T}_{a_k^{\mathcal{D}}}, \bar{\mathbb{T}}_{\bar{a}_k^{\mathcal{D}}}] = \sum_{\mathcal{B},\, k(\mathcal{B})=k} \mathfrak{K}(\mathcal{B},\mu) \prod_{\rho=1}^{C(\mathcal{B})} \delta_{a\bar{a}}^{\mathcal{B}_{(\rho)}} , \qquad (2.13)$$

where the sum runs over all *the* D*-colored graphs* $\mathcal{B}$ *with* $k(\mathcal{B}) = k$ *white vertices, and* ρ *runs over the connected components* $\mathcal{B}_{(\rho)}$ *of* $\mathcal{B}$.

- *a* random tensor *a collection of* N^D *complex random variables whose joint distribution is trace invariant.*

Somewhat abusively we will also call from now on the coefficient of a trace invariant operator in the expansion (2.13), $\mathfrak{K}(\mathcal{B},\mu)$, a cumulant. The expansion in trace invariants can be written directly as a sum over D permutations $\tau^{\mathcal{D}}$:

$$\kappa_{2k}[\mathbb{T}_{a_1^{\mathcal{D}}}, \bar{\mathbb{T}}_{\bar{a}_1^{\mathcal{D}}} \dots \mathbb{T}_{a_k^{\mathcal{D}}}, \bar{\mathbb{T}}_{\bar{a}_k^{\mathcal{D}}}] = \sum_{\tau^{\mathcal{D}}} \mathfrak{K}\big(\mathrm{Gr}(\tau^{\mathcal{D}}),\mu\big) \left(\prod_{c=1}^{D} \prod_{j=1}^{k} \delta_{a_j^c \bar{a}^c_{\tau^{(c)}(j)}} \right) .$$

The invariants one can build out of a tensor (and its complex conjugate) are the observables of tensor models. Following our previous discussion (and Eq. (2.8)), it is enough to study the trace invariants $\mathrm{Tr}_{\mathcal{B}}(\mathbb{T},\bar{\mathbb{T}})$, as they constitute a complete set in the space of invariants.

We are often interested in studying the behavior of random tensors (i.e. we are interested in evaluating the expectations of trace invariants) when the size N goes to infinity. This is done by first expanding the joint moments into cumulants, and then expanding the cumulants themselves into graphs. In order for the end result to make sense in the $N \to \infty$ limit the cumulants *must scale with* N. In Chapter 6 we will discuss in great detail the possible scalings that lead to well defined large N limits.

Example 2.2 (The Gaussian distribution). The simplest example of a trace invariant probability distribution is the normalized Gaussian distribution of covariance σ^2:

$$d\mu_{\frac{\sigma^2}{N^{D-1}} 1^{\mathcal{D}}}(\mathbb{T},\bar{\mathbb{T}}) = e^{-N^{D-1}\frac{1}{\sigma^2}\sum \bar{\mathbb{T}}_{\bar{a}^{\mathcal{D}}}\, \delta_{\bar{a}^{\mathcal{D}} a^{\mathcal{D}}}\, \mathbb{T}_{a^{\mathcal{D}}}} \prod_{a^{\mathcal{D}}} \left(\frac{N^{D-1}}{\sigma^2} \frac{d\bar{\mathbb{T}}_{a^{\mathcal{D}}} d\mathbb{T}_{a^{\mathcal{D}}}}{2\pi\imath} \right) , \qquad (2.14)$$

where $1^{\mathcal{D}}$ denotes the identity operator $1^{\mathcal{D}} : \otimes_{c\in\mathcal{D}} \mathcal{H}_c \to \otimes_{c\in\mathcal{D}} \mathcal{H}_c$. The generating function of the moments of this measure is:

$$Z(J,\bar{J};\mu_{\frac{\sigma^2}{N^{D-1}}1^{\mathcal{D}}}) = e^{\frac{\sigma^2}{N^{D-1}} \sum \bar{J}_{\bar{a}^{\mathcal{D}}} \delta_{\bar{a}^{\mathcal{D}} a^{\mathcal{D}}} J_{a^{\mathcal{D}}}} .$$

The only nontrivial cumulant of the Gaussian distribution is the second cumulant (associated to the unique graph with two vertices $\mathcal{B}^{(2)}$):

$$\kappa_2(\mathbb{T}_{a^{\mathcal{D}}}, \bar{\mathbb{T}}_{\bar{a}^{\mathcal{D}}}) = \frac{\sigma^2}{N^{D-1}} \delta_{\bar{a}^{\mathcal{D}} a^{\mathcal{D}}} \Rightarrow \begin{cases} \mathfrak{K}(\mathcal{B}^{(2)}, \mu_{\frac{\sigma^2}{N^{D-1}}1^{\mathcal{D}}}) = \frac{\sigma^2}{N^{D-1}} \\ \mathfrak{K}(\mathcal{B}, \mu_{\frac{\sigma^2}{N^{D-1}}1^{\mathcal{D}}}) = 0 , \quad \forall \mathcal{B} \neq \mathcal{B}^{(2)} \end{cases} ,$$

and, as expected, it scales with N.

Although the cumulants of the Gaussian distribution are very simple, the computation of the expectations of the connected trace invariant is rather nontrivial: the Gaussian expectation of an invariant,

$$\big\langle \mathrm{Tr}_{\mathcal{B}}(\mathbb{T},\bar{\mathbb{T}})\big\rangle_{\mu_{\frac{\sigma^2}{N^{D-1}}\mathbb{1}^{\mathcal{D}}}} ,$$

involves sums over the indices of the tensors which then combine with the explicit scaling with N of $\mathfrak{K}(\mathcal{B}^{(2)}, \mu_{\frac{\sigma^2}{N^{D-1}}\mathbb{1}^{\mathcal{D}}})$.

We postpone to Section 6.2 in Chapter 6 the computation of the moments of the Gaussian distribution. For now we just quote the end result:

- for any connected D-colored graph $\mathcal{B}$ with $2k(\mathcal{B})$ vertices, there exist two nonnegative integers, $\Omega(\mathcal{B})$ and $R(\mathcal{B})$ such that:

$$\lim_{N\to\infty} N^{-1+\Omega(\mathcal{B})} \big\langle \mathrm{Tr}_{\mathcal{B}}(\mathbb{T},\bar{\mathbb{T}})\big\rangle_{\mu_{\frac{\sigma^2}{N^{D-1}}\mathbb{1}^{\mathcal{D}}}} = \sigma^{2k(\mathcal{B})} R(\mathcal{B}) .$$

  The integer $\Omega(\mathcal{B})$ is called the *convergence order* of $\mathcal{B}$.

- the scaling with N in Eq. (2.14) is the *unique scaling* such that:
  - the convergence order $\Omega(\mathcal{B})$ is positive *for all* $\mathcal{B}$.
  - *for all* $\mathcal{B}$, there exists an *infinite* family of invariants (graphs $\mathcal{B}'$) such that $\Omega(\mathcal{B}) = \Omega(\mathcal{B}')$.

In the case of matrices, $D = 2$, denoting $\mathcal{B}^{(2k)}$ the bicolored cycles with $2k$ white vertices we have the well known values:

$$\Omega(\mathcal{B}^{(2k)}) = 0 , \qquad R(\mathcal{B}^{(2k)}) = \frac{1}{k+1}\binom{2k}{k} .$$

Chapter 3

Generalities on edge colored graphs

Random matrices are studied by a number of methods: eigenvalue decomposition [88], orthogonal polynomials [59] or combinatorial methods. Tensors do not possess a satisfactory notion of eigenvalues and neither the eigenvalues decomposition nor the orthogonal polynomials techniques can be applied directly. In the case of tensors one needs to rely heavily on combinatorial techniques.

It should be clear by now that D-colored graphs will be playing a central role in tensor models. In this chapter we discuss such graphs in detail. We first discuss the cellular structure of such graphs and show that they are dual to vertex colored triangulations in any dimension. We subsequently introduce the notions of open D-colored graphs and boundary graphs, and relate them to triangulations with boundary. Finally, we associate to any D-colored graph a positive integer number, the *degree*, which will play a distinguished role in tensor models.

3.1 Faces, bubbles and the D-complex

A tensor model for a tensor of rank D is studied in perturbation theory using a Feynman graph expansion. The Feynman graphs are made of invariants (which are D-colored graphs) connected by "propagators", that is, a new class of edges. Thus, the Feynman graphs of a tensor model are $(D+1)$-colored graphs. We will be assigning a new color, which we call 0, to this new class of edges. The edges of color 0 are special and will play a distinguished role later on, and for this reason we always draw them as dashed edges (as opposed to the edges of colors $1,\dots D$ which we draw as solid edges). However, from the point of view of combinatorics, the edges of color 0 are no different from the others.

We shift D to $(D+1)$ and we denote a connected $(D+1)$-colored graph (having edges of colors $0,1,\ldots D$) by $\mathcal{G}$. We denote $k(\mathcal{G}) = \frac{1}{2}|\mathcal{V}(\mathcal{G})|$ the number of white vertices of $\mathcal{G}$, $E^c(\mathcal{G}) = |\mathcal{E}^c(\mathcal{G})|$ the number of edges of color c of $\mathcal{G}$, and $E(\mathcal{G}) = |\mathcal{E}(\mathcal{G})|$ the total number of edges of $\mathcal{G}$. Of course:

$$E^c(\mathcal{G}) = k(\mathcal{G}) \,, \qquad E(\mathcal{G}) = (D+1)k(\mathcal{G}) \,.$$

Due to the coloring of the edges, the edge-colored graphs have a very rich topological structure. The notions defined in Definition 3.1 and Definition 3.2 are exemplified in Figure 3.1.

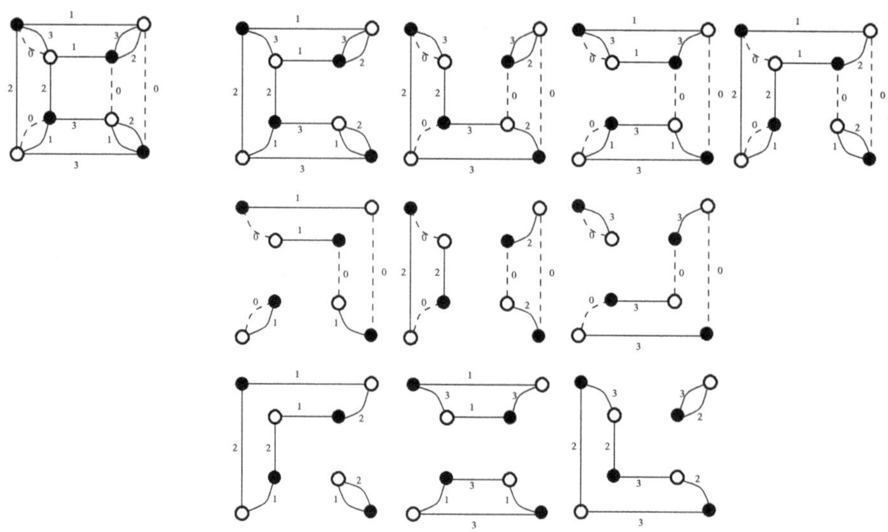

Figure 3.1: Faces and bubbles of a graph.

Definition 3.1. *The* faces *of the $(D+1)$-colored graph $\mathcal{G}$ are its maximal connected subgraphs made of edges of two fixed (distinct) colors.*

The faces are cycles of edges of alternating color. We denote the set of faces of $\mathcal{G}$ made of edges of colors c and c' by $\mathcal{F}^{(c,c')}(\mathcal{G})$ (and $F^{(c,c')}(\mathcal{G}) = |\mathcal{F}^{(c,c')}(\mathcal{G})|$ their number), respectively, $\mathcal{F}(\mathcal{G})$ the set of all faces of $\mathcal{G}$ (and $F(\mathcal{G}) = |\mathcal{F}(\mathcal{G})|$ their number).

Definition 3.2. *For $d \le D$, The* d-cells, *or* d-bubbles *of the $(D+1)$-colored graph $\mathcal{G}$ are its maximal connected subgraphs made of edges with d fixed (distinct) colors.*

The 0-bubbles are the subgraphs of $\mathcal{G}$ having no edges; that is, they are the vertices of $\mathcal{G}$. The 1-bubbles are its edges. The 2-bubbles are its faces. A $(D+1)$-colored graph $\mathcal{G}$ has 0-bubbles, 1-bubbles, 2-bubbles and so on up to D-bubbles.

A vertex belongs to a unique d-bubble with colors $c_1, \ldots c_d$. The vertex v_j, all the vertices $\bar{v}_{\tau^{(c_1)}(j)}$, $\bar{v}_{\tau^{(c_2)}(j)}$ up to $\bar{v}_{\tau^{(c_d)}(j)}$ and all the edges $(v_j, \bar{v}_{\tau^{(c_1)}(j)})$, $(v_j, \bar{v}_{\tau^{(c_2)}(j)})$ up to $(v_j, \bar{v}_{\tau^{(c_d)}(j)})$ belong to the same bubble with colors $c_1, \dots c_d$. The bubbles themselves are d-colored graphs as defined in Definition 2.1. If the $(D+1)$-colored graph $\mathcal{G}$ is defined by the permutations:

$$\tau^{\{0\} \cup \mathcal{D}} = (\tau^{(0)}, \tau^{(1)}, \dots \tau^{(D)}) ,$$

then the d bubbles of colors $\mathcal{C} \subset \{0\} \cup \mathcal{D}$ are the connected components of the d-colored graph defined by the permutations $\tau^{\mathcal{C}} = \{\tau^{(c)} | c \in \mathcal{C}\}$.

A $(d-1)$-bubble with colors $\{c_1, \dots, c_{d-1}\}$ is a subgraph of a unique d-bubble with colors $\{c_1, \dots, c_{d-1}, c_d\}$. Starting from this remark, after some work [93, 90], one eventually finds that any $(D+1)$-colored graph is a D-dimensional cellular complex, hence it is a D-dimensional topological space. All the topological information is encoded in the colors of the edges. A $(D+1)$-colored graph has therefore a well defined Poincaré dual (in the sense of D-dimensional topological spaces), which turns out to be a D-dimensional finite abstract simplicial pseudo manifold [69, 138, 93, 90].

Before presenting in detail this dual D-dimensional pseudo manifold, it is worth mentioning that the initial tensor models proposed [3, 153, 82] failed to give the promised theory of higher dimensional random spaces because they generated only 2-complexes and not higher dimensional cellular complexes. This seemingly benign point effectively halted the development of tensor models for twenty years: instead of searching for a theory of random tensors generating only higher dimensional cellular complexes (which *a posteriori* is clearly the right thing to do), initial efforts [56] concentrated on trying to associate some gluing of higher dimensional simplices to the 2-complexes generated. However, as not every 2-complex is a subcomplex of a higher dimensional complex (which is rather obvious in general and is true also for the subclass of 2-complexes generated by the original tensor models), this strategy is of limited interest. No matter how hard one tries, the ensuing gluing is not some D-dimensional Poincaré dual of the graph 2-complex, as such an object simply does not exist. The "associated" gluing is just an ad hoc construction whose topology is not encoded in the graph, therefore useless.

This is the first important problem solved by the modern tensor models presented in this book.

3.2 The dual triangulation

The $(D+1)$-colored graphs are dual to genuine D-dimensional vertex colored triangulations.

Definition 3.3. *A bipartite, vertex colored D-dimensional triangulation Δ is a gluing of D-simplices such that:*

- *only pairs of positive and negative oriented D-simplices are glued along $(D-1)$-simplices.*

- *all the vertices (0-simplices) have a label $0, 1, \dots D$ such that all the $D+1$ vertices belonging to the same D simplex have distinct labels.*

Because all d-simplices of a triangulation are subsimplices in (several) D-simplices, all the $d+1$ vertices of any d-simplex have distinct colors.

It is important to note that any polyhedral gluing can be transformed into a vertex colored triangulation by a barycentric subdivision. Indeed, under barycentric subdivision all the cells are supplemented with their barycenter and any polyhedron is decomposed into simplices having exactly one vertex of the original polyhedral gluing, one barycenter of a 1-cell, one barycenter of a 2-cell, and so on. Coloring the barycenter of the cells of equal dimension with the same color yields the vertex colored triangulation.

The local topological information in a triangulation is encoded in the *links* of its simplices [117]. The link of a simplex σ in the triangulation Δ is the triangulation composed of simplices τ such that $\tau \cap \sigma = \emptyset$ and $\tau \cup \sigma$ is a simplex of Δ. The link is the gluing of the complements of σ in all the D-simplices to which it belongs.

For instance, for $D = 3$, a $3+1$ colored graph is dual to a gluing Δ of tetrahedra. If σ is a vertex, its complement in a tetrahedron is a triangle. The link of σ is then the surface obtained by gluing together these complementary triangles, i.e. the boundary of the neighborhood of the vertex (see Figure 3.3). The neighborhood of the vertex (called the *star* [117] of that vertex) is the topological cone over the link.

As already mentioned, the $(D+1)$-colored graphs $\mathcal{G}$ are dual to bipartite, vertex colored triangulations [70, 90, 105]. The construction of Δ from $\mathcal{G}$ (and of $\mathcal{G}$ from Δ) are self-evident. Starting from $\mathcal{G}$ we obtain Δ as follows:

- for all vertices v (resp. $\bar{v}$) in $\mathcal{V}(\mathcal{G})$, take a positive (resp. negative) oriented D-simplex Δ_v (resp. $\Delta_{\bar{v}}$). Label its vertices $\mathbf{v}_v^{(0)}$, $\mathbf{v}_v^{(1)}$, ... $\mathbf{v}_v^{(D)}$ (resp. $\mathbf{v}_{\bar{v}}^{(0)}$, $\mathbf{v}_{\bar{v}}^{(1)}$, ... $\mathbf{v}_{\bar{v}}^{(D)}$), hence:

$$\Delta_v = \{\mathbf{v}_v^{(0)}, \mathbf{v}_v^{(1)}, \dots \mathbf{v}_v^{(D)}\} , \qquad \Delta_{\bar{v}} = \{\mathbf{v}_{\bar{v}}^{(D)}, \dots \mathbf{v}_{\bar{v}}^{(1)}, \mathbf{v}_{\bar{v}}^{(0)}\} .$$

- for an edge of color c connecting v and $\bar{v}$, glue Δ_v and $\Delta_{\bar{v}}$ along the $(D-1)$-simplex complementary to (i.e. not containing) the vertices $\mathbf{v}_v^{(c)}$ respectively $\mathbf{v}_{\bar{v}}^{(c)}$:

$$\{\mathbf{v}_v^{(0)}, \mathbf{v}_v^{(1)}, \dots \mathbf{v}_v^{(D)}\} \setminus \{\mathbf{v}_v^{(c)}\} , \qquad \{\mathbf{v}_{\bar{v}}^{(D)}, \dots \mathbf{v}_{\bar{v}}^{(1)}, \mathbf{v}_{\bar{v}}^{(0)}\} \setminus \{\mathbf{v}_{\bar{v}}^{(c)}\} ,$$

  by identifying the vertices with the same label $\mathbf{v}_v^{(c_1)} \equiv \mathbf{v}_{\bar{v}}^{(c_1)}$, $\forall c_1 \neq c$.

Conversely, starting from Δ:

- draw the connectivity graph of the top simplices (obtained by drawing a vertex for each D-simplex, and connecting two vertices by an edge if the corresponding simplices share a $(D-1)$-simplex).

- color each edge of this graph by the complement in the set $\{0, 1, \dots D\}$ of the labels of the vertices of the $(D-1)$-simplex it represents.

For $d \geq 2$, the d-bubbles of $\mathcal{G}$ are d-colored graphs. A d-bubble with colors $c_1, \dots c_d$ represents the gluing of subsimplices with vertices $\{\mathbf{v}_v^{(c_1)}, \dots \mathbf{v}_v^{(c_d)}\}$, therefore it represents the link of a simplex with vertices $\{\mathbf{v}_v^{(c)} | \, c \neq c_1, \dots c_d\}$.

Let us explain in more detail the cases $D = 2$ and $D = 3$.

$D = 2$. The graphs with $2+1 = 3$ colors represent triangulated surfaces [93]. Each vertex of $\mathcal{G}$ represents a triangle with colored vertices, and each edge of $\mathcal{G}$ represents a gluing of two triangles. An example is presented in Figure 3.2.

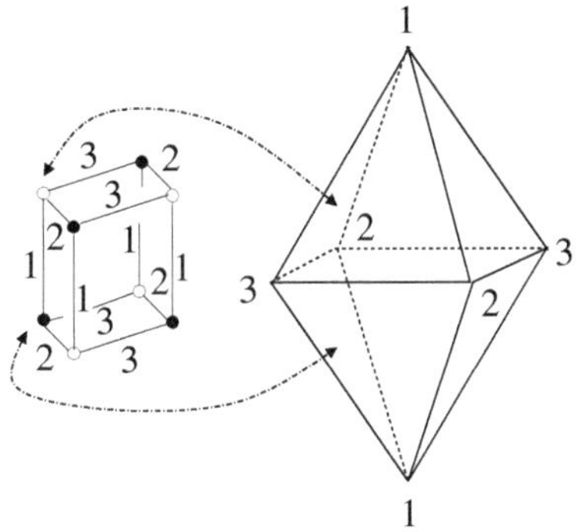

Figure 3.2: Triangulated surfaces and 3-colored graphs.

$D = 3$. The $(3+1)$-colored graphs are topological three-dimensional pseudo manifolds [93]. The black and white vertices of the graph correspond to tetrahedra (3-simplices) with colored vertices. In Figure 3.3(a) we represented a piece of a vertex colored three-dimensional triangulation and its associated $(3+1)$-colored graph.

The link of the vertex with color 0 is the triangulated surface associated with the subgraph (bubble) with colors 1, 2, and 3, as represented in Figure 3.3(b).

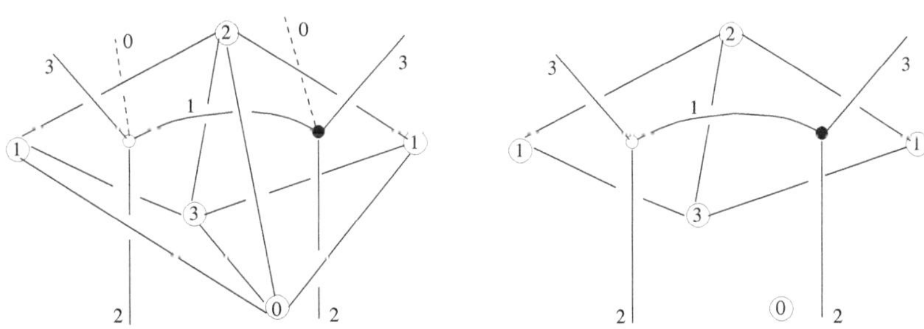

(a) Vertex colored triangulation. (b) The link of a vertex.

Figure 3.3: The triangulation associated with a $(3+1)$-colored graph $\mathcal{G}$.

Alternatively, a graph with $3+1$ colors can be reconstructed by first considering its bubbles $\mathcal{B}$ with colors 1, 2 and 3 and then connecting them via edges of color 0. This is done in two steps: one first decorates every vertex by a halfedge of color 0 and subsequently one connects these halfedges into edges.

Each bubble by itself, being a graph with three colors, represents a triangulated surface. The vertices of the corresponding triangulation are colored 1, 2 and 3. Adding the halfedges of color 0 consists of taking the topological cone over this surface by placing a central vertex with color 0. The topological cone with apex $\mathbf{v}^{(0)}$ over a d dimensional triangulation Δ is the $(d+1)$ dimensional triangulation with boundary consisting of the new vertex $\mathbf{v}^{(0)}$, all the simplices $\sigma \in \Delta$ and all the *coned* simplices $\mathbf{v}^{(0)} \cup \sigma$, $\forall \sigma \in \Delta$.

In Figure 3.4 we represent the topological cone over one triangle, which is a tetrahedron.

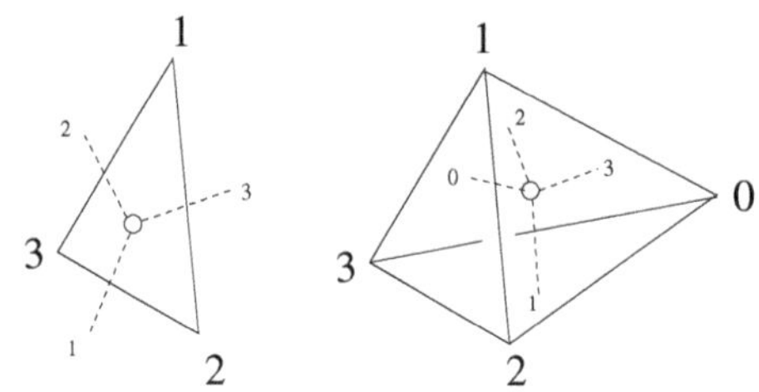

Figure 3.4: Coning of a triangle.

The cone over the surface in Figure 3.2 is represented in Figure 3.5.

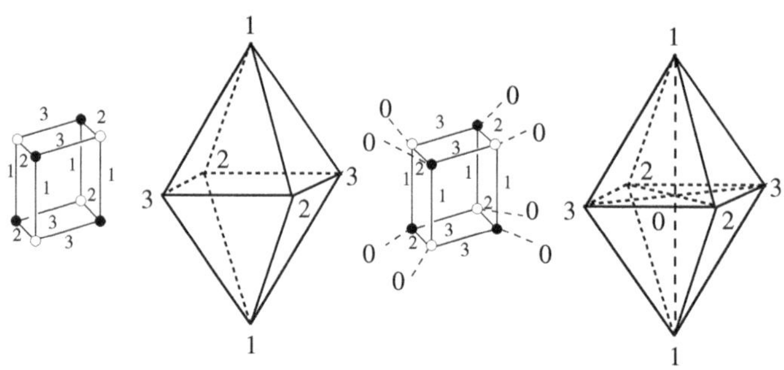

Figure 3.5: Coning of a surface.

Finally, connecting the halfedges of color 0 into edges represents the gluing of the cones along their boundary triangles (which have vertices of colors 1, 2 and 3).

Remark 3.1. We have the following remark.

Simplicial manifolds. A D-dimensional pseudo manifold is a manifold if the neighborhood of any point is topologically a D-ball. Any D-ball is the cone over a $(D-1)$-dimensional sphere. As the neighborhood of a vertex in Δ (the star) is the cone over the link of the vertex, it follows that a $(D+1)$-colored graph is dual to a manifold if and only if all its links are spheres.

Consider a connected $(D+1)$-colored graph $\mathcal{G}$ for $D=3$. The links of the vertices (i.e. the three bubbles) represent surfaces. The neighborhoods of the vertices are cones over these surfaces. The Euler character of a bubble is $2k(\mathcal{B})-E(\mathcal{B})+F(\mathcal{B}) = 2-2g_{\mathcal{B}}$. Summing over the bubbles and taking into account that every face belongs to two bubbles, every edge belongs to three bubbles, every vertex belongs to four bubbles and $E(\mathcal{G}) = 4k(\mathcal{G})$, we get:

$$2k(\mathcal{G}) - E(\mathcal{G}) + F(\mathcal{G}) - B(\mathcal{G}) = -\sum_{\mathcal{B}} g_{\mathcal{B}} ,$$

where $B(\mathcal{G})$ is the total number of 3-bubbles and the sum on the right hand side runs over all 3-bubbles. In particular, a three-dimensional pseudo manifold is a manifold if and only if its Euler characteristic is zero.

Before concluding this section let us remark that the fact that the graphs we are considering are bipartite (which is equivalent to the orientability of Δ) plays a secondary role in this construction. What is crucial is that a colored graph represents a unique vertex colored triangulation. Dropping the bipartiteness condition allows one to consider the graph presented in Figure 3.6. As it represents a gluing of four triangles and any two triangles share exactly one edge one might be tempted to interpret it as a gluing pattern of four triangles bounding a tetrahedron. However, this is wrong: the vertex colored gluing associated with this graph has the topology of a real projective plane $\mathbb{R}P^2$.

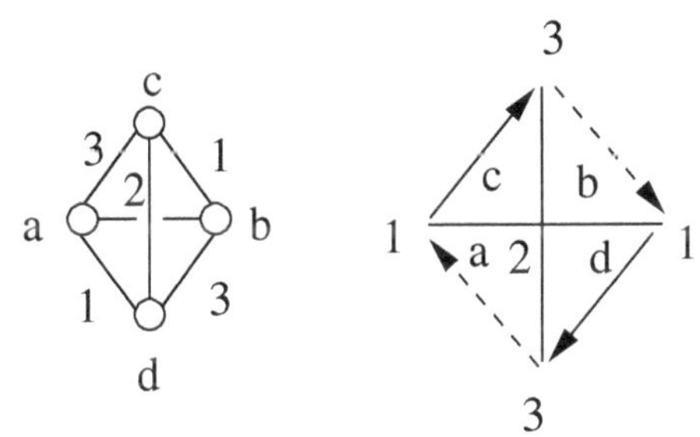

Figure 3.6: An edge colored, nonbipartite graph.

3.3 Open graphs and the boundary graph

So far we have discussed closed D or $(D+1)$-colored graphs. We will now define open $(D+1)$-colored graphs, that is, graphs having some external halfedges. They are dual to triangulations with boundary.

Definition 3.4. *A bipartite* open *edge* $(D+1)$*-colored graph is a graph* $\mathcal{G} = \big(\mathcal{V}(\mathcal{G}), \mathcal{E}(\mathcal{G})\big)$ *with vertex set* $\mathcal{V}(\mathcal{G})$ *and edge set* $\mathcal{E}(\mathcal{G})$ *such that:*

- *$\mathcal{V}(\mathcal{G})$ is bipartite, i.e. it is written as the disjoint union of white and black vertices $\mathcal{V}(\mathcal{G}) = \mathcal{V}^w(\mathcal{G}) \cup \mathcal{V}^b(\mathcal{G})$, such that for any edge $e \in \mathcal{E}(\mathcal{G})$ then $e = (v, \bar{v})$ with $v \in \mathcal{V}^w(\mathcal{G})$ and $\bar{v} \in \mathcal{V}^b(\mathcal{G})$.*
- *The edge set is partitioned into $(D+1)$ subsets $\mathcal{E}(\mathcal{G}) = \bigcup_{c=0}^{D} \mathcal{E}^c(\mathcal{G})$, where $\mathcal{E}^c(\mathcal{G}) = \{e^c = (v, \bar{v})\}$ is the subset of edges with color c.*
- *The white (black) vertices are further partitioned in two categories:*
  - internal *vertices. They are $(D+1)$-valent and all the edges incident to an internal vertex have distinct colors.*
  - external *vertices. They are 1-valent and the edge incident to an external vertex has color 0.*

The set of edges of color 0 is then naturally partitioned into internal *edges connecting internal vertices and* external *edges connecting internal vertices with external vertices, $\mathcal{E}^0(\mathcal{G}) = \mathcal{E}^0_{\text{int}}(\mathcal{G}) \cup \mathcal{E}^0_{\text{ext}}(\mathcal{G})$. All the edges of color $c \neq 0$ are internal.*

We call the internal vertices of $\mathcal{G}$ hooked to the external edges boundary vertices. *Their number of course equals the number of external vertices of $\mathcal{G}$.*

It is easy to see that the numbers of black and white vertices of each type must be equal. Most of the time we forget the external vertices of open graphs and consider the external edges as *halfedges* decorating the boundary vertices of $\mathcal{G}$.

We denote the number of white internal vertices of an open graph $\mathcal{G}$ by $k_{\text{int}}(\mathcal{G})$ and the number of white external (and boundary) vertices by $k_{\text{ext}}(\mathcal{G})$. We emphasize that the $k_{\text{ext}}(\mathcal{G})$ boundary vertices of $\mathcal{G}$ are some of the $k_{\text{int}}(\mathcal{G})$ internal vertices of $\mathcal{G}$.

Remark 3.2. We have the following remark.

Counting open graphs. We are interested in counting the number of open graphs with labeled vertices. When counting open graphs, it is more convenient to consider graphs with labeled vertices such that the internal and external vertices are labeled independently. Let us say that the internal vertices are labeled:

$$v_1, \bar{v}_1, \dots v_{k_{\text{int}}}, \bar{v}_{k_{\text{int}}} \,,$$

and the external ones are labeled:

$$w_1, \bar{w}_1, \dots w_{k_{\text{ext}}}, \bar{w}_{k_{\text{ext}}} \,.$$

In order to count the total number of $(D+1)$-colored open graphs with vertices labeled in this way, we count $\binom{k_{\text{int}}}{k_{\text{ext}}}$ choices of the white (resp. black) boundary vertices, $k_{\text{ext}}!$ relabelings of the white (resp. black) external vertices and $[k_{\text{int}}!]^D (k_{\text{int}} - k_{\text{ext}})!$ choices for the internal edges to obtain a total of:

$$\frac{[k_{\text{int}}!]^{D+2}}{(k_{\text{int}} - k_{\text{ext}})!} \,,$$

open (not necessarily connected) $(D+1)$-colored graphs.

This coning can also be understood in the following manner. There are $[k_{\text{int}}!]^{D+1}$ edge $D+1$-colored graphs with k_{int} labeled vertices. For each of these graphs there are $\binom{k_{\text{int}}}{k_{\text{ext}}}$ choices of edges of color 0 that we can cut into two halfedges. Observe that every open graph is obtained $k_{\text{ext}}!$ times (once for each of the pairings of its external halfedges into edges). Hence there are:

$$\frac{[k_{\text{int}}!]^{D+2}}{(k_{\text{int}}-k_{\text{ext}})!k_{\text{ext}}!k_{\text{ext}}!},$$

graphs with k_{int} labeled vertices and k_{ext} external halfedges. We finally add univalent external vertices at the end of the external halfedges which are labeled in $[k_{\text{ext}}!]^2$ distinct ways ($k_{\text{ext}}!$ ways for the black and $k_{\text{ext}}!$ ways for the white vertices).

Definition 3.5. *We call an open graph with* $2k_{\text{ext}}$ *external vertices a* $2k_{\text{ext}}$-open graph.

Two examples of open $(3+1)$-colored graphs are presented in Figure 3.7 on the left. Both graphs have four external halfedges (or four external vertices).

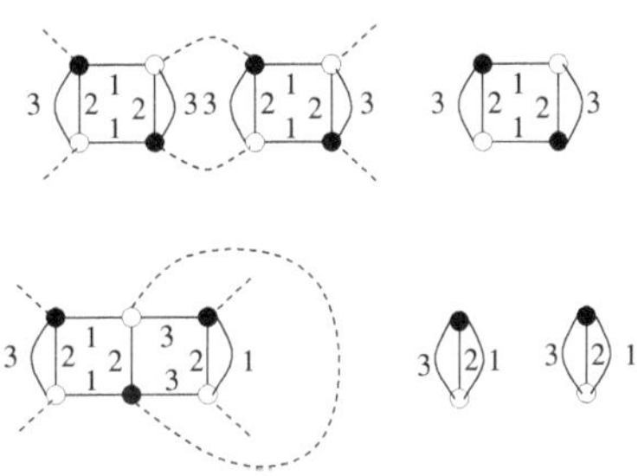

Figure 3.7: Open graphs and their boundary graphs.

Faces are still defined according to Definition 3.1 as subgraphs with two colors. For open $(D+1)$-colored graphs, they fall into two categories. Either (like for closed graphs) they are bicolored cycles of edges, in which case they contain only internal edges and we call them *internal faces*, or they are chains of edges, in which case they necessarily contain two external halfedges and we call them *external faces*.

All the faces (c_1, c_2) with both colors distinct from zero are internal. The faces with colors $(0, c)$ can be either internal of external. We denote the set of internal faces of colors $(0, c)$ of $\mathcal{G}$ by $\mathcal{F}^{(0,c)}_{\text{int}}(\mathcal{G})$ (and its cardinal by $F^{(0,c)}_{\text{int}}(\mathcal{G})$) and the set of external faces of colors $(0, c)$ by $\mathcal{F}^{(0,c)}_{\text{ext}}(\mathcal{G})$ (and its cardinal by $F^{(0,c)}_{\text{ext}}(\mathcal{G})$).

The external faces $f \in \mathcal{F}^{(0,c)}_{\text{ext}}$ necessarily start and end with an external halfedge of color 0, hooked each to a boundary vertex $u(f)$ and $\bar{u}(f)$ of $\mathcal{G}$.

Definition 3.6. *The* boundary *graph of* $\mathcal{G}$ *is the D-colored graph with vertex set* $\mathcal{V}(\partial\mathcal{G})$ *and edge set* $\mathcal{E}(\partial\mathcal{G}) = \cup_{c=1}^{D}\mathcal{E}^c(\partial\mathcal{G})$ *such that:*

- *the vertex set* $\mathcal{V}(\partial\mathcal{G})$ *is the set of boundary vertices of* $\mathcal{G}$.

- *the set of edges of color c, $\mathcal{E}^c(\partial\mathcal{G})$, is:*

$$\mathcal{E}^c(\partial\mathcal{G}) = \bigcup_{f\in\mathcal{F}^{(0,c)}_{\text{ext}}(\mathcal{G})} \Big\{\big(u(f),\bar{u}(f)\big)\Big\},$$

that is, for each external face $(0,c)$ of $\mathcal{G}$ we draw an edge of color c in $\partial\mathcal{G}$ connecting the two boundary vertices belonging to that face.

On the right in Figure 3.7 we represented the boundary graphs $\partial\mathcal{G}$ of the two graphs $\mathcal{G}$. The boundary graph is a D-colored graph, with edges of colors $1,\dots D$.

Note that in the second example in Figure 3.7, in spite of the fact that $\mathcal{G}$ itself is connected, the boundary graph $\partial\mathcal{G}$ can be disconnected.

Open $(D+1)$-colored graphs are dual to D-dimensional triangulations Δ with a boundary. The boundary graph $\partial\mathcal{G}$ is dual to the boundary of the triangulation, $\partial\Delta$. This can be seen as follows: every boundary vertex in $\mathcal{G}$ (which is a vertex also in $\partial\mathcal{G}$) represents a boundary D-simplex in Δ (which contributes a $(D-1)$-simplex to $\partial\Delta$). The D-colored graph $\partial\mathcal{G}$ encodes the connectivity of these $(D-1)$-simplices.

Let us consider the case $D=3$, exemplified in Figure 3.8.

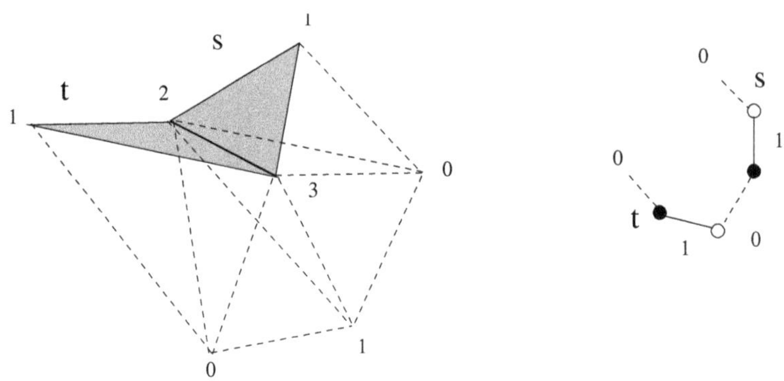

Figure 3.8: Triangulation with boundary.

The two shaded triangles $\{\mathbf{v}_s^{(1)},\mathbf{v}^{(2)},\mathbf{v}^{(3)}\}$ and $\{\mathbf{v}_t^{(1)},\mathbf{v}^{(2)},\mathbf{v}^{(3)}\}$ belong to the boundary of the triangulation, and are glued along their common edge $\{\mathbf{v}^{(2)},\mathbf{v}^{(3)}\}$. The edge $\{\mathbf{v}^{(2)},\mathbf{v}^{(3)}\}$ belongs to several tetrahedra in the triangulation (four tetrahedra in our example). Let us consider more closely these tetrahedra. The triangle $\{\mathbf{v}_s^{(1)},\mathbf{v}^{(2)},\mathbf{v}^{(3)}\}$ belongs to the tetrahedron:

$$v_s = \{\mathbf{v}_1^{(0)},\mathbf{v}_s^{(1)},\mathbf{v}^{(2)},\mathbf{v}^{(3)}\},$$

which is represented in $\mathcal{G}$ as a vertex v_s with an external halfedge of color 0. The edge $\{\mathbf{v}^{(2)},\mathbf{v}^{(3)}\}$ is then shared between v_s and another tetrahedron, say

$$\bar{v}_1 = \{\mathbf{v}_1^{(0)},\mathbf{v}_1^{(1)},\mathbf{v}^{(2)},\mathbf{v}^{(3)}\},$$

glued to v_s by the triangle $\{\mathbf{v}_1^{(0)}, \mathbf{v}^{(2)}, \mathbf{v}^{(3)}\}$. It follows that $\bar{v}_1$ is the vertex of $\mathcal{G}$ connected to v_s by the edge of color 1. Now $\{\mathbf{v}^{(2)}, \mathbf{v}^{(3)}\}$ will be shared between $\bar{v}_1$ and

$$v_2 = \{\mathbf{v}_2^{(0)}, \mathbf{v}_1^{(1)}, \mathbf{v}^{(2)}, \mathbf{v}^{(3)}\} ,$$

the tetrahedron glued to v_1 along the triangle $\{\mathbf{v}_1^{(1)}, \mathbf{v}^{(2)}, \mathbf{v}^{(3)}\}$, hence represented in $\mathcal{G}$ by the vertex v_2 connected to $\bar{v}_1$ by an edge of color 0. Finally, $\{\mathbf{v}^{(2)}, \mathbf{v}^{(3)}\}$ also belongs to the tetrahedron,

$$\bar{v}_t = \{\mathbf{v}_2^{(0)}, \mathbf{v}_t^{(1)}, \mathbf{v}^{(2)}, \mathbf{v}^{(3)}\} ,$$

which is represented in $\mathcal{G}$ by the vertex $\bar{v}_t$ connected to v_2 by the edge of color 1.

Observe that, if the $(D+1)$-colored graph $\mathcal{G}$ consists in a D-colored graph $\mathcal{B}$ decorated by halfedges of color 0, then $\mathcal{B} = \partial\mathcal{G}$. In this case taking the boundary is the inverse operation of the coning, as exemplified in Figure 3.9.

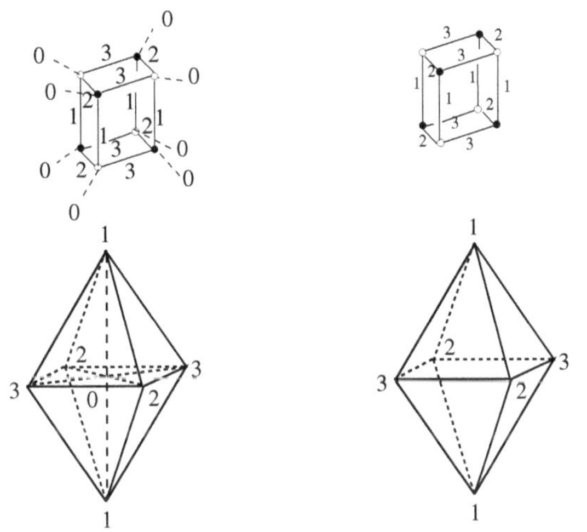

Figure 3.9: Taking the boundary versus coning.

3.3.1 The contraction of edges of color 0

The boundary graph $\partial\mathcal{G}$ can also be obtained by an iterative procedure of *contraction* of the internal edges of color 0 (see Figure 3.10).

Definition 3.7. *Choose* $e^0 = (v, \bar{v})$ an internal edge of color 0 *in* $\mathcal{G}$. *We define the graph* $\mathcal{G}$ *contracted by* e^0, *which we denote* $\mathcal{G}_{/e^0}$, *as the graph obtained from* $\mathcal{G}$ *by deleting* e^0 *(and all other edges* $e^c = (v, \bar{v})$ *connecting the same vertices* v *and* $\bar{v}$ *if they exist) together with the end vertices* v *and* $\bar{v}$, *and reconnecting the remaining edges (which were hooked to* v *and* $\bar{v}$ *in* $\mathcal{G}$*) respecting the colorings.*

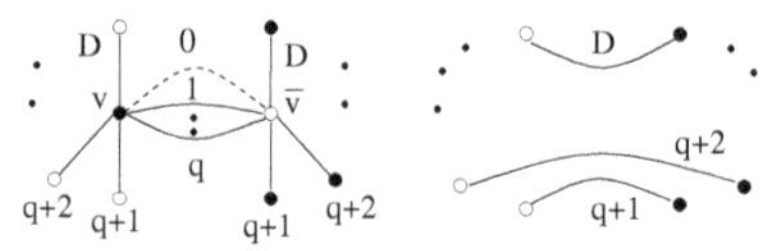

Figure 3.10: The contraction of an edge of color 0.

The contraction of all the internal edges of color 0 in the graphs represented in Figure 3.7 is represented in Figure 3.11. Remark that the end graph obtained in both cases is the boundary graph decorated by external halfedges. This is in fact true for any graph [18].

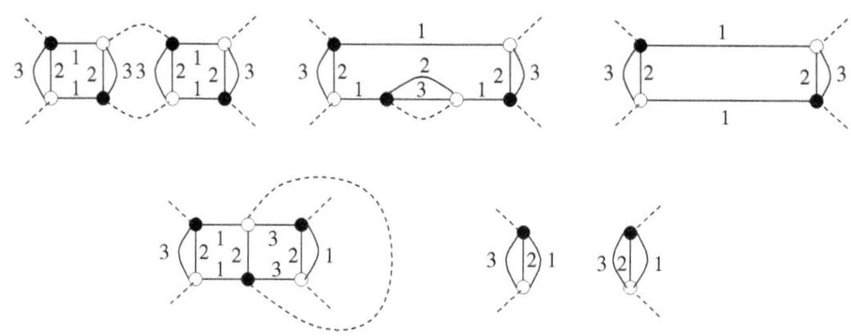

Figure 3.11: The contraction of all the internal edges of color 0 in a graph.

Theorem 3.1. *Contracting all the internal edges of color* 0 *of* $\mathcal{G}$ *in an arbitrary order we obtain the graph* $\partial\mathcal{G} \cup \mathcal{E}^0_{\text{ext}}(\partial\mathcal{G})$*, the boundary graph* $\partial\mathcal{G}$ *decorated by an external halfedge of color* 0 *on each of its vertices.*

Proof. We denote $\mathcal{G}^{(s)}$ the graph obtained after contracting s internal edges of color 0. The contraction of the edge $e^0 = (v, \bar{v})$ leads to $\mathcal{G}^{(s+1)} = \mathcal{G}^{(s)}_{/e^0}$.

Vertices. The graph $\mathcal{G}^{(s+1)}$ can be disconnected. However, while it has two vertices fewer than $\mathcal{G}^{(s)}$, it has the same boundary vertices as $\mathcal{G}^{(s)}$ as boundary vertices cannot be deleted by the contraction.

Edges. The graph $\mathcal{G}^{(s+1)}$ has one internal edge fewer for each color, not only 0 but also all the other $c \neq 0$. However, it has the same external halfedges as $\mathcal{G}^{(s)}$ because external halfedges cannot be deleted by the contraction.

Faces. The graph $\mathcal{G}^{(s+1)}$ might have fewer internal faces than $\mathcal{G}^{(s)}$. However, it has the same external faces because the external faces cannot be modified by the contraction.

The last statement is proved as follows. If there exist edges $e^{c_1}, \dots e^{c_q}$ connecting the same vertices v and $\bar{v}$ as e^0, then the q internal faces of colors $(0, c_1)$, $(0, c_2)$ up to $(0, c_q)$ formed only by the edges $\{e^0, e^{c_1}\}$, $\{e^0, e^{c_2}\}$ up to $\{e^0, e^{c_q}\}$ are deleted. All the other internal (resp. external) faces of colors $(0, c)$ containing e^0 (hence with $c \neq c_1, \dots c_q$) are circuits (resp. chains) of edges with alternating colors of length at least four (resp.

three, not counting the external halfedges for the external faces). Under the contraction their length decreases by two: the edge e^0 and an edge of color c, hence they become circuits (resp. chains) of edges with alternating colors 0 and c of length at least two (resp. one). They are thus internal (resp. external) faces in the new graph $\mathcal{G}^{(s+1)}$.

The final graph $\mathcal{G}^{(s_{\max})}$ obtained after contracting all the internal edges of color 0 is such that:

- $\mathcal{G}^{(s_{\max})}$ has only boundary vertices (otherwise it would still contain internal edges of color 0). Therefore the vertices of $\mathcal{G}^{(s_{\max})}$ are exactly the boundary vertices of $\mathcal{G}$.

- $\mathcal{G}^{(s_{\max})}$ has no more internal edges of color 0. It only has external halfedges of color 0.

- $\mathcal{G}^{(s_{\max})}$ has no more internal faces of colors $(0,c)$ (otherwise it would still contain internal edges of color 0). It only has external faces of colors $(0,c)$ which moreover are one to one to the external faces of $\mathcal{G}$.

In conclusion, each face f of colors $(0,c)$ of $\mathcal{G}^{(s_{\max})}$ corresponds to an external face $\tilde{f}$ of $\mathcal{G}$, must contain exactly one edge of color $c \neq 0$ and must connect the boundary vertices $u(\tilde{f})$ and $\bar{u}(\tilde{f})$.

□

Theorem 3.2. *For any open, connected, $(D+1)$-colored graph $\mathcal{G}$:*

$$(D-1)B(\mathcal{G}) - (D-1)E^0(\mathcal{G}) + \sum_{c\in\mathcal{D}} F_{\text{int}}^{(0,c)}(\mathcal{G}) \leq D - 2(D-1)k(\partial\mathcal{G}) - C(\partial\mathcal{G}) ,$$

where

- *$B(\mathcal{G})$ is the number of bubbles with colors $\mathcal{D} = \{1,\dots D\}$, $E^0(\mathcal{G})$ is the total number of edges (internal and external halfedges) of color 0 and $F_{\text{int}}^{(0,c)}(\mathcal{G})$ is the number of internal faces of color $(0,c)$ of $\mathcal{G}$.*

- *$k(\partial\mathcal{G})$ and $C(\partial\mathcal{G})$ are the total numbers of white vertices and of connected components of $\partial\mathcal{G}$.*

Proof. We will show that, upon contracting an edge of color 0, the quantity:

$$\begin{aligned} Q(s) =& D - C(\mathcal{G}^{(s)}) + (D-1)\big[B(\mathcal{G}^{(s)}) - C(\mathcal{G}^{(s)})\big] \\ &- (D-1)E^0(\mathcal{G}^{(s)}) + \sum_c F_{\text{int}}^{(0,c)}(\mathcal{G}^{(s)}) , \end{aligned}$$

is strictly increasing. As

$$Q(0) = (D-1)B(\mathcal{G}) - (D-1)E^0(\mathcal{G}) + \sum_c F_{\text{int}}^{(0,c)}(\mathcal{G}) ,$$

and

$$Q(s_{\max}) = D - C(\partial\mathcal{G}) - 2(D-1)\, k(\partial\mathcal{G}) ,$$

because the end graph is $\mathcal{G}^{(s_{\max})} = \partial\mathcal{G} \cup \mathcal{E}^0_{\text{ext}}(\partial\mathcal{G})$, this will prove the lemma.

Suppose that, when going from $\mathcal{G}^{(s)}$ to $\mathcal{G}^{(s+1)}$, we contract the edge $e^0 = (v, \bar{v})$ and v and $\bar{v}$ are also connected by the edges $e^{c_1}, \dots e^{c_q}$. Then there are two cases:

$v \in \mathcal{B}_1$ *and* $\bar{v} \in \mathcal{B}_2$. The two vertices v and $\bar{v}$ belong to two distinct bubbles of colors $\mathcal{D}$. In this case no other edge e^c with $c \neq 0$ can connect v and $\bar{v}$. The two bubbles $\mathcal{B}_1$ and $\mathcal{B}_2$ necessarily belong to the same connected component of $\mathcal{G}^{(s)}$. The number of connected components of $\mathcal{G}^{(s)}$ does not change under this contraction, $C(\mathcal{G}^{(s+1)}) = C(\mathcal{G}^{(s)})$ (one can always build a spanning tree in the connected component of $\mathcal{G}^{(s)}$ starting from e^0 and the edges of color 1, e^1_v and $e^1_{\bar{v}}$ hooked to v and $\bar{v}$ and completing to a spanning tree, which remains a spanning tree in the corresponding connected complement of $\mathcal{G}^{(s+1)}$).

The two bubbles $\mathcal{B}_1, \mathcal{B}_2 \subset \mathcal{G}^{(s)}$ are collapsed into a unique bubble of $\mathcal{G}^{(s+1)}$ thus $B(\mathcal{G}^{(s+1)}) = B(\mathcal{G}^{(s)}) - 1$. The number of edges of color 0 decreases by 1, $E^0(\mathcal{G}^{(s+1)}) = E^0(\mathcal{G}^{(s)}) - 1$, and the number of internal faces of color $(0, c)$ does not change $F^{(0,c)}_{\text{int}}(\mathcal{G}^{(s+1)}) = F^{(0,c)}_{\text{int}}(\mathcal{G}^{(s+1)})$, hence:

$$\begin{aligned} Q(s+1) &= D - C(\mathcal{G}^{(s+1)}) + (D-1)\big[B(\mathcal{G}^{(s+1)}) - C(\mathcal{G}^{(s+1)})\big] \\ &\quad - (D-1)E^0(\mathcal{G}^{(s+1)}) + \sum_c F^{(0,c)}_{\text{int}}(\mathcal{G}^{(s+1)}) \\ &= D - C(\mathcal{G}^{(s)}) + (D-1)\big[B(\mathcal{G}^{(s)}) - C(\mathcal{G}^{(s)})\big] \\ &\quad - (D-1)E^0(\mathcal{G}^{(s)}) + \sum_c F^{(0,c)}_{\text{int}}(\mathcal{G}^{(s)}) = Q(s) . \end{aligned}$$

$v, \bar{v} \in \mathcal{B}$. The two vertices v and $\bar{v}$ belong to the same bubble of colors $\mathcal{D}$. In this case the number of connected components of $\mathcal{G}^{(s)}$ can increase upon contracting e^0. As each of the new $D - q$ edges (one for each color $c \neq 0, c_1, \dots c_q$) must belong to some connected component of $\mathcal{G}^{(s+1)}$, we have $C(\mathcal{G}^{(s+1)}) - C(\mathcal{G}^{(s)}) \leq D - q - 1$. Moreover, if one of these edges belongs to a connected component created by the contraction, then it certainly belongs to a new bubble of colors 1, 2 up to D created by this contraction. Hence $C(\mathcal{G}^{(s+1)}) - C(\mathcal{G}^{(s)}) \leq B(\mathcal{G}^{(s+1)}) - B(\mathcal{G}^{(s)})$. As before, $E^0(\mathcal{G}^{(s+1)}) = E^0(\mathcal{G}^{(s)}) - 1$, but the q internal faces of colors $(0, c_1)$, $(0, c_2)$ up to $(0, c_q)$ are deleted, hence:

$$\begin{aligned} Q(s+1) &= D - C(\mathcal{G}^{(s+1)}) + (D-1)\big[B(\mathcal{G}^{(s+1)}) - C(\mathcal{G}^{(s+1)})\big] \\ &\quad - (D-1)E^0(\mathcal{G}^{(s+1)}) + \sum_c F^{(0,c)}_{\text{int}}(\mathcal{G}^{(s+1)}) \\ &\geq D - C(\mathcal{G}^{(s)}) - (D-q-1) \\ &\quad + (D-1)\big[B(\mathcal{G}^{(s)}) - C(\mathcal{G}^{(s)})\big] \\ &\quad - (D-1)E^0(\mathcal{G}^{(s)}) + D - 1 + \sum_c F^{(0,c)}_{\text{int}}(\mathcal{G}^{(s)}) - q = Q(s) , \end{aligned}$$

In both cases $Q(s+1) \geq Q(s)$.

□

3.3.2 The composition of D-colored graphs

Several connected D-colored graphs can be composed to yield a D-colored graph by adding edges of color 0 and taking the boundary.

In detail, consider q connected D-colored graphs labeled $\mathcal{B}_1, \dots \mathcal{B}_q$, and r couples of vertices $(v_i, \bar{v}_i)$ with $v_i \in \bigcup_{s=1}^q \mathcal{V}^w(\mathcal{B}_s)$ and $\bar{v}_i \in \bigcup_{s=1}^q \mathcal{V}^b(\mathcal{B}_s)$.

Definition 3.8. *The composed graph $(\mathcal{B}_1 \uplus \cdots \uplus \mathcal{B}_q)/\{(v_1, \bar{v}_1), \dots (v_r \bar{v}_r)\}$ is the boundary graph of the $(D+1)$ colored graph $\mathcal{G}$ defined as follows:*

- *the vertices of $\mathcal{G}$ are all the vertices of $\mathcal{B}_1, \dots \mathcal{B}_q$, $\mathcal{V}^w(\mathcal{G}) = \bigcup_{s=1}^q \mathcal{V}^w(\mathcal{B}_s)$ and $\mathcal{V}^b(\mathcal{G}) = \bigcup_{s=1}^q \mathcal{V}^b(\mathcal{B}_s)$.*
- *the edges of color $c \neq 0$ of $\mathcal{G}$ are all the edges of $\mathcal{B}_1, \dots \mathcal{B}_q$, $\mathcal{E}^c(\mathcal{G}) = \bigcup_{s=1}^q \mathcal{E}^c(\mathcal{B}_s)$.*
- *the internal edges of color 0 of $\mathcal{G}$ are $\mathcal{E}^0_{\text{int}}(\mathcal{G}) = \{(v_i, \bar{v}_i) | i = 1 \dots r\}$.*
- *all the remaining D-valent vertices, $\bigcup_{s=1}^q \mathcal{V}(\mathcal{B}_s) \setminus \{v_i, \bar{v}_i | i = 1 \dots r\}$, are boundary vertices of $\mathcal{G}$, hence are decorated by an external halfedge of color 0.*

Note that q can be exactly 1, in which case $\mathcal{G}$ is just $\mathcal{B}_1$ decorated by some edges (and external halfedges) of color 0. A subtle point is that the composition is associative only as long as it is defined. Indeed, if one considers $v_{12}, v_{13} \in \mathcal{V}(\mathcal{B}_1)$ and $\bar{v}_{12} \in \mathcal{V}(\mathcal{B}_2), \bar{v}_{13} \in \mathcal{V}(\mathcal{B}_3)$, then

$$\left(\left[(\mathcal{B}_1 \uplus \mathcal{B}_2)/\{(v_{12}, \bar{v}_{12})\}\right] \uplus \mathcal{B}_3\right)/\{(v_{13}, \bar{v}_{13})\}$$
$$= (\mathcal{B}_1 \uplus \mathcal{B}_2 \uplus \mathcal{B}_3)/\{(v_{12}, \bar{v}_{12}), (v_{13}, \bar{v}_{13})\} ,$$

but the expression:

$$\left(\mathcal{B}_1 \uplus \left[(\mathcal{B}_2 \uplus \mathcal{B}_3)/\{(v_{13}, \bar{v}_{13})\}\right]\right)/\{(v_{12}, \bar{v}_{12})\} ,$$

is not defined.

From the point of view of the dual triangulation, the couple of vertices v_i and $\bar{v}_i$ are two $(D-1)$-simplices. Adding and then contracting an edge of color 0 connecting v_i and $\bar{v}_i$ comes to removing the two simplices and coherently regluing together the two resulting boundaries.

Two particular cases of this composition will play an important role.

The composition of two graphs via an edge. (See Figure 3.12). If $v_1 \in \mathcal{B}_1$ and $v_2 \in \mathcal{B}_2$, then $(\mathcal{B}_1 \uplus \mathcal{B}_2)/\{(v_1, \bar{v}_2)\}$ is a connected graph having $k(\mathcal{B}_1) + k(\mathcal{B}_2) - 1$ white vertices.

The addition of an edge on a graph. (See Figure 3.13). If $v_1, \bar{v}_1 \in \mathcal{B}$, then:

$$\mathcal{B}/\{v_1, \bar{v}_1\} ,$$

is a (possibly disconnected) D-colored graph having $k(\mathcal{B}) - 1$ white vertices. If

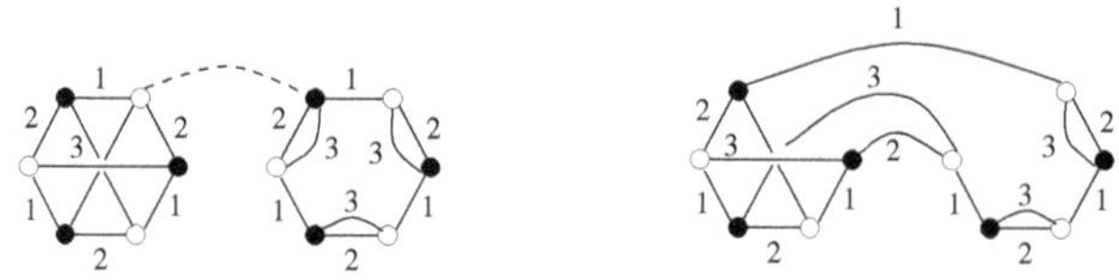

Figure 3.12: The composition of two graphs via an edge.

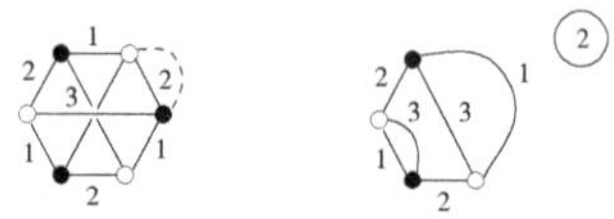

Figure 3.13: The addition of an edge on a graph.

$\mathcal{B}/(v_1, \bar{v}_1)$ has several connected components we denote them $\left[\mathcal{B}/\{v_1, \bar{v}_1\}\right]_{(\rho)}$. Observe that the edges of colors $c \neq 0$ which connect v_1 and $\bar{v}_1$ in $\mathcal{B}$ are deleted. If one wants to keep track of these edges, one can represent them as degenerate colored graphs consisting of a unique edge closing onto itself (as exemplified in Figure 3.13).

The effect of these two moves on the dual triangulation is represented (for $D = 3$) in Figure 3.14. The graphs $\mathcal{B}_1$ and $\mathcal{B}_2$ represent surfaces. The gluing $(\mathcal{B}_1 \uplus \mathcal{B}_2)/\{(v_1, \bar{v}_2)\}$ comes to deleting the triangles corresponding to v_1 and $\bar{v}_2$ on $\mathcal{B}_1$ and $\mathcal{B}_2$ and glue coherently the two surfaces along the boundaries thus created (Figure 3.14(a)). The second operation (Figure 3.14(b)) is essentially the same thing, just that this time the two triangles belong to the same surface.

(a) Gluing of surfaces. (b) Contraction of surfaces.

Figure 3.14: Gluing and contraction of surfaces for tensors of rank $D = 3$.

The topology of the surfaces changes under these moves (in the example of Figure 3.14(b) a planar surface becomes a genus one surface). The first operation is just surgery on the surfaces. The second one has a more involved topological interpretation, and it can lead to an increase of the genus, a splitting of the surface into several connected components or it can preserve the topology.

For $D = 2$, the graphs $\mathcal{B}$ are bicolored cycles. The gluing of two cycles of lengths p and q always leads to a cycle of length $p + q$.

3.4 Combinatorial maps and D-colored graphs

In tensor models and quantum field theory [104], combinatorial maps play a very important role.

Definition 3.9. *A* combinatorial map *is:*

- *A finite set $\mathcal{S}$ of halfedges (or darts) of even cardinality.*
- *A permutation σ on $\mathcal{S}$.*
- *An involution α on $\mathcal{S}$ with no fixed points (a.k.a. a "pairing" of halfedges). The involution α can either be seen as a permutation on $\mathcal{S}$ having only cycles of length 2, or as a list of pairs of halfedges.*

The *vertices* of a combinatorial map are the cycles of σ, and the permutation σ encodes the "next halfedge" when turning clockwise around the vertices. The involution α encodes the pairs of halfedges that are connected into edges.

A combinatorial map has a well defined notion of face obtained by traveling along the vertices from a halfedge to the next halfedge and along the edges from vertex to vertex, that is:

Definition 3.10. *The faces of a combinatorial map are the cycles of the permutation $\alpha\sigma$.*

To each combinatorial map we associate a graph obtained by collapsing the cycles of σ to vertices. That is, we define the equivalence relation on the elements of $\mathcal{S}$, $h_1 \sim h_2$ if h_1 and h_2 belong to the same cycle of σ and we consider the graph with vertex set and edge set given by the cosets $\mathcal{V} = \mathcal{S}_{/\sim}$ and $\mathcal{E} = \alpha_{/\sim}$. Note that this graph can have multiple edges (i.e. several edges connecting the same pair of vertices) as well as edges starting and ending on the same vertex, ("self loops", or "tadpoles" in physics literature). Several combinatorial maps correspond to the same graph. In Figure 3.15 we represented three combinatorial maps associated to the same graph, defined by:

$$\mathcal{S} = \{1,2,3,4,5,6,7,8\}\,, \qquad \sigma = (1,2,3,4)(5,6)(7,8)$$
$$\alpha = \begin{cases} (1,2)(3,8)(4,5)(6,7) \\ (1,2)(3,8)(4,6)(5,7) \\ (1,3)(2,8)(4,5)(6,7) \end{cases} .$$

A combinatorial map is a 2-complex (with two cells given by the faces) hence it is an embedding of a graph into a two dimensional surface. A combinatorial map therefore has a well defined Euler characteristic.

Definition 3.11. *The Euler character χ of a connected combinatorial map (equivalently its genus g) with V vertices, E edges and F faces is the alternated sum:*

$$\chi = V - E + F = 2 - 2g\,.$$

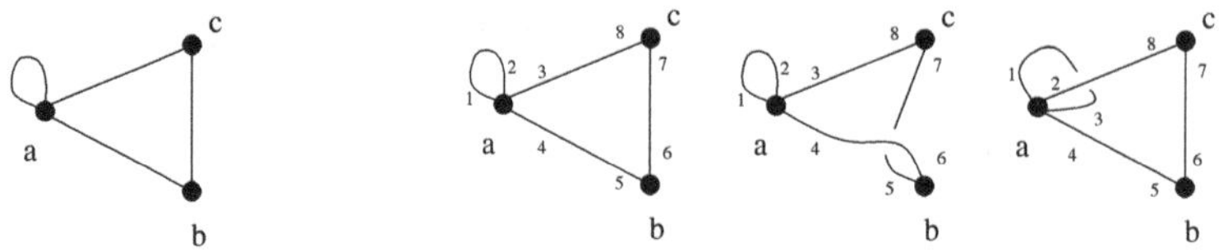

Figure 3.15: Combinatorial maps and associated graph.

The genus of a map is a positive number, and it is an integer if the embedding surface is orientable. The first statement can be proved as follows: one can contract $V-1$ edges forming a tree in the map to obtain a "rosette" map, having $E-V+1$ edges, a single vertex, and the same number of faces. Then one can delete edges as long as they separate distinct faces, obtaining a "super rosette" map having $E-V+1-F+1$ remaining edges, a single vertex and a single face. As this final number of edges is nonnegative, we conclude that $V-E+F\leq 2$, hence $g\geq 0$.

3.4.1 Jackets of colored graphs

Let us go back to $(D+1)$-colored graphs $\mathcal{G}$. We are interested in finding embeddings of $\mathcal{G}$ in two dimensional surfaces, that is, combinatorial maps whose associated graph is $\mathcal{G}$. The graph $\mathcal{G}$ yields a well defined set of halfedges: a vertex has one halfedge for each color[1]:

$$\mathcal{S}(\mathcal{G}) = \Big(\bigcup_{v_j\in\mathcal{V}^w(\mathcal{G})}\{v_j^0,v_j^1,\dots,v_j^D\}\Big)\bigcup\Big(\bigcup_{\bar v_j\in\mathcal{V}^b(\mathcal{G})}\{\bar v_j^0,\bar v_j^1,\dots,\bar v_j^D\}\Big)\,.$$

The involution α is fixed by the permutations $\tau^{(c)}$ (where we also have now a permutation $\tau^{(0)}$ encoding the edges of color 0),

$$\alpha(v_j^c) = \bar v^c_{\tau^{(c)}(j)}\,.$$

However, the only information we have on σ is that all its cycles are of the form:

$$(v_j^{c_0},\dots,v_j^{c_D})(\bar v_j^{c_D},\dots\bar v_j^{c_0})\,,$$

but nothing fixes the order in which the colors $c_0,\dots c_D$ should be listed.

Choosing the *same* ordering at all the vertices (encoded in a cyclic permutation π over the colors $0,1,\dots D$) leads to $D!$ combinatorial maps canonically associated to $\mathcal{G}$. We call them the *jackets* of $\mathcal{G}$ [94, 102, 96] and we denote a jacket by $\mathcal{J}$. The jackets corresponding to the cycles π and π^{-1} are in fact one and the same (up to an overall orientation flip) because, as we will see in a moment, they have the same set of faces. An example is presented in Figure 3.16, where from left to right the jackets correspond to the permutations $(0,1,2,3),(0,3,2,1)$, respectively $(0,1,3,2),(0,2,3,1)$ and $(0,3,1,2),(0,2,1,3)$.

[1]When performing this construction for D-colored graphs $\mathcal{B}$ that represent trace invariants, the halfedges can be associated to the indices indices $a^{\mathcal{D}}$ and $\bar a^{\mathcal{D}}$ of the tensors $\mathbb{T}_{a^{\mathcal{D}}}$ and $\bar{\mathbb{T}}_{\bar a^{\mathcal{D}}}$.

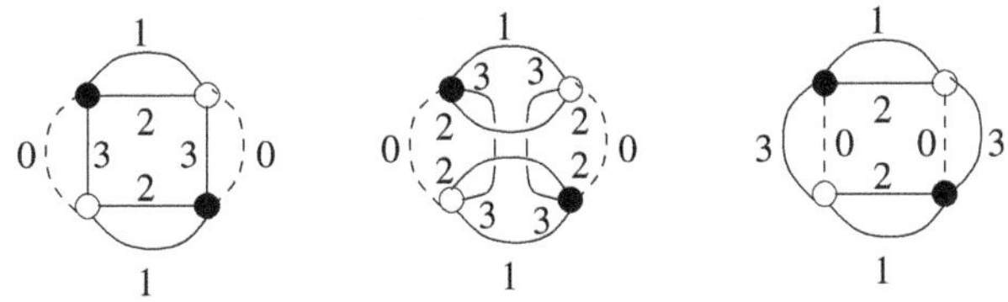

Figure 3.16: Jackets.

Let us consider the jacket $\mathcal{J}$ in which the successor of the color c_1 when going around a vertex is the color c_2 (that is $\pi(c_1) = c_2$):

$$\sigma^{\mathcal{J}}(v_j^{c_1}) = v_j^{c_2} , \qquad \sigma^{\mathcal{J}}(\bar{v}_j^{c_2}) = \bar{v}_j^{c_1} .$$

Note that, as the combinatorial map $\mathcal{J}$ is bipartite, the faces must all have even length. Let us track the face of $\mathcal{J}$ that starts at $v_j^{c_1}$. We have:

$$v_j^{c_1} \to \sigma^{\mathcal{J}}(v_j^{c_1}) \to \alpha\sigma^{\mathcal{J}}(v_j^{c_1}) \to \sigma^{\mathcal{J}}\alpha\sigma^{\mathcal{J}}(v_j^{c_1}) \to \alpha\sigma^{\mathcal{J}}\alpha\sigma^{\mathcal{J}}(v_j^{c_1})\dots$$
$$v_j^{c_1} \to v_j^{c_2} \to \bar{v}^{c_2}_{\tau^{(c_2)}(j)} \to \bar{v}^{c_1}_{\tau^{(c_2)}(j)} \to v^{c_1}_{\left[\tau^{c_1}\right]^{-1}\tau^{(c_2)}(j)} \dots .$$

It follows that the faces (that are cycles of $\alpha\sigma$) starting at a half edge v^{c_1} of color c_1 are the cycles of the permutation $\left[\tau^{(c_1)}\right]^{-1}\tau^{(c_2)}$. But these are nothing but the subgraphs with edges of colors c_1 and c_2 of the graph $\mathcal{G}$:

Proposition 3.1. *The faces of the jacket $\mathcal{J}$ (corresponding to the cycle π) seen as a combinatorial map are exactly the faces of colors $(\pi^p(1), \pi^{p+1}(1))$,$p = 0, \dots D$ of the $(D+1)$-colored graph $\mathcal{G}$.*

Being combinatorial maps, the jackets are dual to surfaces embedded in the triangulation dual to the $(D+1)$-colored graph. From a topological standpoint these surfaces are Heegaard splitting surfaces [148] for $D = 3$, and they provide a generalization of such surfaces in higher dimensions.

3.4.2 The degree

Due to the bipartiteness of $\mathcal{G}$, the jackets $\mathcal{J}$ are orientable hence have integer genus.

Definition 3.12. *The* degree *[94, 96] of a connected $(D+1)$ colored graph is the half sum of the genera of its jackets:*

$$\omega(\mathcal{G}) = \frac{1}{2}\sum_{\mathcal{J}} g(\mathcal{J}) ,$$

where $g(\mathcal{J})$ denotes the genus of the jacket $\mathcal{J}$. In particular the degree is a positive *number.*

The reduced degree *of a graph is:*

$$\delta(\mathcal{G}) = \frac{2}{(D-1)!}\omega(\mathcal{G}) \, .$$

We will see in a moment that the reduced degree is an integer.

The 1/2 in the above definition is conventional. The degree thus defined is still an integer, as the jackets corresponding to π and π^{-1} are identical. The reason for which the degree plays such an important role in tensor models is that it yields a counting of the number of the faces of a graph. To see how this comes about, let us count the faces belonging to a jacket from its genus. Denoting $k(\mathcal{G})$ the number of white vertices, $F^{(c,c')}(\mathcal{G})$ the number of faces of colors (c,c') and $F(\mathcal{G})$ the total number of faces of a connected $(D+1)$-colored graph $\mathcal{G}$ we have:

$$2k(\mathcal{G}) - (D+1)k(\mathcal{G}) + \sum_{p=0}^{D} F^{\left(\pi^p(1),\pi^{p+1}(1)\right)}(\mathcal{G}) = 2 - 2g(\mathcal{J}) \, .$$

Taking into account that a face (with colors (c,c')) belongs to $2(D-1)!$ jackets (those such that $\pi(c) = c'$ and those such that $\pi(c') = c$) and that there are $D!$ jackets in total we get:

$$-D!(D-1)k(\mathcal{G}) + 2(D-1)!F(\mathcal{G}) = 2D! - 2\sum_{\mathcal{J}} g(\mathcal{J})$$

$$\Rightarrow \boxed{F(\mathcal{G}) = D + \frac{D(D-1)}{2}k(\mathcal{G}) - \frac{2}{(D-1)!}\omega(\mathcal{G})} \, . \tag{3.1}$$

This formula plays a crucial role in tensor models. We note already that, as the degree is positive, Eq. (3.1) provides an upper bound for the number of faces of a graph linear in the number of its vertices. This is ultimately the reason why the $1/N$ expansion of tensor models exists. In particular, Eq. (3.1) shows that the reduced degree of a graph is an integer.

Let us denote $\mathcal{B}_{(\rho)}$ the connected subgraphs with colors $\{1,\dots D\}$ of a $(D+1)$ colored graph $\mathcal{G}$. As they are graphs with D colors, the degree $\omega(\mathcal{B}_{(\rho)})$ of each of them is well defined. It turns out that the sum of these degrees is bounded by the degree of $\mathcal{G}$.

Lemma 3.1. *Let $\mathcal{G}$ be a closed connected $D+1$ colored graph and $\mathcal{B}_{(\rho)}$ its connected subgraphs with colors $\{1,\dots D\}$. Then:*

$$\omega(\mathcal{G}) \geq D\sum_{\rho}\omega(\mathcal{B}_{(\rho)}) \, .$$

Proof. There exists a D to 1 correspondence between the cycles π over $(D+1)$ elements $\{0,\dots D\}$ and the cycles $\hat{\pi}$ over D elements $\{1,\dots D\}$:

$$\pi \to \hat{\pi} \qquad \hat{\pi}(c) = \begin{cases} \pi(c) & \text{if } c \neq \pi^{-1}(0) \\ \pi^2(c) & \text{if } c = \pi^{-1}(0) \end{cases} \, .$$

It follows that every jacket $\cup_\rho J_{(\rho)}$ of $\cup_\rho \mathcal{B}_{(\rho)}$ is a submap of D jackets $\mathcal{J}$ of $\mathcal{G}$, obtained by erasing the edges of color 0 in $\mathcal{J}$.

The Euler characteristic of a map with several connected components labeled by μ (each of genus g_μ) is:

$$\chi = V - E + F = \sum_\mu (2 - 2g_\mu) .$$

When deleting an edge in a map:

- either the number of connected components of the map remains constant, in which case the sum of their genera can not increase,
- or the number of connected components of the map increases by one (and the number of faces also increases by one), in which case the sum of the genera of the connect components is constant.

It follows that:

$$D \sum_\rho \sum_{J_{(\rho)}} g(J_{(\rho)}) \le \sum_J g(J) .$$

□

Chapter 4

The classification of edge colored graphs

In this chapter we classify the edge colored graphs in terms of their degree. It is well known that the family of combinatorial maps of fixed genus is infinite and exponentially bounded. It turns out that the same holds for colored graphs, where the degree plays the role of the genus: the family of graphs of fixed degree is infinite and exponentially bounded. We present here a detailed characterization of these families, performed initially in [107].

4.1 Melonic graphs

We first discuss the graphs of degree zero, which we call *melonic graphs*. Melonic graphs play a distinguished role in tensor models. For now we will discuss only their properties relevant for the classification of graphs in terms of the degree, and reserve the entire Chapter 5 to a more detailed presentation of this family.

We can define melonic graphs starting from "cut sets" of edges (two particle reducibility edges in the physics language). Although this is not how melonic graphs have been historically introduced, the advantage of this point of view is that it provides for free a notion of "elimination" of melonic subgraphs in larger graphs.

Definition 4.1. *A* rooted $(D+1)$ *colored graph is a colored graph with a marked edge.*

As we are only dealing with bipartite graphs, marking an edge suffices for the rooting. For non bipartite graphs one would need moreover to choose an orientation of the marked edge.

From a combinatorial standpoint it is much more convenient to deal with rooted colored graphs rather than unrooted colored graphs, because rooted graphs (and rooted combinatorial maps) can be enumerated more easily. All the closed $(D+1)$-colored graphs we are dealing with in this chapter are rooted.

A rooted graph $\mathcal{G}$ can be *opened* by cutting the marked edge into two halfedges. We denote the graph thus obtained $\mathrm{op}(\mathcal{G})$. The graph $\mathrm{op}(\mathcal{G})$ is not an open graphs in the sense of our Definition 3.4, as the halfedges are allowed to have any color, not only 0.

Definition 4.2. *A* 2^c-open colored graph *is a graph with colored edges having two external halfedges of color c that becomes a closed $(D+1)$-colored graph when reconnecting the halfedges into an edge of color c. Such a graph is canonically rooted at the external halfedge incident at the white vertex.*

The graph obtained by reconnecting the two external halfedges of a 2^c-open colored graph $\mathcal{G}$ is canonically rooted at the new edge (which has color c). It is called the *closure* of $\mathcal{G}$ and is denoted $\mathrm{cl}(\mathcal{G})$. We will often ignore the superscript c and call a 2^c-open graph simply a 2-open graph.

Example 4.1. The graph $\mathcal{G}^{(2)}$ having two vertices and $(D+1)$ parallel edges can be rooted by marking any of its edges. In turn, each of these rooted graphs can be opened to obtain the 2^c-open graph $\mathcal{G}^{(2),c}$ having two external halfedges of color c and D parallel internal edges. Of course, $\mathrm{cl}(\mathcal{G}^{(2),c}) = \mathcal{G}^{(2)}$, with the edge of color c marked.

From the onset we include among the closed colored graphs the *ring graph* consisting of a unique edge of color c closed onto itself and having no vertex (see Figure 4.1). The *ring graph has degree zero*, as it has no vertices but it has D faces, one for each color $c' \in \{0, \dots D\} \setminus \{c\}$. The opening of the ring graph leads to the *one edge graph* having only one edge and no vertex.

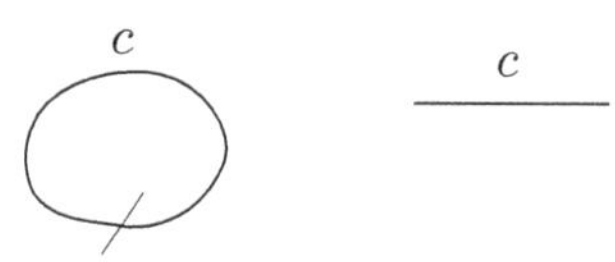

Figure 4.1: The closed ring graph and the open one edge graph.

We will be switching back and forth between closed rooted graphs and the associated 2-open graphs. An important notion for closed graphs is the notion of 2-*edge cut.*

Definition 4.3. *A closed graph is called* 2-connected *if it cannot be disconnected by erasing a vertex.*

A 2-edge-cut *of a 2-connected closed graph $\mathcal{G} = (\mathcal{V}, \mathcal{E})$ is a pair of edges $\{e, e'\}$ such that the graph $\mathcal{G} - \{e, e'\} = (\mathcal{V}, \mathcal{E} \setminus \{e, e'\})$ is not connected (this is represented in Figure 4.2 on the left hand side).*

A simple cycle *of a graph $\mathcal{G}$ is a cycle visiting each vertex of $\mathcal{G}$ at most once.*

In particular closed, connected $(D+1)$-colored graphs are 2-connected. Note that in a colored graph, the two edges in a 2-edge-cut have the same color. A 2-edge-cut is alternatively characterized by the following lemma.

Lemma 4.1. *Let $\mathcal{G}$ be a 2-connected graph. Then $\{e, e'\}$ is a 2-edge-cut in $\mathcal{G}$ if and only if any simple cycle visiting e also visits e'.*

Proof. Let $e = \{x, \bar{y}\}$ and let C be a simple cycle visiting e. Then $\mathcal{G} - \{e, e'\}$ has two connected components, one containing x and the other containing $\bar{y}$. But $C - e$ is a path from x to $\bar{y}$ and e' is the only edge connecting the two components in $\mathcal{G} - \{e\}$.

Conversely, if $\{e, e'\}$ is not a 2-edge-cut, then there exists a path in $\mathcal{G} - \{e, e'\}$ between any two vertices, and in particular between the endpoints of e.

□

The same edge can belong to several 2-edge-cuts.

Lemma 4.2. *Let $\{e, e'\}$ and $\{e, e''\}$ be two 2-edge-cuts in a 2-connected graph $\mathcal{G}$. Then $\{e', e''\}$ is also a 2-edge-cut of $\mathcal{G}$. Moreover, if two oriented cycles visit e in the same direction, then they both visit e' and e'' in the same order after e.*

Proof. Any cycle visiting e' also visits e (since $\{e, e'\}$ is a 2-edge-cut) and thus also e'' (since $\{e, e''\}$ is a 2-edge-cut), and we conclude by Lemma 4.1. Next, suppose there exist two oriented cycles $C_1 = (e, p_1, e', p_1', e'', p_1'')$ and $C_2 = (e, p_2, e'', p_2', e', p_2'')$ that visit e in the same direction. But then the path p_2 would connect the two connected components of $\mathcal{G} - \{e, e'\}$ without visiting e or e'.

□

Collecting together all the edges which, when paired, form 2-edge cuts leads to *proper cut-sets* of edges.

Definition 4.4. *A* proper cut-set *of a 2-connected graph $\mathcal{G}$ is a maximal set $\mathrm{C_{ut}}$ of edges such that any 2 edges of $\mathrm{C_{ut}}$ form a 2-edge-cut.*

In view of the previous lemma, an edge can belong to at most one proper cut-set. We define the *cut-set* of an edge as the unique proper cut-set to which the edge belongs if it exists, or the edge itself otherwise. All the edges in a proper cut-set of a $(D+1)$-colored graph have the same color.

Lemma 4.3. *Let $\mathcal{G} = (\mathcal{V}, \mathcal{E})$ be a colored graph and let $\mathrm{C_{ut}}$ be a cut-set of $\mathcal{G}$. Then there exists a unique way to cyclically arrange the edges of $\mathrm{C_{ut}}$ as $(e_0, \dots, e_\ell)$ and a unique partition $\mathcal{V}_0, \dots, \mathcal{V}_\ell$ of $\mathcal{V}$ such that $\mathcal{E} = \mathrm{C_{ut}} \cup \mathcal{E}_{\mathcal{V}_0} \cup \dots \cup \mathcal{E}_{\mathcal{V}_\ell}$ (where $\mathcal{E}_{\mathcal{V}_0}$ denotes the set of all edges with both end vertices in $\mathcal{V}_0$) and for all $i = 0, \dots, \ell$, e_i connects a black vertex of $\mathcal{V}_i$ to a white vertex of $\mathcal{V}_{i+1}$ (with indices taken modulo $\ell + 1$).*

Proof. Let e_0 be an edge of $\mathrm{C_{ut}}$. Since $\mathcal{G}$ is 2-connected, there exists a simple cycle C visiting e_0, and this cycle also visits the other edges in $\mathrm{C_{ut}}$. Orient this cycle so that e_0 is visited from its black to its white endpoint and let $(e_0, e_1, \dots, e_\ell)$ describe the cyclic arrangement of the edges of $\mathrm{C_{ut}}$ along this oriented cycle. Then $\{e_i, e_{i+1}\}$ forms a 2-edge-cut and one of the components of $\mathcal{G} - \{e_i, e_{i+1}\}$ contains the part of C visiting $e_{i+2}, \dots, e_{i-1}$. Let $\mathcal{V}_i$ denote the vertex set of the other component. Then the $\mathcal{V}_i$s are disjoint and form a partition of $\mathcal{V}$ and the other required properties are immediate. The uniqueness follows from the fact that any other cycle C' has to visit the edges of $\mathrm{C_{ut}}$ in the same order as C (from Lemma 4.2).

□

By cutting the edges in a cut-set one decomposes a graph into 2-open graphs.

Definition 4.5. *The set* $\mathfrak{O}_{\mathrm{C_{ut}}} = \{\mathcal{G}_0, \ldots, \mathcal{G}_\ell\}$ *of* 2-open components *of a cut-set* $\mathrm{C_{ut}}$ *is the set of 2-open connected graphs obtained by cutting each edge of the cut-set into two halfedges.*

The set $\mathfrak{C}_{\mathrm{C_{ut}}}$ *of* closed components *of* $\mathrm{C_{ut}}$ *is, by extension, the set of rooted colored graphs* $\mathrm{cl}(\mathcal{G}_i)$.

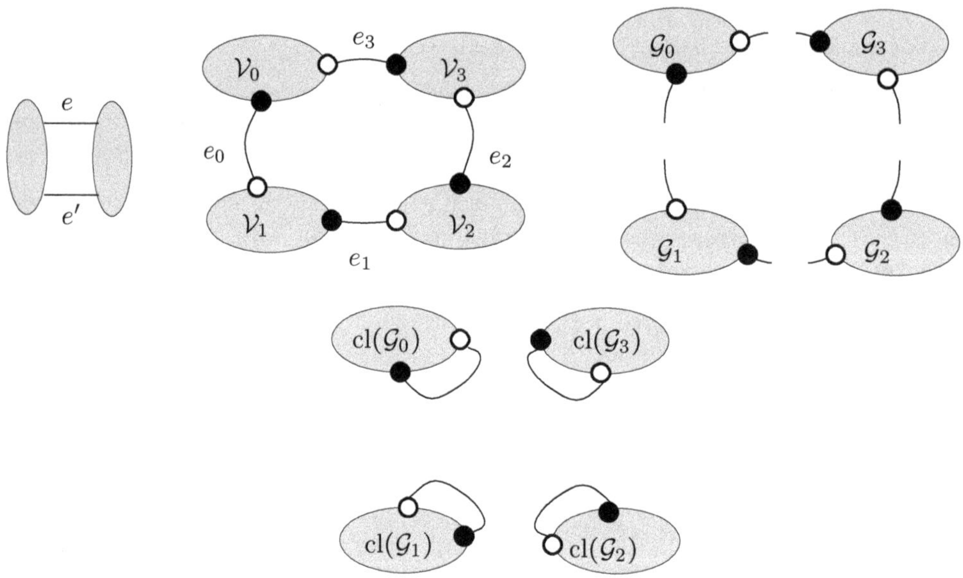

Figure 4.2: A 2-edge-cut, a proper set-cut and the corresponding 2-open and closed components.

These various definitions above are illustrated in Figure 4.2. With the notation of the Lemma 4.3, the vertex set of $\mathcal{G}_i$ is $\mathcal{V}_i$ and the two halfedges of $\mathcal{G}_i$ arise from e_i and e_{i+1} respectively.

Lemma 4.4. *Let* $\mathrm{C_{ut}}'$ *and* $\mathrm{C_{ut}}''$ *be two distinct cut-sets in a colored graph* $\mathcal{G}$. *Then there exists a 2-open component* $\mathcal{G}'$ *of* $\mathrm{C_{ut}}'$ *containing* $\mathrm{C_{ut}}''$, *and a 2-open component* $\mathcal{G}''$ *of* $\mathrm{C_{ut}}''$ *containing* $\mathrm{C_{ut}}'$.

Proof. Assume $\mathrm{C_{ut}}''$ is not contained in any of the 2-open components of $\mathrm{C_{ut}}'$, and let e_i'', e_j'' be two edges of $\mathrm{C_{ut}}''$ appearing in two different components $\mathcal{G}_p'$ and $\mathcal{G}_q'$ of $\mathrm{C_{ut}}'$. Then any cycle visiting e_i'' also visits e_j'', and thus e_p' and e_{p+1}', hence $\mathrm{C_{ut}}'$ and $\mathrm{C_{ut}}''$ are not disjoint.

□

We are now in the position to define the melonic graphs inductively.

Definition 4.6. *(see Figure 4.3) A 2^c-open colored graph $\mathcal{G}$ is a:*

- melonic graph *if:*
  - *either it is the one edge graph,*
  - *or all the 2-open components of the cut-set of the root edge of* $\mathrm{cl}(\mathcal{G})$ *are prime melonic graphs.*
- a prime melonic graph *if by cutting* all *the edges incident to the vertices x and y of $\mathcal{G}$, which carry the external halfedges of $\mathcal{G}$, into pairs of halfedges, one obtains D distinct $2^{c'}$-open graphs, one for each $c' \neq c$, which are melonic graphs. Observe that if x and y are connected by an edge of color c' in $\mathcal{G}$ then this edge is cut twice (once for x and once for y), and the resulting open graph is the one edge graph, which is melonic.*

A melonic graph is called nontrivial *if it is not the one edge graph, and it is called trivial if it is.*

A closed $D+1$ colored graph is melonic *if it is the closure of some 2^c-open melonic graph. In particular, the ring graph is melonic.*

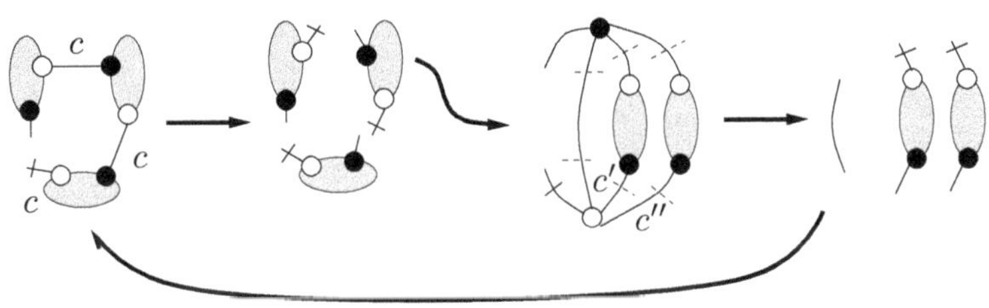

Figure 4.3: Illustration of the definition of melonic graphs.

For example, the closed graph $\mathcal{G}^{(2)}$ and the 2^c-open graphs $\mathcal{G}^{(2),c}$ of Example 4.1 are melonic. We call the graphs $\mathcal{G}^{(2),c}$ *fundamental melons*. We define a *proper sub-melon* of a 2-open colored graph $\mathcal{G}$ as a 2^c-open subgraph of $\mathcal{G}$ which is a fundamental melon $\mathcal{G}^{(2),c}$. Let $\mathcal{G} \smallsetminus \mathcal{G}^{(2),c}$ denote the 2-open colored graph obtained by the *removal* of a proper sub-melon $\mathcal{G}^{(2),c}$ in $\mathcal{G}$, i.e. by removing $\mathcal{G}^{(2),c}$ and gluing together the two resulting unmatched halfedges of color c to form a new edge. Conversely, let $\mathcal{G}[e^c \leftarrow \mathcal{G}^{(2),c}]$ denote the 2-open colored graph obtained by the *insertion* of a fundamental melon $\mathcal{G}^{(2),c}$ at an edge e^c (or at the nonroot external halfedge) of $\mathcal{G}$, i.e. by cutting e^c into two halfedges and gluing these halfedges to those of $\mathcal{G}^{(2),c}$ in the unique way that makes the result bipartite.

Lemma 4.5. *A graph is melonic if and only if it can be obtained from the one edge graph by cutting edges and inserting $\mathcal{G}^{(2),c}$ respecting the colors and the bipartiteness.*

Proof. The only melonic graphs with no vertices are the one edge graphs of various colors. From Definition 4.6 it is immediate that the only 2-open melonic graphs with two vertices are the $\mathcal{G}^{(2),c}$s themselves, which can be seen as the insertion of a $\mathcal{G}^{(2),c}$ at the unique edge of the one edge graph.

Consider now a melonic graph $\mathcal{G}$ having at least four vertices. The root halfedge of $\mathcal{G}$ is incident to some vertex, say x. As $\mathcal{G}$ has at least four vertices, the cut set of its root has at least one open component, therefore there exists a nontrivial prime melonic subgraph of $\mathcal{G}$, $\mathcal{G}_1$, whose external halfedges are incident one to the vertex x and the other one to some other vertex y. It follows that $\mathcal{G}$ is obtained from a fundamental melon $\mathcal{G}^{(2),c}$ with vertices x and y by inserting melonic graphs having at least two fewer vertices at its edges.

The converse is trivial.

□

The original definition of melonic graphs [30] was given in terms of this iterative insertion procedure. The elimination and insertion of a melon are defined identically for closed (rooted) $(D+1)$-colored graphs.

Theorem 4.1. *[93, 90] We have:*

- *Let $\mathcal{G}$ a closed $(D+1)$-colored graph, and $\mathcal{G}^{(2),c}$ a proper submelon. Then the degrees of $\mathcal{G}$ and $\mathcal{G}\setminus\mathcal{G}^{(2),c}$ are the same.*
- *The degree of the ring graph is zero.*
- *A rooted closed colored graph is melonic if and only if it has degree 0.*

Proof. The first two items follow directly from Eq. (3.1), as the removal of a proper submelon decreases the number of vertices by 2 and the number of faces by $\frac{D(D-1)}{2}$. It follows that closed melonic graphs have degree 0. The converse statement requires more work.

Consider a 2-edge-cut $\{e,e'\}$ in a connected closed $D+1$ colored graph $\mathcal{G}$. As any cycle visiting e also visits e', all the faces that contain e also contain e'. It follows that, by cutting e and e' and reconnecting the resulting halfedges in the only other way that keeps the result bipartite, $\mathcal{G}$ splits into two connected components $\mathcal{G}_1$ and $\mathcal{G}_2$ such that:

$$k(\mathcal{G}_1)+k(\mathcal{G}_2)=k(\mathcal{G})\ ,\qquad F(\mathcal{G}_1)+F(\mathcal{G}_2)=F(\mathcal{G})+D\ ,$$

and, by Eq. (3.1),

$$\omega(\mathcal{G}_1)+\omega(\mathcal{G}_2)=\omega(\mathcal{G})\ . \tag{4.1}$$

Let $\mathcal{G}$ be any connected closed $(D+1)$-colored graph having at least two vertices. Denoting $F_s(\mathcal{G})$ the number of faces of $\mathcal{G}$ of length $2s$ (that is having $2s$ vertices), and taking into account that every vertex belongs to $\binom{D+1}{2}$ faces we obtain:

$$\sum_{s\geq 1} sF_s(\mathcal{G})=\frac{D(D+1)}{2}k(\mathcal{G})\ .$$

On the other hand, Eq. (3.1) can be rewritten as:

$$\sum_{s\geq 1} F_s(\mathcal{G})-D+\frac{2}{(D-1)!}\omega(\mathcal{G})=\frac{D(D-1)}{2}k(\mathcal{G})\ ,$$

and, eliminating $k(\mathcal{G})$ between these two equations, we obtain:

$$(D+1)\frac{2}{(D-1)!}\omega(\mathcal{G}) + 2F_1(\mathcal{G}) = \\ = D(D+1) + \sum_{s\geq 2}[s(D-1)-(D+1)]F_s(\mathcal{G}) \,, \qquad (4.2)$$

and the right hand side of this equation is at least $D(D+1)$ for all $D \geq 3$.

Consider $\mathcal{G}$ a closed graph of degree 0 having at least two vertices. Then $\mathcal{G}$ must have at least $\frac{D(D+1)}{2}$ faces with length 2 (that is having exactly two vertices). Let (c, c') be the colors of a face having only two vertices v and $\bar{v}$. The vertices v and $\bar{v}$ are connected by an edge of color c and an edge of color c'. If v and $\bar{v}$ are connected by a unique edge of color c'' for all $c'' \neq c, c'$, then the graph is $\mathcal{G}^{(2)}$, hence it is melonic. Choose a color $c_1 \neq c, c'$ such that the edges of color c_1 hooked to v and $\bar{v}$ are different. We denote them $e_1^{c_1}$ and $e_2^{c_1}$. As $\mathcal{G}$ has degree 0 all its jackets are planar, including the jacket $\pi(c) = c_1, \pi(c_1) = c'$, thus $\{e_1^{c_1}, e_2^{c_1}\}$ is a two edge cut. By cutting $\{e_1^{c_1}, e_2^{c_1}\}$ and reconnecting the halfedges in the only other way that keeps the result bipartite, $\mathcal{G}$ splits into two nontrivial connected components $\mathcal{G}_1$ and $\mathcal{G}_2$, both of degree 0 and both having strictly fewer vertices than $\mathcal{G}$. We conclude by induction.

□

At this point one can understand why melonic graphs are so different from planar graphs. For $D = 2$, $(2+1)$-colored graphs have a unique jacket and the degree reduces to the genus. However, Eq. (4.2) behaves very differently for $D = 2$. Indeed, in this case the coefficient of $F_2(\mathcal{G})$ is *negative*. Thus, while for $D \geq 3$ the graphs of degree zero must have faces of length 2, the same does not hold for $D = 2$: one can build planar graphs with three colors having no such face (see an example in Figure 4.4). In turn, the faces of length 2 lead to the inductive construction of the melonic graphs. Planar graphs do not obey such an inductive construction, and in particular they are more numerous than the melonic graphs.

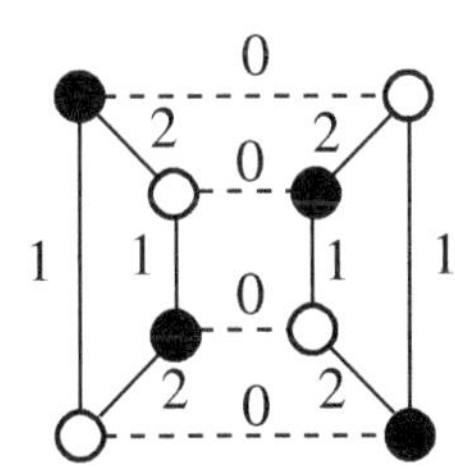

Figure 4.4: A planar graph having only faces of length 4.

4.2 The melonic core

Given an arbitrary graph $\mathcal{G}$, one can identify (and eliminate) all its melonic subgraphs.

Definition 4.7. *A* proper 2-open subgraph *of a 2-open colored graph $\mathcal{G}$ is a connected 2-open subgraph of $\mathcal{G}$. By extension a proper 2-open subgraph of a closed, rooted, colored graph $\mathcal{G}$ is a proper 2-open subgraph of* $\mathrm{op}(\mathcal{G})$. *In particular,* $\mathrm{op}(\mathcal{G})$ *is the largest proper 2-open subgraph of $\mathcal{G}$.*

Two proper 2-open subgraphs of a 2-open colored graph $\mathcal{G}$ are *totally disjoint* if their edge sets are disjoint and their external halfedges belong to different edges of $\mathcal{G}$.

The following lemma is illustrated in Figure 4.5.

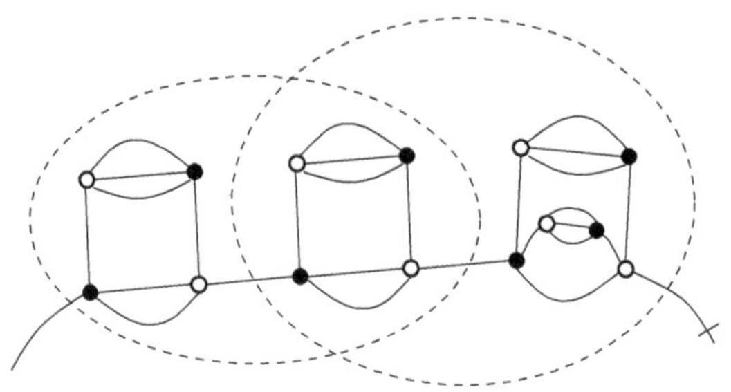

Figure 4.5: Two nondisjoint melonic subgraphs of a 2-open colored graph.

Lemma 4.6. *Let $\mathcal{M}_1$ and $\mathcal{M}_2$ be two nontrivial 2-open proper melonic subgraphs of a rooted, closed, colored graph $\mathcal{G}$. If $\mathcal{M}_1$ and $\mathcal{M}_2$ are not totally disjoint, then their union is a 2-open melonic subgraph of $\mathcal{G}$.*

Proof. First observe that if $\mathcal{M}$ is a nontrivial proper 2-open melonic subgraph of $\mathcal{G}$ with halfedges $h \in e$ and $h' \in e'$ then the edges e and e' of $\mathcal{G}$ form a 2-edge-cut of $\mathcal{G}$, unless $e = e' = r$ the root of $\mathcal{G}$, in which case $\mathcal{M}$ is $\mathrm{op}(\mathcal{G})$. This is because each vertex of $\mathcal{M}$ has the same degree in $\mathcal{M}$ and in $\mathcal{G}$ and therefore all the edges of $\mathcal{G}$ incident to a vertex of $\mathcal{M}$ are in $\mathcal{M}$, except for e and e'.

Now if $\mathcal{M}_1$ and $\mathcal{M}_2$ are two nontrivial 2-open melonic subgraphs of $\mathcal{G}$ with halfedges $h_1 \in e_1$ and $h_1' \in e_1'$, and $h_2 \in e_2$ and $h_2' \in e_2'$, respectively, then the two cuts $\{e_1, e_1'\}$ and $\{e_2, e_2'\}$:

- either belong to two different cut-sets: in this case, in view of Lemma 4.4, either $\mathcal{M}_1 \subset \mathcal{M}_2$ or $\mathcal{M}_2 \subset \mathcal{M}_1$ (or $\mathcal{M}_1$ and $\mathcal{M}_2$ are totally disjoint but this contradicts the hypothesis),
- or belong to the same cut-set: in this case, since $\mathcal{M}_1$ and $\mathcal{M}_2$ are not totally disjoint, then two of their external halfedges are matched into an edge of $\mathcal{G}$. It follows that there are two edges $e_1'' \in \{e_1, e_1'\}$ and $e_2'' \in \{e_2, e_2'\}$ such that $\mathcal{M}_1 \cup \mathcal{M}_2$ is the component of $\mathcal{G} - \{e_1'', e_2''\}$ not containing the root. In particular, the open components of the cut-set of the root of $\mathrm{cl}(\mathcal{M}_1 \cup \mathcal{M}_2)$ are the union of the open components of the cut-sets of the roots of $\mathrm{cl}(\mathcal{M}_1)$ and $\mathrm{cl}(\mathcal{M}_2)$ and they are all prime melonic graphs, so that $\mathcal{M}_1 \cup \mathcal{M}_2$ is melonic.

□

Definition 4.8. *A* maximal melonic subgraph *of a 2-open colored graph $\mathcal{G}$ is a proper 2-open melonic subgraph, which is maximal for inclusion.*

Lemma 4.7. *The maximal melonic subgraphs of $\mathcal{G}$ are totally disjoint.*

Proof. This is an immediate consequence of Lemma 4.6.

□

As the maximal melonic subgraphs are totally disjoint, they can be identified and eliminated independently.

Definition 4.9. *The* core *of a 2-open colored graph $\mathcal{G}$ is the graph $\hat{\mathcal{G}}$ obtained from $\mathcal{G}$ by deleting each of its maximal 2-open melonic subgraphs and gluing the resulting pairs of halfedges to form new edges. The core of a closed rooted colored graph $\mathcal{G}$ is the closure of the core of* op($\mathcal{G}$).

Observe that if an external halfedge of a 2-open graph $\mathcal{G}$ belongs to a melonic subgraph then the external halfedge is cut into two and re-glued. If both the external halfedges of $\mathcal{G}$ belong to a same melonic subgraph then $\mathcal{G}$ is melonic and the core is the one edge graph.

Definition 4.10. *A closed rooted colored graph is* melon-free *if it does not contain a proper sub-melon. The ring graph is a melon-free graph, and in view of Lemma 4.5 it is in fact the only melon-free graph of degree 0.*

By construction the core of a 2-open colored graph $\mathcal{G}$ is melon-free. The following lemma is trivial.

Lemma 4.8. *The core $\hat{\mathcal{G}}$ of a rooted colored graph $\mathcal{G}$ is the unique melon-free graph that can be obtained from $\mathcal{G}$ by a sequence of melon removals.*

The core of any nontrivial, closed, rooted melonic graph is the ring graph.

Looking back at the definition of the core $\hat{\mathcal{G}}$ of a 2-open colored graph $\mathcal{G}$, we see that for every edge $e = \{v, \bar{v}\}$ with color c in $\hat{\mathcal{G}}$,

- either there is a maximal melonic subgraph $\mathcal{G}_e$ in $\mathcal{G}$ whose halfedges have color c and respectively point to v and $\bar{v}$,
- or there is an edge $\{v, \bar{v}\}$ with color c in $\mathcal{G}$ and this edge is not involved in any nontrivial melonic subgraph of $\mathcal{G}$, and in this case, by convention we set $\mathcal{G}_e$ to be the one edge graph.

Similarly, if x is the vertex to which the white halfedge $h_\circ$ (resp. black halfedge $h_\bullet$) of $\hat{\mathcal{G}}$ is hooked, then:

- either there is a maximal nontrivial melonic subgraph $\mathcal{G}_\circ$ (resp. $\mathcal{G}_\bullet$) of $\mathcal{G}$ whose white (resp. black) halfedge is the white (resp. black) halfedge of $\mathcal{G}$,
- or the white (resp. black) halfedge of $\mathcal{G}$ is hooked to the vertex x and this halfedge is not involved in any melonic subgraph of $\mathcal{G}$, and in this case we set $\mathcal{G}_\circ$ (resp. $\mathcal{G}_\bullet$) by convention to be the one edge graph.

Definition 4.11. *The* core decomposition *of a rooted colored graph $\mathcal{G}$ is the list*

$$\left(\hat{\mathcal{G}}; (\mathcal{G}_\circ, \mathcal{G}_\bullet, \mathcal{G}_{e_1}, \dots, \mathcal{G}_{e_{\ell-1}})\right),$$

where $e_1, \dots, e_{\ell-1}$ is a canonical list of the edges of $\hat{\mathcal{G}}$.

This definition is illustrated in Figure 4.6. The following theorem is the main result of this section.

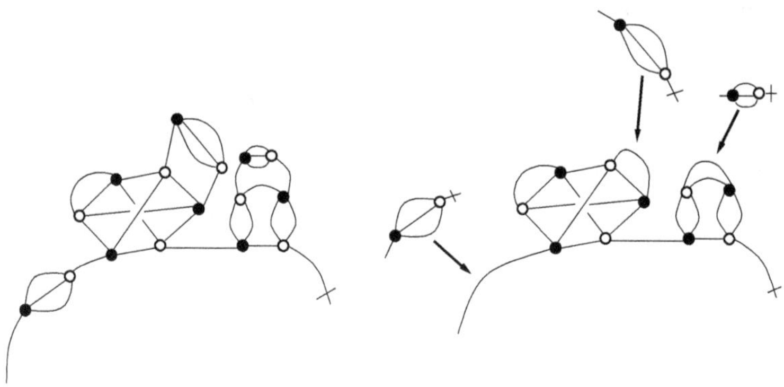

Figure 4.6: The maximal melonic subgraphs of a 2-open colored graph and its core decomposition (only the nontrivial melonic subgraphs are represented).

Theorem 4.2 (Core decomposition). *The core decomposition is one-to-one between:*

- *a rooted, closed, colored graph $\mathcal{G}$ with $2k(\mathcal{G})$ vertices and degree $\omega(\mathcal{G})$,*
- *lists*
$$\left(\hat{\mathcal{G}}; (\mathcal{G}_0, \mathcal{G}_1, \dots, \mathcal{G}_{(D+1)k(\mathcal{G}')})\right)$$
where:
  - *$\hat{\mathcal{G}}$ is a melon-free rooted colored graph with $2k(\hat{\mathcal{G}})$ vertices and degree $\omega(\hat{\mathcal{G}}) = \omega(\mathcal{G})$,*
  - *for all $i = 0, 1, \dots, (D+1)k(\hat{\mathcal{G}})$, $\mathcal{G}_i$ is a (possibly trivial) 2-open melonic graph with $2k(\mathcal{G}_i)$ vertices,*

such that $k(\hat{\mathcal{G}}) + \sum_{i=0}^{(D+1)k(\hat{\mathcal{G}})} k(\mathcal{G}_i) = k(\mathcal{G})$.

Proof. The core decomposition is clearly injective since all the components of $\mathcal{G}$ have been kept in the decomposition as well as the correspondence between edges of the core and subgraphs. Conversely, any such pair yields a rooted colored graph $\mathcal{G}$ by substitution of the $\mathcal{G}_i$ in $\hat{\mathcal{G}}$ and all substituted melonic subgraphs become totally disjoint maximal melonic subgraphs in $\mathcal{G}$, so that $\hat{\mathcal{G}}$ is the core of $\mathcal{G}$ and the core decomposition gives back the $\mathcal{G}_i$s.

The relation between $k(\mathcal{G})$, $k(\hat{\mathcal{G}})$ and $k(\mathcal{G}_i)$ follows from the remark that the core decomposition is a partition of the vertex set of $\mathcal{G}$, while the fact that a rooted colored graph and its core have the same degree follows from Eq. (4.1).

□

4.3 Chains

The set of cores of a fixed degree is not finite. This is due to the presence of *chains* of $(D-1)$-dipoles (to be defined precisely here) of arbitrary length. It follows that, in order to provide a useful classification of graphs at a fixed degree, we need to refine further the core decomposition by identifying and removing maximal chains.

Definition 4.12. *A* $(D-q)$-dipole *in a closed* $(D+1)$ *colored graph is a couple of vertices connected by exactly* $D-q$ *parallel edges* $e_1, \dots e_{D-q}$, none of which is the root *of the graph.*

A $(D-q)$*-dipole is attached to the rest of the graph by* $q+1$ *pairs of halfedges of the same color. It is possible that one of these halfedges belongs to the root edge. Moreover, it is possible that two of these halfedges are in fact matched to form the root edge (see Figure 4.7).*

A $(D-1)$-dipole of colors (c_1, c_2) *is a* $(D-1)$*-uple of parallel edges attached to the rest of the graph by two halfedges of color* c_1 *and two halfedges of color* c_2*. (This definition is slightly different from the one found in [69].)*

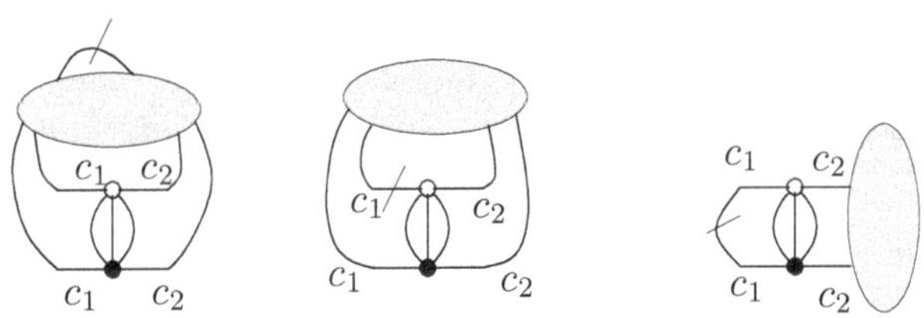

Figure 4.7: Examples of 3-dipoles in $(4+1)$-colored graphs.

Dipoles can join together to form chains of dipoles (as in Figure 4.8). The chains of $(D-1)$ dipoles are especially important: as we will see, the degree of a graph does not depend on the length of the chains, hence all the melonic cores, which only differ by the length of the internal chains of $(D-1)$ dipoles, have the same degree.

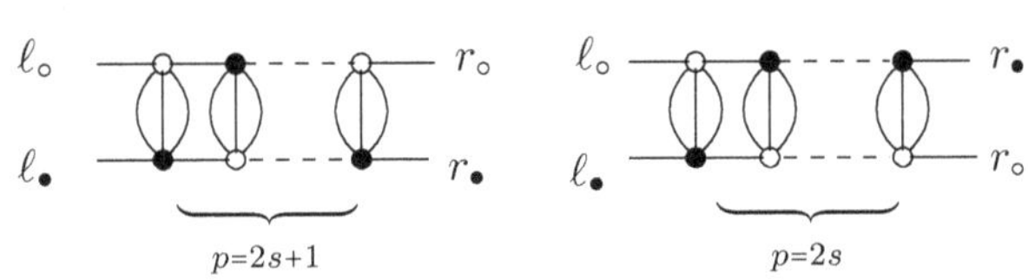

Figure 4.8: Two chains, with odd and even length, respectively (with $D=4$).

One would like to identify the maximal (i.e. vertex disjoint) chains of $(D-1)$ dipoles in a core. The case $D = 3$ is slightly subtle and requires a refinement of the naive definition of a chain.

Definition 4.13. *Let a* chain *in a rooted colored graph $\mathcal{G}$ (see Figure 4.8) be a configuration made of:*

- *two* left external halfedges $\ell_\circ$ *and* $\ell_\bullet$ having the same color,
- *two* right external halfedges $r_\bullet$ *and* $r_\circ$ having the same color,
- $2p$ internal vertices, $p \geq 1$, *forming a sequence* $d_1, \dots, d_p$ *of* $(D-1)$*-dipoles,*

such that:

- *the white (resp. black) vertex of d_1 is incident to $\ell_\circ$ (resp. $\ell_\bullet$)*
- *the white (resp. black) vertex of d_p is incident to $r_\circ$ (resp. $r_\bullet$)*
- *the dipole d_i and d_{i+1} share two edges for each $i = 1, \dots, p-1$*
- *the root of $\mathcal{G}$ is not one of the internal edges of the chain (neither an edge of a dipole, nor an edge connecting two consecutive dipoles). Observe, however, that the root edge can contain one of the external halfedges $\ell_\circ, \ell_\bullet, r_\bullet, r_\circ$ of the chain, or it can consist of* any matching *of a pair of the external halfedges: $(\ell_\circ, \ell_\bullet), (\ell_\circ, r_\bullet), (\ell_\bullet, r_\circ)$ and $(r_\circ, r_\bullet)$ (see Figure 4.9).*
- *the external halfedges $\ell_\circ, \ell_\bullet$ and respectively $r_\circ, r_\bullet$ are* not matched together by a nonroot edge *(i.e. the two external halfedges at the same end of the chain are not part of the same* nonroot *edge in $\mathcal{G}$, as in this case the chain would be a melonic subgraph). On the contrary, $\ell_\circ$ and $r_\bullet$ and respectively $\ell_\bullet$ and $r_\circ$ can be matched together (see Figure 4.9), by a nonroot edge.*

A chain is proper *if it contains at least four internal vertices (or equivalently at least two $(D-1)$-dipoles). A chain is* maximal *if it is not a subgraph of a larger chain.*

We sometimes call a configuration of consecutive dipoles such that the two left external halfedges $\ell_\circ$ and $\ell_\bullet$ have different colors (and in this case, due to color constraints on each dipole, $r_\circ$ and $r_\bullet$ also have different colors) a *twister*. Twisters will not play any role in the sequel: we will regard them as peculiar configurations of $D-1$ dipoles. In particular, a twister is not a chain.

With this definition we have the following important results.

Lemma 4.9. *A chain can be extended to a* unique *maximal chain.*

Proof. Let us consider a chain and its two external left halfedges $\ell_\circ$ and $\ell_\bullet$. If the two halfedges are incident to a $(D-1)$ dipole in $\mathcal{G}$ and neither of the two belongs to the root edge, then the chain can be *extended to the left* by adding the $(D-1)$ dipole. The chain can similarly be *extended to the right*.

Extending maximally the chain, one obtains the same maximal chain, *irrespective of the order* in which the left/right extensions are performed. □

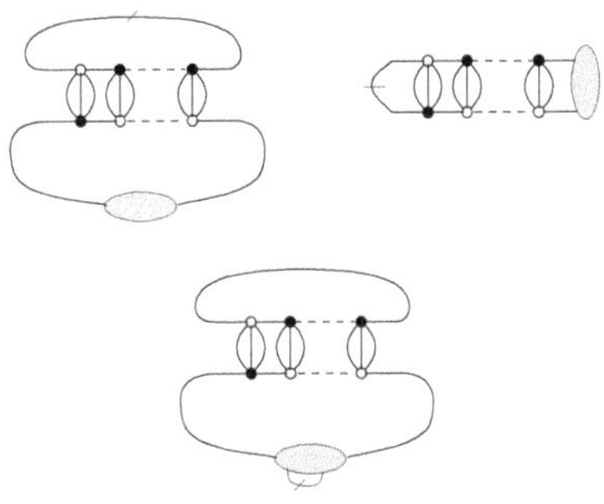

Figure 4.9: Chains with external halfedges matched.

A comment is important at this point. Observe that Lemma 4.9 works because we have allowed a pair of left and right halfedges to be matched by a nonroot edge, but it would not work otherwise. Indeed, consider the case at the bottom of Figure 4.9. If the chain can be extended only if the left and right external halfedges do not match, then one obtains two *different* maximal extensions (including respectively the last dipole on the left or on the right) of the same chain.

Lemma 4.10. *If $D \geq 4$, two distinct maximal chains in a rooted melon-free colored graph cannot share an internal vertex. If $D = 3$, two distinct maximal proper chains in a melon-free rooted colored graph cannot share an internal vertex.*

Proof. Assume first that the rooted colored graph $\mathcal{G}$ contains two maximal chains that share a $(D-1)$-dipole. But then these chains are the maximal left/right extension of the dipole, hence coincide.

Now if two chains share a vertex but no $(D-1)$-dipoles then this vertex must belong to two distinct $(D-1)$-dipoles. Parallel edge count shows that this is not possible if $D \geq 4$.

As illustrated by the right hand side of Figure 4.10, for $D = 3$ a vertex u can belong to two 2-dipoles $u - v$ and $u - w$. But if $u - v$ belongs to a proper chain then w has to belong to the same chain (since the chain has at least four internal vertices), hence there exists a vertex w' that is connected to w by a pair of (nonroot) edges and to u by at least one (nonroot) edge. Applying the same reasoning to the $u - w$ dipole, we conclude that the graph reduces to a double cycle of length 4 (on the right-hand side in Figure 4.10).

As $u - v, w - w'$ is a proper chain, the root edge must be one of the edges connecting u and w or v and w'. But this would make the root an internal edge of the proper chain $u - w, v - w'$, which is impossible.

□

4.3.1 Classification of chains

There are two main types of chains, depending on the way the external edges of the chain are connected by faces of $\mathcal{G}$.

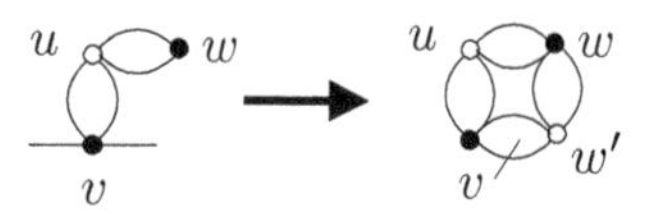

Figure 4.10: A pair of nondisjoint, nonproper chains for $D = 3$.

Let us call the external colors of the chain $\{c_1, c_2\}$ (which can coincide), and the external colors of the first dipole (from the left) in the chain $\{c_1, c'\}$. The two left halfedges $\ell_\circ$ and $\ell_\bullet$ belong to the same face with colors $\{c_1, c\}$ for all $c \neq c'$. The last face, with colors $\{c_1, c'\}$, travels horizontally to the next dipole. Iterating we are in one of the two cases:

- either the face $\{c_1, c'\}$ *does not travel horizontally* through the entire chain, that is there exists a dipole in the chain such that its right external color is neither c_1 nor c'
- or the face $\{c_1, c'\}$ *travels horizontally* along the chain all the way to the right halfedges.

Furthermore, the chains can have an even or odd number of dipoles, hence the chains are in one of the two cases:

- either the chain has an odd number of dipoles, hence $l_\circ$ and $r_\circ$ are both on the top of the chain, and $l_\bullet$ and $r_\bullet$ are on both the bottom (or the reverse)
- or the chain has an even number of dipoles, hence $l_\circ$ and $r_\bullet$ are both on the top of the chain, and $r_\circ$ and $l_\bullet$ are both on the bottom (or the reverse).

Correspondingly the chains are classified into:

Broken chains A chain with external color (c_1, c_2) is *broken* if for all $c \neq c_1$, $\ell_\circ$ and $\ell_\bullet$ are in the same face of color (c, c_1) (and in this case $r_\circ$ and $r_\bullet$ are in the same face of color (c, c_2) for all $c \neq c_2$).

They are subdivided further into chains with equal external colors $c_2 = c_1$ and chains with distinct external colors $(c_2 \neq c_1)$. Furthermore, for each of the two cases the chain can have an even or an odd number of dipoles.

Broken chains can have any number, (larger than two) even or odd, of dipoles. This situation is represented by the first two columns on the left in Figure 4.11.

Unbroken chains Chains that are not broken are *unbroken*. Let us consider separately chains of external colors $(c_1, c_2 \neq c_1)$ and (c_1, c_1):

- External colors $(c_1, c_2 \neq c_1)$: $\ell_\circ$ has color c_1 and belongs to a face of color (c_1, c_2) which does not contain $\ell_\bullet$. This face travels horizontally and leaves the chain through $r_\circ$ (which has color c_2). The chain has an odd number of dipoles. This situation is represented by the third column (from the left) in Figure 4.11.

- External color (c_1, c_1): again there is a face containing $\ell_\circ$ and not $\ell_\bullet$ (since the chain is not broken), which has to travel horizontally and leave the chain via $r_\bullet$. There is only one such face traveling horizontally between $\ell_\circ$ and $r_\bullet$, with color (c_1, c_2) (and a parallel face of color (c_1, c_2) goes through $\ell_\bullet$ and $r_\circ$). The color c_2 is referred to as the secondary color of the unbroken chain. The chain has an even number of dipoles. This situation is represented by the rightmost column in Figure 4.11.

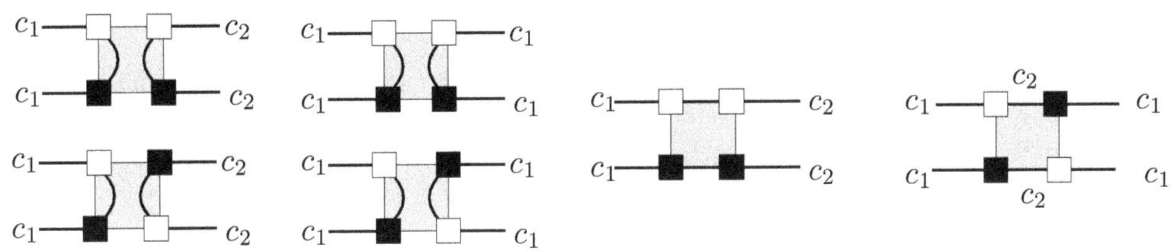

Figure 4.11: The four types of chain-vertices: each big gray square is to be considered a chain-vertex. Black and white little squares serve to recall bipartiteness constraints: the chains with an odd number of dipoles have squares of the same color on top and at the bottom, while the chains with an even number of dipoles have squares with different colors on top and at the bottom.

4.4 Schemes

Definition 4.14. *A rooted* scheme *(see Figure 4.12) is a rooted graph with colored edges having two types of vertices:*

- *Regular black and white vertices of degree $D+1$, each incident to one edge of each color.*
- *Chain-vertices of one of the following four types:*
  - *Broken chain-vertices of color (c_1, c_2), with $c_2 \neq c_1$. There are two types of such chain-vertices, according to the placement of the black and white squares.*
  - *Broken chain-vertices of color (c_1, c_1). There are two types of such chain-vertices, according to the placement of the black and white squares.*
  - *Unbroken chain-vertices of color (c_1, c_2), $c_2 \neq c_1$. There is only one such chain-vertex, with both squares on top (and at the bottom) of the same color.*
  - *Unbroken chain-vertices of color (c_1, c_1), with secondary color c_2. There is only one such chain-vertex, with the squares on top (and at the bottom) of different colors.*

A scheme is reduced *if:*

- *it is melon free,*
- *does not contain any proper chain of* $(D-1)$*-dipoles,*
- *the left (and right) halfedges of any chain vertex are not matched together into a nonroot edge (they can however be matched in the root edge),*
- *the left (or right) halfedges of any chain vertex are* not matched both at the same time *to the left (or right) halfedges of any other chain vertex or* $(D-1)$*-dipole.*

A reduced scheme can contain isolated $D-1$ dipoles (i.e. $D-1$ dipoles that are not part of any chain). It turns out that it will be convenient in the sequel to identify all these isolated $D-1$ dipoles in a scheme. Remark that for $D=3$ these isolated dipoles are not necessarily vertex disjoint, but they are for $D \geq 4$.

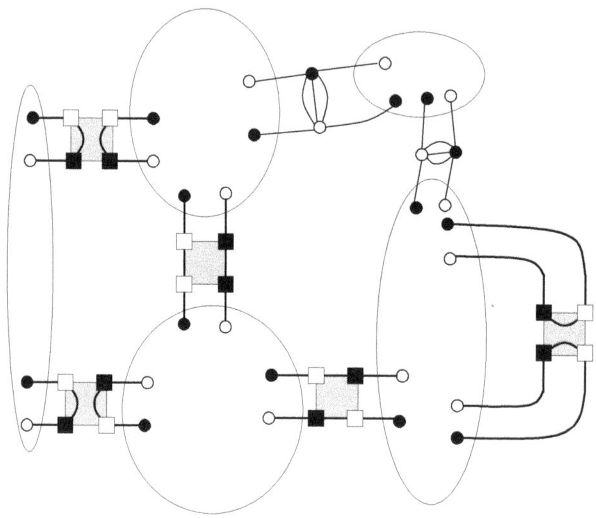

Figure 4.12: An example of scheme (the chain vertices, $(D-1)$ and $(D-2)$-dipoles separate the scheme into components that are free of chain vertices, $(D-1)$ and $(D-2)$-dipoles).

The *scheme* $\tilde{\mathcal{G}}$ *of a core* $\hat{\mathcal{G}}$ is obtained by replacing each maximal proper chain of $\hat{\mathcal{G}}$ by the corresponding *chain-vertex*. Since maximal proper chains are vertex disjoint this can be done independently for each chain. Observe that in the case $D=3$, this operation would not have been well defined if single $(D-1)$-dipoles were considered as chains. This is why we restrict our attention to maximal *proper* chains.

By construction, the scheme of a melon free colored graph is reduced. Observe moreover that we do not associate any chain-vertex to a twister: even if a core contains a twister made of many $(D-1)$-dipoles, these dipoles remains as such in the scheme. In particular, a reduced scheme can contain twisters.

The following theorem is trivial.

Theorem 4.3. *There is a bijection between the set of melon-free rooted colored graphs* $\hat{\mathcal{G}}$ *with* $2k(\hat{\mathcal{G}})$ *vertices and the set of pairs* $(\tilde{\mathcal{G}};(c_1,\ldots,c_q))$ *where* $\tilde{\mathcal{G}}$ *is a reduced scheme with* q *chain-vertices* $x_1,\ldots,x_q$*, and* c_i *is a chain of the same type as* x_i*, such that the total number of vertices in* $\tilde{\mathcal{G}}$ *and the chains* $c_1,\ldots,c_q$ *is* $2k(\hat{\mathcal{G}})$.

The chain-vertices we have introduced allow us to keep track in $\tilde{\mathcal{G}}$ of the faces of the melon-free graph that are not entirely included in a chain.

Lemma 4.11. *Let* $\hat{\mathcal{G}}$ *and* $\hat{\mathcal{G}}'$ *be two melon-free rooted colored graphs with the same scheme* $\tilde{\mathcal{G}}$*. Then* $\hat{\mathcal{G}}$ *and* $\hat{\mathcal{G}}'$ *have the same degree.*

Proof. A chain can be extended into a longer chain represented by a similar chain-vertex by the addition of a pair of $(D-1)$-dipoles (as in Figure 4.13). For unbroken chains the external colors of the two new $(D-1)$ dipoles are constrained, while for broken chains they are not. Any two chains represented by similar chain-vertices are related by a number of such extensions.

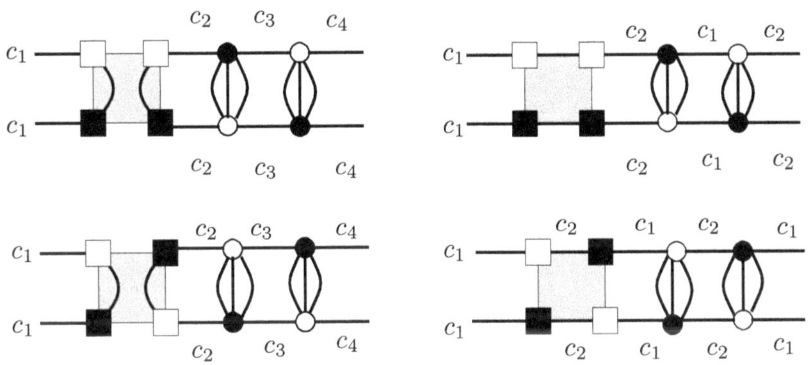

(a) Broken chains.

(b) Unbroken chains.

Figure 4.13: The extension of chains.

In all cases, the number of vertices increases by four. For the faces, we count them for broken (Figure 4.13(a)) and unbroken (Figure 4.13(b)) chains separately:

Broken chains. The number of internal faces increases by (Figure 4.13(a)):

- $2\binom{D-1}{2}=(D-1)(D-2)$ faces of length 2, of colors (c,c') with $c,c'\neq c_2,c_3$ for the first dipole and $c,c'\neq c_3,c_4$ for the second.
- $(D-1)$ faces of colors (c_2,c), $c\neq c_3$ closed through the chain vertex and through one of the parallel edges of the leftmost dipole.
- we have two cases:
  - $c_2=c_4$, $(D-1)$ faces of length 4 containing the vertices of the new dipoles and of colors (c,c') with $c,c'\neq c_2=c_4$.

  - $c_2 \neq c_4$, $(D-2)$ faces of length 4 containing the vertices of the new dipoles and of colors (c, c') with $c, c' \neq c_2, c_4$ and another face of colors (c_2, c_3), closed through the chain vertex and through one of the parallel edges of the rightmost dipole.

Unbroken chains. The number of internal faces increases by (Figure 4.13(b)):

- $2\binom{D-1}{2} = (D-1)(D-2)$ faces of length 2, of colors (c, c') with $c, c' \neq c_1, c_2$ for the two dipoles.
- we have two cases:
  - Chains with external colors (c_1, c_2) with $c_2 \neq c_1$:
    (i) $(D-1)$ faces of colors (c_2, c), $c \neq c_1$ closed through the chain vertex and through one of the parallel edges the leftmost dipole.
    (ii) $(D-1)$ faces of length 4 and colors (c_1, c) with $c \neq c_2$ containing the vertices of the new dipoles.
  - Chains with external colors (c_1, c_1):
    (i) $(D-1)$ faces of colors (c_1, c), $c \neq c_2$ closed through the chain vertex and through one of the parallel edges the leftmost dipole.
    (ii) $(D-1)$ faces of length 4 and colors (c_2, c) with $c \neq c_1$ containing the vertices of the new dipoles.

In all cases the number of faces increases by $D(D-1)$ and, as the number of vertices increases by four, the degree is unaffected.

□

This lemma allows us to define the *degree of a reduced scheme*, as the common degree of all cores that have it as scheme.

4.5 Schemes of fixed degree

All the graphs corresponding to a scheme have the same degree (the degree of the scheme). Furthermore, unlike the set of cores of fixed degree, the set of schemes of fixed degree is finite. The reduced degree of a graph (or core or scheme) is defined as:

$$\delta(\mathcal{G}) \equiv \frac{2}{(D-1)!}\omega(\mathcal{G}) .$$

Theorem 4.4. *The number of reduced schemes with a fixed (reduced) degree is finite.*

Proof. The proof of this is divided into two parts. First we analyze the iterative elimination of chain-vertices, $(D-1)$-dipoles, and in the cases $D \geq 4$, $(D-2)$-dipoles in a scheme, to prove the following result:

Proposition 4.1. *The number of chain-vertices, $(D-1)$, and for $D \geq 4$, $(D-2)$-dipoles in a reduced scheme of reduced degree δ is bounded by 19δ.*

Proof. See Section 4.5.1. □

Once this result is granted, we observe that the minimal realization of any chain-vertex consists of at most four $(D-1)$-dipoles, so that there is an injective map from the reduced schemes of reduced degree δ into colored graphs of reduced degree δ with at most 76δ $(D-1)$- and $(D-2)$-dipoles.

We then establish the following result:

Proposition 4.2. *For $D = 3$, the number of colored graphs with fixed degree and a fixed number of 2-dipoles is finite. For $D \geq 4$, the number of colored graphs with fixed degree and fixed numbers of $(D-1)$-dipoles and $(D-2)$-dipoles is finite.*

Proof. See Section 4.5.2. □

Theorem 4.4 is an immediate consequence of these two propositions. □

4.5.1 Proof of Proposition 4.1

As a preliminary we analyze the effect that the deletion of a single $(D-q)$-dipole has on the degree and subsequently extend the analysis to the deletion of a chain-vertex. The conclusion of this analysis is that the deletion of a $(D-q)$-dipole:

- either separates the graph into $q+1$ connected components, in which case the degree is distributed among the connected components,
- or it separates the graph into less than $q+1$ connected components, in which case the degree strictly decreases.

Similarly, for chains, the deletion:

- either separates the graph into two connected components, in which case the degree is distributed among the connected components,
- or the deletion does not separate the graph, in which case the degree strictly decreases.

As a consequence, we can bound the number of non separating deletions in terms of the degree.

Analysis of a $(D-q)$-dipole removal

Let us define more precisely the *removal* of a $(D-q)$-dipole (with $1 \leq q \leq D-2$) of a colored graph $\mathcal{G}$: assuming the nonparallel edges of the dipole (the edges *not connecting* the two vertices of the dipole) have colors $c_0, c_1, \dots c_q$, we delete the two vertices of the dipole and their incident halfedges and form one new edge for each color $c_0, c_1, \dots c_q$ with the remaining pairs of halfedges.

By construction, the number of vertices decreases by two and the number of edges decreases by $(D+1)$ at a deletion. In order to track the variation of the degree under this move we need to consider more precisely the variation of the number of faces. This is somewhat involved, as it depends on the number of connected components the graph separates into upon removal of the $(D-q)$-dipole and also on whether couples of new edges (c_a, c_b) belong to a same face or not.

Connected components and faces after a dipole removal. We denote:

$$\mathcal{G}_1, \mathcal{G}_2, \dots \mathcal{G}_C ,$$

the C connected components obtained after the removal of the $(D-q)$-dipole, $1 \leq C \leq q+1$. As the removal of the dipole deletes two vertices we have:

$$k(\mathcal{G}) = k(\mathcal{G}_1) + k(\mathcal{G}_2) + \dots + k(\mathcal{G}_C) + 1 .$$

We denote t_1 the number of new edges belonging to $\mathcal{G}_1$, t_2 the number of new edges belonging to $\mathcal{G}_2$, and so on. We have:

$$t_1 + t_2 + \dots + t_C = q + 1 .$$

Without loss of generality we can assume that the colors of the t_1 new edges belonging to $\mathcal{G}_1$ are $c_0, \dots c_{t_1-1}$, the colors of the t_2 new edges belonging to $\mathcal{G}_2$ are $c_{t_1}, \dots c_{t_1+t_2-1}$ and so on.

The faces affected by the $(D-q)$-dipole removal are the ones containing at least one of its vertices. They fall into three categories:

- Faces with colors (c, c') such that $\{c, c'\} \cap \{c_0, c_1, \dots c_q\} = \emptyset$. For each of the $\binom{D-q}{2}$ choices of such pairs, exactly one face of degree 2 (made of two parallel edges of the dipole) is deleted by the removal of the dipole:

$$F^{(c,c')}(\mathcal{G}) = F^{(c,c')}(\mathcal{G}_1) + F^{(c,c')}(\mathcal{G}_2) + \dots + F^{(c,c')}(\mathcal{G}_C) + 1 .$$

- Faces with colors (c_i, c), with $c_i \in \{c_0, c_1, \dots c_q\}$ and $c \notin \{c_0, c_1, \dots c_q\}$. For each of the $(D-q)(q+1)$ choices of such pairs, exactly one face is incident to the dipole: this face has length at least four in $\mathcal{G}$ and the dipole removal reduces its length by two, so that the number of faces with these colors is unchanged:

$$F^{(c_i,c)}(\mathcal{G}) = F^{(c_i,c)}(\mathcal{G}_1) + F^{(c_i,c)}(\mathcal{G}_2) + \dots + F^{(c_i,c)}(\mathcal{G}_C) .$$

- Faces with colors (c_i, c_j) with $\{c_i, c_j\} \subset \{c_0, c_1, \dots c_q\}$. For each of the $\binom{q+1}{2}$ choices of such colors, either one or two faces are incident to the dipole. In this case we distinguish between two possibilities:

  - **Type** a. The four edges of color c_i and c_j belong to the same cycle (c_i, c_j) in $\mathcal{G}$. Upon removal of the $(D-q)$-dipole the cycle (c_i, c_j) splits into two disjoint cycles:

$$F^{(c_i,c_j)}(\mathcal{G}) = F^{(c_i,c_j)}(\mathcal{G}_1) + F^{(c_i,c_j)}(\mathcal{G}_2) + \dots + F^{(c_i,c_j)}(\mathcal{G}_C) - 1 .$$

  This is presented in Figure 4.14.

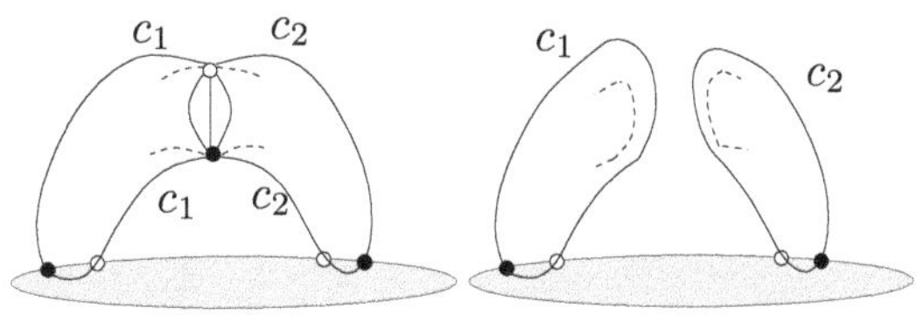

Figure 4.14: A face that splits by deleting a dipole.

– **Type** b. The four edges of color c_i and c_j belong to two distinct cycles (c_i, c_j) in $\mathcal{G}$. Upon removal of the $(D-q)$-dipole the two cycles (c_i, c_j) are merged into a unique cycle:

$$F^{(c_i,c_j)}(\mathcal{G}) = F^{(c_i,c_j)}(\mathcal{G}_1) + F^{(c_i c_j)}(\mathcal{G}_2) + \cdots + F^{(c_i,c_j)}(\mathcal{G}_C) + 1 \, .$$

This is presented in Figure 4.15.

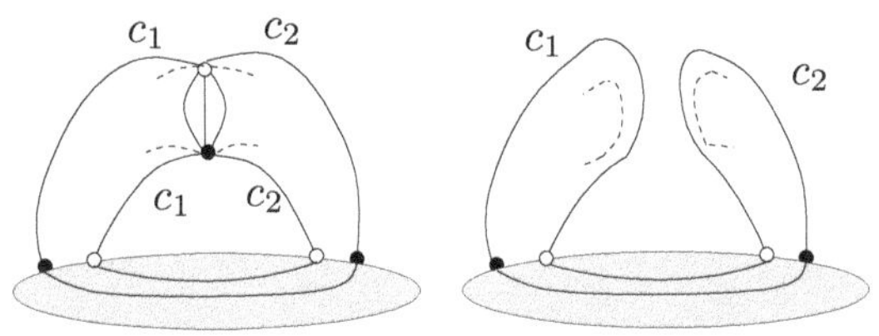

Figure 4.15: Two faces that merge by deleting a dipole.

A third possibility, namely that the two pairs of edges of color (c_i, c_j) belong to the same face before the removal and the face does not split, does not exist because the graph $\mathcal{G}$ is bipartite.

We are now in position to describe the global effect of the removal of a dipole on the degree.

Case I. Completely separating $(D-q)$*-dipoles.* We first consider the case in which the removal of $(D-q)$-dipoles splits the graph into $q+1$ connected components each containing exactly one new edge ($C = q+1$ with the notation above): we refer to such a dipole as *completely separating.* We illustrate this case in Figure 4.16.

In this case, all the faces of colors (c_i, c_j) with $\{c_i, c_j\} \subset \{c_0, c_1, \dots c_q\}$ are of Type a, hence:

$$\begin{aligned} F(\mathcal{G}) &= F(\mathcal{G}_1) + F(\mathcal{G}_2) + \cdots + F(\mathcal{G}_{q+1}) + \binom{D-q}{2} - \binom{q+1}{2} \\ &= F(\mathcal{G}_1) + F(\mathcal{G}_2) + \cdots + F(\mathcal{G}_{q+1}) + \frac{1}{2} D(D-2q-1) \, . \end{aligned}$$

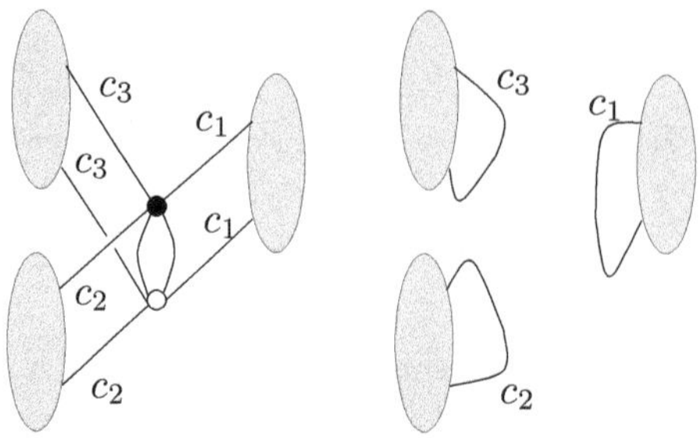

Figure 4.16: The decomposition at a completely separating $(D-2)$-dipole.

From Eq. (3.1) we get:

$$\begin{aligned}\delta(\mathcal{G}) &= \frac{1}{2}D(D-1)k(\mathcal{G}) + D - F(\mathcal{G}) \\ &= \frac{1}{2}D(D-1)\Big(k(\mathcal{G}_1) + k(\mathcal{G}_2) + \dots + k(\mathcal{G}_{q+1}) + 1\Big) + D \\ &\quad - \Big(F(\mathcal{G}_1) + F(\mathcal{G}_2) + \dots + F(\mathcal{G}_{q+1}) + \frac{1}{2}D(D-2q-1)\Big) \\ &= \delta(G_1) + \delta(G_2) + \dots + \delta(G_{q+1}) \,.\end{aligned}$$

In other terms the degree is distributed between the $q+1$ connected components.

Case II. Non completely separating $(D-q)$-dipoles. We now consider the remaining cases ($1 \le C \le q$): we refer to such a dipole as *non completely separating.*

All the faces (c_i, c_j) with $c_i \in \{c_0, \dots c_{t_1-1}\}$ and $c_j \notin \{c_0, \dots c_{t_1-1}\}$ are of Type a. On the contrary, the faces (c_i, c_j) with $\{c_i, c_j\} \subset \{c_0, \dots c_{t_1-1}\}$ can be either of Type a or of Type b.

Let us denote r_1 the number of faces of Type b, $0 \le r_1 \le \binom{t_1}{2}$ in $\mathcal{G}_1$, r_2 the number of faces of Type b in $\mathcal{G}_2$, and so on. We have:

$$\begin{aligned}F(\mathcal{G}) =& F(\mathcal{G}_1) + F(\mathcal{G}_2) + \dots + F(\mathcal{G}_C) + \binom{D-q}{2} - \binom{q+1}{2} \\ &+ 2r_1 + 2r_2 + \dots + 2r_C \\ =& F(\mathcal{G}_1) + F(\mathcal{G}_2) + \dots + F(\mathcal{G}_C) + \frac{1}{2}D(D-2q-1) \\ &+ 2r_1 + 2r_2 + \dots + 2r_C \,.\end{aligned}$$

Using again Eq. (3.1) we get:

$$\begin{aligned}\delta(\mathcal{G}) =& \frac{1}{2}D(D-1)k(\mathcal{G}) + D - F(\mathcal{G}) \\ =& \frac{1}{2}D(D-1)\Big(k(\mathcal{G}_1) + k(\mathcal{G}_2) + \dots + k(\mathcal{G}_C) + 1\Big) + D\end{aligned}$$

$$
\begin{aligned}
&- \Big(F(\mathcal{G}_1) + F(\mathcal{G}_2) + \cdots + F(\mathcal{G}_C) \\
&\qquad + \frac{1}{2} D(D - 2q - 1) + 2r_1 + 2r_2 + \cdots + 2r_C \Big) \\
&= \delta(\mathcal{G}_1) + \delta(\mathcal{G}_2) + \cdots + \delta(\mathcal{G}_C) \\
&\quad + D(q + 1 - C) - 2r_1 - 2r_2 - \cdots - 2r_C \,.
\end{aligned}
$$

In other terms the variation of the degree through the removal depends on the structure of the incident faces. Observe that, as $r_i \le t_i$:

$$
\begin{aligned}
&D(q+1-C) - 2r_1 - 2r_2 - \cdots - 2r_C \ge D(q+1-C) - t_1(t_1-1) - \ldots t_C(t_C-1) \\
&= (D - t_1)(t_1 - 1) + \ldots (D - t_C)(t_C - 1) \,,
\end{aligned}
$$

as $t_1 + \cdots + t_C = q + 1 \le D - 1$. For non completely separating deletions we have $t_i \ge 1$ and at least one of them (say t_1) is in fact at least two, hence

$$
(D - t_1)(t_1 - 1) + \ldots (D - t_C)(t_C - 1) \ge D - t_1 \ge 1 \,.
$$

We now make the various possibilities more explicit in the cases of $(D-1)$- and $(D-2)$-dipoles:

Non separating $(D-1)$-dipoles. We have $q = 1$, $C = 1$ and $t_1 = 2$. Depending on the value of r_1 we thus distinguish two cases:

- Case II.A: the face (c_0, c_1) is of Type a, hence $r_1 = 0$ and:
$$\delta(\mathcal{G}) = \delta(\mathcal{G}_1) + D \,.$$
- Case II.B: the face (c_0, c_1) is of Type b, $r_1 = 1$ and:
$$\delta(\mathcal{G}) = \delta(\mathcal{G}_1) + D - 2 \,.$$

Observe that for all $D \ge 3$ the degree strictly decreases under the removal of a non separating $(D-1)$-dipole.

Non completely separating $(D-2)$-dipoles. We have $q = 2$ and there are two possible values for C, and a total of six possible cases:

- $C = 1$, $t_1 = 3$ and:
  - $r_1 = 0$, $\delta(\mathcal{G}) = \delta(\mathcal{G}_1) + 2D$.
  - $r_1 = 1$, $\delta(\mathcal{G}) = \delta(\mathcal{G}_1) + 2D - 2$.
  - $r_1 = 2$, $\delta(\mathcal{G}) = \delta(\mathcal{G}_1) + 2D - 4$.
  - $r_1 = 3$, $\delta(\mathcal{G}) = \delta(\mathcal{G}_1) + 2D - 6$.

  we call this a *non separating $(D-2)$-dipole deletion.*
- $C = 2$, $t_1 = 1$ (hence $r_1 = 0$), $t_2 = 2$ and
  - $r_2 = 0$, $\delta(\mathcal{G}) = \delta(\mathcal{G}_1) + \delta(\mathcal{G}_2) + D$.
  - $r_2 = 1$, $\delta(\mathcal{G}) = \delta(\mathcal{G}_1) + \delta(\mathcal{G}_2) + D - 2$.

  we call this a *partially separating $(D-2)$-dipole deletion.*

As expected, for all $D \ge 4$, the degree strictly decreases under the removal of a non separating or partially separating $(D-2)$dipole.

Chain-vertex removal vs $(D-1)$-dipole removal

The removal of a chain-vertex consists in deleting this vertex and the incident halfedges and creating two new edges by joining two by two the remaining halfedges that arise from the same extremity of the chain-vertex. The removal of a chain-vertex in a scheme can alternatively be performed in the following equivalent way:

- Replace the chain-vertex by its minimal length chain representation: this yields a scheme $\mathcal{G}^+$ with same degree.
- Remove one of the $(D-1)$-dipole of the inserted chain: one of the three cases above applies.
- Eliminate the melons that might have been created: these operations do not affect the degree.

This last procedure, although slightly more complex *a priori*, allows us to built on the case analysis already carried out for the $(D-1)$-dipole removals.

Case I. Separating chain-vertex. The removal of a chain-vertex separates the graph $\mathcal{G}$ into two components $\mathcal{G}_1$ and $\mathcal{G}_2$ if and only if the deletion of any $(D-1)$-dipole of the equivalent chain separates the graph $\mathcal{G}^+$ into two components $\mathcal{G}_1^+$ and $\mathcal{G}_2^+$. In such a case,

$$\delta(\mathcal{G}) = \delta(\mathcal{G}^+) = \delta(\mathcal{G}_1^+) + \delta(\mathcal{G}_2^+) = \delta(\mathcal{G}_1) + \delta(\mathcal{G}_2) \,.$$

Observe that in this case, the chain-vertex can represent indifferently an unbroken or a broken chain.

Case II. Nonseparating chain-vertex, two resulting faces. This case is similar to a Case II.A removal of a $(D-1)$-dipole: the removal of the chain-vertex does not separate the graph $\mathcal{G}$ and in the resulting graph $\mathcal{G}'$ the two new edges belong to two different (c_0, c_1)-cycles. Then the removal of the chain-vertex is equivalent to a Case II.A removal of $(D-1)$-dipole in the graph $\mathcal{G}^+$, followed by some melon deletions:

$$\delta(\mathcal{G}) = \delta(\mathcal{G}^+) = \delta(\mathcal{G}') + D \,,$$

whether the chain-vertex represents an unbroken or a broken chain.

Case III. Nonseparating chain-vertex, single resulting face. In this case the broken and unbroken chains behave differently:

- *Case III.a: unbroken chain.* This case is similar to a Case II.B removal of a $(D-1)$-dipole: the removal of the chain-vertex does not separate the graph $\mathcal{G}$ but in the resulting graph $\mathcal{G}'$ the two new edges belong to a same (c_0, c_1)-cycle. If the chain-vertex represents an unbroken chain then the removal of the chain-vertex is equivalent to a Case II.B removal of $(D-1)$-dipole in the graph $\mathcal{G}^+$, followed by some melon deletions:

$$\delta(\mathcal{G}) = \delta(\mathcal{G}^+) = \delta(\mathcal{G}') + D - 2 \,.$$

- *Case III.b: broken chain.* The removal of the chain-vertex does not separate the graph $\mathcal{G}$ and in the resulting graph $\mathcal{G}'$ the two new edges belong to the same (c_0, c_1)-cycle, but the chain-vertex represents a broken chain: in this case the removal is equivalent to a Case II.A removal of $(D-1)$-dipole in the graph $\mathcal{G}^+$, followed by some melon deletions:

$$\delta(\mathcal{G}) = \delta(\mathcal{G}^+) = \delta(\mathcal{G}') + D \, .$$

Iterative deletion of chain-vertices, $(D-1)$-dipoles and $(D-2)$-dipoles

In this section we introduce an algorithm that allows us to eliminate one by one chain vertices, $(D-1)$-dipoles and, for $D \geq 4$, $(D-2)$-dipoles in a reduced scheme. This will allow us to show that the total number of chain vertices, $(D-1)$-dipoles and $(D-2)$-dipoles in a reduced scheme is bounded linearly in terms of the degree. This algorithm is not unique, and here we will adapt the one presented in [73]. For $D = 3$, the algorithm goes through ignoring the $(D-2)$-dipoles.

The deletions of chain vertices and $(D-1)$-dipoles can either be separating or not. The deletions of $(D-2)$-dipoles can be completely separating, partially separating or non separating. Under any completely separating deletion the degree is distributed among the resulting connected components, while under the non completely separating deletions the degree strictly decreases.

We first perform a maximal number of *non completely separating deletions*, that is non separating deletions of chain vertices and $(D-1)$ dipoles and non separating or only partially separating deletions of $(D-2)$-dipoles. Some examples are represented in Figure 4.17 and Figure 4.18.

Let $\tilde{\mathcal{G}}$ be a reduced scheme of degree $\delta(\tilde{\mathcal{G}})$, and let us denote $\mathfrak{D}(\tilde{\mathcal{G}})$ the total number of chain vertices, $(D-1)$-dipoles and $(D-2)$-dipoles in $\tilde{\mathcal{G}}$. Besides the color $c \in \{0, \dots D\}$, we assign to all the edges in the scheme a new color, say black.

As long we find a non separating chain vertex, a non separating $(D-1)$-dipole *formed by parallel black edges* or a non completely separating $(D-2)$-dipole *formed by parallel black edges* we delete it. We color the new edges created by a deletion with a new color, say blue. Observe that the deletion of a chain vertex having a left and a right halfedge matched into an edge creates only one blue edge. As we delete partially separating $(D-2)$-dipoles the scheme can be disconnected and the deletions can create ring components (see Figure 4.17 and Figure 4.18).

Our aim is to delete at each step only dipoles that were present in the original scheme, hence we only delete dipoles having all the parallel edges colored in black. If the deletion creates new $(D-1)$-dipoles and $(D-2)$-dipoles having one or several of their parallel edges colored in blue, these blue edges *do not count* when identifying dipoles. However, the ring components consisting of one blue edge *count* as connected components when deciding whether a deletion is (completely) separating or not.

Observe that $\mathfrak{D}(\tilde{\mathcal{G}})$ always goes down by one under a chain vertex deletion, but, for $3 \leq D \leq 5$ it can go down by as much as three for the dipole deletions (see Figure 4.17):

- in $D = 3$ the 2-dipoles (i.e. $(D-1)$-dipoles) are not vertex disjoint. Each vertex of a 2-dipole can belong to another 2-dipole, in which case three dipoles are erased by a deletion.

- in $D = 4$ the 3-dipoles and 2-dipoles (i.e. $(D-1)$-dipoles and $(D-2)$-dipoles) and couples of 2-dipoles (i.e. $(D-2)$-dipoles) are not vertex disjoint and again up to three dipoles can be erased in one step.
- in $D = 5$ the 3-dipoles (i.e. $(D-2)$-dipoles) are not vertex disjoint and again up to three dipoles can be erased in one step.

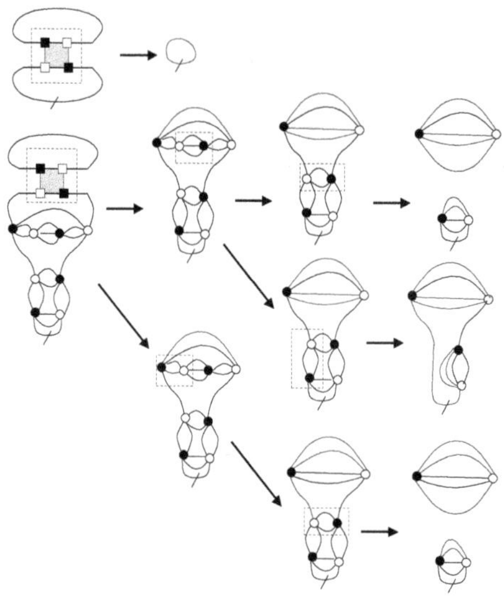

Figure 4.17: Iterated deletions of non separating chain vertices and $(D-1)$-dipoles and non separating or partially separating $(D-2)$-dipoles.

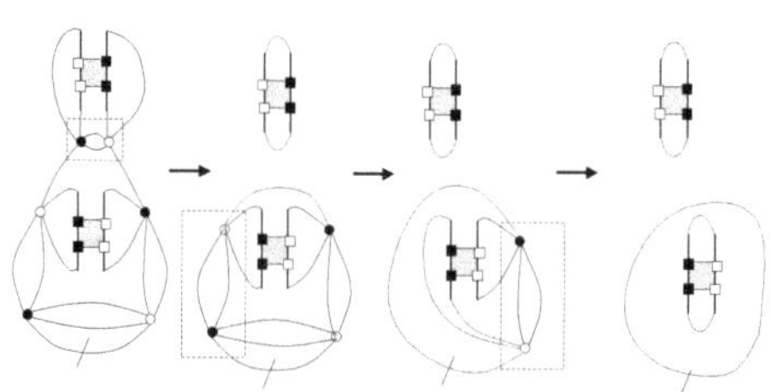

Figure 4.18: Iterated deletions of partially separating dipoles.

We iterate the non completely separating deletions maximally and obtain a scheme $\tilde{\mathcal{G}}'$. Unlike $\tilde{\mathcal{G}}$, $\tilde{\mathcal{G}}'$ in general is neither connected nor reduced (as depicted for instance in Figure 4.17 and Figure 4.18). Furthermore, $\tilde{\mathcal{G}}'$ is not unique, and depends on the sequence of deletions performed.

The degree of $\mathcal{G}'$ is the sum of the degrees of its connected components. Let us denote the maximal number of non completely separating deletions one can perform starting from a reduced scheme by $q^{\text{n.c.s.}}$. We have the following inequalities:

- the degree goes down by at least one for each of these deletions, hence:

$$q^{\text{n.c.s.}} \le \delta(\mathcal{G}') + q^{\text{n.c.s.}} \le \delta(\tilde{\mathcal{G}}) ,$$

- $\mathfrak{D}(\tilde{\mathcal{G}})$ goes down by at most three at each step, hence:

$$\mathfrak{D}(\tilde{\mathcal{G}}) \le \mathfrak{D}(\tilde{\mathcal{G}}') + 3q^{\text{n.c.s.}} \le \mathfrak{D}(\tilde{\mathcal{G}}') + 3\delta(\tilde{\mathcal{G}}) ,$$

- every deletion creates at most three new blue edges, hence the total number of blue edges of $\tilde{\mathcal{G}}'$, $E_{\text{blue}}(\tilde{\mathcal{G}}')$ is bounded by:

$$E_{\text{blue}}(\tilde{\mathcal{G}}') \le 3q^{\text{n.c.s.}} \le 3\delta(\tilde{\mathcal{G}}) .$$

In the scheme $\mathcal{G}'$, all the chain vertices, $(D-1)$ and $(D-2)$-dipoles are (completely) separating. It follows in particular that, for any D, all the remaining $(D-1)$ and $(D-2)$ dipoles *are vertex disjoint.*

In order to conclude that $\mathfrak{D}(\tilde{\mathcal{G}})$ is bounded linearly in terms of $\delta(\tilde{\mathcal{G}})$, it is enough to bound $\mathfrak{D}(\tilde{\mathcal{G}}')$ in terms of $\delta(\mathcal{G}')$ and $E_{\text{blue}}(\tilde{\mathcal{G}}')$. This is slightly subtle because the remaining deletions are all separating, hence they conserve the degree.

All the remaining $(D-1)$- and $(D-2)$-dipoles are vertex disjoint, hence we can delete them together with the remaining chain vertices in any order. We color the new edges created by the deletions in black if *both halfedges* that are matched are black and in blue otherwise. Observe that the number of blue edges can either remain constant or decrease with these deletions. We add marks on the new edges, *keeping track of the multiplicity.* That is, if the two halfedges connected at a step come from edges having, say m_1 and m_2 marks respectively, the new edge will have $m_1 + m_2 + 1$ marks. For each $(D-2)$-dipole deleted (having external colors, say, c_1, c_2 and c_3) we add a copy of the fundamental melon and mark its edges c_1, c_2 and c_3.

As before, these deletions can again create ring components having only one edge and zero vertices. In Figure 4.19 we represent the deletion of a maximal set of nonseparating dipoles and chain vertices, followed by the deletion of all the separating dipole and chain vertices in a reduced scheme.

Let us denote the scheme obtained in this way by $\tilde{\mathcal{G}}''$. By construction $\tilde{\mathcal{G}}''$ has:

- blue and black edges that can be (multiply) marked. The number of blue edges of $\tilde{\mathcal{G}}''$ is at most equal to the number of blue edges of $\tilde{\mathcal{G}}'$:

$$E_{\text{blue}}(\tilde{\mathcal{G}}'') \le E_{\text{blue}}(\tilde{\mathcal{G}}') .$$

- the same degree as $\tilde{\mathcal{G}}'$.
- ring components that can be made either of a blue or of a black edge.
- no $(D-1)$-dipole made of black, unmarked parallel edges.

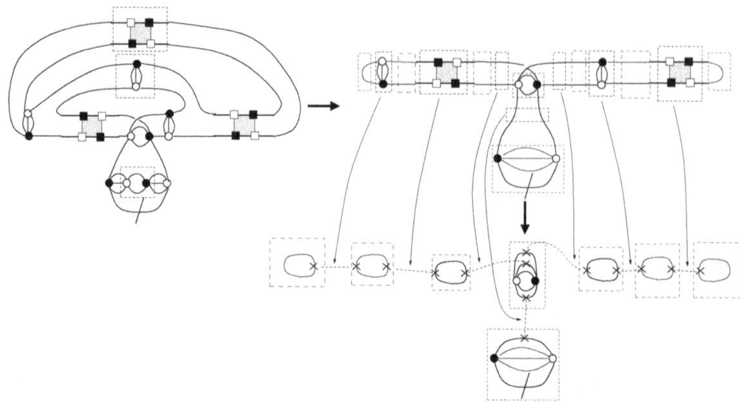

Figure 4.19: Maximal deletion of non separating chain vertices and dipoles, followed by the deletion of the separating chain vertices and dipoles in a reduced scheme.

- no chain vertex.
- all the $(D-2)$-dipoles made of black, unmarked parallel edges are copies of the fundamental melon with three marked edges.
- at least a mark in every connected component.

The connected components of $\tilde{\mathcal{G}}''$ can be seen as the vertices of an abstract graph $\mathfrak{F}$ (represented in Figure 4.19 at the bottom) whose edges correspond:

- either to the chain vertices and to the $(D-1)$-dipoles in $\tilde{\mathcal{G}}'$,
- or to pairs of external halfedges of the $(D-2)$-dipoles in $\tilde{\mathcal{G}}'$.

As all the chain vertices and $(D-1)$- and $(D-2)$-dipoles of $\tilde{\mathcal{G}}'$ are completely separating, $\mathfrak{F}$ is a *forest* (every tree in $\mathfrak{F}$ corresponds to one of the connected components of $\tilde{\mathcal{G}}'$). A subtle point is the following (see Figure 4.19): a pair of external halfedges of a $(D-2)$-dipole in $\tilde{\mathcal{G}}'$ can lead:

- either *to an edge and a vertex* in $\mathfrak{F}$ if the pair is matched to a pair of external halfedges of a chain vertex, $(D-1)$- or $(D-2)$-dipole,
- or *to just an edge* in $\mathfrak{F}$ if it is not.

The following lemma characterizes the components of degree zero having only black edges in $\tilde{\mathcal{G}}''$.

Lemma 4.12. *The components of degree zero having only black edges of $\tilde{\mathcal{G}}''$ and not containing the root:*

- *either are rings with only two marks such that at least one of the marks comes from a $(D-2)$-dipole deletion,*

- *or have et least three marks.*

Proof. The crucial observation is that all the black unmarked edges in $\tilde{\mathcal{G}}''$ are in fact edges that were preset in $\tilde{\mathcal{G}}$, and $\tilde{\mathcal{G}}$ is a reduced scheme.

The components of degree zero of $\tilde{\mathcal{G}}''$ are either ring components or melonic graphs with at least two vertices.

Ring components with one mark can be created by separating deletions only if one either deletes $D-1$ edges in a melon with D edges in $\tilde{\mathcal{G}}$, or one deletes a chain vertex whose left (or right) external halfedges are matched together into an edge in $\tilde{\mathcal{G}}$. Both cases are impossible as $\tilde{\mathcal{G}}$ is reduced.

Ring components with two marks can be created by the deletion of only $(D-1)$-dipoles and chain vertices only if, in $\tilde{\mathcal{G}}$, the two left (or right) external halfedges of a $(D-1)$-dipole or a chain vertex are both matched to the two left (or right) external halfedges of another $(D-1)$-dipole or a chain vertex. This is again impossible, as $\tilde{\mathcal{G}}$ is reduced.

Ring components with three marks can be created, as depicted in Figure 4.20.

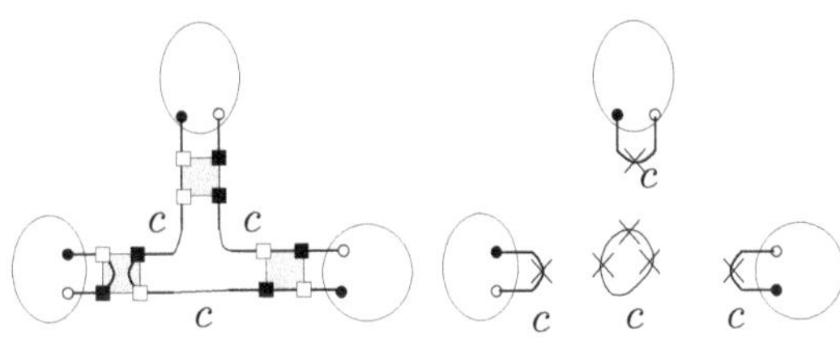

Figure 4.20: Ring components with three marks.

A nontrivial melonic graph with fewer than three marks represent either a melonic subgraph of $\tilde{\mathcal{G}}$ (which is impossible as $\tilde{\mathcal{G}}$ is reduced), or a $(D-1)$-dipole made of black, unmarked parallel edges in $\tilde{\mathcal{G}}''$, which is again impossible.

□

Let us denote:

- $n_+ + 1 \le \delta(\mathcal{G}') + 1$ the number of non root connected components of $\tilde{\mathcal{G}}''$ having positive degree plus the component containing the root and $n_{0,\text{blue}} \le E_{\text{blue}}(\tilde{\mathcal{G}}')$ the number of non root connected components of $\tilde{\mathcal{G}}''$ having degree zero and having at least a blue edge.

- $n^{(2)}_{0,\text{black}}$ the number of connected components of degree zero of $\tilde{\mathcal{G}}''$ having only black edges and only two marks. They are necessarily ring components and at least one of the marks comes from a $(D-2)$-dipole deletion.

- $n^{(3)}_{0,\text{black}}$ the number of connected components of degree zero of $\tilde{\mathcal{G}}''$ having only black edges and having at least three marks.

We also denote $\mathfrak{D}^{(D-1)}$ the number of chain vertices and $D-1$ dipoles of $\tilde{\mathcal{G}}'$ and $\mathfrak{D}^{(D-2)}$ the number of $(D-2)$-dipoles of $\tilde{\mathcal{G}}'$, hence:

$$\mathfrak{D}(\tilde{\mathcal{G}}') = \mathfrak{D}^{(D-1)} + \mathfrak{D}^{(D-2)} .$$

Denoting $E(\mathfrak{F})$ the number of edges of the abstract forest $\mathfrak{F}$, we have:

$$\begin{aligned} \mathfrak{D}^{(D-1)} + 3\mathfrak{D}^{(D-2)} &= E(\mathfrak{F}) , \\ 1 + n_+ + n_{0,\text{blue}} + n^{(2)}_{0,\text{black}} + n^{(3)}_{0,\text{black}} &\geq E(\mathfrak{F}) + 1 , \\ 2n^{(2)}_{0,\text{black}} + 3n^{(3)}_{0,\text{black}} &\leq 2E(\mathfrak{F}) , \\ n^{(2)}_{0,\text{black}} &\leq 3\mathfrak{D}^{(D-2)} , \end{aligned}$$

where the last inequality follows from the fact that, for all the ring components with a black edge and only two marks, at least one of the marks comes from the deletion of a $(D-2)$-dipole. It follows that:

$$\begin{aligned} 2n^{(2)}_{0,\text{black}} + 3n^{(3)}_{0,\text{black}} &\leq 2n_+ + 2n_{0,\text{blue}} + 2n^{(2)}_{0,\text{black}} + 2n^{(3)}_{0,\text{black}} \Rightarrow \\ n^{(3)}_{0,\text{black}} &\leq 2n_+ + 2n_{0,\text{blue}} , \end{aligned}$$

which further leads to:

$$\begin{aligned} \mathfrak{D}^{(D-1)} + 3\mathfrak{D}^{(D-2)} &= E(\mathfrak{F}) \leq 3n_+ + 3n_{0,\text{blue}} + n^{(2)}_{0,\text{black}} \Rightarrow \\ \mathfrak{D}^{(D-1)} + 3\mathfrak{D}^{(D-2)} &\leq 3n_+ + 3n_{0,\text{blue}} + 3\mathfrak{D}^{(D-2)} \Rightarrow \\ \mathfrak{D}^{(D-1)} &\leq 3n_+ + 3n_{0,\text{blue}} . \end{aligned}$$

On the other hand, as only the positive degree components or the components with blue edges can be univalent in $\mathfrak{F}$, $\mathfrak{D}^{(D-2)}$ is bounded by the maximal number of three valent vertices in a forest with exactly $1 + n_+ + n_{0,\text{blue}}$ univalent vertices, that is:

$$\mathfrak{D}^{(D-2)} \leq n_+ + n_{0,\text{blue}} - 1 ,$$

i.e. the number of such vertices in a binary tree. Thus:

$$\mathfrak{D}(\tilde{\mathcal{G}}') \leq 4n_+ + 4n_{0,\text{blue}} - 1 \leq 4\delta(\tilde{\mathcal{G}}') + 4E_{\text{blue}}(\tilde{\mathcal{G}}') ,$$

which finally leads us to:

$$\mathfrak{D}(\tilde{\mathcal{G}}) \leq 3\delta(\tilde{\mathcal{G}}) + \mathfrak{D}(\tilde{\mathcal{G}}') \leq 3\delta(\tilde{\mathcal{G}}) + 4\delta(\tilde{\mathcal{G}}') + 4E_{\text{blue}}(\tilde{\mathcal{G}}') \leq 19\delta(\tilde{\mathcal{G}}) ,$$

as $\delta(\tilde{\mathcal{G}}') \leq \delta(\tilde{\mathcal{G}})$ and $E_{\text{blue}}(\tilde{\mathcal{G}}') \leq 3\delta(\tilde{\mathcal{G}})$.

This completes the proof of Proposition 4.1.

4.5.2 Proof of Proposition 4.2

For any connected closed $(D+1)$-colored graph $\mathcal{G}$, denoting $F_s(\mathcal{G})$ the number of faces of $\mathcal{G}$ of length $2s$ (that is having $2s$ vertices), and taking into account that every vertex belongs to $\binom{D+1}{2}$ faces we have:

$$\sum_{s\geq 1} sF_s(\mathcal{G}) = \frac{D(D+1)}{2}k(\mathcal{G}) \,, \tag{4.3}$$

and Eq. (3.1) can be rewritten as:

$$\sum_{s\geq 1} F_s(\mathcal{G}) - D + \delta(\mathcal{G}) = \frac{D(D-1)}{2}k(\mathcal{G}) \,.$$

Eliminating $k(\mathcal{G})$ between these two equations, we obtain:

$$(D+1)\,\delta(\mathcal{G}) + 2F_1(\mathcal{G}) = D(D+1) + \sum_{s\geq 2}\left[s(D-1)-(D+1)\right]F_s(\mathcal{G}) \,, \tag{4.4}$$

and the right hand side is at least $D(D+1)$ for all $D\geq 3$.

The case $D=3$.

For $D=3$, we are interested in graphs with fixed number of 2-dipoles, or equivalently, with a fixed number of faces of degree 2. In view of Eq. (4.4), the number of faces of degree 6 or more in such a graph satisfies:

$$\sum_{s\geq 3} F_s(\mathcal{G}) \leq \sum_{s\geq 2} 2(s-2)F_s(\mathcal{G}) \leq (D+1)\delta(\mathcal{G}) + 2F_1(\mathcal{G}) \,,$$

i.e. this number is finite. Moreover the number of vertices incident to a face of degree 6 or more is finite:

$$\sum_{s\geq 3} 2sF_s(\mathcal{G}) = \sum_{s\geq 3} 2[s-2+2]F_s(\mathcal{G}) \leq 5(D+1)\delta(\mathcal{G}) + 10F_1(\mathcal{G}).$$

However, there could *a priori* be an arbitrary number of faces of degree 4 (that is infinitely many vertices incident only to faces of degree 4), since the coefficient of F_2 in Eq. (4.4) is zero for $D=3$. Let us rule this possibility out.

Let us count the number of vertices that there can be at distance at most 3 of a vertex on a face of degree not equal to 4. The vertices belonging to a face of degree not equal to 4 either belong to a face of degree 2 or to a face of degree larger than or equal to 6. According to our previous remark, the number of such vertices is at most:

$$5(D+1)\delta(\mathcal{G}) + 12F_1(\mathcal{G}) \,.$$

Since all vertices have degree 4, the number of vertices at distance at most 3 of such a vertex is $1+4+4^2+4^3=85$. Therefore if a regular colored graph $\mathcal{G}$ has more than $[5(D+1)\delta(\mathcal{G}) + 12F_1(\mathcal{G})]85$ vertices, then it contains a vertex v such that all vertices at distance less than 3 of v belong *only* to faces of length 4.

Now take an arbitrary jacket of $\mathcal{G}$: for instance the one corresponding to the cycle $(0,1,2,3)$. Then the faces of color $(0,1)$, $(1,2)$, $(2,3)$ and $(3,0)$ of $\mathcal{G}$ are faces of the resulting map, which is thus locally a regular square grid around v (up to distance 3 at least). Then the fact that faces of color $(1,3)$ and $(0,2)$ also have length 4 implies that this map is in fact a four by four toroidal grid. In particular $\mathcal{G}$ has only finitely many vertices.

We conclude that there are only finitely many colored graphs with a fixed number of 2-dipoles, hence Proposition 4.2 is proved for $D = 3$.

The case $D \geq 4$.

The proof of Proposition 4.2 is similar for all $D \geq 4$. Consider a $(D+1)$-colored graph $\mathcal{G}$ of reduced degree $\delta(\mathcal{G})$ with $2k(\mathcal{G})$ vertices, having $t_1(\mathcal{G})$ $(D-1)$-dipoles and $t_2(\mathcal{G})$ $(D-2)$-dipoles. We will show that the number of such graphs is finite. The bound we establish on the number of such graphs is not tight and can be improved with minimal effort, but it is sufficient for our purpose.

Let us count faces of degree 2 according to whether they belong to a $(D-1)$-dipole, a $(D-2)$-dipole or none of these two:

$$F_1(\mathcal{G}) \leq t_1(\mathcal{G})\binom{D-1}{2} + t_2(\mathcal{G})\binom{D-2}{2} + \alpha(D)k(\mathcal{G}) \,, \tag{4.5}$$

where:

- $\alpha(4) = 0$ as, for $D = 4$, all the faces with two vertices must belong to a $(D-1)$- or a $(D-2)$-dipole (i.e. a 3- or a 2-dipole).
- $\alpha(5) = 3$ as, for $D = 5$, a vertex not belonging to a $(D-1)$- or $(D-2)$-dipole can belong to at most three 2-dipoles.
- $\alpha(6) = 6$ as, for $D = 6$, a vertex not belonging to a $(D-1)$- or $(D-2)$-dipole belongs to the largest number of faces of degree 2 when it belongs to two 3-dipoles i.e. to six faces of degree 2.
- $\alpha(D) = \frac{(D-3)(D-4)}{2} + 6$, for all $D \geq 7$ as, in this case, a vertex not belonging to a $(D-1)$- or $(D-2)$-dipole belongs to the largest number of faces of degree 2 when it belongs to a $(D-3)$-dipole and a 4-dipole.

On the one hand, the bound (4.5) together with Eq. (4.4) gives:

$$\begin{aligned}
&\sum_{s\geq 2}[(D-1)s - D - 1]F_s(\mathcal{G}) \\
&\quad \leq (D+1)\delta(\mathcal{G}) - D(D+1) \\
&\qquad + 2t_1(\mathcal{G})\binom{D-1}{2} + 2t_2(\mathcal{G})\binom{D-2}{2} + 2\alpha(D)k(\mathcal{G}) \,.
\end{aligned}$$

On the other hand, together with Eq. (4.3) it yields:

$$\left[\frac{D(D+1)}{2} - \alpha(D)\right]k(\mathcal{G}) \leq \sum_{s\geq 2} sF_s + t_1(\mathcal{G})\binom{D-1}{2} + t_2(\mathcal{G})\binom{D-2}{2} . \tag{4.6}$$

Eliminating $k(\mathcal{G})$ between these two equations and reordering we get:

$$\sum_{s\geq 2}\left[\left(D-1-\frac{4\alpha(D)}{D(D+1)-2\alpha(D)}\right)s-D-1\right]F_s(\mathcal{G})$$
$$\leq (D+1)\delta - D(D+1) + \left(2+\frac{4\alpha(D)}{D(D+1)-2\alpha(D)}\right)t_1(\mathcal{G})\binom{D-1}{2}$$
$$+\left(2+\frac{4\alpha(D)}{D(D+1)-2\alpha(D)}\right)t_2(\mathcal{G})\binom{D-2}{2}.$$

The coefficient of F_s on the left hand side is:

- for $D=4$: $3s-5$.
- for $D=5$: $\frac{7}{2}s-6$.
- for $D=6$: $\frac{21}{5}s-7$.
- for $D\geq 7$: $\frac{3(D-4)(D+1)}{4(D-3)}(s-2)+\frac{(D-6)(D+1)}{2(D-3)}$.

In particular, this coefficient is strictly positive for $s\geq 2$ so that we get for the previous equation an upper bound for each $F_s(\mathcal{G})$, $s\geq 2$ depending only on D, $\delta(\mathcal{G})$, $t_1(\mathcal{G})$ and $t_2(\mathcal{G})$, and there is a maximal value of s, depending again only on D, $\delta(\mathcal{G})$, $t_1(\mathcal{G})$ and $t_2(\mathcal{G})$ for which F_s can be nonzero.

On the other hand:

$$\frac{D(D+1)}{2}-\alpha(D)-\begin{cases}10\,, & D=4\\ 12\,, & D=5\\ 15\,, & D=6\\ 4(D-3)\,, & D\geq 7\end{cases}$$

is always positive, so that from Eq. (4.6) we finally get an upper bound on $k(\mathcal{G})$ depending only on D, $\delta(\mathcal{G})$, $t_1(\mathcal{G})$ and $t_2(\mathcal{G})$.

This completes the proof of Proposition 4.2.

4.6 Exact enumeration

4.6.1 Melonic graphs and cores

In view of Theorem 4.2 we will need, in order to enumerate colored graphs, the generating function of melonic graphs.

Proposition 4.3 (*See e.g.* [30]). *The generating function $T(z)$ of rooted closed melonic graphs (and 2-open melonic graphs) with respect to the number of black vertices is the unique power series solution of the equation:*

$$T(z)=1+zT(z)^{D+1}.$$

Proof. Let $M(z)$ be the generating function of prime melonic graphs. Then the inductive definition of melonic graphs and prime melonic graphs immediately translate into the equations:

$$T(z) = 1 + \sum_{i\geq 1} M(z)^i = \frac{1}{1 - M(z)}$$

and

$$M(z) = zT(z)^D.$$

□

Corollary 4.1 (*e.g.* [30]). *The generating series* $T(z)$ *admits the power series expansion*

$$T(z) = \sum_{k\geq 0} \frac{1}{(D+1)k+1} \binom{(D+1)k+1}{k} z^k .$$

It has a dominant singularity at $z_0 = D^D/(1+D)^{1+D}$ *and the following singular expansion in a slit domain around* z_0

$$T(z) = \frac{1+D}{D} - \sqrt{2\frac{D+1}{D^3}}\sqrt{1 - z/z_0} + O(1 - z/z_0) . \qquad (4.7)$$

In particular $T(z_0) = (1+D)/D$ *and* $z_0 T(z_0)^{D+1} = 1/D$.

Proof. The first expansion follows immediately from Proposition 4.3 using Lagrange inversion formula. The singular expansion is a direct instance of the standard theory of singularity analysis of simple trees generating functions [71], chapter VII.4.

□

Proposition 4.4. *Let* $\hat{\mathcal{G}}$ *be a rooted melon-free graph with* $2k(\hat{\mathcal{G}})$ *vertices, and thus* $(D+1)k(\hat{\mathcal{G}})$ *edges. The generating function* $H_{\hat{\mathcal{G}}}(z)$ *of rooted colored graphs with core* $\hat{\mathcal{G}}$ *with respect to the number of black vertices is*

$$H_{\hat{\mathcal{G}}}(z) = z^{k(\hat{\mathcal{G}})} T(z)^{(D+1)k(\hat{\mathcal{G}})+1}.$$

Proof. This immediately follows from the bijection of Theorem 4.2. The case $k(\hat{\mathcal{G}}) = 0$ corresponds to the ring graph, which is the core of the melonic graphs.

□

4.6.2 Chains of $(D-1)$-dipoles and schemes

In view of Theorem 4.3, in order to enumerate cores in terms of schemes, we will need several chain generating series, depending on whether the chain is broken or not, whether its external edges have identical color or not and whether its white squares are on the same side or not. Recall that a proper chain has at least four internal vertices.

Unbroken chains. Let us fix two colors $c_1 \neq c_2$. There is exactly one (c_1, c_2)-unbroken chain with $2k$ vertices, for $k \geq 1$, so their generating function $U(u)$ with respect to the number of $(D-1)$-dipoles, and the generating function $U^+(u)$ of the proper ones are

$$U(u) = \frac{u}{1-u}, \qquad U^+(u) = \frac{u^2}{1-u}.$$

The external edges have different colors if the number of dipoles is odd, and equal colors if it is even.

The generating functions for (c_1, c_2)-unbroken proper chains with left extremity of color c_1 and right extremity of color c_2, or c_1, are thus respectively:

$$U^+_{\neq;\substack{\square\square\\\blacksquare\blacksquare}}(u) = \frac{u^3}{1-u^2}, \qquad U^+_{\neq;\substack{\square\blacksquare\\\blacksquare\square}}(u) = 0,$$
$$U^+_{=;\substack{\square\blacksquare\\\blacksquare\square}}(u) = \frac{u^2}{1-u^2}, \qquad U^+_{=;\substack{\square\square\\\blacksquare\blacksquare}}(u) = 0,$$

where the squares in subscript track the white and black little squares of the chain-vertex corresponding to the chain.

Arbitrary chains. Let us fix one color c_1. The generating function of arbitrary nonempty and arbitrary proper chains of dipoles with left external color c_1 are:

$$A(u) = \frac{Du}{1-Du}, \qquad A^+(u) = \frac{(Du)^2}{1-Du}.$$

A nonempty chain with external colors (c_1, c_1) consists of a nonempty chain not reusing color c_1 followed by a $(D-1)$-dipole with right external color c_1 and a possibly empty chain with external colors (c_1, c_1):

$$\begin{aligned} A_=(u) &= \frac{Du}{1-(D-1)u} \cdot u \cdot (1 + A_=) = \frac{Du^2}{1-(D-1)u} \frac{1}{1 - \frac{Du^2}{1-(D-1)u}} \\ &= \frac{Du^2}{(1+u)(1-Du)} = \frac{Du^2(1-u)(1+Du)}{(1-u^2)(1-D^2u^2)}. \end{aligned}$$

The chains with an even number of dipoles correspond to the even powers of u, while the ones with an odd number of dipoles correspond to the odd powers of u, hence:

$$A_{=,\substack{\square\blacksquare\\\blacksquare\square}}(u) = \frac{Du^2(1-Du^2)}{(1-u^2)(1-D^2u^2)},$$
$$A_{=;\substack{\square\square\\\blacksquare\blacksquare}}(u) = \frac{Du^2(D-1)u}{(1-u^2)(1-D^2u^2)}.$$

Now fix a second color $c_2 \neq c_1$. A nonempty chain with external colors (c_1, c_2) is an arbitrary nonempty chain that does not have equal external colors:

$$A_{\neq}(u) = \frac{1}{D}[A(u) - A_=(u)] = \frac{u}{(1+u)(1-Du)} = \frac{u(1-u)(1+Du)}{(1-u^2)(1-D^2u^2)}.$$

Again, the chains with an even number of dipoles correspond to the even powers of u, while the ones with an odd number of dipoles to the odd powers of u, hence:

$$A_{\neq;\substack{\square\blacksquare\\\blacksquare\square}}(u) = \frac{u(D-1)u}{(1-u^2)(1-D^2u^2)},$$
$$A_{\neq;\substack{\square\square\\\blacksquare\blacksquare}}(u) = \frac{u(1-Du^2)}{(1-u^2)(1-D^2u^2)}.$$

Broken chains. Let us fix two colors c_1 and c_2 (maybe equal). A proper broken chain with external colors (c_1, c_2) is an arbitrary proper chain that is not unbroken. If $c_1 = c_2$, all the D possible second colors for the unbroken chain have to be considered:

$$\begin{aligned}
B^+_{=;\substack{\square\blacksquare\\\blacksquare\square}}(u) &= A_{=;\substack{\square\blacksquare\\\blacksquare\square}}(u) - DU^+_{=;\substack{\square\blacksquare\\\blacksquare\square}}(u) = \frac{Du^2(1-Du^2)}{(1-u^2)(1-D^2u^2)} - D\frac{u^2}{1-u^2}\\
&= \frac{D^2(D-1)u^4}{(1-u^2)(1-D^2u^2)},\\
B^+_{=;\substack{\square\square\\\blacksquare\blacksquare}}(u) &= A_{=;\substack{\square\square\\\blacksquare\blacksquare}}(u) - DU^+_{=;\substack{\square\square\\\blacksquare\blacksquare}}(u)\\
&= \frac{D(D-1)u^3}{(1-u^2)(1-D^2u^2)},\\
B^+_{\neq;\substack{\square\blacksquare\\\blacksquare\square}}(u) &= A_{\neq;\substack{\square\blacksquare\\\blacksquare\square}}(u) - U^+_{\neq;\substack{\square\blacksquare\\\blacksquare\square}}(u)\\
&= \frac{(D-1)u^2}{(1-u^2)(1-D^2u^2)},\\
B^+_{\neq;\substack{\square\square\\\blacksquare\blacksquare}}(u) &= A_{\neq;\substack{\square\square\\\blacksquare\blacksquare}}(u) - u - U^+_{\neq;\substack{\square\square\\\blacksquare\blacksquare}}(u) = \frac{u(1-Du^2)}{(1-u^2)(1-D^2u^2)} - u - \frac{u^3}{1-u^2}\\
&= \frac{D(D-1)u^3}{(1-u^2)(1-D^2u^2)}.
\end{aligned}$$

This is summarized in Figure 4.21.

Proposition 4.5. *Let $\mathcal{S}$ be a reduced scheme with $2k(\mathcal{S})$ black and white vertices, and with:*

- $\mathbf{U}_{\neq;\substack{\square\square\\\blacksquare\blacksquare}}$ *unbroken chain-vertices with different external colors (and consequently having little squares of the same color on the top).*
- $\mathbf{U}_{=;\substack{\square\blacksquare\\\blacksquare\square}}$ *unbroken chain-vertices with equal external colors (and consequently having little squares of different colors on the top).*
- $\mathbf{B}_{\neq;\substack{\square\square\\\blacksquare\blacksquare}}$ *broken chain-vertices with different external colors and squares of the same color on the top.*
- $\mathbf{B}_{=;\substack{\square\square\\\blacksquare\blacksquare}}$ *broken chain-vertices with equal external colors and squares of the same color on the top.*

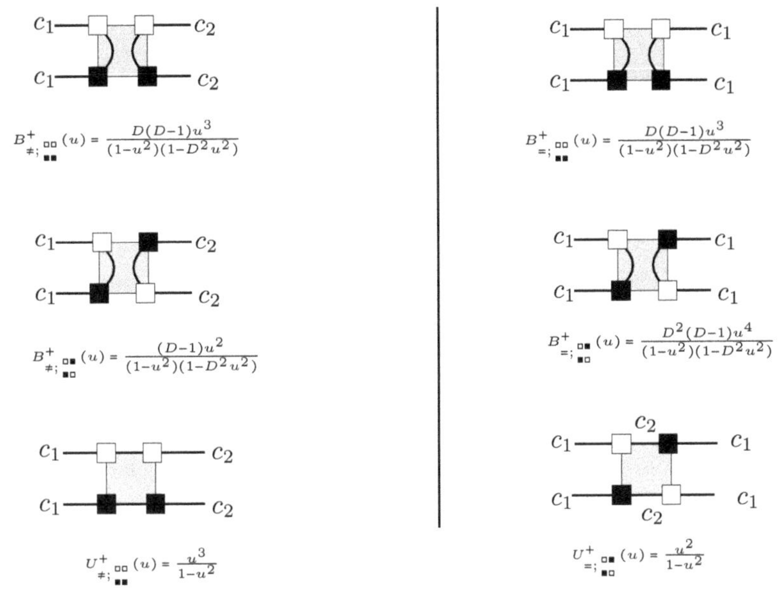

Figure 4.21: The generating series of the chain vertices.

- $\mathbf{B}_{\neq;\substack{\square\blacksquare\\\blacksquare\square}}$ *broken chain-vertices with different external colors and squares of different colors on the top.*
- $\mathbf{B}_{=;\substack{\square\blacksquare\\\blacksquare\square}}$ *broken chain-vertices with equal external colors and squares of different colors on the top.*

The generating function $G_{\mathcal{S}}(u)$ of rooted melon-free graphs with scheme $\mathcal{S}$ with respect to the number of black vertices is:

$$\begin{aligned}G_{\mathcal{S}}(u) &= u^{k(\mathcal{S})}\left[U^+_{\neq;\substack{\square\square\\\blacksquare\blacksquare}}(u)\right]^{\mathbf{U}_{\neq;\substack{\square\square\\\blacksquare\blacksquare}}}\left[U^+_{=;\substack{\square\blacksquare\\\blacksquare\square}}(u)\right]^{\mathbf{U}_{=;\substack{\square\blacksquare\\\blacksquare\square}}}\\ &\quad\times\left[B^+_{\neq;\substack{\square\square\\\blacksquare\blacksquare}}(u)\right]^{\mathbf{B}_{\neq;\substack{\square\square\\\blacksquare\blacksquare}}}\left[B^+_{=;\substack{\square\square\\\blacksquare\blacksquare}}(u)\right]^{\mathbf{B}_{=;\substack{\square\square\\\blacksquare\blacksquare}}}\left[B^+_{\neq;\substack{\square\blacksquare\\\blacksquare\square}}(u)\right]^{\mathbf{B}_{\neq;\substack{\square\blacksquare\\\blacksquare\square}}}\left[B^+_{=;\substack{\square\blacksquare\\\blacksquare\square}}(u)\right]^{\mathbf{B}_{=;\substack{\square\blacksquare\\\blacksquare\square}}}\\ &= \frac{P_{\mathcal{S}}(u)}{(1-u^2)^{\mathbf{U}+\mathbf{B}}(1-D^2u^2)^{\mathbf{B}}},\end{aligned}$$

where $\mathbf{U} = \mathbf{U}_{\neq;\substack{\square\square\\\blacksquare\blacksquare}} + \mathbf{U}_{=;\substack{\square\blacksquare\\\blacksquare\square}}$, $\mathbf{B} = \mathbf{B}_{\neq;\substack{\square\square\\\blacksquare\blacksquare}} + \mathbf{B}_{=;\substack{\square\square\\\blacksquare\blacksquare}} + \mathbf{B}_{\neq;\substack{\square\blacksquare\\\blacksquare\square}} + \mathbf{B}_{=;\substack{\square\blacksquare\\\blacksquare\square}}$ *and $P_{\mathcal{S}}(u)$ is the monomial:*

$$\begin{aligned}P_{\mathcal{S}}(u) &= (D-1)^{\mathbf{B}}\; D^{\mathbf{B}_{\neq;\substack{\square\square\\\blacksquare\blacksquare}}+\mathbf{B}_{=;\substack{\square\square\\\blacksquare\blacksquare}}+2\mathbf{B}_{=;\substack{\square\blacksquare\\\blacksquare\square}}}\\ &\quad\times u^{k(\mathcal{S})+2\mathbf{U}_{=;\substack{\square\blacksquare\\\blacksquare\square}}+3\mathbf{U}_{\neq;\substack{\square\square\\\blacksquare\blacksquare}}+3\mathbf{B}_{\neq;\substack{\square\square\\\blacksquare\blacksquare}}+3\mathbf{B}_{=;\substack{\square\square\\\blacksquare\blacksquare}}+2\mathbf{B}_{\neq;\substack{\square\blacksquare\\\blacksquare\square}}+4\mathbf{B}_{=;\substack{\square\blacksquare\\\blacksquare\square}}}.\end{aligned}$$

Proof. This follows immediately from the bijection in Theorem 4.3. □

4.6.3 The enumeration of rooted colored graph of fixed degree

Putting together Theorem 4.2, Theorem 4.3 and Theorem 4.4 we obtain the enumeration of the edge colored graphs of fixed degree.

Theorem 4.5. *Let $\delta \geq 1$. The generating function of rooted colored graphs with root edge of color* 0 *and reduced degree δ with respect to the number of black vertices is*

$$H_\delta^0(z) = T(z) \sum_{\mathcal{S} \in \mathbf{S}_\delta^0} G_{\mathcal{S}}\left(zT(z)^{D+1}\right) ,$$

where the sum is over the finite set $\mathbf{S}_\delta^0$ of reduced schemes with degree δ and root edge of color 0.

The first values can be computed explicitly. Let us take the example of the first nontrivial schemes of reduced degree $\delta = D-2$. The scheme must have a non separating

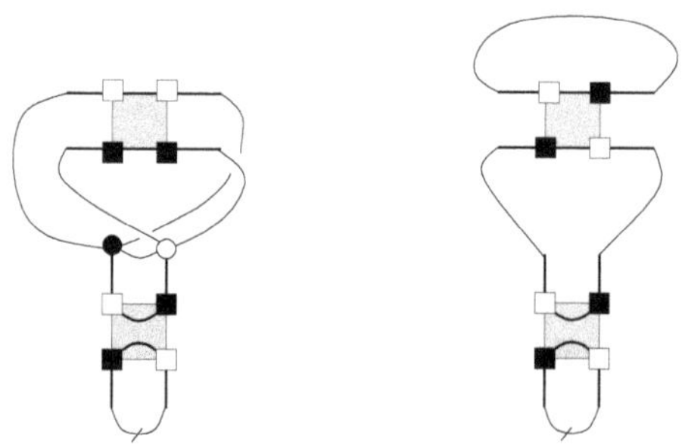

Figure 4.22: The "lollipop" schemes.

unbroken chain-vertex and a single resulting face after removal of the chain-vertex (or a non separating $(D-1)$-dipole Case II.B). It can possess a separating chain-vertex, a separating $(D-1)$ dipole, or a partially separating $D-2$ dipole or not. It thus must be of one of the two types presented in Figure 4.22.

Let us analyze first the vertical arm in Figure 4.22. Including the degenerate configurations it can be empty, consist of one dipole, or consist of a (broken or unbroken) chain vertex. Fixing the incoming color (bottom color of the arm in Figure 4.22) to be c_1, allowing all the possible outgoing colors c_2, and taking into account that if the chain vertex is unbroken and has external colors (c_1, c_1) then it can have any secondary color $c_2 \neq c_1$, the generating function of the vertical arm in Figure 4.22 is:

$$\begin{aligned} & 1 + B^+_{=;\substack{\square\blacksquare \\ \blacksquare\square}}(u) + DU^+_{=;\substack{\square\blacksquare \\ \blacksquare\square}}(u) + B^+_{=;\substack{\square\square \\ \blacksquare\blacksquare}}(u) + D[u + B^+_{\neq;\substack{\square\square \\ \blacksquare\blacksquare}}(u) + U^+_{\neq;\substack{\square\square \\ \blacksquare\blacksquare}}(u) + B^+_{\neq;\substack{\square\blacksquare \\ \blacksquare\square}}(u)] \\ & = 1 + \frac{Du^2(1-Du^2)}{(1-u^2)(1-D^2u^2)} + \frac{D(D-1)u^3}{(1-u^2)(1-D^2u^2)} + \\ & + D\left[\frac{u(1-Du^2)}{(1-u^2)(1-D^2u^2)} + \frac{(D-1)u^2}{(1-u^2)(1-D^2u^2)}\right] \end{aligned}$$

$$= 1 + \frac{Du^2}{(1+u)(1-Du)} + \frac{Du}{(1+u)(1-Du)} = \frac{1}{1-Du} .$$

The generating function of the lollipop schemes are therefore:

$$T(z)\left[\frac{1}{1-Du} \cdot u \cdot \binom{D}{2} \cdot \frac{u}{1-u^2}\right]_{u=zT(z)^{D+1}} ,$$
$$T(z)\left[\frac{1}{1-Du} \cdot D \cdot \frac{u^2}{1-u^2}\right]_{u=zT(z)^{D+1}} ,$$

where we counted the fact that the non separating chain vertex can be reduced to a unique dipole for the leftmost scheme. Putting the two together we obtain:

$$H^0_{D-2}(z) = T(z)\left[\frac{1}{1-Du}\left[\binom{D}{2} + D\right]\frac{u^2}{1-u^2}\right]_{u=zT(z)^{D+1}}$$
$$= T(z)\frac{D(D+1)}{2}\frac{z^2T(z)^{2D+2}}{1-z^2T(z)^{2D+2}}\frac{1}{1-DzT(z)^{D+1}} ,$$

reproducing the result of [112]. In principle it is more efficient to list all reduced schemes of $\bigcup_{\delta\geq 1}\mathbf{S}^0_\delta$ according to their size, rather than degree by degree; however, computing all the terms by hand is quite tedious.

Chapter 5

Melonic graphs

The graphs of genus zero, the planar graphs, yield the leading order of the $1/N$ expansion of matrix models. They are replaced in the case of tensor models by the melonic graphs [30]. In Chapter 4 we saw (and used) only some general properties of the melonic graphs: their two equivalent iterative definitions (Definition 4.6 and Lemma 4.5) and the fact that they have degree zero (Theorem 4.1).

In this chapter we will study in more detail this family of graphs and especially the random geometry they represent. This study requires a number of prerequisites.

The edge colored graphs represent topological triangulations. In order to associate a geometry to a topological triangulation, one needs to endow it with a metric, i.e. a distance function. For abstract triangulations there are two natural candidate distance functions.

Equilateral. One can assign the same length to all the 1-simplices in the triangulation (i.e. consider the associated equilateral triangulation). With this metric, the distance between two vertices of the triangulation is the number of 1-simplices in (one of) the shortest path(s) of 1-simplices connecting them. In the case of the invariant tensor models, which are the object of this book, this distance function is natural: the invariant tensor models endow a graph with an "amplitude", which turns out to be (see Chapter 12) exactly the Einstein Hilbert action discretized on this metric triangulation.

Dual equilateral. Alternatively, one can choose to use the graph distance of the edge colored graph itself. This time the distance is computed between vertices of the graph (corresponding to barycenters of the D-simplices in the triangulation) and equals the number of graph edges in (one of) the shortest path(s) connecting the two graph vertices.

These two notions of distance are quite different for individual cases. Consider the two-dimensional triangulation represented in Figure 5.1. We choose a vertex (in the triangulation, and respectively the dual graph) that we consider as root and we measure the distances (in blue in Figure 5.1) from any other vertex to it. Observe that for the case represented in the figure, no triangulation vertex is at a distance of more than two from the distinguished triangulation vertex in the equilateral metric, while in the dual equilateral metric, one graph vertex is at a distance of four from the distinguished graph vertex. This example shows that triangulations that are "compact" in the equilateral metric can be quite "extended" in the dual equilateral metric. The most one can hope for is that the two metrics become equivalent in the limit of large graphs.

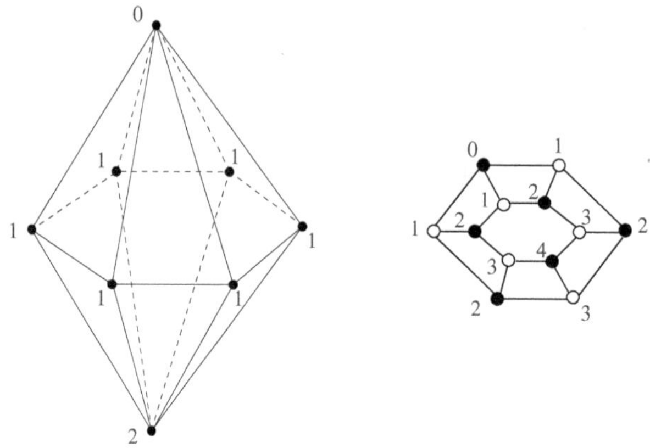

Figure 5.1: Equilateral and dual equilateral metric assignments.

Once the metric of a triangulation is assigned, one constructs an infinitely refined random metric space as follows. One chooses a sequence of finite but increasingly larger triangulations by selecting for each k a triangulation with $2k$ simplices with some probability. In the case of the invariant tensor models all the melonic graphs with $2k$ vertices are assigned the same amplitude, hence it is natural to choose this finite triangulation *uniformly*. Taking the $k \to \infty$ limit one obtains a limit object that is an infinitely refined random metric space. Of course, in order to keep the total volume fixed, one needs to send at the same time the edge length to zero.

The limit random metric space obtained by this procedure is strongly dependent not only on the family of triangulations one chooses (dual to melons in our case) but also on the distance function used (equilateral in our case). We will show that under these assumptions the melonic family converges to the *continuous random tree* (or "branched polymers" in the physics literature).

5.1 Quartic melonic graphs

The (closed) melonic graphs with D-colors and four vertices are called *quartic melons.* They are the simplest D colored graphs with more than two vertices and we will encounter them in many instances in the sequel. They consist of two pairs of vertices such that the vertices in a pair are connected by an edge for each color in $\mathcal{D} \setminus \{c\}$, and the pairs are connected in-between them by edges of color c (see for instance Figure 5.2 for some examples). We denote the quartic melonic graphs by $\mathcal{B}^{(4)}_{\{c\}}$ (where c is the color of the edges connecting the two pairs). The trace invariant operator associated to $\mathcal{B}^{(4)}_{\{c\}}$ is:

$$\delta^{\mathcal{B}^{(4)}_{\{c\}}}_{a\bar{a}} = \left(\delta_{a_1^{\mathcal{D}\setminus\{c\}}\bar{a}_1^{\mathcal{D}\setminus\{c\}}}\right) \delta_{a_1^c\bar{a}_2^c}\delta_{a_2^c\bar{a}_1^c} \left(\delta_{a_2^{\mathcal{D}\setminus\{c\}}\bar{a}_2^{\mathcal{D}\setminus\{c\}}}\right) .$$

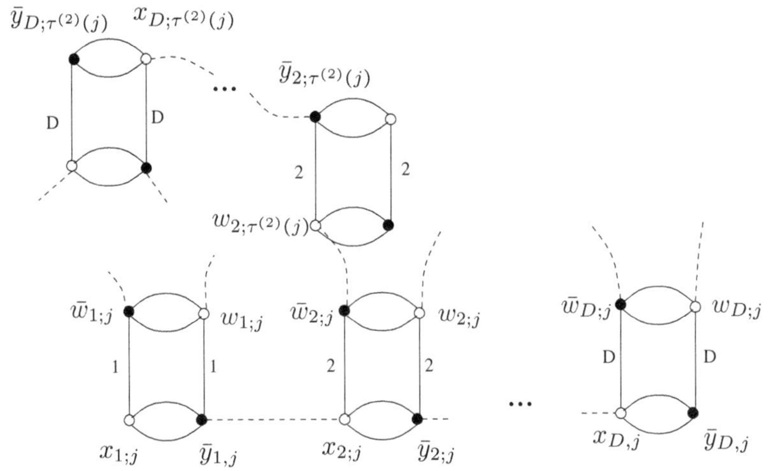

Figure 5.2: Obtaining any edge colored graph as a composition of quartic melonic graphs.

The quartic melonic graphs have a very important property: they generate upon composition (recall Definition 3.8) all the edge colored graphs.

Theorem 5.1. *Any D-colored graph can be obtained as a composition (as defined in Definition 3.8) of quartic melons.*

Proof. Consider a D-colored graph $\mathcal{B}$ with $2k$ vertices encoded in D permutations $\tau^{\mathcal{D}}$:

$$\mathcal{V}(\mathcal{B}) = \{v_j, \bar{v}_j | j = 1, \dots k\} \qquad \mathcal{E}^c(\mathcal{B}) = \{(v_j, \bar{v}_{\tau^{(c)}(j)}) | j = 1, \dots k\} , \ \forall c .$$

For each $j = 1, \dots k$, we define D quartic melons, one for each color c:

$$\begin{aligned} \mathcal{B}^{(4)}_{\{c\};j} \ : \mathcal{V}\left(\mathcal{B}^{(4)}_{\{c\};j}\right) &= \{x_{c;j}, \bar{y}_{c;j}, w_{c;j}, \bar{w}_{c;j}\} \\ \mathcal{E}^c\left(\mathcal{B}^{(4)}_{\{c\};j}\right) &= \{(x_{c;j}, \bar{w}_{c;j}), (w_{c;j}, \bar{y}_{c;j})\} \\ \mathcal{E}^{c'}\left(\mathcal{B}^{(4)}_{\{c\};j}\right) &= \{(x_{c;j}, \bar{y}_{c;j}), (w_{c;j}, \bar{w}_{c;j})\} , \ \forall c' \neq c , \end{aligned}$$

and we set $v_j = x_{1;j}$ and $\bar{v}_j = \bar{y}_{D;j}$. We have:

$$\mathcal{B} = \left(\uplus_{c;j} \ \mathcal{B}^{(4)}_{\{c\};j}\right) / \left\{ \bigcup_{j=1}^{k} \left[\cup_{c=1}^{D} \{(w_{c;\tau^{(c)}(j)}, \bar{w}_{c;j})\} \cup \left[\cup_{c=1}^{D-1} \{(x_{j,c+1}, \bar{y}_{c,j})\}\right] \right] \right\} .$$

To see this, consider the open $(D+1)$ colored graph made of the $\mathcal{B}^{(4)}_{c;j}$s and the edges of color 0:

$$\bigcup_{j=1}^{k}\left[\cup_{c=1}^{D}\{(w_{c;\tau^{(c)}(j)},\bar{w}_{c;j})\}\cup\left[\cup_{c=1}^{D-1}\{(x_{j,c+1},\bar{y}_{c,j})\}\right]\right].$$

The boundary vertices of $\mathcal{G}$ are $\bigcup_{j=1}^{k}\{x_{1;j},\bar{y}_{D;j}\}$, which we relabeled v_j and $\bar{v}_j$. Consider the external face of colors $(0,c)$ originating at $v_j = x_{1;j}$. This face necessarily steps through $x_{c;j}$, $\bar{w}_{c,j}$,$w_{c;\tau^{(c)}(j)}$, $\bar{y}_{c;\tau^{(c)}(j)}$ from which it steps to $\bar{y}_{D;\tau^{(c)}(j)} = \bar{v}_{\tau^{(c)}(j)}$.

□

5.2 Melonic graphs and colored, rooted, $(D+1)$-ary trees

The 2^c-open melonic graphs with $D+1$ colors are one to one with *colored rooted* $(D+1)$*-ary trees.*

Definition 5.1. *A* colored, rooted $(D+1)$-ary tree *is a tree such that:*

- *the tree has only* $(D+2)$*-valent and univalent vertices and a univalent vertex is the root vertex,*
- *the edges of the tree have a color* $c\in\mathcal{D}\cup\{0\}$*,*
- *all the descendants of a* $(D+2)$*-valent vertex (in the genealogy induced by the root vertex) are connected to that vertex by edges of different colors.*

The branches of color c *of a colored rooted* $(D+1)$*-ary tree are its maximally connected subtrees (subgraphs) having only edges of color* c.

With the exception of the root univalent vertex, we routinely forget the rest of the univalent vertices and consider the edges connecting them to the tree as halfedges. By extension, we refer to the edge connecting the root univalent vertex to the rest of the tree as the root halfedge of the tree (and we keep in mind that the root halfedge ends in the root univalent vertex).

The following construction is illustrated in Figure 5.3. Let $\mathcal{G}$ be a 2^c-open melonic graph, hence rooted at the external halfedge (of color c) incident to the white vertex. If $\mathcal{G}$ is the graph $\mathcal{G}^{(2),c}$, we associate it to the tree with a unique $(D+2)$-valent vertex having a root halfedge of color c and one descendant halfedge for each color $c'\in(\mathcal{D}\cup\{0\})\setminus\{c\}$.

If $\mathcal{G}\neq\mathcal{G}^{(2),c}$ we build the colored, rooted $(D+1)$-ary tree associated to $\mathcal{G}$ recursively as follows:

- We list the open components $\{\mathcal{G}_0,\dots\mathcal{G}_\ell\}$ of the cut set of the root r of $\mathrm{cl}(\mathcal{G})$ in the order they are encountered starting from the root halfedge of $\mathcal{G}$. Each $\mathcal{G}_i$ is a prime melonic subgraph of $\mathcal{G}$.

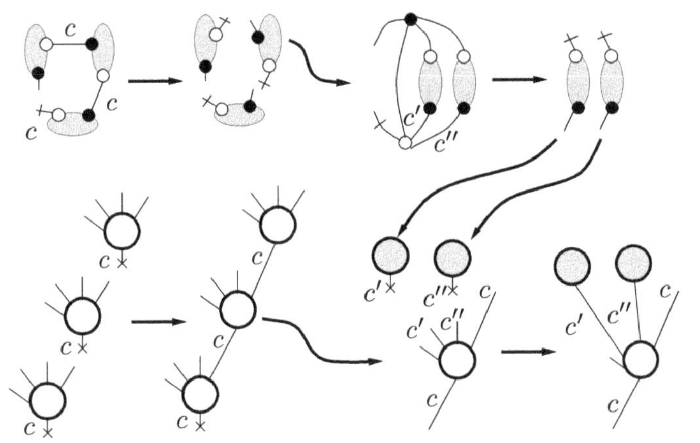

Figure 5.3: From melonic graphs to trees.

- Corresponding to each $\mathcal{G}_i$ we draw an abstract $(D+2)$-valent vertex $\mathbf{V}(\mathcal{G}_i)$ with a root halfedge of color c having $(D+1)$ descendant halfedges (one for each color). We build the *branch of color* c of $\mathcal{G}$ by connecting the root halfedge of $\mathbf{V}(\mathcal{G}_i)$ to the descendant halfedge of color c of $\mathbf{V}(\mathcal{G}_{i-1})$, and declare the root halfedge of $\mathbf{V}(\mathcal{G}_0)$ as the root of the branch.
- For each $\mathcal{G}_i$ we denote x_i and $\bar{y}_i$ the vertices hooked to its external halfedges (which have color c). For all $c' \neq c$, we connect the descendant halfedge of color c' of $\mathbf{V}(\mathcal{G}_i)$ to the root halfedge of the branch of color c' associated to the nonempty melonic graph (with external halfedges of color c') obtained by cutting the edges of color c' in $\mathcal{G}_i$ incident to the vertices x_i and $\bar{y}_i$.

The $(D+2)$ valent vertex $\mathbf{V}(\mathcal{G}_i)$ can be canonically associated to the two vertices x_i and $\bar{y}_i$ of $\mathcal{G}_i$, which are hooked to its external halfedges.

In light of Lemma 4.5, the rooted tree associated with a melon can also be interpreted as follows. Each melonic graph $\mathcal{G}$ is obtained from a fundamental melon $\mathcal{G}^{(2),c}$ (with vertices x and $\bar{y}$) by iterated insertions of fundamental melons on its edges and at the nonroot external halfedge. The vertex $\mathbf{V}$ of the tree hooked to the root halfedge represents the fundamental melon $\mathcal{G}^{(2),c}$, its descendants of colors different from c represent the fundamental melons subsequently inserted on its edges and its descendant of color c represents the fundamental melon inserted at its nonroot external halfedge. Observe that the vertices x and $\bar{y}$ of any fundamental melon in this insertion procedure become the external vertices of prime melonic subgraphs of $\mathcal{G}$.

We thus obtain the following correspondences:

Tree vertex $\mathbf{V}$ in the associated $(D+1)$-ary tree $\Leftrightarrow$	
Fundamental melon $\mathcal{G}^{2,(c)}$ inserted iteratively $\Leftrightarrow$	(5.1)
Couples of external vertices $x, \bar{y}$ of prime melonic subgraphs	

5.3 The melonic balls

As we saw in Section 3.2, all the colored graphs are dual to triangulations. Melonic graphs are dual to a specific class of triangulations, which we call by extension melonic.

Let us consider a fundamental melon $\mathcal{G}^{(2),c}$ and let us denote its vertices x and $\bar{y}$. This fundamental melon is dual to a vertex-colored triangulation with boundary (see Figure 5.4). Following the construction in Chapter 3, the dual triangulation consists of two D-simplices:

$$\Delta_x = \{\mathbf{v}_x^{(0)}, \mathbf{v}_x^{(1)}, \dots \mathbf{v}_x^{(D)}\}, \qquad \Delta_{\bar{y}} = \{\mathbf{v}_{\bar{y}}^{(D)}, \dots \mathbf{v}_{\bar{y}}^{(1)}, \mathbf{v}_{\bar{y}}^{(0)}\},$$

glued along the $(D-1)$-simplices

$$\{\mathbf{v}_x^{(0)}, \mathbf{v}_x^{(1)}, \dots \mathbf{v}_x^{(D)}\} \setminus \{\mathbf{v}_x^{(c_1)}\}, \ \{\mathbf{v}_{\bar{y}}^{(D)}, \dots \mathbf{v}_{\bar{y}}^{(1)}, \mathbf{v}_{\bar{y}}^{(0)}\} \setminus \{\mathbf{v}_{\bar{y}}^{(c_1)}\} \quad \forall c_1 \neq c,$$

by identifying the vertices with the same color.

This gluing represents the coherent identification of two D simplices along D of their boundary simplices, hence topologically it is a D-ball. The boundary of this D-ball is triangulated by the two remaining $(D-1)$ simplices (dual to the halfedges of color c):

$$\{\mathbf{v}_x^{(0)}, \mathbf{v}_x^{(1)}, \dots \mathbf{v}_x^{(D)}\} \setminus \{\mathbf{v}_x^{(c)}\} \ \{\mathbf{v}_{\bar{y}}^{(D)}, \dots \mathbf{v}_{\bar{y}}^{(1)}, \mathbf{v}_{\bar{y}}^{(0)}\} \setminus \{\mathbf{v}_{\bar{y}}^{(c)}\},$$

whose vertices are pairwise identified $\mathbf{v}_x^{(c')} = \mathbf{v}_{\bar{y}}^{(c')}, c' \neq c$, as represented in Figure 5.4.

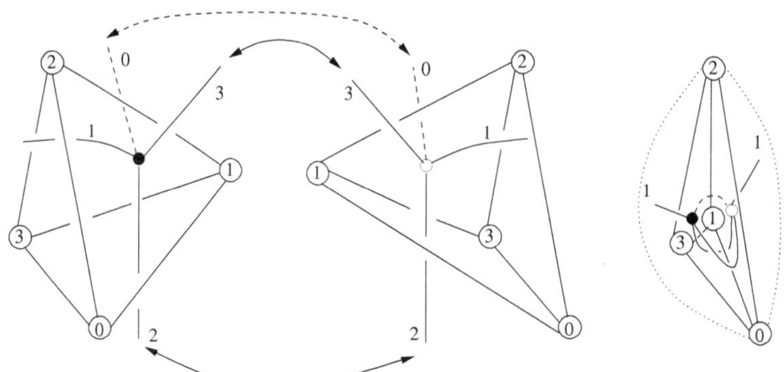

Figure 5.4: The melon ball. The dotted lines on the rightmost figure are drawn in order to emphasize that the triangulation is made of two tetrahedra with vertices 0, 1, 2, 3, sharing the triangles 012, 013 and 123 and its boundary is formed by two triangles (one on the right and one on the left), both with vertices 023.

We call this ball the *elementary melon ball.* The elementary melon ball dual to $\mathcal{G}^{(2),c}$ has D vertices on the boundary ($\mathbf{v}^{(c')}, c' \neq c$) and one interior vertex $\mathbf{v}^{(c)}$. Any melonic graph can be obtained from a fundamental melon by repeated insertions of fundamental melons. The insertion of a melon $\mathcal{G}^{(2),c_1}$ on an edge of a graph amounts to splitting a $(D-1)$-simplex σ of color c_1 in the triangulation and inserting an elementary melon ball by gluing a boundary $(D-1)$ simplex of the ball on each of the two copies of σ. This adds exactly one new interior vertex $\mathbf{v}^{(c_1)}$ to the triangulation, the interior vertex of $\mathcal{G}^{(2),c_1}$. Observe that the boundary vertices of the first fundamental melon ball, $\mathbf{v}^{(c')}, c' \neq c$ will always remain the boundary vertices of the triangulation.

Melonic graphs represent the coherent insertion of fundamental melon balls into fundamental melon balls, hence they represent topological balls themselves. We therefore call the triangulations dual to melonic graphs *melonic balls*.

The melonic graphs and the dual melonic balls are then related by:

Fundamental melon $\mathcal{G}^{2,(c)}$ inserted iteratively $\Leftrightarrow$
Interior vertex $\mathbf{v}^{(c)}$ in the melon ball

which together with the correspondences in (5.1) lead to:

Fundamental melon $\mathcal{G}^{2,(c)}$ inserted iteratively $\Leftrightarrow$	
Tree vertex $\mathbf{V}$ in the associated $(D+1)$-ary tree $\Leftrightarrow$	(5.2)
Interior vertex $\mathbf{v}^{(c)}$ in the melon ball $\Leftrightarrow$	
Couples of external vertices $x, \bar{y}$ of prime melonic subgraphs	

5.4 Random melons and branched polymers

The local geometrical features of a metric space (be it random or not) are encoded in its geodesic curves: one can completely explore the geometry using (an infinity of) free falling observers that follow the geodesic flow.

Studying the geodesic curves in random metric spaces is of course very difficult [121] and in most cases one needs to contend with studying some of their less sensitive, more global, features. In particular, a good amount of information can be gained by studying the dimension of the emergent random geometry. There are several notions of dimension of a (pseudo) manifold, which are all equal for the case of fixed smooth manifolds. In the case of random metric spaces however, these different dimensions often turn out to be different.

The Hausdorff dimension. Beyond the topological dimension, the simplest notion of dimension of a space is the Hausdorff dimension. For a metric space X, the Hausdorff dimension $d_H(X)$ captures how the volume of an infinitesimal ball scales with respect to its radius. Let X be a metric space with distance function $\rho(\cdot,\cdot)$. For every subset $U \subset X$ we define the diameter of U as:

$$\mathrm{di}(U) = \sup\{\rho(x,y) | x,y \in U\} .$$

Let us denote $\{U_i\}$ a countable cover of X, $X \subset \bigcup_i U_i$, and let us define $H^\delta_d(X)$ as:

$$H^\delta_d(X) = \inf_{\substack{\{U_i\} \\ X \subset \bigcup_i U_i,\ \mathrm{di}(U_i)<\delta}} \Big\{ \sum_i \big[\mathrm{di}(U_i)\big]^d \Big\} .$$

The smaller the δ, the less coverings are included and the infimum increases. The Hausdorff measure $H_d(X)$ of X is:

$$H_d(X) = \lim_{\delta\to 0} H_d^\delta(X)\ .$$

Finally the Hausdorff dimension of X, $d_H(X)$ is:

$$d_H(X) = \inf_{H_d(X)=0} \{d\}\ .$$

In spite of this rather cumbersome definition, the Hausdorff dimension is relatively straightforward. Intuitively, the minimal number of balls Ball_δ of radius δ required to cover X scales like $\frac{1}{\text{Volume}(\text{Ball}_\delta)}$, hence:

$$H_d^\delta(X) \sim \frac{1}{\text{Volume}(\text{Ball}_\delta)}\delta^d\ , \quad H_d(X) = \lim_{\delta\to 0}\frac{1}{\text{Volume}(\text{Ball}_\delta)}\delta^d\ ,$$
$$d_H(X) = \inf_{\lim_{\delta\to 0}\frac{\delta^d}{\text{Volume}(\text{Ball}_\delta)}=0} \{d\}\ ,$$

that is, the volume of an infinitesimal ball scales with the radius like:

$$\text{Volume}(\text{Ball}_\delta) \sim \delta^{d_H}\ .$$

For instance, the number of balls of radius δ required to cover the flat Euclidean $\mathbb{R}^D$ space scales like $\frac{1}{\delta^D}$, hence $d_H = D$. However, it is clear that some work must be done in order to extract the average Hausdorff dimension of the emergent random geometry of melonic D-balls.

The spectral dimension. Another notion of dimension of a metric space X is the spectral dimension $d_S(X)$. This is the dimension of X as seen by a random walk and as such it encodes more refined features of X than the Hausdorff dimension $d_H(X)$.

Consider a random walk in X and denote $P(t)$ the probability for the walker to return to its starting point after the diffusion time t. The spectral dimension of X, $d_S(X)$ is defined as:

$$d_S(X) = -2\frac{d\log P(t)}{d\log t}\ ,$$

that is $P(t) \sim \frac{1}{t^{d_S/2}}$. For the simple example of the Euclidean $\mathbb{R}^D$, the return probability is given by the heat kernel at coincident points, hence $d_S = D = d_H$, but this is not true in general.

In the rest of this chapter we will show that the average values of these two dimensions for melonic balls are:

$$\langle d_H\rangle = 2\ , \qquad \langle d_S\rangle = \frac{4}{3}\ ,$$

which are the average Hausdorff and spectral dimensions of branched polymers [111].

5.4.1 The Hausdorff dimension

We first need to understand in more detail the distance function induced by the equilateral metric for melonic balls. This is somewhat involved and relies heavily on the use of the $(D+1)$-ary tree associated with the melonic graphs.

The depth in a melonic triangulation

To each $(D+2)$-valent vertex $\mathbf{V}$ of a colored rooted $(D+1)$-ary tree we can associate a canonical label [92], which is a word over the alphabet $0, \ldots D$.

Definition 5.2. *For a given tree vertex* $\mathbf{V}$*, its word is constructed by listing (left to right) all the colors one encounters on the edges, when going from the root univalent vertex to the vertex in question.*

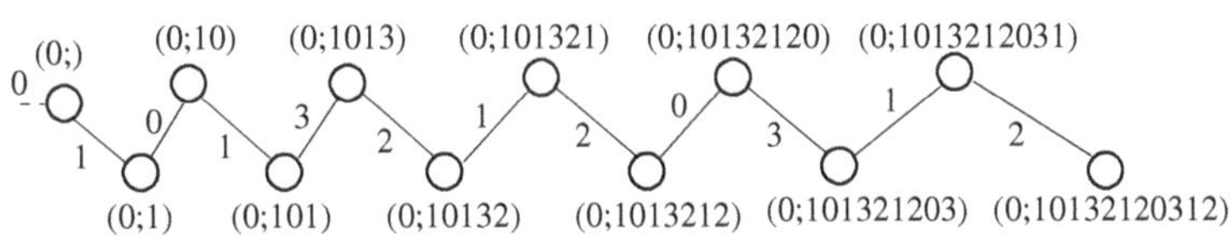

Figure 5.5: The words associated to some vertices of a colored rooted $(D+1)$–ary tree.

As all the words of the vertices in a tree start with the color of the root halfedge, we separate the first letter in the word by a semi column. An example is presented in Figure 5.5. Using the equivalences in (5.2), these words can also be used to label the interior vertices in the melonic ball, the fundamental melonic graphs in the iterative insertion procedure or the couples of external vertices of prime melonic subgraphs, that is:

$$\boxed{\begin{array}{l} \text{Fundamental melon } \mathcal{G}^{2,(c)} \text{ inserted iteratively } \Leftrightarrow \\ \text{Tree vertex } \mathbf{V} \text{ in the associated } (D+1)\text{-ary tree } \Leftrightarrow \\ \text{Interior vertex } \mathbf{v}^{(c)} \text{ in the melon ball } \Leftrightarrow \\ \text{Couples of external vertices } x, \bar{y} \text{ of prime melonic subgraphs } \Leftrightarrow \\ \text{Word } (\,;\ldots c) \text{ over } \{0,\ldots D\} \end{array}} \tag{5.3}$$

In an equilateral triangulation, the distance between two triangulation vertices is the minimal number of 1-simplices (edges of the triangulation) in a simple path (i.e. a path of 1-simplices visiting each vertex at most once) from one vertex to another.

The following construction is illustrated in Figure 5.6.

Consider first the fundamental melon $\mathcal{G}^{(2),c}$ corresponding to the tree with a unique $(D+2)$ valent vertex $\mathbf{V}$ with root halfedge of color c. The word of this vertex is $(c;)$. The elementary melon ball associated to it is a triangulation with $D+1$ vertices $\mathbf{v}^{(0)}, \ldots \mathbf{v}^{(D)}$. The vertices $\mathbf{v}^{(c')}, c' \neq c$ belong to the boundary of the elementary melon ball. The vertex $\mathbf{v}^{(c)}$, the interior vertex of the elementary melon ball, will be declared the root vertex of the triangulation and is denoted $\mathbf{x}_0 \equiv \mathbf{v}^{(c)}$.

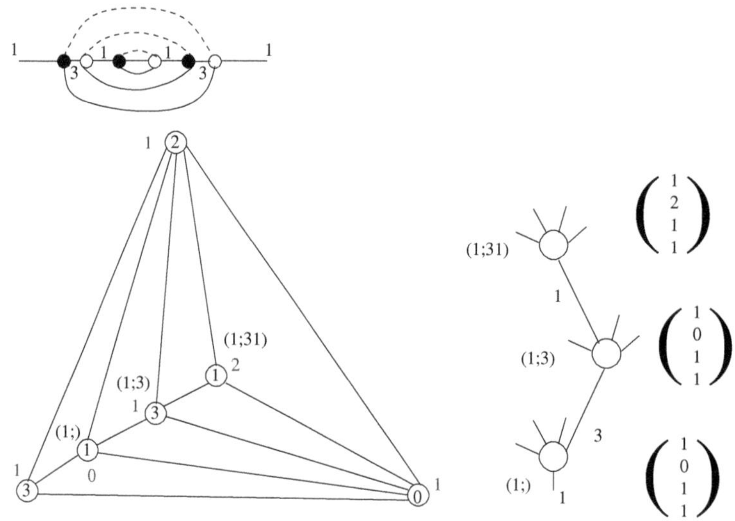

Figure 5.6: A melonic graph, the dual triangulation and the corresponding tree. The depths of the triangulation vertices are indicated in blue. The words of the tree vertices (and associated interior vertices of the triangulation) are indicated, as well as the distance arrays.

We are interested in evaluating the distance from any other vertex $\mathbf{x}$ of the triangulation to the root vertex $\mathbf{x}_0$, which we call the *depth* of $\mathbf{x}$. To this end we introduce an auxiliary structure, the *distance array* of $\mathbf{V}$, which is an array with $(D+1)$ entries listing (in the order of the colors) the depths of each of the triangulation vertices of its corresponding elementary melon ball. The depth of $\mathbf{v}^{(c)} \equiv \mathbf{x}_0$ is zero, while the depths of the other vertices $\mathbf{v}^{(c')}, c' \neq c$ is 1 (see Figure 5.6):

$$\mathfrak{da}[\mathbf{V}] = \begin{pmatrix} \mathfrak{da}[\mathbf{V}]_0 \\ \vdots \\ \mathfrak{da}[\mathbf{V}]_D \end{pmatrix} \qquad \mathfrak{da}[(c;)]_{c'} = \begin{cases} 0, & c' = c \\ 1, & c' \neq c \end{cases} .$$

Consider now the insertion of a fundamental melon $\mathcal{G}^{(2),c_1}$ in this elementary melon ball. It consists of splitting the D simplex $\{\mathbf{v}^{(0)}, \dots \mathbf{v}^{(D)}\} \setminus \{\mathbf{v}^{(c_1)}\}$ and inserting a melon ball in its place. The D boundary vertices of $\mathcal{G}^{(2),c_1}$ are identified with the vertices of $\{\mathbf{v}^{(0)}, \dots \mathbf{v}^{(D)}\} \setminus \{\mathbf{v}^{(c_1)}\}$, while the interior vertex of $\mathcal{G}^{(2),c_1}$, which we label $\mathbf{x}_1$, becomes a new interior vertex in the triangulation. This vertex is connected by one edge to each of the boundary vertices of $\mathcal{G}^{(2),c_1}$. The tree vertex corresponding to this fundamental melon has word $(c; c_1)$.

All the boundary vertices of $\mathcal{G}^{(2)c_1}$ were already present in the triangulation, hence their depths are already known. For the interior vertex there are two cases:

- if $c_1 \neq c$, then $\mathbf{x}_0$ is one of the boundary vertices of $\mathcal{G}^{(2),c_1}$, and the interior vertex $\mathbf{x}_1$ has depth 1 (as in Figure 5.6).

- if $c_1 = c$, then $\mathbf{x}_0$ is not among the boundary vertices of $\mathcal{G}^{(2),c_1}$, and the depth of $\mathbf{x}_1$ is 2.

This is expressed by the formula:

$$\mathfrak{da}[(c;c_1)]_{c'} = \begin{cases} \min_{c''\neq c'} \left\{ \mathfrak{da}[(c;c_1)]_{c''} \right\} + 1 \,, & c' = c_1 \\ \\ \mathfrak{da}[(c;)]_{c'} \,, & c' \neq c_1 \end{cases} .$$

This generalizes for the subsequent insertions: when inserting a fundamental melon with corresponding word $(c;c_1,\dots c_i)$, its boundary vertices of color $c' \neq c_i$ are the vertices of color c' of the fundamental melon with word $(c;c_1,\dots c_{i-1})$, (already present in the triangulation) and keep their depth. Any path from $\mathbf{x}_0$ to the new interior vertex $\mathbf{x}_i$ must pass through one of these boundary vertices, hence the depth of $\mathbf{x}_i$ is the minimum of the depths of the boundary vertices plus 1:

$$\mathfrak{da}[(c;c_1\dots c_i)]_{c'} = \begin{cases} \min_{c''\neq c'} \left\{ \mathfrak{da}[(c;c_1\dots c_{i-1})]_{c''} \right\} + 1 \,, & c' = c_i \\ \\ \mathfrak{da}[(c;c_1\dots c_{i-1})]_{c'} \,, & c' \neq c_i \end{cases} .$$

The last remark we need to make is that, except the vertices of the first elementary melon ball, all the other triangulation vertices are inserted one at a time. The melon with word $(c;c_1)$ inserts the interior vertex $\mathbf{x}_1$, the melon with word $(c;c_1c_2)$ inserts the interior vertex $\mathbf{x}_2$ and so on (see Figure 5.6). Furthermore, the interior root vertex $\mathbf{x}_0$ can be canonically associated to the melon with word $(c;)$. Thus the words label not only the tree vertices, but also all the interior vertices of the triangulation (that is all the triangulation vertices except the boundary vertices of the first melon). The *depth of the vertex* $\mathbf{x}_i$ with word $(c;c_1,\dots c_i)$ is the entry c_i in the corresponding distance array:

$$\Lambda[(c;c_1,\dots c_i)] \equiv \mathfrak{da}[(c;c_1\dots c_i)]_{c_i} \,.$$

The computation of the depth of a vertex $\mathbf{x}_i$ with word $(c;c_1,\dots c_i)$ involves all its predecessors, i.e. the vertices with words $(c,c_1,\dots c_k), k < i$. For example the distance arrays of the vertex $\mathbf{x} = (0;10132120312)$ and all its predecessors are presented in Table 5.1 (the depth is indicated in bold in the distance array).

w	0	1	0	1	3	2	1	2	0	3	1	2
$\Lambda(w)$	0	1	2	2	2	3	3	3	3	4	4	4
$\mathfrak{da}[w]$	$\begin{pmatrix}\mathbf{0}\\1\\1\\1\end{pmatrix}$	$\begin{pmatrix}0\\\mathbf{1}\\1\\1\end{pmatrix}$	$\begin{pmatrix}\mathbf{2}\\1\\1\\1\end{pmatrix}$	$\begin{pmatrix}2\\\mathbf{2}\\1\\1\end{pmatrix}$	$\begin{pmatrix}2\\2\\1\\\mathbf{2}\end{pmatrix}$	$\begin{pmatrix}2\\2\\\mathbf{3}\\2\end{pmatrix}$	$\begin{pmatrix}2\\\mathbf{3}\\3\\2\end{pmatrix}$	$\begin{pmatrix}2\\3\\\mathbf{3}\\2\end{pmatrix}$	$\begin{pmatrix}\mathbf{3}\\3\\3\\2\end{pmatrix}$	$\begin{pmatrix}3\\3\\3\\\mathbf{4}\end{pmatrix}$	$\begin{pmatrix}3\\\mathbf{4}\\3\\4\end{pmatrix}$	$\begin{pmatrix}3\\4\\\mathbf{4}\\4\end{pmatrix}$

Table 5.1: Calculation of distance array and depth.

There is a second method to compute the depth of a vertex starting from its associated word. We call the *depth of the letter* c_i in the word $(c;c_1,\dots c_n)$ the depth of the vertex with word $(c; c_1, \dots c_i)$. The depth of a letter in a word can only increase (as the depth is one of the entries of the distance array and all the entries in the array can only increase). We split the word into disjoint adjacent subwords τ_r comprised of groups of letter of the same depth, r.

The depth of the first letter, c is always 0, and any other letter will have a strictly larger depth, hence the word τ_0 is always (c). As long as the letter c does not appear in the word, the distance array remains constant (having 1 for all the entries different from c, and 0 for the entry c), and the depth is 1. Thus the second word is $\tau_1 = (c_1, \dots c_{a_1})$ with $c_1, \dots c_{a_1} \neq c$, $c_{a_1+1} = c$.

Starting with $r \geq 2$, the algorithm changes. Consider the letters of depth 2. The distance array of $(c)(c_1 \dots c_{a_1})(c)$ has 2 in the entry c and 1 in all the other entries. The first time a letter appears in the word its corresponding entry in the distance array passes to 2, and subsequent repetitions of the same letter do not modify the depth (this includes the first letter, c, of τ_2). The depth changes from 2 to 3 and the word is decomposed as $(c)(c_1, \dots c_{a_1})(c_{a_1+1} \dots c_{a_2})(c_{a_2+1})$ only when:

- there is only one entry equal to 1 in the distance array, that is the letters $(c_{a_1+1} \dots c_{a_2})$ are all the letters $\{0, 1 \dots D\}$ except exactly one.
- the letter c_{a_2+1} is exactly the letter missing in $(c_{a_1+1} \dots c_{a_2})$.

Denoting W_{D+1} the set of words in which each letter $\{0, \dots D\}$ appears *at least once*, the condition is rewritten as:

$$\tau_2 \notin W_{D+1} \ , \quad \tau_2 c_{a_2+1} \in W_{D+1} \ .$$

By this algorithm any word is decomposed as:

$$\tau_0 \tau_1 \tau_2 \dots \tau_q \ ,$$

where τ_q might be incomplete, and the depth of this word is exactly $\Lambda[w] = q$. For example the word:

$$(0; 10132120312) = (0)(1)(013)(2120)(312) \ ,$$

has depth 4.

Let us recapitulate where we stand. Every interior vertex of the triangulation $\mathbf{x}$ is one to one to a vertex in the tree, $\mathbf{V}$, and has a word $w = (c; c_1, \dots c_n)$. The vertex $\mathbf{x}$ has a depth $\Lambda[w]$. On the other hand, the distance in the tree between $\mathbf{V}$ and the root univalent vertex of the tree is $n + 1$, that is the number of letters of the word w. In general there is no simple formula relating n and $\Lambda[w]$. However, on average, the two are proportional. This is a consequence of the following two lemmas.

Lemma 5.1. *We have:*

$$S_q^n = \sum_{n_1, \dots n_q \geq 1}^{n_1 + \dots n_q = n} \frac{n!}{n_1! \dots n_q!} = \sum_{0 \leq r \leq q} (-1)^{q-r} \binom{q}{r} r^n \ .$$

Proof. We note that:

$$q^n = \sum_{n_1,\dots n_q \geq 0}^{n_1+\dots n_q=n} \frac{n!}{n_1!\dots n_q!} = \sum_{r=1}^{q} \binom{q}{r} \sum_{n_1,\dots n_r \geq 1}^{n_1+\dots n_r=n} \frac{n!}{n_1!\dots n_r!} = \sum_{r=1}^{q} \binom{q}{r} S_r^n \,.$$

Let us define L as a lower triangular $q \times q$ matrix with nonzero entries: $L_{rs} = \binom{r}{s}$ for $q \geq r \geq s \geq 1$. Then, the S_q^n's are the solution of the system of equations:

$$r^n = \sum_s L_{rs} S_s^n \,, \quad \text{for all } r \text{ such that } 1 \leq r \leq q \,.$$

The lemma follows by noting that the lower triangular $q \times q$ matrix P, with entries:

$$P_{rs} = (-1)^{r-s} \binom{r}{s} \,, \quad \text{for} \quad q \geq r \geq s \geq 1 \,,$$

is the inverse of L, as it can be seen by considering the appropriate monomials in the Cauchy product of the two series:

$$(1+x)^r = \sum_{s=0}^{r} \binom{r}{s} x^s \,, \qquad \frac{x^t}{(1+x)^{t+1}} = \sum_{s \geq t} \binom{s}{t} (-1)^{s-t} x^s \,.$$

□

Lemma 5.2. *Let $c_1, \dots, c_n$ be a sequence of random variables uniformly drawn from the alphabet $\{0, \dots D\}$, and denote $w = 0c_1 \dots c_n$. One has:*

$$\lim_{n\to\infty} \frac{1}{n+1} \Lambda(w) = \Lambda_\Delta \,, \qquad \Lambda_\Delta^{-1} = (D+1) \sum_{0 \leq r \leq D} (-1)^{D-r} \binom{D}{r} \frac{r}{(D+1-r)^2} \,.$$

Proof. Let us denote the length of a word τ by $|\tau|$. The constant Λ_Δ^{-1} is the average length of a subword τ_i for infinitely long words w. Recalling the definition of the τ_i, one notes that $|\tau_0| = 1$ and $|\tau_1|$ is special: $|\tau_1| = (-1 + \textit{length of the sequence of letters that ends the first time the letter } 0 \textit{ appears})$. The probability P_n that this sequence ends after n letters is:

$$P_1 = \frac{1}{D+1} \,, \qquad P_{n+1} = (1 - P_n) P_1 \,,$$

that is $P_n = (1 - P_1)^{n-1} P_1$ and the average length of τ_1 is:

$$\langle |\tau_1| \rangle = \sum_{n \geq 1} (n-1)(1-P_1)^{n-1} P_1 = \frac{1 - P_1}{P_1} = D \,.$$

As $\langle |\tau_1| \rangle$ is finite, Λ_Δ^{-1} is the average length of τ_i, for $i \geq 2$. Let us denote P_q^n the probability that after n draws one has obtained a sequence of letters with exactly q distinct colors. It follows that:

$$\langle |\tau_i| \rangle = \sum_{n \geq 1} (n-1) P_D^{n-1} P_1 = \frac{1}{D+1} \sum_{n \geq 0} n P_D^n \,.$$

The probability of any fixed sequence of results on n draws is $\frac{1}{(D+1)^n}$. The number of sequences of n letters having n_1 times the color c_1, n_2 times the color c_2 up to n_q times the color c_q is the multinomial coefficient, hence, using Lemma 5.1:

$$P_q^n = \frac{1}{(D+1)^n}\binom{D+1}{q}\sum_{n_1,\dots n_q\geq 1}^{n_1+\dots n_q=n}\frac{n!}{n_1!\dots n_q!}$$
$$= \frac{1}{(D+1)^n}\binom{D+1}{q}\sum_{0\leq r\leq q}(-1)^{q-r}\binom{q}{r}r^n,$$

therefore:

$$\langle|\tau_i|\rangle = \sum_{n\geq 0}\frac{nP_D^n}{D+1} = \sum_{n\geq 0}\frac{n}{(D+1)^n}\sum_{0\leq r\leq D}(-1)^{D-r}\binom{D}{r}r^n$$
$$= \sum_{0\leq r\leq D}(-1)^{D-r}\binom{D}{r}\sum_{n\geq 0}n\left(\frac{r}{D+1}\right)^n$$
$$= (D+1)\sum_{0\leq r\leq D}(-1)^{D-r}\binom{D}{r}\frac{r}{(D+1-r)^2}.$$

□

In particular for $D = 2, 3, 4$ we get $\Lambda_\Delta^{-1} = 9/2, 22/3, 125/12$, respectively.

On average, the depth of a vertex $\mathbf{x}$ in a melonic D–ball is, up to the constant rescaling by Λ_Δ, the tree distance between its associated vertex $\mathbf{V}$ and the root of the $(D+1)$-ary tree. As the $(D+1)$-ary trees are one to one to the melonic graphs, it follows that the statistical ensemble of uniformly distributed random melonic balls, with distance function given by the depth, is similar to the statistical ensemble of uniformly distributed random trees with the natural graph distance.

Random Trees

The statistical ensemble of uniformly distributed random trees with the graph distance function is well known in the literature. The infinitely refined metric space obtained in continuous limit of this ensemble (in a precise sense we will specify below) is known as the continuum random tree (CRT).

Galton–Watson trees. A Galton–Watson process is a stochastic process $\{Z_k\}$, evolving according to $Z_{k+1} = \sum_{j=1}^{Z_k}\xi_j^{(k)}$ where the $\xi_j^{(k)}$ are independent identically distributed random variables taking values in $\mathbb{Z}_+$. We denote by μ the probability distribution of one ξ.

Colloquially, k is the generation number, Z_k is the number of offspring in the kth generation, while μ is the offspring distribution, so that μ_j is the probability that a member has j children. Such a process is called *critical* if its mean is one: $\sum_{j=0}^{\infty} j\,\mu_j = 1$. The variance of the distribution is $\sigma = \sum_{j=0}^{\infty} j(j-1)\,\mu_j$. Obviously, if $Z_1 = 1$, the process has one initiator (ancestor) and any instance of the process can be mapped to a rooted tree.

A *Galton–Watson tree* is a tree generated by this process (with a root halfedge hooked to Z_1). We are interested in Galton–Watson trees conditioned to have a total of n progeny, that is, trees T_n with n vertices (not counting the root univalent vertex).

The crucial fact about critical Galton–Watson trees is that they can be obtained as simply generated trees. The offspring distribution μ induces a distribution for T_n on the set of all (plane) trees with n vertices. Specifically we associate to every tree a weight $\Pi_\mu(T_n) := \prod_{v\in T_n} \mu_{k_v}$, where v are the vertices of T_n (distinct from the univalent root vertex) and k_v is the number of offspring of v. We then pick at random plane trees with probabilities proportional to this weight:

$$P(T_n) = \frac{1}{\sum_{T_n} \Pi_\mu(T_n)} \Pi_\mu(T_n) \ .$$

Continuum Random Tree. Let us consider a rooted plane tree T_n with n interior vertices (i.e. distinct from the root).

We denote $d_{T_n}(v,u)$ the graph distance in T_n between the vertices v and u (i.e. the number of edges in the unique path in the tree joining v and u). We also denote $d_{T_n}(v)$ the distance from the vertex v to the root. As it is embedded in the plane, T_n has an associated contour walk obtained by starting from the root and turning clockwise around the tree passing from one vertex to the next in unit time-steps. We denote $v(i)$ the tree vertex reached by this walk at the time step i. Of course $v(0)$ is just the root of the tree. The contour walk can be encoded in a piecewise linear function $f(t)$ obtained by setting $f(i) = d_{T_n}(v(i))$ and using a linear interpolation between time steps. We illustrate this in Figure 5.7. By construction the journey ends at time step

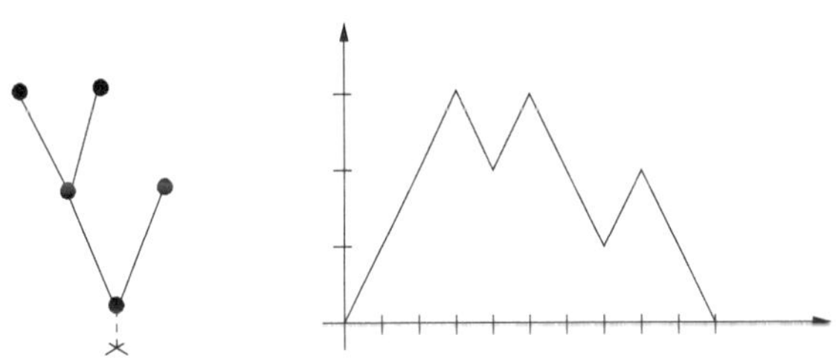

Figure 5.7: A tree and its associated contour walk.

$2n$, with $f(2n) = 0$ and $f(t) > 0$ for $0 < t < 2n$. The plane trees are one to one with such piecewise linear functions f.

A real continuous function $f(t)$, such that $f(0) = f(1) = 0$ and $f(t) > 0$ for $0 < t < 1$, encodes a rooted *real tree* $\mathcal{T}_f$. The intuitive picture is as follows. One reconstructs $\mathcal{T}_f$ by "spraying glue" under the graph of f and squashing the interval $[0,1]$ to a point. More formally, we set:

$$\forall s < t \in [0,1] \ , \qquad m_f(s,t) \equiv \inf_{s\le r\le t} f(r) \ ,$$

and we define the equivalence relation:

$$s \underset{f}{\sim} t \qquad \Longleftrightarrow \qquad f(s) = f(t) = m_f(s,t) \ .$$

Then the rooted real tree is the quotient: $\mathcal{T}_f = [0,1]/\underset{f}{\sim}$ and the distance on the real tree is given by:

$$d_f(s,t) = f(s) + f(t) - 2m_f(s,t) .$$

The branching vertices of the real tree are those values of s in $[0,1]$ that are equivalent to at least two other values.

A particular function f one can consider is the Brownian excursion. It is defined as follows. The Wiener process is a stochastic process W_t (that is a random variable for every time t) such that $W_0 = 0$, $t \to W_t$ is almost surely continuous, W_t has independent increments and $W_t - W_s$ is distributed on a normal distribution of mean 0 and variance $\sigma^2 = t - s$ for $s \le t$. The normalized Brownian excursion e_t is a Wiener process conditioned to be positive for $0 < t < 1$ and to be at 0 at time 1. It is formally represented by the path integral measure:

$$d\mu_e = \frac{1}{Z}\left[dq(t)\right]\Big|_{q(0)=q(1)=0;q(t)>0} e^{-\frac{1}{2}\int_0^1 [\dot{q}(t)]^2 dt} ,$$

with Z a normalization constant.

Definition 5.3. *The continuum random tree (CRT) $(\mathcal{T}_{2e}, d_{2e})$ is the random tree associated to twice a normalized Brownian excursion $2e$, endowed with the metric d_{2e}.*

Gromov-Hausdorff topology and convergence. We will consider a sequence of random metric spaces (the trees endowed with the graph distance). In order to discuss convergence for such sequences we need to define the "space of metric spaces" along with an appropriate topology. This is provided by the Gromov–Hausdorff topology on the space of isometry classes of compact metric spaces.

Consider a metric space (E, d_E). The Hausdorff distance between two compact sets, K_1 and K_2, in E is:

$$d_E^{\text{Haus}}(K_1, K_2) = \inf\{r | K_1 \subset K_2^r, K_2 \subset K_1^r\} ,$$

where $K_i^r = \cup_{x \in K_i} B_E(x, r)$ is the union of open balls of radius r centered on the points of K_i.

Now, given two compact metric spaces (E_1, d_1) and (E_2, d_2), the Gromov–Hausdorff distance between them is:

$$d_{\text{GH}}(E_1, E_2) = \inf\left\{d_E^{\text{Haus}}(\phi_1(E_1), \phi_2(E_2))\right\} ,$$

where the infimum is taken over all the metric spaces E and all the isometric embeddings ϕ_1 and ϕ_2 from (E_1, d_1) and (E_2, d_2) into (E, d_E).

It turns out that $\mathbb{K}$, the set of all isometry classes of compact metric spaces, endowed with the Gromov–Hausdorff distance d_{GH}, is a complete metric space in its own right. Therefore, one may study the convergence (in distribution) of $\mathbb{K}$–valued random variables.

The continuum limit of trees. Let us consider the family of trees T_n generated by a critical Galton Watson process with variance σ. We regard these trees as metric spaces $(T_n, d_{T_n}/\sqrt{n/\sigma})$. In order to explore the limit $n \to \infty$ we look at sequences of trees with increasing number of vertices n, such that the element T_n of the sequence at any given n is chosen randomly with probability $P(T_n)$. One thing to note is that as n gets large, one would like to ensure convergence to some compact metric space (rather than just letting the structure get infinitely large). This requires the precise n dependent rescaling of metric $d_{T_n}/\sqrt{n/\sigma}$. The seminal result of Aldous [2] concerning the $n \to \infty$ limit is:

Theorem 5.2 (Aldous). *Under the uniform distribution, the family of trees associated to a critical Galton–Watson process with variance σ converges in the Gromov–Hausdorff topology on compact metric spaces to the continuum random tree:*

$$\left(T_n, \frac{d_{T_n}}{\sqrt{n/\sigma}}\right) \xrightarrow[n\to\infty]{} (\mathcal{T}_{2e}, d_{2e}) .$$

While the Gromov–Hausdorff topology appears quite abstract, convergence in the Gromov–Hausdorff topology is a consequence of *any* convergence of the sequence T_n embedded within some common metric space E.

This result rests on two points:

- Skorohod's representation theorem states that there exists a metric space Ω within which T_n (for all n) and $\mathcal{T}_{2e}$ can be embedded such that the image of d_{T_n} approaches the image of d_{2e} almost surely as $n \to \infty$;
- under the uniform distribution:

$$\left(\frac{d_{T_n}(s_1 n, s_2 n)}{\sqrt{n/\sigma}}\right)_{s_1,s_2\in[0,1]} \xrightarrow[n\to\infty]{} \big(d_{2e}(s_1,s_2)\big)_{s_1,s_2\in[0,1]} . \tag{5.4}$$

Importantly, the Gromov–Hausdorff distance between T_n and $\mathcal{T}_{2e}$ is bounded from above by:

$$\sup\left\{\left(d_{T_n}(s_1 n, s_2 n)/\sqrt{n/\sigma}\right) - d_{2e}(s_1,s_2)\right)\ ,\ s_1,s_2 \in [0,1]\right\} .$$

hence, using (5.4), the convergence is ensured.

The continuum limit of melons

We denote a random melonic D–ball with n (internal) vertices by M_n, and the set of its vertices by $m_n = \{\mathbf{x}_0, \dots \mathbf{x}_{n-1}\}$. Each $\mathbf{x}_r$ is labeled by a word, w_r. We denote $d_{m_n}(r_1,r_2)$ the graph distance (in the melonic ball M_n) between $\mathbf{x}_{r_1}$ and $\mathbf{x}_{r_2}$, hence $d_{m_n}(0,r)$ is just the depth $\Lambda(w_r)$. We have the following result:

Theorem 5.3. *Under the uniform distribution, the family of melonic D-balls converges in the Gromov-Hausdorff topology on compact metric spaces to the continuum random tree:*

$$\left(m_n, \frac{d_{m_n}}{\Lambda_\Delta \sqrt{\frac{(D+1)n}{D}}} \right) \longrightarrow_{n\to\infty} (\mathcal{T}_{2e}, d_{2e}) \, .$$

A detailed proof of this theorem is presented in [106]. Instead of reproducing this slightly technical proof here we will just emphasize some of the more important points.

Galton Watson trees. The melonic D-balls are represented by $(D+1)$-ary trees T_n, where n denotes the number of $(D+2)$-valent (internal) vertices of the tree. Forgetting the univalent vertices, such trees are generated by the critical Galton Watson process with nonzero binomial offspring distribution on the first $D+2$ weights

$$\mu_j = \binom{D+1}{j} \frac{1}{(D+1)^j} \left(\frac{D}{D+1} \right)^{D+1-j} \, , \; 0 \le j \le D+1 \, ,$$

which is critical with variance $\sigma = D/(D+1)$. A subtlety to note is that this distribution on the offspring for combinatorial trees induces a distribution of offspring

$$\mu_{c_1,\dots c_j} = \frac{1}{(D+1)^j} \left(\frac{D}{D+1} \right)^{D+1-j}$$

for trees with colored edges, as the binomial coefficient counts the number of ways to choose the colorings $\{c_1, \dots c_j\}$ of the offspring. As the sum of the number of offspring of all the vertices is $n-1$, it follows that this Galton Watson process corresponds to the uniform distribution on colored rooted $(D+1)$-ary trees with n vertices.

Metric. The distance between two vertices in the melonic ball can be well estimated from their depth. Consider two vertices r_1 and r_2 with words wu_0u and wv_0v, that is the two words have w in common, $u_0 \neq v_0$ are the first letters at which the two words differ and u and v denote the remaining letters in the words. Using the triangle inequality [106] one can show that:

$$\left| d_{m_n}(r_1, r_2) - \Lambda(u_0 u) - \Lambda(v_0 v) \right| \le 6 \, .$$

As the distance in the triangulation is well estimated by the depth, and on average the depth is proportional with just the tree distance, one can trade, up to a rescaling, d_{m_n} for d_{T_n} and invoke Aldous's result.

In particular, the rescaling of the distance function by $\sqrt{n}$ implies that in the $n \to \infty$ limit the diameter of the melonic triangulation scales like the inverse of the square root of its volume, hence the emergent Hausdorff dimension is $d_H = 2$.

5.4.2 The spectral dimension

The small scale geometry of a triangulation is well encoded in the Laplacian operator (or the Dirac operator). The Laplacian and its associated heat kernel can be understood in terms of random walks, that is diffusion processes, in the triangulation. The behavior of a random walk on a graph depends strongly on the connectivity of this graph: the more distinct "short" simple paths exist between two vertices, the higher the probability to jump from one vertex to the other. For this reason the study of random walks in generic graphs is highly nontrivial.

Diffusion in a melon

We shall study here the diffusion (random walk) directly in the rooted melonic graphs. This means that, from the point of view of the triangulation, the walk jumps from the barycenter of a simplex to the barycenter of one of its neighboring simplices, i.e. we endow the triangulation with the dual equilateral metric. This is of course a different process from a random walk in the triangulation itself, which would jump between triangulation vertices along triangulation edges. Our choice is justified by the fact that a diffusion process on the vertices of the melonic balls themselves is difficult to analyze. While for individual triangulations the two processes can be very different, for typical triangulations (and, most importantly, on average) they are expected to present the same rough characteristics.

For rooted melonic graphs, the appropriate diffusion process generates a return / transit probability, where transit refers to the process of traversing from one external vertex to the other[1]. In this section we derive the equation expressing these probabilities for a melonic graph as a function of the ones of its constituent submelons.

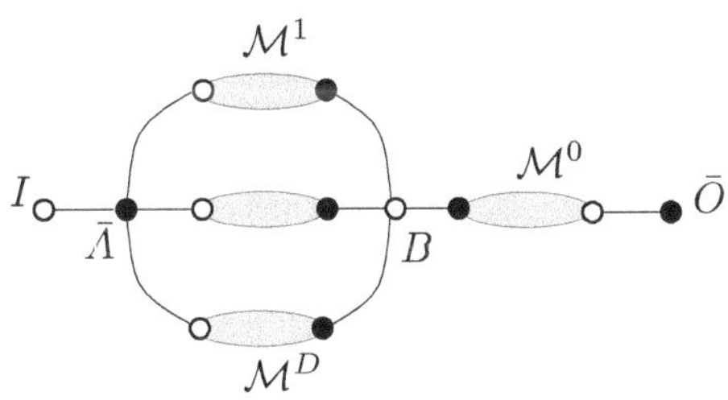

Figure 5.8: A rooted melonic graph $\mathcal{M}$ with submelons $\mathcal{M}^i$.

Consider a rooted melonic graph $\mathcal{M}$, with external color c. We add two univalent vertices I and $\bar{O}$ at the end of its external halfedges, keeping the graph bipartite, see Figure 5.8 (where the external color is chosen as $c = 0$). The graph is rooted at the external vertex $\bar{O}$ (the univalent end vertex of the root halfedge). We consider the

[1]To be more precise, a diffusion process is characterized by the diffusion equation along with appropriate boundary conditions. For the rooted melonic graphs, it is appropriate to choose cyclic boundary conditions. This allows the interpretation of the process as occurring on a closed manifold, that is the closure of the melonic graph.

prime melonic graphs $\mathcal{G}_i, i = 1, q$ in the cut set of the root. We denote the melonic graph $\mathcal{G}_1 \cup \dots \mathcal{G}_{p-1}$ by $\mathcal{M}^0$. Let us call the vertices carrying the external edges of $\mathcal{G}_p$ by $\bar{A}$ and B. The melonic graph obtained by cutting the edges of color $c_1 \neq c$ incident at $\bar{A}$ and B is denoted $\mathcal{M}^{c_1}$. We decorate the external halfedges of any melonic graph by vertices, and we denote generically the external vertices by $\circ$ and $\bullet$ (for instance, for $\mathcal{M}$ itself, $\circ = I$ and $\bullet = \bar{O}$).

Consider a random walk on the melon $\mathcal{M}$. If the walker is at one of the external vertices $\circ = I$ or $\bullet = \bar{O}$, then it steps with probability one to its unique neighbor. If the walker is at any of the internal $(D+1)$-valent vertices of $\mathcal{M}$, it steps with probability $\frac{1}{D+1}$ to one of its $D+1$ neighbors.

Let us denote by $P^1_{\mathcal{M}}(t)$ the 2×2 matrix encoding the first return/first transit probabilities, in time t, for the melonic graph $\mathcal{M}$. Its four elements are:

- $P^{1;\circ\circ}_{\mathcal{M}}(t) = P^{1;II}_{\mathcal{M}}(t)$ is the probability that the walker starts from the external point $\circ = I$ and returns for the first time to $\circ = I$ at time t without touching the external point $\bullet = \bar{O}$ in the intervening time. For any $\mathcal{M}$, $P^{1;\circ\circ}_{\mathcal{M}}(0) = 1$.

- $P^{1;\circ\bullet}_{\mathcal{M}}(t) = P^{1;I\bar{O}}_{\mathcal{M}}(t)$ is the probability that the walker starts from the external point $\circ = I$ and reaches for the first time $\bullet = \bar{O}$ at time t, without touching I a second time. For any $\mathcal{M}$, $P^{1;\circ\bullet}_{\mathcal{M}}(0) = 0$.

- $P^{1;\bullet\circ}_{\mathcal{M}}(t) = P^{1;\bar{O}I}_{\mathcal{M}}(t)$ is the probability that the walker starts from the external point $\bullet = \bar{O}$ and reaches for the first time $\circ = I$ at time t without touching $\bar{O}$ a second time. For any $\mathcal{M}$, $P^{1;\bullet\circ}_{\mathcal{M}}(0) = 1$.

- $P^{1;\bullet\bullet}_{\mathcal{M}}(t) = P^{1;\bar{O}\bar{O}}_{\mathcal{M}}(t)$ is the probability that the walker starts from the external point $\bullet = \bar{O}$ and returns for the first time to $\bullet = \bar{O}$ at time t without touching the external point $\circ = I$ in the intervening time. For any $\mathcal{M}$, $P^{1;\bullet\bullet}_{\mathcal{M}}(0) = 1$.

Let us denote by $\mathcal{G}^{(0),c}$ the (melonic) graph made by a unique edge of color c connecting the external points. Its first return/first transit probability matrix is: $P^1_{\mathcal{G}^{(0),c}}(t) = \begin{pmatrix} 0 & \delta_{t,1} \\ \delta_{t,1} & 0 \end{pmatrix}$.

The return/transit probability matrix $P_{\mathcal{M}}(t)$ is defined similarly to the first return/first transit probability matrix, except that the walker is allowed any trajectory between the end points. Consider a generic path from the external vertex X to the external vertex Y. It can be decomposed as a word on the alphabet I and $\bar{O}$ starting with X and ending with Y (like XY, XIY, $X\bar{O}Y$, etc.), where each pair of consecutive letters represent a walk from the first to the second letter that does not touch either external point in the intervening time. Denoting by w_q the words over I and $\bar{O}$ of length q, and by $w_q(i)$ the i'th letter of w_q, the return/transit probability in time t is:

$$P^{XY}_{\mathcal{M}}(t) = \delta^{XY}\delta_{t,0} + P^{1;XY}_{\mathcal{M}}(t) + \sum_{q=1}^{\infty}\sum_{w_q}\sum_{t_0+\dots+t_q=t} P^{1;Xw_q(1)}_{\mathcal{M}}(t_0) \\ \times P^{1;w_q(1)w_q(2)}_{\mathcal{M}}(t_1)\dots P^{1;w_q(q-1)w_q(q)}_{\mathcal{M}}(t_{q-1})P^{1;w_q(q)Y}_{\mathcal{M}}(t_q)\,.$$

This equation simplifies when translated in terms of the generating functions:

$$P_{\mathcal{M}}^{1;XY}(y) = \sum_{t\geq 0} y^t \, P_{\mathcal{M}}^{1;XY}(t) \,, \qquad P_{\mathcal{M}}^{XY}(y) = \sum_{t\geq 0} y^t \, P_{\mathcal{M}}^{XY}(t) \,,$$

$$P_{\mathcal{M}}^{XY}(y) = \delta^{XY} + P_{\mathcal{M}}^{1;XY}(y) + \sum_{q=1}^{\infty} \sum_{w_q} P_{\mathcal{M}}^{1;X w_q(1)}(y) \times P_{\mathcal{M}}^{1;w_q(1)w_q(2)}(y) \dots P_{\mathcal{M}}^{1;w_q(q-1)w_q(q)}(y) P_{\mathcal{M}}^{1;w_q(q)Y}(y) \,.$$

Taking into account that $w_q(i) \in \{I, \bar{O}\}$, we obtain the matrix equation:

$$P_{\mathcal{M}}(y) = 1 + P_{\mathcal{M}}^1(y) + [P_{\mathcal{M}}^1(y)]^2 + \dots = \frac{1}{1 - P_{\mathcal{M}}^1(y)} \,,$$

or in detail:

$$\begin{pmatrix} P_{\mathcal{M}}^{\circ\circ}(y) & P_{\mathcal{M}}^{\circ\bullet}(y) \\ P_{\mathcal{M}}^{\bullet\circ}(y) & P_{\mathcal{M}}^{\bullet\bullet}(y) \end{pmatrix} = \frac{1}{[1 - P_{\mathcal{M}}^{1;\circ\circ}(y)][1 - P_{\mathcal{M}}^{1;\bullet\bullet}(y)] - P_{\mathcal{M}}^{1;\bullet\circ}(y) P_{\mathcal{M}}^{1;\circ\bullet}(y)} \times \begin{pmatrix} 1 - P_{\mathcal{M}}^{1;\bullet\bullet}(y) & P_{\mathcal{M}}^{1;\circ\bullet}(y) \\ P_{\mathcal{M}}^{1;\bullet\circ}(y) & 1 - P_{\mathcal{M}}^{1;\circ\circ}(y) \end{pmatrix} .$$

The first return/first transit probability matrix of $\mathcal{M}$ can be computed in terms of the first return/first transit probability matrices of its sub-melons $\mathcal{M}^c$. A walk starting at I and returning for the first time to I at time t, which does not touch $\bar{O}$ in the intervening time (that is a walk contributing to $(1, II)$), can be further decomposed as a word over $I, \bar{A}, B, \bar{O}$ (see Figure 5.8). Starting with I, the walker first steps to $\bar{A}$. Once at $\bar{A}$:

- the walker may travel out from $\bar{A}$ and return without hitting B.
- the walker may traverse to B. If the walker is at B at some time t, then:
  - the walker may travel out (also in the direction of $\bar{O}$, although never hitting it) and return.
  - the walker may traverse back to $\bar{A}$.

These sub-paths may occur any number of times before eventually the walker returns to $\bar{A}$ a last time and steps back to I. Similarly, a walk contributing to $(1, IO)$ starts from I, passes through $\bar{A}$, then $\bar{A}$ and B a number of times, before finally going to B a last time and traversing to $\bar{O}$. Thus:

$$\begin{aligned} (1, II) &= I \, \bar{A} \dots \bar{A} \, I \,, \\ (1, I\bar{O}) &= I \, \bar{A} \dots \bar{A} B \dots B \, \bar{O} \,, \\ (1, \bar{O}I) &= \bar{O} \, B \dots B \bar{A} \dots \bar{A} \, I \,, \\ (1, \bar{O}\bar{O}) &= \bar{O}\bar{O} \text{ or } \bar{O} \, B \dots B \, \bar{O} \,, \end{aligned}$$

where the dots signify an arbitrary string of $\bar{A}$ and B. The walks from $\bar{O}$ to $\bar{O}$ are special: the walks indicated as by $\bar{O}\bar{O}$ start from $\bar{O}$, go into the melon $\mathcal{M}^0$ and return to $\bar{O}$ without ever touching B.

The first return/first transit probabilities between $\bar{A}$ and B may be written in terms of the first return/first transit probabilities for the sub-melons: $\mathcal{M}^1, \dots, \mathcal{M}^D$ and $\mathcal{M}^0$. Indeed, for a random walk to go from say $\bar{A}$ to $\bar{A}$ it needs to choose one of the sub-melons $\mathcal{M}^1, \dots \mathcal{M}^D$ and go from $\bullet$ to $\bullet$ in it. Thus:

$$
\begin{aligned}
P^{1;\bar{A}\bar{A}}(t) &= \frac{1}{D+1}\Big(P^{1,\bullet\bullet}_{\mathcal{M}^1}(t) + \dots + P^{1,\bullet\bullet}_{\mathcal{M}^D}(t)\Big) , \\
P^{1;\bar{A}B}(t) &= \frac{1}{D+1}\Big(P^{1,\bullet\circ}_{\mathcal{M}^1}(t) + \dots + P^{1,\bullet\circ}_{\mathcal{M}^D}(t)\Big) , \\
P^{1;B\bar{A}}(t) &= \frac{1}{D+1}\Big(P^{1,\circ\bullet}_{\mathcal{M}^1}(t) + \dots + P^{1,\circ\bullet}_{\mathcal{M}^D}(t)\Big) , \\
P^{1;BB}(t) &= \frac{1}{D+1}\Big(P^{1,\circ\circ}_{\mathcal{M}^1}(t) + \dots + P^{1,\circ\circ}_{\mathcal{M}^D}(t) + P^{1,\circ\circ}_{\mathcal{M}^0}(t)\Big) ,
\end{aligned}
$$

where again one notes that the first return probability for B to B is special, as the walk may also go into the melon $\mathcal{M}^0$. Denoting by $P^1(t)$ the first return/first transit probability matrix between $\bar{A}$ and B, one has:

$$
P^1(t) = \frac{1}{D+1} P^1_{\mathcal{M}^1}(t) + \dots + \frac{1}{D+1} P^1_{\mathcal{M}^D}(t) + \frac{1}{D+1} \begin{pmatrix} P^{1,\circ\circ}_{\mathcal{M}^0}(t) & 0 \\ 0 & 0 \end{pmatrix} .
$$

Furthermore, the first return/first transit probabilities between B and $\bar{O}$ are:

$$
P^{1;\bar{O}B}(t) = P^{1,\bullet\circ}_{\mathcal{M}^0}(t) , \quad P^{1;B\bar{O}}(t) = \frac{1}{D+1} P^{1,\circ\bullet}_{\mathcal{M}^0}(t) , \quad P^{1;\bar{O}\bar{O}}(t) = P^{1,\bullet\bullet}_{\mathcal{M}^0}(t) .
$$

Denoting by w_q the words of q letters over $\bar{A}$ and B, the walks of first return/first transit in $\mathcal{M}$ decompose as:

$$
\begin{aligned}
P^{1,\circ\circ}_{\mathcal{M}}(t) =& \frac{1}{D+1}\delta_{t,2} + \frac{1}{D+1} P^{1;\bar{A}\bar{A}}(t-2) + \\
& + \frac{1}{D+1} \sum_{q=1}^{\infty} \sum_{w_q} \sum_{t_0 + \dots + t_q = t-2} P^{1;\bar{A}w_q(1)}(t_0) \\
& \qquad \times P^{1;w_q(1)w_q(2)}(t_1) \dots P^{1;w_q(q-1)w_q(q)}(t_{q-1}) P^{1;w_q(q)\bar{A}}(t_q) , \\
P^{1,\circ\bullet}_{\mathcal{M}}(t) =& \sum_{t_0+t_1=t-1} P^{1;\bar{A}B}(t_0) P^{1;B\bar{O}}(t_1) \\
& + \sum_{q=1}^{\infty} \sum_{w_q} \sum_{t_0 + \dots + t_{q+1} = t-1} P^{1;\bar{A}w_q(1)}(t_0) \\
& \qquad \times P^{1;w_q(1)w_q(2)}(t_1) \dots P^{1;w_q(q)B}(t_q) P^{1;B\bar{O}}(t_{q+1})\Big) , \\
P^{1,\bullet\circ}_{\mathcal{M}}(t) =& \frac{1}{D+1} \sum_{t_0+t_1=t-1} P^{1;\bar{O}B}(t_0) P^{1;B\bar{A}}(t_1)
\end{aligned}
$$

$$
\begin{aligned}
&+\frac{1}{D+1}\sum_{q=1}^{\infty}\sum_{w_q}\sum_{t_0+\cdots+t_{q+1}=t-1} P^{1;\bar{O}B}(t_0) \\
&\qquad \times P^{1;Bw_q(1)}(t_1)\dots P^{1;w_q(q)\bar{A}}(t_{q+1}) \,, \\
P^{1,\bullet\bullet}_{\mathcal{M}}(t) =& P^{1;\bar{O}\bar{O}}(t) + \sum_{t_0+t_1=t} P^{1;\bar{O}B}(t_0)P^{1;B\bar{O}}(t_1) + \\
&+ \sum_{t_0+t_1+t_2=t} P^{1;\bar{O}B}(t_0)P^{1;BB}(t_1)P^{1;B\bar{O}}(t_2) \\
&+\sum_{q=1}^{\infty}\sum_{w_q}\sum_{t_0+\cdots+t_{q+2}=t} P^{1;\bar{O}B}(t_0) \\
&\qquad \times P^{1;Bw_q(1)}(t_1)\dots P^{1;w_q(q)B}(t_{q+1})P^{1;B\bar{O}}(t_{q+2}) \,.
\end{aligned}
$$

As one might expect, $P^{1;\bar{O}\bar{O}}_{\mathcal{M}}(t)$ requires special consideration. The first term, $P^{1;\bar{O}\bar{O}}(t)$, represents the walks that start from $\bar{O}$ and end in $\bar{O}$ without touching B, while the second represents the walks that hit B exactly once. The equations simplify again for generating functions:

$$
\begin{aligned}
P^{1,\circ\circ}_{\mathcal{M}}(y) =& \frac{1}{D+1}y^2 + \frac{1}{D+1}y^2 P^{1;\bar{A}\bar{A}}(y) \\
&+\frac{1}{D+1}y^2\sum_{q=1}^{\infty} P^{1;\bar{A}w_q(1)}(y)P^{1;w_q(1)w_q(2)}(y)\dots P^{1;w_q(q)\bar{A}}(y) \,, \\
P^{1,\circ\bullet}_{\mathcal{M}}(y) =& yP^{1;\bar{A}B}(y)P^{1;B\bar{O}} + y\sum_{q=1}^{\infty} P^{1;\bar{A}w_q(1)}(y) \\
&\qquad \times P^{1;w_q(1)w_q(2)}(y)\dots P^{1;w_q(q)B}(y)P^{1;B\bar{O}}(y) \,, \\
P^{1,\bullet\circ}_{\mathcal{M}}(y) =& \frac{1}{D+1}yP^{1;\bar{O}B}(y)P^{1;B\bar{A}}(y) + \frac{1}{D+1}y\sum_{q=1}^{\infty} P^{1;\bar{O}B}(y) \\
&\qquad \times P^{1;Bw_q(1)}(y)P^{1;w_q(1)w_q(2)}(y)\dots P^{1;w_q(q)\bar{A}}(y) \,, \\
P^{1,\bullet\bullet}_{\mathcal{M}}(y) =& P^{1;OO}(y) + P^{1;\bar{O}B}(y)P^{1;B\bar{O}}(y) + P^{1;\bar{O}B(y)}P^{1;BB}(y)P^{1;B\bar{O}}(y) \\
&+\sum_{q=1}^{\infty} P^{1;\bar{O}B}(y)P^{1;Bw_q(1)}(y) \\
&\qquad \times P^{1;w_q(1)w_q(2)}(y)\dots P^{1;w_q(q)B}(y)P^{1;B\bar{O}}(y) \,.
\end{aligned}
$$

Again, these equations can be written in a compact matrix form. Denoting $\sigma = \begin{pmatrix} 0 & 1 \\ 1 & 0 \end{pmatrix}$ we have:

$$
\begin{aligned}
P^{1}_{\mathcal{M}}(y) =& \begin{pmatrix} 0 & 0 \\ 0 & P^{1;\bar{O}\bar{O}}(y) \end{pmatrix} + \begin{pmatrix} y & 0 \\ 0 & P^{1;\bar{O}B}(y) \end{pmatrix} \\
&\times \left[\sigma\frac{1}{1-P^{1}(y)}\sigma\right]\begin{pmatrix} \frac{y}{D+1} & 0 \\ 0 & P^{1;B\bar{O}}(y) \end{pmatrix} .
\end{aligned}
$$

Substituting the various probabilities as functions of the submelons one gets a recursive equation along with an initial condition:

$$P^1_{\mathcal{M}}(y) = \begin{pmatrix} 0 & 0 \\ 0 & P^{1,\bullet\bullet}_{\mathcal{M}^0}(y) \end{pmatrix} + \frac{1}{D+1} \begin{pmatrix} y & 0 \\ 0 & P^{1,\bullet\circ}_{\mathcal{M}^0}(y) \end{pmatrix}$$

$$\times \sigma \frac{1}{1 - \frac{1}{D+1} \left[P^1_{\mathcal{M}^1}(y) + \dots P^1_{\mathcal{M}^D}(y) + \begin{pmatrix} P^{1,\circ\circ}_{\mathcal{M}^0}(t) & 0 \\ 0 & 0 \end{pmatrix} \right]} \sigma \begin{pmatrix} y & 0 \\ 0 & P^{1,\circ\bullet}_{\mathcal{M}^0}(y) \end{pmatrix} ,$$

$$P^1_{\mathcal{G}^{(0)},c}(y) = \begin{pmatrix} 0 & y \\ y & 0 \end{pmatrix} \equiv y\sigma \, . \tag{5.5}$$

In principle, Eq. (5.5) solves the problem of determining the first return/first transit probabilities for an arbitrary melon. For example, for $\mathcal{G}^{(2),0}$ one gets:

$$P^1_{\mathcal{G}^{(2)},0}(y) = \frac{1}{D+1} \begin{pmatrix} y & 0 \\ 0 & y \end{pmatrix} \sigma \frac{1}{1 - \frac{D}{D+1} \begin{pmatrix} 0 & y \\ y & 0 \end{pmatrix}} \sigma \begin{pmatrix} y & 0 \\ 0 & y \end{pmatrix}$$

$$= \frac{\frac{1}{D+1} y^2}{1 - \frac{D^2}{(D+1)^2} y^2} \begin{pmatrix} 1 & \frac{D}{D+1} y \\ \frac{D}{D+1} y & 1 \end{pmatrix} .$$

Defining the auxiliary 2×2 matrices $E^{ab}_{\alpha\beta} = \delta^a_\alpha \, \delta^b_\beta$ with $a, b, \alpha, \beta = 1, 2$ simplifies further Eq. (5.5):

$$P^1_{\mathcal{M}} = E^{22} P^1_{\mathcal{M}^0} E^{22} + \left(E^{12} y + E^{22} P^1_{\mathcal{M}^0} E^{11} \right)$$

$$\times \frac{1}{D + 1 - \sum_{c=1}^D P^1_{\mathcal{M}^c} - E^{11} P^1_{\mathcal{M}^0} E^{11}} \left(y E^{21} + E^{11} P^1_{\mathcal{M}^0} E^{22} \right) ,$$

$$P^1_{\mathcal{G}^{(0)},c}(y) = y \, \sigma \, .$$

By induction one trivially obtains that the first return/first transit matrix is symmetric $P^{1,\bullet\circ}_{\mathcal{M}}(y) = P^{1,\circ\bullet}_{\mathcal{M}}(y)$.

Simple melons. The recursion in Eq. (5.5) is somewhat difficult to handle. In order to proceed we will restrict ourselves from now on to *simple melons*, which are prime melonic graphs $\mathcal{M}$ (with external halfedges of color 0) such that all the melonic graphs obtained by cutting the edges incident to the vertices of $\mathcal{M}$ hooked to the external halfedges are themselves prime melonic graphs. In terms of the associated $(D+1)$–ary tree, a simple melon corresponds to a tree such that all the branches of color c consist of exactly one edge. For simple melons, $\mathcal{M}^0$ in Eq. (5.5) is empty and the recursion reduces to:

$$P^1_{\mathcal{M}} = y^2 \sigma \frac{1}{D + 1 - \sum_{c=1}^D P^1_{\mathcal{M}^c}} \sigma \, , \quad P^1_{\mathcal{G}^{(0)},c}(y) = y\sigma \, .$$

Lemma 5.3. *$P^1_{\mathcal{M}} = a + b\sigma$ for all simple melons $\mathcal{M}$, where $a, b \in \mathbb{R}$ and a implicitly multiplies the 2×2 identity matrix.*

Proof. For the elementary melon $P^1_{\mathcal{G}(0),c} = y\sigma$, so it is of the claimed form. The lemma follows by induction, taking into account that:

$$\sigma(\alpha+\beta\sigma)^{-1}\sigma = \sigma\Big(\frac{\alpha}{\alpha^2-\beta^2} - \frac{\beta}{\alpha^2-\beta^2}\sigma\Big)\sigma = \frac{\alpha}{\alpha^2-\beta^2} - \frac{\beta}{\alpha^2-\beta^2}\sigma \,.$$

□

Furthermore, as $\sigma(a+b\sigma)\sigma = a+b\sigma$, it follows that the recursion may be rewritten as:

$$P^1_{\mathcal{M}} = y^2 \frac{1}{D+1-\sum_{c=1}^D P^1_{\mathcal{M}^c}} \,,$$

and that all the $P^1_{\mathcal{M}}$ may be diagonalized simultaneously in the basis:

$$\frac{1}{\sqrt{2}}\begin{pmatrix}1\\1\end{pmatrix}, \frac{1}{\sqrt{2}}\begin{pmatrix}1\\-1\end{pmatrix},$$

that is:

$$P^1_{\mathcal{M}} = \frac{1}{2}\begin{pmatrix}1 & 1\\ 1 & -1\end{pmatrix}\begin{pmatrix}\lambda^1_{\mathcal{M};1} & 0\\ 0 & \lambda^1_{\mathcal{M};2}\end{pmatrix}\begin{pmatrix}1 & 1\\ 1 & -1\end{pmatrix}.$$

For the eigenvalues the recursion takes the simple form:

$$\lambda^1_{\mathcal{M};\frac{1}{2}} = \frac{y^2}{D+1-\sum_c \lambda^1_{\mathcal{M}^c;\frac{1}{2}}} \,, \qquad \lambda^1_{\mathcal{G}(0),c;\frac{1}{2}} = \pm y \,,$$

which can be restated in terms of a single $\lambda^1_{\mathcal{M}}(y)$:

$$\lambda^1_{\mathcal{M}}(y) = \frac{y^2}{D+1-\sum_c \lambda^1_{\mathcal{M}^c}(y)} \,, \quad \lambda^1_{\mathcal{G}(0),c}(y) = y \,; \qquad \lambda^1_{\mathcal{M};\frac{1}{2}} = \lambda^1_{\mathcal{M}}(\pm y) \,,$$

which (up to the powers of y and the presence of colors) is the recursion for branched polymers of [111].

Random walks on random graphs

Let us recapitulate where we stand. The generating function of the first return/ first transit probability matrix of a melon is related by a simple equation to the generating function of the return/ transit probability matrix of the melon:

$$P^1_{\mathcal{M}}(y) = \sum_{t\geq 0} y^t\, P^1_{\mathcal{M}}(t) \,, \quad P_{\mathcal{M}}(y) = \sum_{t\geq 0} y^t\, P_{\mathcal{M}}(t) \,, \quad P_{\mathcal{M}}(y) = \frac{1}{1-P^1_{\mathcal{M}}(y)} \,.$$

Furthermore, for simple melons, the generating function of the first return/first transit probability matrix obeys a simple recursion:

$$P^1_{\mathcal{M}} = y^2\sigma \frac{1}{D+1-\sum_{c=1}^D P^1_{\mathcal{M}^c}}\sigma \,, \quad P^1_{\mathcal{G}(0),c}(y) = y\sigma \,,$$

which can be translated into a scalar recursion on the eigenvalues of the first return/first transit probability matrix.

In order to get from this recursion to the average spectral dimension of the emergent melonic geometry we need to perform quite a bit of work [111]. The considerations that follow apply to any statistical ensemble of graphs.

The statistical ensemble. This discussion follows the analysis of [111]. We will discuss nonbipartite graphs (the bipariteness changes only irrelevant details in the discussion below). Our aim is to derive the emergent spectral dimension of some family of connected graphs (like the melonic family for instance). Of course, this family should contain arbitrarily large graphs.

We first consider a graph $\mathcal{G}$ with a fixed number of vertices, n. We assume that no edge starts and ends on the same vertex, but several edges might connect the same two vertices. Let us denote $D_{\mathcal{G}}$ the $n \times n$ Laplacian matrix of the graph $\mathcal{G}$, $\lambda_1 < \lambda_2 < \lambda_3 \dots$ its eigenvalues and $d_1, d_2 \dots$ the dimension of the corresponding eigenspaces. For any connected graph we have $\lambda_1 = 0, d_1 = 1$ hence the return probability of a random walk on $\mathcal{G}$ (averaged over its vertices) is:

$$P_{\mathcal{G}}(t) = \frac{1}{n}\operatorname{Tr}(e^{-tD_{\mathcal{G}}}) = \frac{1}{n}\sum_{i\geq 1} d_i e^{-\lambda_i} \approx \frac{1}{n} + d_2 e^{-t\lambda_2} \,.$$

Let us denote $P_n(t)$ the return probability averaged over the set of graphs with n vertices (that is the average return probability in the canonical ensemble):

$$P_n(t) \approx \frac{1}{n} + \frac{1}{N(n)}\sum_{\mathcal{G}} d_2 e^{-t\lambda_2} \,,$$

where the sum runs over the connected graphs $\mathcal{G}$ with n vertices belonging to our family and $N(n)$ denotes the number of such graphs. The second term above is very complicated, and a direct computation is impossible. However, it is relatively straightforward to propose an ansatz. For small t, that is as long as the walk did not have the time to explore the finite size effects on the graphs, one expects a power law in $t^{-d_S/2}$, while for large t one expects an exponential suppression. Finally, the finite size effects should be felt after a diffusion time that scales with the number of vertices. We are thus led to the ansatz:

$$P_n(t) \approx \frac{1}{n} + \frac{a}{t^{d_S/2}} e^{-\frac{t}{n^{\Delta}}} \,,$$

where Δ is called the gap exponent. It is convenient to pass to the generating function of $P_n(t)$:

$$P_n(y) = \sum_{t\geq 0} y^t P_n(t) \approx \frac{1}{n}\frac{1}{1-y} + \frac{a}{(1-ye^{-n^{-\Delta}})^{1-d_S/2}} \,,$$

where we have used:

$$\sum_t t^{\alpha} x^t \approx \int^{\infty} dt\, t^{\alpha} x^t = \int^{\infty} dt\, t^{\alpha} e^{t\ln[1-(1-x)]}$$

$$\underset{x\to 1}{\sim} \int^{\infty} dt\, t^{\alpha} e^{-t(1-x)} \sim \frac{1}{(1-x)^{1+\alpha}} \,.$$

The spectral dimension is captured by the nonanalytic part of $P_n(y)$. However, care must be taken: for $n \to \infty$ the relevant singularity is at $y = 1$, exactly on top of the simple pole contribution coming from the zero mode of the Laplacian. It is thus quite difficult to extract the spectral dimension in the canonical ensemble.

In order to disentangle the contribution of the pole at $y = 1$ from the physically interesting singularity at $ye^{-n^{-\Delta}} = 1$ one needs to work a bit more. It turns out that the two contributions can be more readily separated in the grand canonical ensemble. If one considers graphs with arbitrarily many vertices the two singularities of $P_n(y)$ have very different fates. The generating function of the average return probability in the grand canonical ensemble is:

$$\mathbf{Q}(z,y) = \sum_n z^n N(n) P_n(y) \approx \frac{1}{1-y} \sum_n \frac{1}{n} z^n N(n) + a \sum_n \frac{z^n N(n) e^{n^{-\Delta}(1-d_S/2)}}{(e^{n^{-\Delta}} - y)^{1-d_S/2}} \ .$$

Note first that the grand canonical ensemble makes sense only if the family of graphs in question is exponentially bounded, that is $N(n) \sim n^{\gamma-2} z_0^{-n}$ for some constant z_0. This is the case for melons for example, for which $\gamma = 1/2$.

The first term in $\mathbf{Q}(z,y)$ behaves like:

$$\frac{1}{1-y}\left(1 - \frac{z}{z_0}\right)^{4-\gamma} ,$$

and survives in the grand canonical ensemble as a simple pole at $y = 1$. The second term behaves very differently for $y \approx 1$. Subtracting the pole and truncating at the first nontrivial order in $1/n$ it becomes:

$$a \sum_n \left(\frac{z}{z_0}\right)^n n^{\gamma-2} \frac{1}{(n^{-\Delta} + 1 - y)^{1-d_S/2}}$$
$$= a \sum_n \left(\frac{z}{z_0}\right)^n n^{\gamma-2+\Delta(1-d_S/2)} \frac{1}{\left[1 + n^{\Delta}(1-y)\right]^{1-d_S/2}} ,$$

which exhibits the asymptotic behavior as $z \to z_0$:

$$\int^{\infty} dx\, e^{-x[1-z/z_0]} x^{\gamma-2+\Delta(1-d_S/2)} \frac{1}{[1 + (1-y)x^{\Delta}]^{1-d_S/2}}$$
$$\sim \left(1 - \frac{z}{z_0}\right)^{1-\gamma-\Delta(1-d_S/2)} \Phi\left(\frac{1-y}{\left(1 - \frac{z}{z_0}\right)^{\Delta}}\right)$$

and the function Φ is analytic for $y = 1$.

In conclusion, in order to determine the spectral dimension of a family of graphs one needs to prove (starting from the study of random walks on the graphs) that the average generating function of the return probability in the grand canonical ensemble is written for $y < 1$ and $z \to z_0$ as:

$$\mathbf{Q}(z,y) = \frac{1}{(1-y)}\left(1 - \frac{z}{z_0}\right)^{\delta} + \left(1 - \frac{z}{z_0}\right)^{\beta} f\left(\frac{(1-y)}{(1 - z/z_0)^{\alpha}}\right) , \qquad (5.6)$$

to identify the exponents β and α and γ (the critical exponent of $N(n)$) and to derive the spectral dimension from the equations:

$$\alpha = \Delta\,, \qquad \beta = 1 - \gamma - \Delta(1 - d_S/2)\,. \tag{5.7}$$

We now apply this procedure to melons.

The spectral dimension of melons. In the grand canonical ensemble of melons the return probability $\mathbf{Q}(z,y)$ is replaced by a two by two return/transit probability matrix,

$$\mathbf{Q}(z,y) = \sum_{\mathcal{M}} z^{k(\mathcal{M})} P_{\mathcal{M}}(y)\,,$$

where the sum runs over all the melons and the number of *simple* melons with k white vertices is:

$$\sum_{\mathcal{M},k(\mathcal{M})=k} = \frac{1}{Dk+1}\binom{Dk+1}{k} \sim k^{\gamma-2} z_0^{-k}\,,$$

with $\gamma = 1/2$ and $z_0 = (D-1)^{D-1}/D^D$. Like $P_{\mathcal{M}}(y)$, the matrix $\mathbf{Q}(z,y)$ is diagonal in the basis $\frac{1}{\sqrt{2}}\begin{pmatrix}1\\1\end{pmatrix}, \frac{1}{\sqrt{2}}\begin{pmatrix}1\\-1\end{pmatrix}$, and we can study only one of its eigenvalues. Recall that:

$$\lambda^1_{\mathcal{M}}(y) = \frac{y^2}{D+1-\sum_{c=1}^{D}\lambda^1_{\mathcal{M}^c}(y)}\,, \qquad \lambda^1_{\mathcal{G}^{(0)},c}(y) = y\,,$$

and denoting by abuse of notation the eigenvalue of $P_{\mathcal{M}}(y)$ by $P_{\mathcal{M}}(y)$ we have:

$$P_{\mathcal{M}}(y) = \frac{1}{1-\lambda^1_{\mathcal{M}}}\,.$$

It is convenient to define:

$$h_{\mathcal{M}}(y) = \frac{1}{(1-y)}\frac{1}{P_{\mathcal{M}}(y)} = \frac{1}{(1-y)}(1-\lambda^1_{\mathcal{M}})$$

and, denoting the corresponding eigenvalue of $\mathbf{Q}(z,y)$ by $Q(z,y)$ we have:

$$Q(z,y) = \sum_{\mathcal{M}} z^{k(\mathcal{M})}\frac{1}{(1-y)}\frac{1}{h_{\mathcal{M}}(y)}\,,$$

$$h_{\mathcal{M}}(y) = \frac{1+y+\sum_{c=1}^{D} h_{\mathcal{M}^c}(y)}{1+(1-y)\sum_{c=1}^{D} h_{\mathcal{M}^c}(y)}\,, \qquad h_{\mathcal{G}^{(0)},c}(y) = 1\,.$$

In order to compute the spectral dimension we need to show that Q takes the form of Eq. (5.6) and identify α and β (recall that $\gamma = 1/2$).

Pole contribution. The function $Q(z,y)$ has a simple pole at $y = 1$. At $y = 1$, the recursion and initial condition become:

$$h_{\mathcal{M}}(1) = 2 + \sum_{c=1}^{D} h_{\mathcal{M}^c}(1) \quad \text{and} \quad h_{\mathcal{G}^{(0)},c}(1) = 1\,,$$

whose solution is $h_{\mathcal{M}}(1) = (D+1)k(\mathcal{M}) + 1$, hence:

$$\sum_{\mathcal{M}} z^{k(\mathcal{M})} \frac{1}{h_{\mathcal{M}}(1)} = \sum_{k} z^k k^{\gamma-3} z_0^{-k} ,$$

which converges for $z < z_0$.

The non pole part. The spectral dimension is captured by the remaining no-pole part of the generating function. In order to identify it, let us define:

$$\widetilde{Q}(z,y) = -\frac{d}{dy}(1-y)Q(z,y) = -\sum_{\mathcal{M}} z^{k(\mathcal{M})} \frac{d}{dy}\frac{1}{h_{\mathcal{M}}(y)}$$
$$= \sum_{n\geq 0} \frac{(y-1)^n}{n!} \widetilde{Q}_n(z) ,$$

where

$$\widetilde{Q}_n(z) = \frac{d^n}{dy^n}\widetilde{Q}(z,y)|_{y=1} = -\sum_{\mathcal{M}} z^{k(\mathcal{M})} \frac{d^{n+1}}{dy^{n+1}} \frac{1}{h_{\mathcal{M}}(y)}\Big|_{y=1} .$$

Remark that from Eq. (5.6), we expect that:

$$\widetilde{Q}(z,y) \sim \left(1 - \frac{z}{z_0}\right)^{\beta} \sum_{p} c_p \left(\frac{1-y}{(1-z/z_0)^{\Delta}}\right)^{p} , \tag{5.8}$$

where the series in p is convergent. Remark that, if $\beta = 0$, the term corresponding to $p = 0$, should be understood as $\ln(1 - z/z_0)$ rather than $\left(1 - \frac{z}{z_0}\right)^0$. Denoting:

$$h_{\mathcal{M}}^{(n)} = \frac{d^n}{dy^n} h_{\mathcal{M}}(y)\Big|_{y=1} ,$$

one can prove that [111, 106]:

$$\widetilde{Q}_n(z) - \sum_{r=1}^{n+1} (-1)^{r+1}\, r! \sum_{\substack{a_1,\dots,a_{n+1} \\ \sum a_j = r\, ;\, \sum j a_j = n+1}}$$
$$\frac{(n+1)!}{(1!)^{a_1}\dots[(n+1)!]^{a_{n+1}} a_1! \dots a_{n+1}!} \sum_{\mathcal{M}} z^{k(\mathcal{M})} \frac{\prod_j \left[h_{\mathcal{M}}^{(j)}\right]^{a_j}}{\left[h_{\mathcal{M}}^{(0)}\right]^{r+1}} .$$

We introduce the notation:

$$H^{(n_1,\dots n_p)}(z) := \sum_{\mathcal{M}} z^{k(\mathcal{M})}\, h_{\mathcal{M}}^{(n_1)} h_{\mathcal{M}}^{(n_2)} \dots h_{\mathcal{M}}^{(n_p)} .$$

In order to derive the asymptotic behavior of $\widetilde{Q}_n(z)$ we first note that:

$$\left((D+1)z\frac{\partial}{\partial z} + 1\right)^{r+1} \left(\sum_{\mathcal{M}} z^{k(\mathcal{M})} \frac{\prod_j (h_{\mathcal{M}}^{(j)})^{a_j}}{(h_{\mathcal{M}}^{(0)})^{r+1}}\right)$$

$$
\begin{aligned}
&= \sum_{\mathcal{M}} z^{k(\mathcal{M})} \Big[(D+1)k(\mathcal{M}) + 1 \Big]^{r+1} \frac{\prod_j \big(h_{\mathcal{M}}^{(j)}\big)^{a_j}}{\big(h_{\mathcal{M}}^{(0)}\big)^{r+1}} \\
&= \sum_{\mathcal{M}} z^{k(\mathcal{M})} \prod_j \big(h_{\mathcal{M}}^{(j)}\big)^{a_j} = H^{(1^{a_1} \dots, (n+1)^{a_{n+1}})} , \qquad i^{a_i} \equiv \underbrace{i, \dots i}_{a_i} .
\end{aligned}
$$

The asymptotic behavior of the term of order r in the expansion of $\widetilde{Q}_n(z)$ can be obtained by integrating $r+1$ times the asymptotic behavior of $H^{(1^{a_1} \dots, (n+1)^{a_{n+1}})}$. This asymptotic behavior is exhibited in the following lemma [106]:

Lemma 5.4. *For all $n_1, \dots n_p$, for $z \to z_0$ the following asymptotic behavior holds:*

$$
\sum_{\mathcal{M}} z^{k(\mathcal{M})} h_{\mathcal{M}}^{(n_1)} h_{\mathcal{M}}^{(n_2)} \dots h_{\mathcal{M}}^{(n_p)} \sim \left(1 - \frac{z}{z_0}\right)^{\frac{1}{2} - p - \frac{3}{2}(n_1 + n_2 + \dots + n_p)} .
$$

In particular

$$
H^{(1^{a_1} \dots, (n+1)^{a_{n+1}})} \sim \left(1 - \frac{z}{z_0}\right)^{\frac{1}{2} - r - \frac{3}{2}(n+1)} .
$$

The proof of this lemma is quite technical and will not be reproduced here. Substituting this leading order behavior and integrating $r+1$ times, one obtains

$$
\widetilde{Q}_0(z) \sim \log(1 - z/z_0) , \qquad \widetilde{Q}_n(z) \sim (1 - z/z_0)^{-\frac{3}{2}n} , \quad \forall n > 0 ,
$$

thus

$$
\tilde{Q}(z, y) \approx a_0 \log(1 - z/z_0) + \sum_{p \geq 1} a_p \left(\frac{1 - y}{(1 - z/z_0)^{3/2}} \right)^p .
$$

Comparing this equation with Eq. (5.8), it follows that $\beta = 0$ and $\alpha = 3/2$. The gap exponent and the spectral dimension are then obtained from Eq. (5.7) as:

$$
\Delta = \alpha = \frac{3}{2} , \qquad d_S = 2 \frac{\Delta + \gamma + \beta - 1}{\Delta} = \frac{4}{3} .
$$

Chapter 6

The universality theorem

In this chapter we initiate the study of random tensors. For pedagogical purposes we will first discuss in some detail random matrices and introduce some of the notions also relevant for tensors. We then discuss the Gaussian distribution for random tensors, and in particular compute its moments to leading order in $\frac{1}{N}$. Finally, we go to the core of this chapter, namely the study of general trace invariant measures.

It turns out that, in order for the large N limit to exist, the cumulants (Definition 2.3) $\mathfrak{K}(\mathcal{B},\mu)$ of a trace invariant measure must scale with N. We show that, assuming that the scaling with N of $\mathfrak{K}(\mathcal{B},\mu)$ is uniform (i.e. depends only on the roughest features of $\mathcal{B}$), a nontrivial large N limit exists only if the cumulants obey a precise scaling bound. We call a measure whose cumulants respect this scaling bound *properly uniformly bounded*. We then show that any properly uniformly bounded trace invariant probability measure becomes Gaussian in the large N limit. This is the *universality theorem* (Theorem 6.3) for tensors, which is the main result of this chapter.

The universality theorem shows that the Gaussian distribution is a very strong attractor for random tensors. In particular, for the case of random matrices such a universality property does not hold. At the end of the day, two important lessons must be drawn from this theorem. First, in order to escape the universality theorem one must consider distributions whose cumulants do not obey a uniform scaling bound. Second, and perhaps more important, in order to discriminate between different tensor measures, one needs to understand in great detail the behavior of the subleading orders in the $1/N$ expansion.

What we do not do in this chapter is to present examples of properly uniformly bounded distributions. Such examples will be presented and discussed at length later on.

6.1 Random matrices

Let us recall Example 2.1. The connected bicolored graphs with $2k$ vertices are cycles with alternating colors. Up to relabelings of the vertices, such graphs are encoded in the couple of permutations $\tau^{(1)}, \tau^{(2)}$:

$$\tau^{(1)} = (1)\dots(k) , \qquad \tau^{(2)} = (1,2,\dots k) ,$$
$$\mathrm{Gr}(\tau^{(1)},\tau^{(2)}) = \Big\{ \{v_j, \bar v_j | j = 1,\dots k\} , \ \bigcup_{c=1}^{2} \{(v_j, \bar v_{\tau^{(c)}(j)}) | j = 1,\dots k\} \Big\} ,$$

with associated trace invariants, Eq. (2.9):

$$\mathrm{Tr}_{\mathrm{Gr}(\tau^{(1)},\tau^{(2)})}(\mathbb{M},\bar{\mathbb{M}}) = \sum_{a,\bar a} \Big(\prod_{j=1}^{k} \delta_{a^1_j \bar a^1_{\tau^{(1)}(j)}} \delta_{a^2_j \bar a^2_{\tau^{(2)}(j)}} \Big) \Big(\prod_{j=1}^{k} \mathbb{M}_{a^1_j a^2_j} \bar{\mathbb{M}}_{\bar a^1_j \bar a^2_j} \Big) =$$
$$= \mathrm{Tr}\Big(\big[\mathbb{M}^\dagger \mathbb{M}\big]^k \Big) .$$

Any invariant function of a generic (i.e. not necessarily Hermitian) matrix can be evaluated starting from these trace invariants, as they fix the spectral measure of $\mathbb{M}^\dagger\mathbb{M}$.

6.1.1 Gaussian distribution of a random matrix

The Gaussian distribution of a non-Hermitian random $N \times N$ matrix $\mathbb{M}$ of covariance 1 is the probability measure for a complex field (see Appendix B):

$$d\mu_{\frac{1}{N}\mathbf{1}^{\{1,2\}}} = e^{-N \sum \mathbb{M}_{a^1a^2} \delta_{a^1\bar a^1} \delta_{a^2 \bar a^2} \bar{\mathbb{M}}_{\bar a^1 \bar a^2}} \prod_{(a^1,a^2)} \Big(N \frac{d\mathbb{M}_{a^1a^2} d\bar{\mathbb{M}}_{a^1a^2}}{2\pi\imath} \Big) . \tag{6.1}$$

The covariance $\frac{1}{N}\mathbf{1}\otimes\mathbf{1} \equiv \frac{1}{N}\mathbf{1}^{\{1,2\}}$ is the identity operator on the N^2 dimensional spaces of $N \times N$ matrices with matrix elements $\frac{1}{N}\delta_{a^{\{1,2\}}\bar a^{\{1,2\}}} \equiv \frac{1}{N}\delta_{a^1\bar a^1}\delta_{a^2\bar a^2}$. The exponent in Eq. (6.1) can alternatively be written in the more familiar form $N\,\mathrm{Tr}(\mathbb{M}^\dagger\mathbb{M})$. Let us first show that the Gaussian expectations of the connected trace invariants obey:

$$\lim_{N\to\infty} \frac{1}{N} \Big\langle \mathrm{Tr}\big[(\mathbb{M}^\dagger\mathbb{M})^k\big] \Big\rangle_{\mu_{\frac{1}{N}\mathbf{1}^{\{1,2\}}}} = \frac{1}{k+1}\binom{2k}{k} .$$

Feynman graphs

In order to compute the Gaussian expectations we will use the Feynman graph representation of Appendix B, adapted to random matrices. We have:

$$\frac{1}{N} \big\langle \mathrm{Tr}\big[(\mathbb{M}^\dagger\mathbb{M})^k\big] \big\rangle_{\mu_{\frac{1}{N}\mathbf{1}^{\{1,2\}}}} =$$
$$= \frac{1}{N} \sum_{a,\bar a} \Big(\prod_{j=1}^{k} \delta_{a^1_j \bar a^1_{\tau^{(1)}(j)}} \delta_{a^2_j \bar a^2_{\tau^{(2)}(j)}} \Big) \Big\langle \prod_{j=1}^{k} \mathbb{M}_{a^1_j a^2_j} \bar{\mathbb{M}}_{\bar a^1_j \bar a^2_j} \Big\rangle_{\mu_{\frac{1}{N}\mathbf{1}^{\{1,2\}}}} ,$$

and from the Definition B.1 of the Gaussian measure:

$$\left\langle \prod_{j=1}^{k} \mathbb{M}_{a_j^1 a_j^2} \bar{\mathbb{M}}_{\bar{a}_j^1 \bar{a}_j^2} \right\rangle_{\mu_{\frac{1}{N}\mathbf{1}^{\{1,2\}}}} = \sum_{\pi \in \mathfrak{S}(k)} \prod_{j=1}^{k} \frac{1}{N} \delta_{a_j^1 \bar{a}^1_{\pi(j)}} \delta_{a_j^2 \bar{a}^2_{\pi(j)}} ,$$

where $\mathfrak{S}(k)$ denotes the set of permutations of k elements. Relabeling the permutation π by $\tau^{(0)}$ we obtain:

$$\frac{1}{N} \sum_{\tau^{(0)} \in \mathfrak{S}} \sum_{a,\bar{a}} \Big(\prod_{j=1}^{k} \delta_{a_j^1 \bar{a}^1_{\tau^{(1)}(j)}} \delta_{a_j^2 \bar{a}^2_{\tau^{(2)}(j)}} \Big) \Big(\prod_{j=1}^{k} \frac{1}{N} \delta_{a_j^1 \bar{a}^1_{\tau^{(0)}(j)}} \delta_{a_j^2 \bar{a}^2_{\tau^{(0)}(j)}} \Big) . \tag{6.2}$$

The sum over $\tau^{(0)}$ is a sum over the pairings of a $\mathbb{M}$ and a $\bar{\mathbb{M}}$. We represent each such pairing as a dashed edge to which we assign the color 0. The sum in Eq. (6.2) becomes then a sum over graphs $\mathcal{G}$ obtained from the graph $\mathrm{Gr}(\tau^{(1)}, \tau^{(2)})$ by adding edges of color 0, that is:

$$\mathcal{G} = \mathrm{Gr}(\tau^{(0)}, \tau^{(1)}, \tau^{(2)}) = \Big\{ \{v_j, \bar{v}_j | j = 1, \dots k\} , \ \bigcup_{c=0}^{2} \{(v_j, \bar{v}_{\tau^{(c)}(j)}) | j = 1, \dots k\} \Big\} .$$

Definition 6.1. *We call a graph with three colors $\mathcal{G}$ a* covering graph *of $\mathcal{B}$ if $\mathcal{G}$ reduces to $\mathcal{B}$ by deleting the edges of color* 0, $\mathcal{G} \smallsetminus \mathcal{E}^0(\mathcal{G}) = \mathcal{B}$.

The Gaussian expectation of $\mathcal{B} = \mathrm{Gr}(\tau^{(1)}, \tau^{(2)})$ is a sum over its covering graphs $\mathcal{G}$. An example of a covering graph $\mathcal{G}$ contributing to the expectation of $\mathrm{Tr}\big[(\mathbb{M}^\dagger \mathbb{M})^3\big]$ is presented in Figure 6.1. The edges of color 0 are drawn outwards such that the colors are encountered in the order 0, 1, 2 when turning clockwise (respectively anti clockwise) around the black (respectively white) vertices. As the edges of color 1 and 2 of $\mathcal{B}$ are

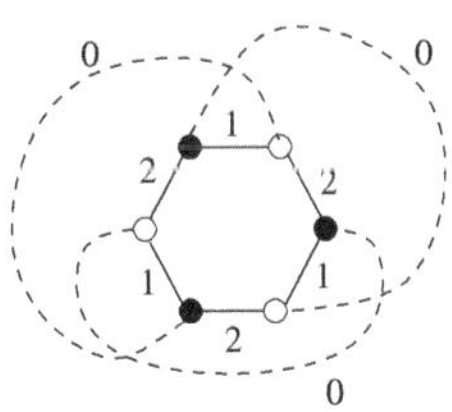

Figure 6.1: A covering graph $\mathcal{G}$ of an observable $\mathcal{B}$.

also the edges of color 1 and 2 of $\mathcal{G}$, we rewrite Eq. (6.2) as:

$$\frac{1}{N} \Big\langle \mathrm{Tr}\big[(\mathbb{M}^\dagger \mathbb{M})^k\big] \Big\rangle_{\mu_{\frac{1}{N}\mathbf{1}^{\{1,2\}}}} = \sum_{\mathcal{G}, \mathcal{G} \smallsetminus \mathcal{E}^0(\mathcal{G}) = \mathcal{B}} \sum_{a,\bar{a}}$$
$$\times \Big(\prod_{c=1}^{2} \prod_{e^c = (v_j, \bar{v}_k) \in \mathcal{E}^c(\mathcal{G})} \delta_{a_j^c \bar{a}_k^c} \Big) \Big(\prod_{e^0 = (v_p, \bar{v}_q) \in \mathcal{E}^0(\mathcal{G})} \frac{1}{N} \delta_{a_p^1 \bar{a}_q^1} \delta_{a_p^2 \bar{a}_q^2} \Big) .$$

In this expression of the Gaussian expectation, the edges of color 0 play a very different role from the edges of colors 1 and 2. While the latter identify one index of the appropriate color, the former not only bring a $1/N$ scaling factor, but also identify *all the indices* of their two end vertices. It follows that an index a^c is identified along an edge of color c, then along an edge of color 0, then along an edge of color c and so on, until the cycle of colors $(0,c)$ closes. Thus we obtain a free sum (yielding a factor N) for every face of colors $(0,c)$. Taking into account that we have exactly k edges of color 0 we get:

$$\frac{1}{N}\Big\langle \operatorname{Tr}\big[(\mathbb{M}^\dagger\mathbb{M})^k\big]\Big\rangle_{\mu_{\frac{1}{N}\mathbb{1}^{\{1,2\}}}} = \sum_{\mathcal{G},\mathcal{G}\setminus\mathcal{E}^0(\mathcal{G})=\mathcal{B}} \frac{1}{N^{k+1}} N^{F^{(0,1)}(\mathcal{G})+F^{(0,2)}(\mathcal{G})} .$$

Remark that the face of colors $(1,2)$, corresponding to the circuit $\mathcal{B}$, that is to the observable itself, does not bring any sum. The graph $\mathcal{G}$ has $2k$ vertices (k black and k white), $3k$ edges (k dashed edges of color 0 and k solid edges for each of the colors 1 and 2) and faces $(F^{(0,1)}(\mathcal{G}) + F^{(0,2)}(\mathcal{G})$ representing free sums and $F^{(1,2)}(\mathcal{G}) = 1$ with no sum).

As $\mathcal{G}$ is a graph with three colors, it is identified with its unique jacket (see Section 3.4.1), that is $\mathcal{G}$ is a combinatorial map corresponding to the cycle $(1,2,3)$ over the colors. The Euler characteristic of $\mathcal{G}$ is:

$$2k - 3k + F^{(0,1)}(\mathcal{G}) + F^{(0,2)}(\mathcal{G}) + 1 = 2 - 2g(\mathcal{G}) ,$$

and the genus of the graph $\mathcal{G}$ is a positive number. The expectation of an invariant is a sum of powers of $1/N$:

$$\frac{1}{N}\Big\langle \operatorname{Tr}\big[(\mathbb{M}^\dagger\mathbb{M})^k\big]\Big\rangle_{\mu_{\frac{1}{N}\mathbb{1}^{\{1,2\}}}} = \sum_{\mathcal{G},\mathcal{G}\setminus\mathcal{E}^0(\mathcal{G})=\mathcal{B}} N^{-2g(\mathcal{G})} ,$$

and higher genera graphs are suppressed. It follows that in the large N limit only graphs $\mathcal{G}$ of genus $g(\mathcal{G}) = 0$ contribute (that is planar graphs). We call such graphs *minimal covering graphs* of $\mathcal{B}$. Equivalently they can be seen as the covering graphs of $\mathcal{B}$ with maximal number of faces $F^{(0,1)}(\mathcal{G}) + F^{(0,2)}(\mathcal{G})$. In the $N \to \infty$ limit we obtain:

$$\lim_{N\to\infty} \frac{1}{N}\Big\langle \operatorname{Tr}\big[(\mathbb{M}^\dagger\mathbb{M})^k\big]\Big\rangle_{\mu_{\frac{1}{N}\mathbb{1}^{\{1,2\}}}} = R_k ,$$

where R_k counts the number of minimal (planar) covering graphs $\mathcal{G}$ of $\mathcal{B}$.

For $k = 1$ there exists a unique covering graph of $\mathcal{B}$. For $k > 1$, we consider the white vertex labeled v_1. The edge of color 0 connects the white vertex v_1 with the black vertex $\bar{v}_{\tau^{(0)}(1)}$. As $\mathcal{G}$ is planar, all the vertices comprised between v_1 and $\bar{v}_{\tau^{(0)}(1)}$ along the cycle $\mathcal{B}$ connect in-between them, as well as the vertices comprised between $\bar{v}_{\tau^{(0)}(1)}$ and v_1, therefore:

$$R_{k+1} = \sum_{p=0}^{k} R_p R_{k-p} ,$$

thus $R_k = \frac{1}{k+1}\binom{2k}{k}$, i.e. R_k are the Catalan numbers. The normalization of the Gaussian is canonical, and not a matter of choice: any other normalization leads either to infinite or to zero expectations in the large N limit.

6.1.2 Invariant probability measures for random matrices

We now consider a trace invariant probability measure ν_N for N^2 random variables, that is a random matrix (see Definition 2.3). In this case $D = 2$ and $\mathcal{D} = \{1, 2\}$. The cumulants of ν_N can be written as:

$$\kappa_{2k}[\mathbb{T}_{a_1^{\mathcal{D}}}, \bar{\mathbb{T}}_{\bar{a}_1^{\mathcal{D}}} \dots \mathbb{T}_{a_k^{\mathcal{D}}}, \bar{\mathbb{T}}_{\bar{a}_k^{\mathcal{D}}}] = \sum_{\mathcal{B},\ k(\mathcal{B})=k} \mathfrak{K}(\mathcal{B}, \nu_N) \prod_{\rho=1}^{C(\mathcal{B})} \delta_{a\bar{a}}^{\mathcal{B}_{(\rho)}}, \tag{6.3}$$

where $\mathcal{B}$ runs over all the bicolored graphs with $2k$ vertices. The graph $\mathcal{B}$ is the disjoint union of its connected components $\mathcal{B}_{(\rho)}$, and each $\mathcal{B}_{(\rho)}$ is a bicolored cycle.

Definition 6.2. *For $D = 2$, the* rescaled cumulants *of ν_N are defined as:*

$$K(\mathcal{B}, \nu_N) \equiv \frac{\mathfrak{K}(\mathcal{B}, \nu_N)}{N^{2-2k(\mathcal{B})-C(\mathcal{B})}},$$

where $k(\mathcal{B})$ is the number of white vertices of $\mathcal{B}$ and $C(\mathcal{B})$ is the number of connected components of $\mathcal{B}$.

When computing the expectation of an invariant, we first expand this expectation into cumulants and then substitute the invariant form of the cumulants, that is we first write the expectation as:

$$\frac{1}{N}\left\langle \mathrm{Tr}\left[(\mathbb{M}^\dagger \mathbb{M})^k\right]\right\rangle_{\nu_N} = \frac{1}{N}\sum_{a,\bar{a}} \delta_{a\bar{a}}^{\mathcal{B}} \sum_{\mathbf{P}} \prod_{\alpha=1}^{|\mathbf{P}|} \kappa_{2k(\alpha)}\left[\mathbb{M}_{a_i^{\mathcal{D}}}, \bar{\mathbb{M}}_{\bar{a}_j^{\mathcal{D}}} | a_i^{\mathcal{D}}, \bar{a}_j^{\mathcal{D}} \in \mathcal{V}(\alpha)\right],$$

where $\mathbf{P}$ runs over the partitions of the set $\mathcal{V} = \{a_j^{\mathcal{D}}, \bar{a}_j^{\mathcal{D}} | j = 1 \dots k\}$ into disjoint bipartite subsets $\mathcal{V}(\alpha)$ (with $|\mathcal{V}(\alpha)| = 2k(\alpha)$), and $|\mathbf{P}|$ denotes the number of parts in the partition. Each of the cumulants $\kappa_{2k(\alpha)}$ is in turn a sum over graphs $\mathcal{B}(\alpha)$ with $2k(\alpha)$ vertices and $C(\mathcal{B}(\alpha))$ connected components. We denote the connected components of $\mathcal{B}(\alpha)$ by $\mathcal{B}_{(\rho)}(\alpha)$, with $\rho = 1, \dots C(\mathcal{B}(\alpha))$, and we get, in terms of rescaled cumulants, the expansion:

$$\kappa_{2k(\alpha)}\left[\mathbb{M}_{a_i^{\mathcal{D}}}, \bar{\mathbb{M}}_{\bar{a}_j^{\mathcal{D}}} | a_i^{\mathcal{D}}, \bar{a}_j^{\mathcal{D}} \in \mathcal{V}(\alpha)\right]$$
$$= \sum_{\mathcal{B}(\alpha),\ k(\mathcal{B}(\alpha))=k(\alpha)} N^{2-2k(\mathcal{B}(\alpha))-C(\mathcal{B}(\alpha))} K(\mathcal{B}(\alpha), \nu_N) \prod_{\rho=1}^{C(\mathcal{B}(\alpha))} \delta_{a\bar{a}}^{\mathcal{B}_{(\rho)}(\alpha)}.$$

The index $\alpha = 1, \dots |\mathbf{P}|$ tracks the cumulant $\kappa_{2k(\alpha)}$ appearing in the expansion of the joint moment. The index $\rho = 1, \dots C(\mathcal{B}(\alpha))$ labels (at fixed $\mathcal{B}(\alpha)$) the connected components $\mathcal{B}_{(\rho)}(\alpha)$ in the expansion of $\kappa_{2k(\alpha)}$ in trace invariants.

When evaluating the expectation of a trace invariant, every term in the sum over $\mathbf{P}$ and $\mathcal{B}(\alpha)$ can be represented by a *doubled graph* $\mathcal{G}$.

Definition 6.3. *A* doubled graph *$\mathcal{G}$ built over $\mathcal{B}$ is obtained as follows:*

- *we draw the observable $\mathcal{B}$. As a matter of convention we flip all the black and all the white vertices of $\mathcal{B}$.*
- *we draw an invariant $\mathcal{B}(\alpha)$ (with connected components $\mathcal{B}_{(\rho)}(\alpha)$) for each $\alpha = 1,\dots|\mathbf{P}|$. Note that $\sum_{\rho=1}^{C(\mathcal{B}(\alpha))} k\big(\mathcal{B}_{(\rho)}(\alpha)\big) = k(\alpha)$ and $\sum_{\alpha=1}^{|\mathbf{P}|} k(\alpha) = k$.*
- *all the original vertices of $\mathcal{B}$ are doubled: every vertex appears once in $\mathcal{B}$ and once in some $\mathcal{B}_{(\rho)}(\alpha)$. We connect every vertex representing a matrix entry $\mathbb{M}$ in $\mathcal{B}$ with the vertex representing* the same matrix entry $\mathbb{M}$ *in the corresponding $\mathcal{B}_{(\rho)}(\alpha)$ by a fictitious dashed edge of color 0.*

Some examples are presented in Figure 6.2. The graph $\mathcal{G}$ is a bipartite 3-colored graph (it is bipartite because we flipped the black and white vertices on $\mathcal{B}$, hence all the edges of color 0 in $\mathcal{G}$ will connect a black and a white vertex).

The sums over partitions $\mathbf{P}$ and invariants $\mathcal{B}(\alpha)$ become a sum over all doubled graphs $\mathcal{G}$ one can build starting from $\mathcal{B}$. We denote the set of these doubled graphs by $\mathrm{Doub}(\mathcal{B})$. The index α takes values $\alpha = 1,\dots|\mathbf{P}|$ and the index ρ takes values, at fixed α, $\rho = 1,\dots C\big(\mathcal{B}(\alpha)\big)$. Observe that if $\mathcal{G} \in \mathrm{Doub}(\mathcal{B})$ then:

$$\mathcal{G} \setminus \mathcal{E}^0(\mathcal{G}) = \mathcal{B} \cup \bigcup_{\alpha=1}^{|\mathbf{P}|} \left(\bigcup_{\rho=1}^{C(\mathcal{B}(\alpha))} \mathcal{B}_{(\rho)}(\alpha) \right) .$$

Starting from a doubled graph $\mathcal{G}$ one readily identifies the observable $\mathcal{B}$, the partition $\mathbf{P}$ and the graphs $\mathcal{B}(\alpha)$ to which it corresponds. The observable $\mathcal{B}$ is the subgraph with colors $1,\dots D$ of $\mathcal{G}$ having no label α while all the other subgraphs with colors $1,\dots D$ of $\mathcal{G}$ represent the various $\mathcal{B}_{(\rho)}(\alpha)$s. The graph $\mathcal{B}(\alpha)$ is the union of the subgraphs with colors $1,\dots D$ sharing the same index α, $\mathcal{B}(\alpha) = \bigcup_\rho \mathcal{B}_{(\rho)}(\alpha)$, hence and $C\big(\mathcal{B}(\alpha)\big)$ is the number of the subgraphs sharing the same index α. Finally, the parts $\mathcal{V}(\alpha)$ of $\mathbf{P}$ are the sets of vertices of $\mathcal{B}(\alpha)$.

This graphical representation applies to all trace invariant measures. We will see in Chapter 7 that the relation between the usual Feynman graphs for perturbed Gaussian measures and these doubled graphs is in fact much more subtle than it might appear at first sight.

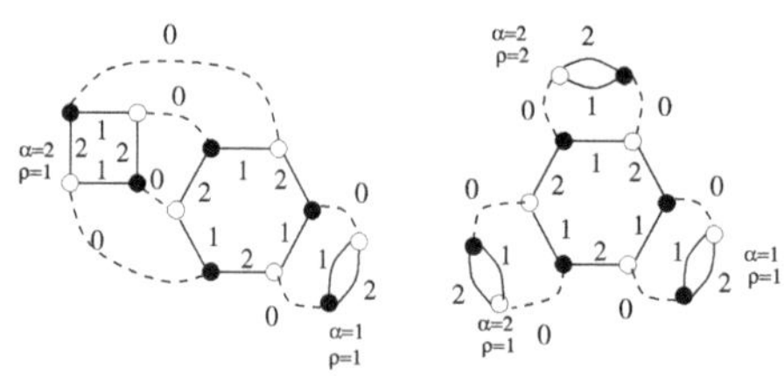

Figure 6.2: Doubled graphs contributing to an observable.

Two doubled graphs contributing to the observable $\mathrm{Tr}\big[(\mathbb{M}^\dagger\mathbb{M})^3\big]$ are given in Figure 6.2. The face $(1,2)$ associated with $\mathcal{B}$ is the one with six vertices, while the faces $(1,2)$ with four and two vertices correspond to various $\mathcal{B}_{(\rho)}(\alpha)$. We include in the figure the labels α and ρ of the various connected components $\mathcal{B}_{(\rho)}(\alpha)$. Thus on the left hand side of Figure 6.2 we represented a contribution from a partition $\mathbf{P}$ of the six vertices into a set with two vertices and a set with four vertices, $k\big(\mathcal{B}(1)\big) = 1$, $k\big(\mathcal{B}(2)\big) = 2$. The invariant for the first set, $\mathcal{B}(1)$ has one connected component $C\big(\mathcal{B}(1)\big) = 1$ with two vertices $k\big(\mathcal{B}_{(1)}(1)\big) = 1$ (that is $\mathcal{B}(1) = \mathcal{B}_{(1)}(1)$). The invariant for the second set, $\mathcal{B}(2)$ has also one connected component $C\big(\mathcal{B}(2)\big) = 1$ but this time with four vertices $k\big(\mathcal{B}_{(1)}(2)\big) = 2$. On the right of Figure 6.2 we presented a contribution coming from *the same* partition $\mathbf{P}$, with $k\big(\mathcal{B}(1)\big) = 1$, $k\big(\mathcal{B}(2)\big) = 2$. The invariant $\mathcal{B}(1)$ has again one connected component $C\big(\mathcal{B}(1)\big) = 1$ with two vertices $k\big(\mathcal{B}_{(1)}(1)\big) = 1$. But this time the invariant $\mathcal{B}(2)$ has two connected components $C\big(\mathcal{B}(2)\big) = 2$, each with two vertices $k\big(\mathcal{B}_{(1)}(2)\big) = 1, k\big(\mathcal{B}_{(2)}(2)\big) = 1$ (and $\mathcal{B}(2) = \mathcal{B}_{(1)}(2) \cup \mathcal{B}_{(2)}(2)$).

To evaluate the contribution of a doubled graph $\mathcal{G}$ to the expectation of an observable one must remember that we first divide the $2k$ vertices of $\mathcal{B}$ among the $|\mathbf{P}|$ parts, and subsequently the $2k(\alpha)$ vertices in each part of $\mathbf{P}$ are further subdivided into $C\big(\mathcal{B}(\alpha)\big)$ connected graphs $\mathcal{B}_{(\rho)}(\alpha)$. As the edges of color 0 connect two copies of the same vertex, the indices of their end points are identical, hence each $e^0 = (v_i, \bar v_j) \in \mathcal{E}^0(\mathcal{G})$ contributes $\delta_{a^1_i \bar a^1_j}\delta_{a^2_i \bar a^2_j}$. The expectation of an invariant observable becomes:

$$
\begin{aligned}
&\frac{1}{N}\nu_N\Big(\mathrm{Tr}\big[(\mathbb{M}^\dagger\mathbb{M})^k\big]\Big)\\
&= \frac{1}{N} \sum_{\substack{\mathcal{G},\ \mathcal{G}\in \mathrm{Doub}(\mathcal{B})\\ \mathcal{G}\setminus\mathcal{E}^0(\mathcal{G}) = \mathcal{B}\cup_{\alpha;\rho}\mathcal{B}_{(\rho)}(\alpha)}} N^{\sum_\alpha\left(2-2k\left(\mathcal{B}(\alpha)\right)-C\left(\mathcal{B}(\alpha)\right)\right)} \prod_\alpha K\big(\mathcal{B}(\alpha),\nu_N\big)\\
&\quad\times \sum_{a,\bar a}\Big(\delta^{\mathcal{B}}_{a\bar a}\prod_{\alpha,\rho}\delta^{\mathcal{B}_{(\rho)}(\alpha)}_{a\bar a}\Big) \prod_{e^0=(v_i,\bar v_j)\in\mathcal{E}^0(\mathcal{G})} \delta_{a^1_i \bar a^1_j}\delta_{a^2_i \bar a^2_j}\ .
\end{aligned}
\tag{6.4}
$$

The total operator $\Big(\delta^{\mathcal{B}}_{a\bar a}\prod_{\alpha,\rho}\delta^{\mathcal{B}_{(\rho)}(\alpha)}_{a\bar a}\Big)$ encodes the invariant associated with all the subgraphs of colors $\{1,\dots D\}$, including both the observable $\mathcal{B}$ and the graphs $\mathcal{B}_{(\rho)}(\alpha)$. Substituting the trace invariant operators $\delta^{\mathcal{B}}_{a\bar a}$ and $\delta^{\mathcal{B}_{(\rho)}(\alpha)}_{a\bar a}$ and taking into account that all the edges of colors 1 and 2 of $\mathcal{G}$ come from either $\mathcal{B}$ or a $\mathcal{B}_{(\rho)}(\alpha)$ we obtain:

$$
\begin{aligned}
&\frac{1}{N} \sum_{\substack{\mathcal{G},\ \mathcal{G}\in \mathrm{Doub}(\mathcal{B})\\ \mathcal{G}\setminus\mathcal{E}^0(\mathcal{G}) = \mathcal{B}\cup_{\alpha;\rho}\mathcal{B}_{(\rho)}(\alpha)}} N^{\sum_\alpha\left(2-2k\left(\mathcal{B}(\alpha)\right)-C\left(\mathcal{B}(\alpha)\right)\right)} \prod_\alpha K\big(\mathcal{B}(\alpha),\nu_N\big)\\
&\quad\times \sum_{a,\bar a}\Big(\prod_{c=1}^{2}\ \prod_{e^c=(v_i,\bar v_j)\in\mathcal{E}^c(\mathcal{G})} \delta_{a^c_i \bar a^c_j}\Big) \prod_{e^0=(v_i,\bar v_j)\in\mathcal{E}^0(\mathcal{G})} \delta_{a^1_i \bar a^1_j}\delta_{a^2_i \bar a^2_j}\ .
\end{aligned}
$$

As in the case of the Gaussian measure treated in Section 6.1.1, the Kronecker δs compose along the faces of colors $(0,1)$ and $(0,2)$ of $\mathcal{G}$, thus:

$$\frac{1}{N}\nu_N\Big(\mathrm{Tr}\big[(\mathbb{M}^\dagger\mathbb{M})^k\big]\Big) = \sum_{\mathcal{G},\,\mathcal{G}\in\mathrm{Doub}(\mathcal{B})} N^{-1-2k+F^{(0,1)}(\mathcal{G})+F^{(0,2)}(\mathcal{G})-\sum_\alpha\left(C\left(\mathcal{B}(\alpha)\right)-2\right)} \times\prod_\alpha K\big(\mathcal{B}(\alpha),\nu_N\big)\,.$$

The doubled graph $\mathcal{G}$ has $4k$ vertices, $2k$ coming from $\mathcal{B}$ and $2k$ coming from all the $\mathcal{B}_{(\rho)}(\alpha)$. It has $1+\sum_\alpha C\big(\mathcal{B}(\alpha)\big)$ faces $(1,2)$, one associated with the observable $\mathcal{B}$, and one for each $\mathcal{B}_{(\rho)}(\alpha)$. Furthermore it has $2k$ edges of color 0, $2k$ edges of color 1 and $2k$ edges of color 2. The Euler characteristic of $\mathcal{G}$ is:

$$4k-6k+1+\sum_\alpha C\big(\mathcal{B}(\alpha)\big)+F^{01}(\mathcal{G})+F^{02}(\mathcal{G}) = 2-2g(\mathcal{G})\,,$$

hence the global scaling with N of a term is $N^{-2g(\mathcal{G})-2\sum_\alpha\left(C\left(\mathcal{B}(\alpha)\right)-1\right)}$. It follows that, assuming that the rescaled cumulants $K\big(\mathcal{B}(\alpha),\nu_N\big)$ are bounded, $\mathcal{G}$ contributes to the expectation of an observable in the large N limit if it is planar and each cumulant $\kappa_{2k(\alpha)}$ contributes exactly one connected invariant $C\big(\mathcal{B}(\alpha)\big)=1$.

In the large N limit only planar graphs $\mathcal{G}$ contribute. However, they involve arbitrary cumulants (as long as the cumulants correspond to connected graphs). Indeed, the only requirement is that the partition $\mathbf{P}$ is *a noncrossing partition*, that is random matrices distributed on ν_N become *free* in the large N limit (see [130]).

6.2 Gaussian distribution for tensors

We are now going to study probability distributions for tensors. We start again with the Gaussian distribution, and then proceed to generic invariant distributions. The graphical representations we introduced for matrices generalize for tensors.

Let us compute the large N limit of the Gaussian expectations:

$$\frac{1}{N}\Big\langle\mathrm{Tr}_{\mathcal{B}}(\mathbb{T},\bar{\mathbb{T}})\Big\rangle_{\mu_{\frac{\sigma^2}{N^{D-1}}1^{\mathcal{D}}}} = \frac{1}{N}\int\prod_{a^{\mathcal{D}}}\Big(\frac{N^{D-1}}{\sigma^2}\frac{d\mathbb{T}_{a^{\mathcal{D}}}d\bar{\mathbb{T}}_{a^{\mathcal{D}}}}{2\pi\imath}\Big) \times e^{-\frac{N^{D-1}}{\sigma^2}\sum_{a^{\mathcal{D}}\bar{a}^{\mathcal{D}}}\mathbb{T}_{a^{\mathcal{D}}}\delta_{a^{\mathcal{D}}\bar{a}^{\mathcal{D}}}\bar{\mathbb{T}}_{\bar{a}^{\mathcal{D}}}}\,\mathrm{Tr}_{\mathcal{B}}(\mathbb{T},\bar{\mathbb{T}})\,. \tag{6.5}$$

As in the case of matrices, such an expectation is evaluated as a sum over contractions. As before, we represent two tensors contracted together (hence contributing a covariance to the expectation) as a dashed edge to which we assign the color 0. The expectation of an observable is then a sum over graphs $\mathcal{G}$, which restrict to $\mathcal{B}$ by erasing the dashed edges of color 0.

Definition 6.4. *A $(D+1)$ colored graph $\mathcal{G}$ is called a* covering graph *of $\mathcal{B}$ if it reduces to $\mathcal{B}$ by erasing the edges of color 0, $\mathcal{G}\smallsetminus\mathcal{E}^0(\mathcal{G})=\mathcal{B}$.*

A covering graph of $\mathcal{B}$ of minimal degree, $\mathcal{G}^{\min}$:

$$\mathcal{G}^{\min} \smallsetminus \mathcal{E}^0(\mathcal{G}^{\min}) = \mathcal{B}\,, \qquad \textit{with} \quad \omega(\mathcal{G}^{\min}) = \min_{\mathcal{G}\,,\mathcal{G}\smallsetminus\mathcal{E}^0(\mathcal{G})=\mathcal{B}} \omega(\mathcal{G})\,,$$

is called a minimal covering graph *of $\mathcal{B}$. We denote $R(\mathcal{B}) \geq 0$ the number of minimal covering graphs of $\mathcal{B}$.*

The convergence order of the observable $\mathcal{B}$, *$\Omega(\mathcal{B})$ is the integer:*

$$\Omega(\mathcal{B}) = \frac{2}{(D-1)!}\omega(\mathcal{G}^{\min}) - \frac{2}{(D-2)!}\omega(\mathcal{B}) \geq 0\,.$$

The fact that the convergence order is positive follows from the remark that, according to Lemma 3.1, $\omega(\mathcal{G}) \geq D\omega(\mathcal{B})$, thus for any covering graph of $\mathcal{B}$:

$$\begin{aligned}\frac{2}{(D-1)!}\omega(\mathcal{G}) - \frac{2}{(D-2)!}\omega(\mathcal{B}) &= \frac{2}{D!}\omega(\mathcal{G}) + \frac{2}{D(D-2)!}\Big[\omega(\mathcal{G}) - D\omega(\mathcal{B})\Big] \\ &\geq \frac{2}{D!}\omega(\mathcal{G}) \geq \frac{2}{(D-1)!}\omega(\mathcal{B}) \geq 0\,. \end{aligned} \tag{6.6}$$

The moments of the Gaussian distribution are computed by the following theorem.

Theorem 6.1. *For any connected D-colored graph $\mathcal{B}$ with $2k(\mathcal{B})$ vertices we have:*

$$\frac{1}{N}\Big\langle \mathrm{Tr}_{\mathcal{B}}(\mathbb{T},\bar{\mathbb{T}})\Big\rangle_{\mu_{\frac{\sigma^2}{N^{D-1}}\mathbf{1}^{\mathcal{D}}}} = \sigma^{2k(\mathcal{B})} \sum_{\mathcal{G},\,\mathcal{G}\smallsetminus\mathcal{E}^0(\mathcal{G})=\mathcal{B}} N^{-\frac{2}{(D-1)!}\omega(\mathcal{G})+\frac{2}{(D-2)!}\omega(\mathcal{B})}\,, \tag{6.7}$$

which is a series in $1/N$. In particular:

$$\lim_{N\to\infty} N^{-1+\Omega(\mathcal{B})}\Big\langle \mathrm{Tr}_{\mathcal{B}}(\mathbb{T},\bar{\mathbb{T}})\Big\rangle_{\mu_{\frac{\sigma^2}{N^{D-1}}\mathbf{1}^{\mathcal{D}}}} = \sigma^{2k(\mathcal{B})}R(\mathcal{B})\,.$$

Proof. By the definition of the Gaussian measure:

$$\frac{1}{N}\Big\langle \mathrm{Tr}_{\mathcal{B}}(\mathbb{T},\bar{\mathbb{T}})\Big\rangle_{\mu_{\frac{\sigma^2}{N^{D-1}}\mathbf{1}^{\mathcal{D}}}} = \frac{1}{N}\sum_{a\bar{a}}\delta^{\mathcal{B}}_{a\bar{a}} \sum_{\tau^{(0)}\in\mathfrak{S}(k)}\prod_{j=1}^{k}\frac{\sigma^2}{N^{D-1}}\delta_{a_j^{\mathcal{D}}\bar{a}^{\mathcal{D}}_{\tau^{(0)}(j)}}\,.$$

Like in the matrix case, the sum becomes a sum over the covering graphs of $\mathcal{B}$. Every face of colors $(0,c)$ in $\mathcal{G}$ brings a free sum (hence a factor N) and every dashed edge generated by the covariance brings a factor $N^{-(D-1)}$ (where we set $\sigma = 1$). The moments of the Gaussian measure become:

$$\begin{aligned}\frac{1}{N}\Big\langle \mathrm{Tr}_{\mathcal{B}}(\mathbb{T},\bar{\mathbb{T}})\Big\rangle_{\mu_{\frac{1}{N^{D-1}}\mathbf{1}^{\mathcal{D}}}} &= \sum_{\mathcal{G},\,\mathcal{G}\smallsetminus\mathcal{E}^0(\mathcal{G})=\mathcal{B}} N^{-1-k(\mathcal{B})(D-1)}N^{\sum_c F^{(0,c)}(\mathcal{G})} \\ &= \sum_{\mathcal{G},\,\mathcal{G}\smallsetminus\mathcal{E}^0(\mathcal{G})=\mathcal{B}} N^{-1-k(\mathcal{B})(D-1)+F(\mathcal{G})-F(\mathcal{B})}\,,\end{aligned}$$

where we used the fact that the faces of colors (c, c') with $c, c' \neq 0$ of $\mathcal{G}$ are exactly the faces of colors (c, c') of $\mathcal{B}$:

$$\sum_c F^{(0,c)}(\mathcal{G}) = F(\mathcal{G}) - F(\mathcal{B}) .$$

The total numbers of faces of $\mathcal{G}$ and $\mathcal{B}$ are computed using Eq. (3.1):

$$F(\mathcal{B}) = (D-1) + \frac{(D-1)(D-2)}{2} k(\mathcal{B}) - \frac{2}{(D-2)!}\, \omega(\mathcal{B}) ,$$
$$F(\mathcal{G}) = D + \frac{D(D-1)}{2} k(\mathcal{G}) - \frac{2}{(D-1)!}\, \omega(\mathcal{G}) ,$$

and, as $k(\mathcal{G}) = k(\mathcal{B})$, we get:

$$\frac{1}{N}\Big\langle \mathrm{Tr}_{\mathcal{B}}(\mathbb{T}, \bar{\mathbb{T}}) \Big\rangle_{\mu_{\frac{1}{N^{D-1}}\mathbb{1}^{\mathcal{D}}}} = \sum_{\mathcal{G},\, \mathcal{G}\setminus\mathcal{E}^0(\mathcal{G}) = \mathcal{B}} N^{-\frac{2}{(D-1)!}\omega(\mathcal{G}) + \frac{2}{(D-2)!}\omega(\mathcal{B})} , \tag{6.8}$$

reproducing Eq. (6.7). The rest of the theorem is trivial.

□

Note that the minimal covering graphs of $\mathcal{B}$ are the covering graphs having the maximal possible number of faces $\sum_c F^{(0,c)}(\mathcal{G})$.

Melonic observables

Theorem 6.1 ensures that in the large N limit the expectation of an observable scales at most like N, and (using Eq. (6.6)) it scales like N only if there exists a melonic graph $\mathcal{G}$ which restricts to $\mathcal{B}$ by erasing the edges of color zero.

Lemma 6.1. *Let $\mathcal{B}$ be a melonic D-colored graph. Then it admits a unique melonic covering graph.*

Proof. We root $\mathcal{B}$ by marking one of its edges (and we root any melonic covering graph $\mathcal{G}$ of $\mathcal{B}$ by marking the same edge). From Section 5.2, $\mathcal{B}$ is represented by a D-ary tree $\mathcal{T}_{\mathcal{B}}$ with $k(\mathcal{B})$ vertices and edges of colors $\{1, \dots D\}$ and $\mathcal{G}$ by a $D+1$-ary tree $\mathcal{T}_{\mathcal{G}}$ with $k(\mathcal{G}) = k(\mathcal{B})$ vertices and edges of colors $\{0, \dots D\}$. Erasing the edges of color 0 in $\mathcal{G}$ comes to erasing the edges and halfedges of color 0 in its associated tree. The only $D+1$-ary tree which reduces to $\mathcal{T}_{\mathcal{B}}$ by this procedure is $\mathcal{T}_{\mathcal{B}}$ decorated by a halfedge of color 0 on each of its vertices.

□

The unique minimal covering graph of a melonic observable $\mathcal{B}$ is obtained by connecting by edges of color 0 the pairs of external vertices of its prime melonic subgraphs. This is readily understood: as $\mathcal{G}$ is melonic, and it stays connected by deleting the edges of color 0, it follows that the cut set of any edge of color 0 is formed by just the edge itself, hence all the melonic subgraphs of $\mathcal{G}$ with external edges of color 0 are empty. The expectation of a melonic observable $\mathcal{B}$ is, in the large N limit:

$$\lim_{N\to\infty} \frac{1}{N} \Big\langle \mathrm{Tr}_{\mathcal{B}}(\mathbb{T}, \bar{\mathbb{T}}) \Big\rangle_{\mu_{\frac{1}{N^{D-1}}1^{\mathcal{D}}}} = 1 \,,$$

reproducing the result of Theorem 6.1 with $\Omega(\mathcal{B}) = 0$ and $R(\mathcal{B}) = 1$.

Taking into account Eq. (6.6) we conclude that the melonic observables are the only observables of convergence order 0 and their expectation at leading order is 1.

Arbitrary observables

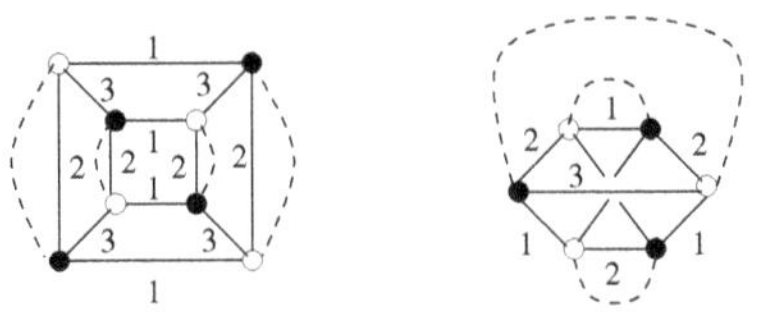

Figure 6.3: Observables of lower order for $D = 3$ and minimal covering graphs.

In the case of matrices, $D = 2$, the minimal covering graphs are exactly the planar graphs with one face of colors 12, which are counted by the Catalan numbers. The situation is more involved for $D \geq 3$. Take for example $D = 3$ and the two invariants depicted in Figure 6.3. They are both of order $\Omega(\mathcal{B}) = 1$ and the number of minimal covering graphs is in both cases $R(\mathcal{B}) = 3$. Lemma 6.1 states that a melonic observable admits a unique minimal covering graph $\mathcal{G}$. The exact counting of the number of minimal covering graphs for more complicated sub families of observables has been performed [28]. However, determining the degree of the minimal covering graphs (hence the order $\Omega(\mathcal{B})$ of an observable), and their number $R(\mathcal{B})$ for an arbitrary observable $\mathcal{B}$ is a difficult problem: we do not know of any systematic way to compute them for a generic $\mathcal{B}$.

However, some necessary conditions for a graph $\mathcal{G}$ to be a minimal covering graph of $\mathcal{B}$ can be given. In particular, an important property of the minimal covering graphs of an arbitrary observable $\mathcal{B}$ is established in Lemma 6.2.

Lemma 6.2. *Let $\mathcal{G}^{\min}$ be a minimal covering graph of the D colored graph $\mathcal{B}$ with D odd (respectively even). Then any two edges of color 0 of $\mathcal{G}^{\min}$, $e_1^0 = (v, \bar{v}), e_2^0 = (w, \bar{w}) \in \mathcal{E}^0(\mathcal{G}^{\min})$ share* at most *$\frac{D-1}{2}$ (respectively $\frac{D}{2}$) faces of colors $(0, c)$ for any c.*

Proof. Let us denote the number of faces of colors $(0, c)$ shared by $e_1^0 = (v, \bar{v})$ and $e_2^0 = (w, \bar{w})$ by q. We build the 4-open graph $\tilde{\mathcal{G}}^{\min}$ obtained from $\mathcal{G}^{\min}$ by cutting the edges e_1^0 and e_2^0 into two pairs of halfedges, as in Figure 6.4.

The boundary graph of $\tilde{\mathcal{G}}^{\min}$, $\partial\tilde{\mathcal{G}}^{\min}$ is a D colored graph with four vertices, hence it necessarily has the structure presented in Figure 6.4 on the right, with q edges connecting v with $\bar{w}$ (respectively $\bar{v}$ and w) and $D - q$ edges connecting v with $\bar{v}$ (respectively $\bar{w}$ and w). Observe that q can be 0.

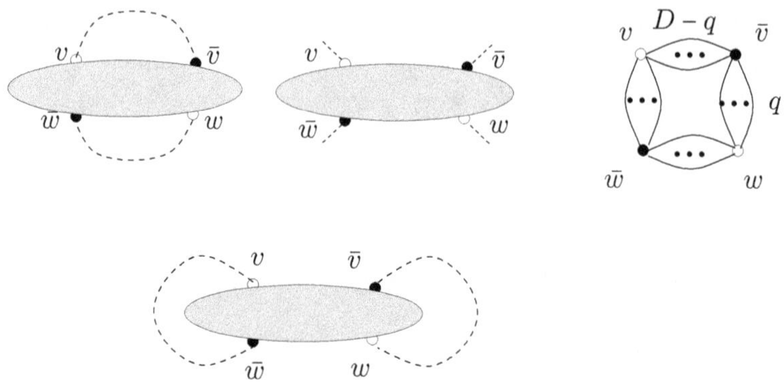

Figure 6.4: A minimal covering graph $\mathcal{G}^{\min}$, the opened graph $\tilde{\mathcal{G}}^{\min}$, the boundary graph $\partial\tilde{\mathcal{G}}^{\min}$ and $\mathcal{G}^{\min,\times}$.

Consider then the graph $\mathcal{G}^{\min,\times}$ obtained from $\tilde{\mathcal{G}}^{\min}$ by reconnecting the halfedges the other way around (forming two new edges of color 0, $(v,\bar{w}),(w,\bar{v})$, like in Figure 6.4 on the second line). It is also a covering graph of $\mathcal{B}$. The numbers of faces of colors $(0,c)$ of $\mathcal{G}^{\min}$ and $\mathcal{G}^{\min,\times}$ are respectively:

$$\sum_c F^{(0,c)}(\mathcal{G}^{\min}) = \sum_c F^{(0,c)}_{\text{int}}(\tilde{\mathcal{G}}^{\min}) + q + 2(D-q) \, ,$$
$$\sum_c F^{(0,c)}(\mathcal{G}^{\min,\times}) = \sum_c F^{(0,c)}_{\text{int}}(\tilde{\mathcal{G}}^{\min}) + D - q + 2q \, ,$$

with $F^{(0,c)}_{\text{int}}(\tilde{\mathcal{G}}^{\min})$ the number of internal faces of colors $(0,c)$ of $\tilde{\mathcal{G}}^{\min}$. As $\mathcal{G}^{\min}$ is a minimal covering graph of $\mathcal{G}$, we have:

$$\sum_c F^{(0,c)}(\mathcal{G}^{\min}) \geq \sum_c F^{(0,c)}(\mathcal{G}^{\min,\times}) \Rightarrow D \geq 2q \, .$$

□

Note that this lemma also holds for $D = 2$.

6.2.1 Uniqueness of the normalization

Theorem 6.2. *The normalization of the Gaussian measure in Eq.* (6.5) *is the* only normalization *such that:*

- *the convergence order* $\Omega(\mathcal{B})$ *is positive for all* $\mathcal{B}$.
- for all $\mathcal{B}$, *there exists an* infinite *family of invariants (graphs* $\mathcal{B}'$*) such that* $\Omega(\mathcal{B}) = \Omega(\mathcal{B}')$.

Proof. Suppose that we choose a normalization of the Gaussian:

$$\Big(\prod_{a^{\mathcal{D}}} N^{\nu} \frac{d\mathbb{T}_{a^{\mathcal{D}}} d\bar{\mathbb{T}}_{a^{\mathcal{D}}}}{2\pi\imath}\Big) e^{-N^{\nu}\sum_{a^{\mathcal{D}}\bar{a}^{\mathcal{D}}} \mathbb{T}_{a^{\mathcal{D}}}\delta_{a^{\mathcal{D}}\bar{a}^{\mathcal{D}}}\bar{\mathbb{T}}_{\bar{a}^{\mathcal{D}}}} ,$$

and denote $\Omega^{(\nu)}(\mathcal{B})$ the order of convergence of $\mathcal{B}$ with this normalization.

We first show that if $\nu \neq D-1$ then either $\Omega^{(\nu)}(\mathcal{B}) \leq 0$ for some $\mathcal{B}$ or there exists a $\mathcal{B}$ such that $\mathcal{B}$ is the only observable whose convergence order is $\Omega^{(\nu)}(\mathcal{B})$. With this normalization of the Gaussian measure, the scaling with N of a term in Eq. (6.7) becomes:

$$N^{k(\mathcal{B})(D-1-\nu)-\frac{2}{(D-1)!}\omega(\mathcal{G})+\frac{2}{(D-2)!}\omega(\mathcal{B})} ,$$

leading to a modified order of convergence:

$$\Omega^{(\nu)}(\mathcal{B}) = \frac{2}{(D-1)!}\omega(\mathcal{G}^{\min}) - \frac{2}{(D-2)!}\omega(\mathcal{B}) - k(\mathcal{B})\big(D-1-\nu\big) .$$

This number is positive for all the melonic observables only if $D-1 \leq \nu$. If $\nu > D-1$, then the order of convergence of the fundamental melon $\mathcal{B}^{(2)}$ is:

$$\Omega^{(\nu)}(\mathcal{B}^{(2)}) = (\nu - D - 1) ,$$

and for any other observable the order of convergence is at least $2(\nu - D - 1)$, hence strictly larger.

Let us now set $\nu = D-1$. The convergence order, (which we will denote $\Omega(\mathcal{B})$, and not $\Omega^{(D-1)}(\mathcal{B})$) is positive for any observable: as we already remarked,

$$\Omega(\mathcal{B}) \geq \frac{2}{(D-1)!}\omega(\mathcal{B}) \geq 0 .$$

In order to establish the second statement, consider any observable $\mathcal{B}$ and the observable $\mathcal{B}'$ obtained by inserting a fundamental melon with $D-1$ colors $\mathcal{B}^{(2),c}$ (formed by two vertices connected by one edge for each color $c' \in \{1,\dots D\} \setminus \{c\}$ and having two external halfedges of color c) on one of the edges of color c of $\mathcal{B}$. Call v and $\bar{v}$ the vertices of this fundamental melon.

Any minimal covering graph $\mathcal{G}'^{\min}$ of $\mathcal{B}'$ is such that the two vertices v and $\bar{v}$ are connected by an edge of color 0. Indeed, any covering graph of $\mathcal{B}'$ in which v and $\bar{v}$ are not connected by an edge of color 0 has two edges of color 0 that share $D-1$ faces of colors $(0,c)$ hence cannot be minimal by virtue of Lemma 6.2. Thus:

$$\Omega(\mathcal{B}') = \frac{2}{(D-1)!} \min_{\substack{\mathcal{G}',\mathcal{G}'\setminus\mathcal{E}^0(\mathcal{G}')=\mathcal{B}' \\ (v,\bar{v})\in\mathcal{E}^0(\mathcal{G}')}} \omega(\mathcal{G}') - \frac{2}{(D-2)!}\omega(\mathcal{B}') .$$

Let us consider the graph $\mathcal{B}'$ and one of its covering graphs $\mathcal{G}'$ such that v and $\bar{v}$ are connected by an edge of color 0 in $\mathcal{G}'$. By deleting v and v' and reconnecting the edges respecting the colors, $\mathcal{B}'$ becomes $\mathcal{B}$ and $\mathcal{G}'$ becomes some covering graph $\mathcal{G}$ of $\mathcal{B}$. All the covering graphs of $\mathcal{B}$ are obtained starting from exactly one such $\mathcal{G}'$.

Moreover, as $\mathcal{B}$ is obtained from $\mathcal{B}'$ by deleting a fundamental melon with $D-1$ colors and $\mathcal{G}$ from $\mathcal{G}'$ by deleting a fundamental melon with D colors, $\omega(\mathcal{B}') = \omega(\mathcal{B})$ and $\omega(\mathcal{G}') = \omega(\mathcal{G})$. Thus:

$$\Omega(\mathcal{B}) = \frac{2}{(D-1)!} \min_{\mathcal{G}, \mathcal{G} \setminus \mathcal{E}^0(\mathcal{G}) = \mathcal{B}} \omega(\mathcal{G}) - \frac{2}{(D-2)!} \omega(\mathcal{B})$$
$$= \frac{2}{(D-1)!} \min_{\substack{\mathcal{G}', \mathcal{G}' \setminus \mathcal{E}^0(\mathcal{G}') = \mathcal{B}' \\ (v,\bar{v}) \in \mathcal{E}^0(\mathcal{G}')}} \omega(\mathcal{G}') - \frac{2}{(D-2)!} \omega(\mathcal{B}') = \Omega(\mathcal{B}') .$$

By inserting fundamental melons with $D-1$ colors arbitrarily on the edges of $\mathcal{B}$ one could build infinitely many graphs $\mathcal{B}'$ with $\Omega(\mathcal{B}') = \Omega(\mathcal{B})$.

□

Different scaling of the Gaussian can make sense only if one decides to look at *subsets* of observables. Consider for instance a tensor with four indices. One can decide to only consider tensor observables in which the tensor effectively behaves like a $N^2 \times N^2$ matrix, that is the indices $(1,2)$ and the indices $(3,4)$ are always contracted between the same tensors. A scaling $\nu = 2$ leads to a well defined large N limit for these observables (this is just the usual large N limit of matrices). However, other tensor observables do not behave well with this scaling: the melonic observables are arbitrarily divergent. The scaling in N^{D-1} of the Gaussian renders *all* the tensor observables convergent in the large N limit.

6.3 Trace invariant tensor measures

Having studied the Gaussian distribution for tensors we now proceed to generic invariant distributions. Recall that the cumulants of such a distribution μ_N are written as:

$$\kappa_{2k}[\mathbb{T}_{a_1^{\mathcal{D}}}, \bar{\mathbb{T}}_{\bar{a}_1^{\mathcal{D}}} \dots \mathbb{T}_{a_k^{\mathcal{D}}}, \bar{\mathbb{T}}_{\bar{a}_k^{\mathcal{D}}}] = \sum_{\mathcal{B},\, k(\mathcal{B})=k} \mathfrak{K}(\mathcal{B}, \mu_N) \prod_{\rho=1}^{C(\mathcal{B})} \delta_{a\bar{a}}^{\mathcal{B}_{(\rho)}} ,$$

where $\mathcal{B}$ runs over all the D-colored graphs with $2k$ labeled vertices and $C(\mathcal{B})$ denotes the number of connected components of $\mathcal{B}$. As already mentioned, the cumulants must scale with N in order for the large N limit to make sense.

Definition 6.5. *We define the* rescaled cumulants *of the distribution* μ_N *as:*

$$K(\mathcal{B}, \mu_N) \equiv \frac{\mathfrak{K}(\mathcal{B}, \mu_N)}{N^{D-2k(\mathcal{B})(D-1)-C(\mathcal{B})}} .$$

Following step by step the discussion of the trace invariant measures for matrices, the expectation of an observable $\mathcal{B}$ with $2k(\mathcal{B})$ vertices is computed by expanding first into cumulants and subsequently substituting the invariant form of the cumulants:

$$\frac{1}{N} \left\langle \mathrm{Tr}_{\mathcal{B}}[\mathbb{T}, \bar{\mathbb{T}}] \right\rangle_{\mu_N} = \frac{1}{N} \sum_{a,\bar{a}} \delta_{a\bar{a}}^{\mathcal{B}} \sum_{\mathbf{P}} \prod_{\alpha=1}^{|\mathbf{P}|} \kappa_{2k(\alpha)}\left[\mathbb{T}_{a_i^{\mathcal{D}}}, \bar{\mathbb{T}}_{\bar{a}_j^{\mathcal{D}}} | a_i^{\mathcal{D}}, \bar{a}_j^{\mathcal{D}} \in \mathcal{V}(\alpha)\right] ,$$

where $\mathbf{P}$ runs over the partitions of the set $\mathcal{V} = \{a_j^{\mathcal{D}}, \bar{a}_j^{\mathcal{D}} | j = 1 \dots k\}$ into disjoint bipartite subsets $\mathcal{V}(\alpha)$ (with $|\mathcal{V}(\alpha)| = 2k(\alpha)$), and $|\mathbf{P}|$ denotes the number of parts in the partition. Again, each of the cumulants $\kappa_{2k(\alpha)}$ is a sum over graphs $\mathcal{B}(\alpha)$ with $2k(\alpha)$ vertices and $C(\mathcal{B}(\alpha))$ connected components. As before, we denote the connected components of $\mathcal{B}(\alpha)$ by $\mathcal{B}_{(\rho)}(\alpha)$, with $\rho = 1, \dots C(\mathcal{B}(\alpha))$, and we get, in terms of rescaled cumulants, the expansion:

$$\kappa_{2k(\alpha)}\left[\mathbb{T}_{a_i^{\mathcal{D}}}, \bar{\mathbb{T}}_{\bar{a}_j^{\mathcal{D}}} | a_i^{\mathcal{D}}, \bar{a}_j^{\mathcal{D}} \in \mathcal{V}(\alpha)\right]$$
$$= \sum_{\substack{\mathcal{B}(\alpha) \\ k\left(\mathcal{B}(\alpha)\right) = k(\alpha)}} N^{D - 2k\left(\mathcal{B}(\alpha)\right)(D-1) - C\left(\mathcal{B}(\alpha)\right)} K\left(\mathcal{B}(\alpha), \mu_N\right) \prod_{\rho=1}^{C\left(\mathcal{B}(\alpha)\right)} \delta_{a\bar{a}}^{\mathcal{B}_{(\rho)}(\alpha)},$$

where again the index $\alpha = 1, \dots |\mathbf{P}|$ tracks the cumulant $\kappa_{2k(\alpha)}$ and the index $\rho = 1, \dots C(\mathcal{B}(\alpha))$ labels the connected components $\mathcal{B}_{(\rho)}(\alpha)$ of $\mathcal{B}(\alpha)$.

The moments of trace invariant measures

As in the case of the invariant measures for matrices, the expectation of a trace invariant is evaluated as a sum over doubled graphs $\mathcal{G}$ generalizing the ones of Definition 6.3.

Definition 6.6. *A* doubled graph $\mathcal{G}$ *built over* $\mathcal{B}$ *is obtained as follows:*

- *we draw the observable* $\mathcal{B}$ *and flip all the black and all the white vertices.*
- *we draw an invariant* $\mathcal{B}(\alpha)$ *with connected components* $\mathcal{B}_{(\rho)}(\alpha)$ *for each* $\alpha = 1, \dots |\mathbf{P}|$*. We have* $\sum_{\rho=1}^{C(\mathcal{B}(\alpha))} k(\mathcal{B}_{(\rho)}(\alpha)) = k(\alpha)$ *and* $\sum_{\alpha=1}^{|\mathbf{P}|} k(\alpha) = k$.
- *we connect every vertex representing a tensor entry* $\mathbb{T}$ *in* $\mathcal{B}$ *with the vertex representing* the same tensor entry $\mathbb{T}$ *in the corresponding* $\mathcal{B}_{(\rho)}(\alpha)$ *by a dashed edge of color* 0.

Again the subgraphs with colors $1 \dots D$ of a doubled graph $\mathcal{G}$ fall in two categories: one of them is $\mathcal{B}$ (and has no label α), while the others, $\mathcal{B}_{(\rho)}(\alpha)$ correspond to the various cumulants. The edges of color 0 of $\mathcal{G}$ connect two copies of the same vertex, hence the indices of their end vertices are identified. Eq. (6.4) generalizes for arbitrary D to:

$$\frac{1}{N} \left\langle \mathrm{Tr}_{\mathcal{B}}(\mathbb{T}, \bar{\mathbb{T}}) \right\rangle_{\mu_N}$$
$$= \frac{1}{N} \sum_{\substack{\mathcal{G},\, \mathcal{G} \in \mathrm{Doub}(\mathcal{B}) \\ \mathcal{G} \setminus \mathcal{E}^0(\mathcal{G}) = \mathcal{B} \cup_{\alpha;\rho} \mathcal{B}_{(\rho)}(\alpha)}} N^{\sum_\alpha \left(D - 2k\left(\mathcal{B}(\alpha)\right)(D-1) - C\left(\mathcal{B}(\alpha)\right)\right)} \prod_\alpha K\left(\mathcal{B}(\alpha), \mu_N\right)$$
$$\times \sum_{a, \bar{a}} \left(\delta_{a\bar{a}}^{\mathcal{B}} \prod_{\alpha, \rho} \delta_{a\bar{a}}^{\mathcal{B}_{(\rho)}(\alpha)} \right) \prod_{e^0 = (v_i, \bar{v}_j) \in \mathcal{E}^0(\mathcal{G})} \left(\prod_{c=1}^{D} \delta_{a_i^c \bar{a}_j^c} \right),$$

which becomes, by substituting the trace invariant operators:

$$\frac{1}{N} \sum_{\substack{\mathcal{G},\, \mathcal{G}\in\mathrm{Doub}(\mathcal{B}) \\ \mathcal{G}\setminus\mathcal{E}^0(\mathcal{G})=\mathcal{B}\cup_{\alpha\rho}\mathcal{B}_{(\rho)}(\alpha)}} N^{\sum_\alpha\left(D-2k\left(\mathcal{B}(\alpha)\right)(D-1)-C\left(\mathcal{B}(\alpha)\right)\right)} \prod_\alpha K\big(\mathcal{B}(\alpha),\mu_N\big)$$

$$\times \sum_{a,\bar{a}} \Big(\prod_{c=1}^{D} \prod_{e^c=(v_i,\bar{v}_j)\in\mathcal{E}^c(\mathcal{G})} \delta_{a_i^c \bar{a}_j^c} \Big) \prod_{e^0=(v_i,\bar{v}_j)\in\mathcal{E}^0(\mathcal{G})} \Big(\prod_{c=1}^{D} \delta_{a_i^c \bar{a}_j^c} \Big) .$$

Like in the case of random matrices, the Kronecker δs compose along the faces with colors $(0,c)$, and the expectation becomes:

$$\frac{1}{N}\big\langle \mathrm{Tr}_{\mathcal{B}}(\mathbb{T},\bar{\mathbb{T}}) \big\rangle_{\mu_N} = \tag{6.9}$$

$$\sum_{\mathcal{G},\, \mathcal{G}\in\mathrm{Doub}(\mathcal{B})} N^{-1-2(D-1)k(\mathcal{B})+\sum_\alpha\left[D-C\left(\mathcal{B}(\alpha)\right)\right]+\sum_{c=1}^{D} F^{(0,c)}(\mathcal{G})} \prod_\alpha K\big(\mathcal{B}(\alpha),\mu_N\big) .$$

As all the faces of colors (c,c') with $c,c'\neq 0$ of $\mathcal{G}$ belong either to $\mathcal{B}$ or to a $\mathcal{B}_{(\rho)}(\alpha)$, the number of faces of color $(0,c)$ of $\mathcal{G}$ is:

$$\sum_{c=1}^{D} F^{(0,c)}(\mathcal{G}) = F(\mathcal{G}) - F(\mathcal{B}) - \sum_{\alpha,\rho} F\big(\mathcal{B}_{(\rho)}(\alpha)\big).$$

The total numbers of faces of $\mathcal{G}$, $\mathcal{B}$ and $\mathcal{B}_{(\rho)}(\alpha)$ are computed using Eq. (3.1):

$$\begin{aligned}
F(\mathcal{G}) &= D + \frac{D(D-1)}{2}k(\mathcal{G}) - \frac{2}{(D-1)!}\,\omega(\mathcal{G}) ,\\
F(\mathcal{B}) &= (D-1) + \frac{(D-1)(D-2)}{2}k(\mathcal{B}) - \frac{2}{(D-2)!}\,\omega(\mathcal{B}) ,\\
F\big(\mathcal{B}_{(\rho)}(\alpha)\big) &= (D-1) + \frac{(D-1)(D-2)}{2}k\big(\mathcal{B}_{(\rho)}(\alpha)\big) - \frac{2}{(D-2)!}\,\omega\big(\mathcal{B}_{(\rho)}(\alpha)\big) ,
\end{aligned}$$

therefore, taking into account that $k(\mathcal{G}) = 2k(\mathcal{B})$ and $k(\mathcal{B}) = \sum_{\alpha,\rho} k\big(\mathcal{B}_{(\rho)}(\alpha)\big)$, we obtain:

$$\begin{aligned}
\sum_{c=1}^{D} F^{(0,c)}(\mathcal{G}) &= 1 + 2(D-1)k(\mathcal{B}) - \sum_{\alpha,\rho}(D-1)\\
&\quad - \left[\frac{2}{(D-1)!}\,\omega(\mathcal{G}) - \frac{2}{(D-2)!}\,\omega(\mathcal{B}) - \sum_{\rho,\alpha} \frac{2}{(D-2)!}\,\omega\big(\mathcal{B}_{(\rho)}(\alpha)\big) \right] ,
\end{aligned}$$

hence finally:

$$\frac{1}{N}\big\langle \mathrm{Tr}_{\mathcal{B}}(\mathbb{T},\bar{\mathbb{T}}) \big\rangle_{\mu_N} = \sum_{\mathcal{G},\, \mathcal{G}\in\mathrm{Doub}(\mathcal{B})} \prod_\alpha K\big(\mathcal{B}(\alpha),\mu_N\big) \tag{6.10}$$

$$\times N^{-D\sum_\alpha \left[C\left(\mathcal{B}(\alpha)\right)-1\right]-\left[\frac{2}{(D-1)!}\,\omega(\mathcal{G})-\frac{2}{(D-2)!}\,\omega\left(\mathcal{B}\right)-\sum_{\rho,\alpha}\frac{2}{(D-2)!}\,\omega\left(\mathcal{B}_{(\rho)}(\alpha)\right)\right]} .$$

As $\mathcal{B}$ and $\mathcal{B}_{(\rho)}(\alpha)$ are all the subgraphs (bubbles) of colors $\{1,\dots D\}$ of the graph $\mathcal{G}$, using Lemma 3.1 the explicit scaling with N of a doubled graph $\mathcal{G}$ in Eq. (6.10) is bounded by:

$$N^{-\frac{2}{D!}\omega(\mathcal{G})-D\sum_\alpha\left[C\left(\mathcal{B}(\alpha)\right)-1\right]} . \tag{6.11}$$

6.3.1 Universality for random tensors

Let us consider a measure μ_N and the behavior of the expectations of the trace invariants as N becomes large. In this limit the expectations can exhibit three kinds of behavior:

- the order of convergence of the observables is not bounded from below. In this case one obtains arbitrarily divergent expectations in the large N limit, and we call them *divergent probability measure.*
- the order of convergence is bounded from below, and only a finite number of observables arise at any given order. We call them *convergent probability measures.*
- the order of convergence is bounded from below and at each order infinitely many observables contribute. We call them *marginal probability measures.*

The division into divergent convergent and marginal probability measures is inspired by physics and parallels the classification of field theories into non renormalizable, super renormalizable and just renormalizable. Among the three possible behaviors we are mostly interested in the marginal probability measures: not much can be said about the divergent distributions, while the convergent ones are trivial. The distributions with an interesting large N limit are the marginal ones.

Note that we have already used this criterion in Theorem 6.2 to identify the canonical scaling of the Gaussian measure.

Definition 6.7. *We say that the trace invariant probability distribution μ_N is* properly uniformly bounded *at large N if the two point rescaled cumulant converges to a finite limit as N goes to infinity:*

$$\lim_{N\to\infty} K(\mathcal{B}^{(2)},\mu_N) = K(\mathcal{B}^{(2)}) < \infty ,$$

and all the other rescaled cumulants are bounded for N large enough:

$$K(\mathcal{B},\mu_N) \le K(\mathcal{B}) ,$$

for some constants $K(\mathcal{B})$.

Definition 6.8. *We say that a random tensor $\mathbb{T}$ distributed with the probability measure μ_N* converges in distribution *to the distributional limit of a Gaussian tensor model of* *covariance* σ^2 *if the large* N *limit of the expectation of any connected trace invariant equals the large* N *Gaussian expectation of the invariant:*

$$\lim_{N\to\infty} N^{-1+\Omega(\mathcal{B})}\left\langle \mathrm{Tr}_{\mathcal{B}}(\mathbb{T},\bar{\mathbb{T}})\right\rangle_{\mu_N} = \sigma^{2k(\mathcal{B})}R(\mathcal{B})\,,$$

with $\Omega(\mathcal{B})$ *and* $R(\mathcal{B})$ *of Definition 6.4.*

The main result of this chapter is that in the large N limit all properly uniformly bounded trace invariant distributions for random tensors become Gaussian. Before proving this result, let us discuss its implications. We have seen that for properly uniformly bounded distributions for random matrices, in the large N limit only planar doubled graphs contribute. However, planar doubled graphs can be arbitrarily complicated and receive contributions from cumulants at arbitrary order. The planar graphs are the subset of doubled graphs, representing all the spherical surfaces. For tensors, the large N limit restricts to melonic graphs that form a more restricted family of graphs than the planar ones (and in particular involves only the two point cumulants). While, fortunately, the melonic graphs represent spheres in any dimension, they are only a subclass of triangulations of the sphere. It should also be noted that, while Gaussian, the large N limit of tensors is nontrivial: the large N covariance $K(\mathcal{B}^{(2)})$ strongly depends on the details of the joint distribution at finite N, and exhibits critical and even multicritical behaviors [31].

Theorem 6.3 (The universality theorem). *Consider* N^D *random variables* $\mathbb{T}_{a^{\mathcal{D}}}$ *whose joint distribution is trace invariant with covariance* $K(\mathcal{B}^{(2)},\mu_N)$ *and is properly uniformly bounded. Then in the large* N *limit the tensor* $\mathbb{T}_{a^{\mathcal{D}}}$ *converges in distribution to a Gaussian tensor of covariance* $K(\mathcal{B}^{(2)}) = \lim_{N\to\infty} K(\mathcal{B}^{(2)},\mu_N)$.

Proof. We start from the expansion of the expectations in doubled graphs, Eq. (6.10) and (6.11). We can restrict from the onset to doubled graphs in which every cumulant $\mathcal{B}(\alpha)$ has only one connected component (as all the other doubled graphs are strictly suppressed with respect to one member of this family). This connect component is labeled $\mathcal{B}_1(\alpha)$.

Melonic observables: direct computation

We first discuss the case when $\omega(\mathcal{G}) = 0$, which implies $\omega(\mathcal{B}) = 0, \omega\big(\mathcal{B}_{(1)}(\alpha)\big) = 0$, hence $\mathcal{B}$ and $\mathcal{B}_{(1)}(\alpha)$ are melonic. Furthermore $\mathcal{G}$ has $4k(\mathcal{B})$ vertices, $2k(\mathcal{B})$ of them belonging to $\mathcal{B}$ and the other $2k(\mathcal{B})$ belonging to the invariants $\mathcal{B}_{(1)}(\alpha)$ and all edges of color 0 connect some vertex in $\mathcal{B}$ with its image in one of the $\mathcal{B}_{(1)}(\alpha)$s.

Every melonic graph can be represented as a tree. We denote $\mathcal{T}_{\mathcal{B}}$ the tree associated to $\mathcal{B}$ rooted at some edge (having thus $k(\mathcal{B})$ tree vertices), and $\mathcal{T}_{\mathcal{G}}$ the tree associated to $\mathcal{G}$ rooted at the same edge (having thus $2k(\mathcal{B})$ tree vertices). All the edges of color 0 in $\mathcal{G}$ connect a vertex of $\mathcal{B}$ with a vertex of some $\mathcal{B}_{(1)}(\alpha)$. It follows that all the tree edges of color 0 in $\mathcal{T}_{\mathcal{G}}$ must connect a tree vertex belonging to $\mathcal{T}_{\mathcal{B}}$ to some tree vertex not belonging to $\mathcal{T}_{\mathcal{B}}$. Thus $\mathcal{T}_{\mathcal{G}}$ is necessarily $\mathcal{T}_{\mathcal{B}}$ decorated by an edge of color 0 (ending in a tree vertex) on all its vertices, hence it is unique. Furthermore, all the $\mathcal{B}_{(1)}(\alpha)$

are represented by only one tree vertex with no descendants in $\mathcal{T}_{\mathcal{G}}$, hence are just the fundamental melon $\mathcal{B}^{(2)}$ representing the two point cumulant.

The expectation of the melonic observables is then:

$$\lim_{N\to\infty} N^{-1+\Omega(\mathcal{B})} \langle \mathrm{Tr}_{\mathcal{B}}(\mathbb{T},\bar{\mathbb{T}}) \rangle_{\mu_N} = \Big(\lim_{N\to\infty} K(\mathcal{B}^{(2)},\mu_N) \Big)^{k(\mathcal{B})} .$$

This result can also be obtained as follows. As all the edges of color 0 connect a vertex of $\mathcal{B}$ with a vertex of some $\mathcal{B}_{(1)}(\alpha)$ and $\mathcal{G}$ is melonic it follows that for every edge of color 0 in $\mathcal{G}$, its cut set must have exactly two edges. Furthermore, one of the two open components of any cut set does not contain any vertex of $\mathcal{B}$, hence it is not incident to any edge of color 0. Thus this component is one of the $\mathcal{B}_{(1)}(\alpha)$s, and it cannot have more than two vertices (as otherwise it would be incident to some edges of color 0). Removing the pairs of edges of color 0 in cut sets and the corresponding fundamental melons $\mathcal{B}_{(1)}(\alpha)$ one must obtain $\mathcal{B}$.

Arbitrary observables

Consider now the expectation of an arbitrary observable $\mathcal{B}$. Taking into account that only doubled graphs in which each cumulant contributes a connected invariant survive in the large N limit, the relevant terms in the expectation are:

$$\frac{1}{N} \langle \mathrm{Tr}_{\mathcal{B}}(\mathbb{T},\bar{\mathbb{T}}) \rangle_{\mu_N} \\ = \sum_{\mathcal{G},\, \mathcal{G}\in \mathrm{Doub}(\mathcal{B})} N^{-\left[\frac{2}{(D-1)!}\,\omega(\mathcal{G}) - \sum_\alpha \frac{2}{(D-2)!}\,\omega\left(\mathcal{B}_{(1)}(\alpha)\right)\right]} \prod_\alpha K\big(\mathcal{B}_{(1)}(\alpha),\mu_N\big) ,$$

where $\mathcal{B}_{(1)}(\alpha)$ is the unique connected component of $\mathcal{B}(\alpha)$.

The doubled graphs $\mathcal{G} \in \mathrm{Doub}(\mathcal{B})$ fall in two categories:

- either $\mathcal{G}$ is some covering graph $\mathcal{G}'$ of $\mathcal{B}$ decorated by a fundamental melon $\mathcal{B}^{(2)}$ on each edge of color 0:
$$\mathcal{G} = \mathcal{G}' \cup \bigcup_{\mathcal{E}^0(\mathcal{G}')} \mathcal{B}^{(2)} .$$
They correspond to partitions $\mathbf{P}$ in two element subsets $|\mathcal{V}(\alpha)| = 2$. As only such partitions contribute in the case of the Gaussian measure, we call them *Gaussian contributions*.

- or $\mathcal{G}$ is not of this type, that is it involves at least one cumulant having more than two points (the partition $\mathbf{P}$ has at least a set with strictly more than two elements: $\exists\alpha, |\mathcal{V}(\alpha)| \geq 4$). We call them *nonGaussian contributions.*

Gaussian contributions: Every cumulant is the fundamental melon, $\mathcal{B}_{(1)}(\alpha) = \mathcal{B}^{(2)}$ (hence $\omega\big(\mathcal{B}_{(1)}(\alpha)\big) = 0$). The graph $\mathcal{G}' \cup \bigcup_{\mathcal{E}^0(\mathcal{G}')} \mathcal{B}^{(2)}$ has $\frac{D(D-1)}{2} k(\mathcal{G}')$ extra faces with respect to $\mathcal{G}'$ (all the faces of colors $0 < c < c'$ made by couples of edges of the various $\mathcal{B}^{(2)}$ insertions) and $k(\mathcal{G}')$ extra white vertices (one for each $\mathcal{B}^{(2)}$), hence by Eq. (3.1)

$\omega(\mathcal{G}' \cup \bigcup_{\mathcal{E}^0(\mathcal{G}')} \mathcal{B}^{(2)}) = \omega(\mathcal{G}')$. The Gaussian contributions are further subdivided into the contributions corresponding to minimal covering graphs $\mathcal{G}^{\min}$ and the other covering graphs, $\mathcal{G}^{\text{n. min}}$. Separating these terms among the terms contributing to the expectation we get:

$$\begin{aligned} &\frac{1}{N}\left\langle \mathrm{Tr}_{\mathcal{B}}(\mathbb{T},\bar{\mathbb{T}})\right\rangle_{\mu_N} \\ &= \sum_{\mathcal{G}^{\min},\mathcal{G}^{\min}\setminus\mathcal{E}^0(\mathcal{G}^{\min})=\mathcal{B}} \Big[K(\mathcal{B}^{(2)},\mu_N)\Big]^{k(\mathcal{B})} N^{-\frac{2}{(D-1)!}\omega(\mathcal{G}^{\min})+\frac{2}{(D-2)!}\omega(\mathcal{B})} \\ &+ \sum_{\mathcal{G}^{\text{n. min}},\mathcal{G}^{\text{n. min}}\setminus\mathcal{E}^0(\mathcal{G}^{\text{n. min}})=\mathcal{B}} \Big[K(\mathcal{B}^{(2)},\mu_N)\Big]^{k(\mathcal{B})} N^{-\frac{2}{(D-1)!}\omega(\mathcal{G}^{\text{n. min}})+\frac{2}{(D-2)!}\omega(\mathcal{B})} \\ &+ \text{nonGaussian}\,. \end{aligned}$$

As $\omega(\mathcal{G}^{\text{n. min}}) > \omega(\mathcal{G}^{\min})$, provided that the nonGaussian terms are strictly smaller in $1/N$, we obtain (recall the Definition 6.4):

$$\lim_{N\to\infty} N^{-1+\Omega(\mathcal{B})}\left\langle \mathrm{Tr}_{\mathcal{B}}(\mathbb{T},\bar{\mathbb{T}})\right\rangle_{\mu_N} = \Big[\lim_{N\to\infty} K(\mathcal{B}^{(2)},\mu_N)\Big]^{k(\mathcal{B})} R(\mathcal{B})\,,$$

reproducing large N expectations of the Gaussian distribution of covariance $K(\mathcal{B}^{(2)}) = \lim_{N\to\infty} K(\mathcal{B}^{(2)},\mu_N)$.

NonGaussian contributions. In order to conclude we prove that the nonGaussian contributions are strictly suppressed with respect to the Gaussian ones. The doubled graph $\mathcal{G}$ has at least a higher order cumulant $\mathcal{B}_{(1)}(\alpha) \neq \mathcal{B}^2$. It is convenient to go back to Eq. (6.9), and write the scaling with N of such a contribution as:

$$-1 - 2(D-1)k(\mathcal{B}) + \sum_{\alpha}[D-1] + \sum_{c=1}^{D} F^{(0,c)}(\mathcal{G})\,,$$

where we restricted again to the case $C\big(\mathcal{B}(\alpha)\big) = 1$, as the other cases are strictly suppressed.

Consider two edges of color 0, $(v,\bar{a})$ and $(a,\bar{v})$ hooked to two vertices v and $\bar{v}$ connected by an edge of color c of $\mathcal{B}_{(1)}(\alpha)$. We will compare the scaling of $\mathcal{G}$ with the one of the graph $\tilde{\mathcal{G}}$ obtained by reconnecting the two edges of color 0 into an edge of color 0, namely $(a,\bar{a})$, with a $\mathcal{B}^{(2)}$ insertion, and reconnecting all the other edges hooked to v and $\bar{v}$ respecting the colors[1] (see Figure 6.5). We consider the two point subgraph $\mathcal{B}^{(2)}$ as coming from a different cumulant in $\tilde{\mathcal{G}}$.

The graph $\tilde{\mathcal{G}}$ is also a doubled graph in $\mathrm{Doub}(\mathcal{B})$. The number of sets in the partition $\tilde{\mathbf{P}}$ associated with $\tilde{\mathcal{G}}$ is the number of sets in $\mathbf{P}$ plus 1. The number of faces respects:

$$\sum_{c} F^{(0,c)}(\tilde{\mathcal{G}}) \geq \sum_{c} F^{(0,c)}(\mathcal{G}) - (D-1)\,,$$

[1] If there are several edges (of colors different from 0) connecting v and $\bar{v}$ in $\mathcal{G}$, we delete them. If $\mathcal{B}_{(1)}(\alpha)$ divides in several connected components under this procedure, we associate a different label α to each of them (i.e. we consider each of them as coming from a different cumulant in $\tilde{\mathcal{G}}$). Both cases give strictly subleading contributions.

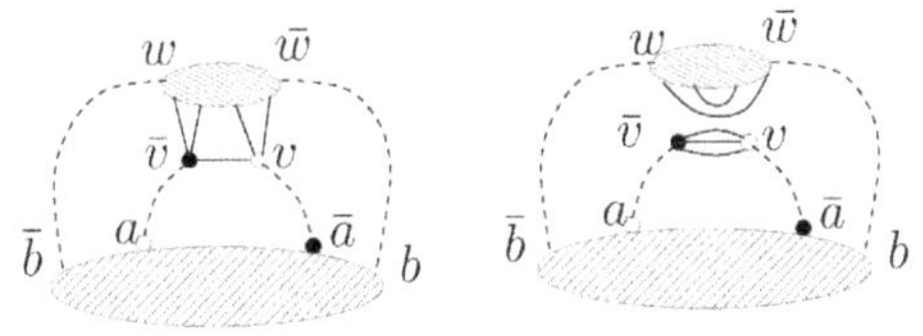

Figure 6.5: The graphs $\mathcal{G}$ and $\tilde{\mathcal{G}}$.

as the face of colors $(0,c)$ cannot be erased by this move, and all the other $D-1$ faces $(0,c')$ touching v and $\bar{v}'$ can at most merge two by two. Thus:

$$\begin{aligned}&-2(D-1)k(\mathcal{B})+(D-1)|\mathbf{P}|+\sum_c F^{(0,c)}(\mathcal{G})\leq\\&\leq -2(D-1)k(\mathcal{B})+(D-1)|\tilde{\mathbf{P}}|+\sum_c F^{(0,c)}(\tilde{\mathcal{G}}),\end{aligned}\tag{6.12}$$

and equality holds only if all the faces of colors $(0,c')$, for all $c'\neq c$ touching v and $\bar{v}$ are merged after this move. Iterating we reduce the order of all the cumulants and obtain in the end a doubled graph representing a covering graph $\mathcal{G}^{\text{final}}$ of $\mathcal{B}$ with two point insertions $\mathcal{B}^{(2)}$ on all edges, $\mathcal{G}^{\text{final}}\cup\bigcup_{\mathcal{E}^0(\mathcal{G}^{\text{final}})}\mathcal{B}^{(2)}$.

At the last step we reduced a four point cumulant connected to the rest of the graph by four edges of color 0 namely $(v,\bar{a})$, $(a,\bar{v})$ and another two edges, say $(b,\bar{w})$ and $(w,\bar{b})$. In order for the inequality in Eq. (6.12) to be saturated (if not the contribution of $\mathcal{G}$ is strictly suppressed with respect to the one of $\mathcal{G}^{\text{final}}\cup\bigcup_{\mathcal{E}^0(\mathcal{G}^{\text{final}})}\mathcal{B}^{(2)}$), it follows that the two edges of $\mathcal{G}^{\text{final}}$, $(a,\bar{a})$ and $(b,\bar{b})$ obtained after eliminating the insertions $\mathcal{B}^{(2)}$ in $\mathcal{G}^{\text{final}}\cup\bigcup_{\mathcal{E}^0(\mathcal{G}^{\text{final}})}\mathcal{B}^{(2)}$ share all the $D-1$ faces of colors $(0,c')$ for $c'\neq c$. Hence from Lemma 6.2 the graph $\mathcal{G}^{\text{final}}$ cannot be minimal.

In all the cases the contribution of $\mathcal{G}$ is strictly suppressed with respect to the one of minimal covering graphs decorated by $\mathcal{B}^{(2)}$ insertions.

□

Two natural questions arise:

- which classes of nonGaussian measures are properly uniformly bounded? This question is addressed in the next chapters, where we show that proper uniform boundedness holds for a very large class of trace invariant measures.
- do there exist marginal probability measures that escape the universality theorem and do not become Gaussian in the large N limit?

6.3.2 Nonuniform scalings

In order to discuss possible alternative scalings leading to a large N limit we will assume for the rest of this section that the rescaled cumulants admit an asymptotic expansion:

$$K(\mathcal{B},\mu_N)=N^{S(\mathcal{B})}K(\mathcal{B})+o(N^{S(\mathcal{B})}).$$

The properly uniformly bounded distributions encompass the $S(\mathcal{B}) \le 0$ case and the slightly more general case when the rescaled cumulants do not admit an asymptotic expansion but only a bound for N large enough.

As divergent measures are not interesting, we are interested in identifying some scaling $S(\mathcal{B}) > 0$ that leads to a marginal measure whose cumulants are boosted with respect to the ones of a properly uniformly bounded distribution but still has an interesting large N limit.

A uniform boosting factor $N^{S(\mathcal{B})}$ cannot work: if $S(\mathcal{B}) = ak(\mathcal{B}) + bC(\mathcal{B})$, $(a, b \ge 0)$, from Eq. (6.10):

$$\frac{1}{N}\left\langle \mathrm{Tr}_{\mathcal{B}}(\mathbb{T},\bar{\mathbb{T}})\right\rangle_{\mu_N} = \sum_{\mathcal{G},\, \mathcal{G}\in \mathrm{Doub}(\mathcal{B})} \left(\prod_{\alpha} K\big(\mathcal{B}(\alpha),\mu_N\big)\right)$$
$$\times N^{-D\sum_\alpha\left[C\left(\mathcal{B}(\alpha)\right)-1\right]-\left[\frac{2}{(D-1)!}\,\omega(\mathcal{G})-\frac{2}{(D-2)!}\,\omega\left(\mathcal{B}\right)-\sum_{\alpha,\rho}\frac{2}{(D-2)!}\,\omega\left(\mathcal{B}_{(\rho)}(\alpha)\right)\right]} ,$$

we conclude that if either a or b are nonzero then the measure is divergent. Indeed, it is enough to choose some melonic observable $\mathcal{B}$ and cumulants corresponding to melonic $\mathcal{B}_{(\rho)}(\alpha)$. The extra divergent factors in N pile up arbitrarily when choosing increasingly larger melonic observables.

The question becomes more interesting for nonuniform boosting factors $N^{S(\mathcal{B})}$, that is if $S(\mathcal{B})$ is allowed to depend on finer details of the graph $\mathcal{B}$. Of course, if one chooses

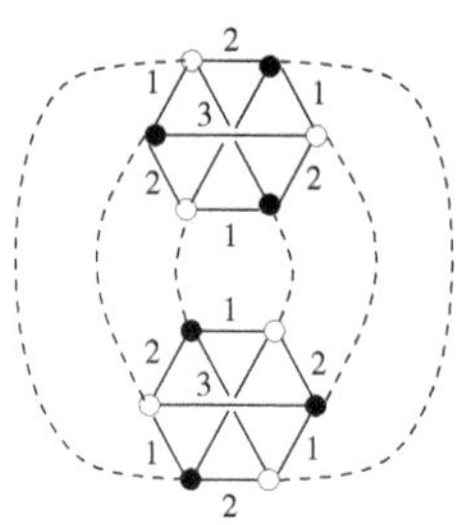

Figure 6.6: A doubled graph scaling like the Gaussian contribution for a measure with boosted scaling of the cumulants.

$S(\mathcal{B})$ too large, one will lose the marginality of the measure. A particular scaling one can consider is to boost each invariant by its Gaussian order of convergence, $S(\mathcal{B}) = \Omega(\mathcal{B})$:

$$\mathfrak{K}(\mathcal{B},\mu_N) = N^{D-2k(\mathcal{B})(D-1)-C(\mathcal{B})+\Omega(\mathcal{B})}K(\mathcal{B},\mu_N) ,$$

where $\Omega(\mathcal{B})$ factors over the connected components $\mathcal{B}_{(\rho)}$ of $\mathcal{B}$,

$$\Omega(\mathcal{B}) = \sum_{\rho}\Omega(\mathcal{B}_{(\rho)}) .$$

The expectation of an observable (rescaled by its Gaussian order of convergence) is written again as a sum over doubled graphs $\mathcal{G}$:

$$N^{-1+\Omega(\mathcal{B})}\left\langle \mathrm{Tr}_{\mathcal{B}}(\mathbb{T},\bar{\mathbb{T}})\right\rangle_{\mu_N}$$

$$= \sum_{\mathcal{G},\, \mathcal{G}\in \mathrm{Doub}(\mathcal{B})} \left(\prod_{\alpha} K(\mathcal{B}(\alpha), \mu_N) \right) N^{\Omega(\mathcal{B}) + \sum_{\alpha,\rho} \Omega\left(\mathcal{B}_{(\rho)}(\alpha)\right)} \tag{6.13}$$
$$\times N^{-D\sum_\alpha \left[C\left(\mathcal{B}(\alpha)\right)-1\right] - \left[\frac{2}{(D-1)!}\omega(\mathcal{G}) - \frac{2}{(D-2)!}\omega\left(\mathcal{B}\right) - \sum_{\rho,\alpha}\frac{2}{(D-2)!}\omega\left(\mathcal{B}_{(\rho)}(\alpha)\right)\right]},$$

where the convergence orders $\Omega(\mathcal{B})$ and $\Omega(\mathcal{B}_{(\rho)}(\alpha))$ are respectively:

$$\Omega(\mathcal{B}) = \frac{2}{(D-1)!} \min_{\mathcal{G}', \mathcal{G}' \setminus \mathcal{E}^0(\mathcal{G}) = \mathcal{B}} \omega(\mathcal{G}') - \frac{2}{(D-2)!}\omega(\mathcal{B})$$
$$\Omega\left(\mathcal{B}_{(\rho)}(\alpha)\right) = \frac{2}{(D-1)!} \min_{\substack{\mathcal{G}_\rho(\alpha) \\ \mathcal{G}_\rho(\alpha)\setminus \mathcal{E}^0\left(\mathcal{G}_\rho(\alpha)\right) = \mathcal{B}_{(\rho)}(\alpha)}} \omega(\mathcal{G}_\rho(\alpha)) - \frac{2}{(D-2)!}\omega(\mathcal{B}_{(\rho)}(\alpha)),$$

with $\mathcal{G}'$ and $\mathcal{G}_\rho(\alpha)$ covering graphs of $\mathcal{B}$ and $\mathcal{B}_{(\rho)}(\alpha)$. Denoting the total scaling of a term in Eq. (6.13) by $N^{-\Lambda(\mathcal{G})}$ and restricting to the terms with $C\left(\mathcal{B}(\alpha)\right) = 1$ we get:

$$\frac{(D-1)!}{2}\Lambda(\mathcal{G}) = \omega(\mathcal{G}) - \min_{\substack{\mathcal{G}' \\ \mathcal{G}'\setminus\mathcal{E}^0(\mathcal{G}')=\mathcal{B}}} \omega(\mathcal{G}') - \sum_{\alpha} \min_{\substack{\mathcal{G}(\alpha) \\ \mathcal{G}(\alpha)\setminus\mathcal{E}^0\left(\mathcal{G}(\alpha)\right)=\mathcal{B}(\alpha)}} \omega(\mathcal{G}(\alpha)) .$$

The measure is marginal if and only if $\Lambda(\mathcal{G}) \geq 0$ for any $\mathcal{G}$. We conjecture this to be the case.

Conjecture 6.1. *Let there be a connected, $(D+1)$-colored graph $\mathcal{G}$. Denote $\mathcal{B}_{(\rho)}$ its subgraphs (D bubbles) with colors $\{1,\dots D\}$. Then there exists a set of covering graphs $\mathcal{G}_\rho$ of $\mathcal{B}_{(\rho)}$, $\mathcal{G}_\rho \setminus \mathcal{E}^0(\mathcal{G}_\rho) = \mathcal{B}_{(\rho)}$ such that:*

$$\omega(\mathcal{G}) \geq \sum_\rho \omega(\mathcal{G}_\rho) .$$

If this conjecture holds the measure admits a large N limit. This would be especially interesting because if the large N limit exists then it is *not* Gaussian. To see this, consider the example of Figure 6.6 consisting of an observable and a cumulant in $D=3$. The minimal covering graphs for both the observable and the cumulant have six faces of colors $(0,c)$, hence degree $\omega(\mathcal{G}(\alpha)^{\min}) = \omega(\mathcal{G}'^{\min}) = 3$. The doubled graph has nine faces of colors $(0,c)$, that is degree $\omega(\mathcal{G}) = 6$, thus $\Lambda(\mathcal{G}) = 0$.

Part II

Random tensor models

Chapter 7

A digest of matrix models

Any invariant probability measure can formally be treated as a perturbed Gaussian measure. In this chapter we present this formal treatment. The results we present here apply to all the invariant measures. The reader should keep in mind that establishing them beyond the formal level is quite difficult but, at least in some cases, can be done [100].

There are many excellent reviews on random matrices both in the mathematics [88] and in the physics [59] literature. We present here only a very quick overview of some selected topics in random matrix theory: the $1/N$ expansion, the continuum limit, the double scaling limit and the Schwinger–Dyson equations.

7.1 Invariant matrix models

A perturbed Gaussian measure for a random matrix is (formally) defined as:

$$\frac{1}{Z(\lambda;\{t_p\})} d\mu_V(\mathbb{M},\mathbb{M}^\dagger) = \frac{1}{Z(\lambda;\{t_p\})} e^{-NV(\mathbb{M},\mathbb{M}^\dagger)} [d\mathbb{M} d\mathbb{M}^\dagger] \,,$$
$$[d\mathbb{M} d\mathbb{M}^\dagger] = \prod_{(a^1,a^2)} \Big(\frac{N}{\lambda} \frac{d\mathbb{M}_{a^1a^2} d\bar{\mathbb{M}}_{a^1a^2}}{2\pi\imath} \Big) \,,$$
$$V(\mathbb{M},\mathbb{M}^\dagger) = \frac{1}{\lambda} \mathrm{Tr}[\mathbb{M}\mathbb{M}^\dagger] + \sum_{p\ge 2} \frac{1}{p} t_p \,\mathrm{Tr}[(\mathbb{M}\mathbb{M}^\dagger)^p] \,, \qquad (7.1)$$

where $Z(\lambda;\{t_p\})$ is a normalization constant that we call the partition function.

This definition is formal because, as a function of the coupling constants λ and t_p, the integral of $d\mu_V(\mathbb{M},\mathbb{M}^\dagger)$ might not converge.

Rigorous results on matrix models can be established (especially when the potential $V(\mathbb{M},\mathbb{M}^\dagger)$ is a convex function [88]) but we will not insist on mathematical rigor here.

Instead we will present the *perturbative expansion* of matrix models which, while not rigorous, allows for a direct contact between matrix models and random surfaces.

The perturbative expansion is a Taylor expansion in the coupling constants t_p. The partition function and the moments of the measure in Eq. (7.1) are expressed as power series in t_p, the so-called *perturbative series.* One can then study the perturbative series and establish a number of results term by term, for every term in such series. Unfortunately the perturbative series are only formal, as they do not converge.

7.2 The $1/N$ expansion and the large N limit

The observables of matrix models are the invariants $\mathrm{Tr}[(\mathbb{M}\mathbb{M}^\dagger)^p]$ themselves, whose expectations can be computed as the derivatives of the partition function $Z(\lambda;\{t_p\})$ with respect to the coupling constants:

$$\begin{aligned}
Z(\lambda;\{t_p\}) &\equiv \int d\mu_V(\mathbb{M},\mathbb{M}^\dagger)\,,\\
\frac{1}{N}\left\langle \mathrm{Tr}[\mathbb{M}\mathbb{M}^\dagger]\right\rangle &= \frac{\int d\mu_V(\mathbb{M},\mathbb{M}^\dagger)\ \frac{1}{N}\mathrm{Tr}[\mathbb{M}\mathbb{M}^\dagger]}{\int d\mu_V(\mathbb{M},\mathbb{M}^\dagger)}\\
&= \lambda + \lambda^2\frac{\partial}{\partial\lambda}\left(\frac{1}{N^2}\ln Z(\lambda;\{t_p\})\right)\,,\\
\forall p\geq 2\,,\quad \frac{1}{N}\left\langle \mathrm{Tr}[(\mathbb{M}\mathbb{M}^\dagger)^p]\right\rangle &= \frac{\int d\mu_V(\mathbb{M},\mathbb{M}^\dagger)\ \frac{1}{N}\mathrm{Tr}[(\mathbb{M}\mathbb{M}^\dagger)^p]}{\int d\mu_V(\mathbb{M},\mathbb{M}^\dagger)}\\
&= -p\frac{\partial}{\partial t_p}\left(\frac{1}{N^2}\ln Z(\lambda;\{t_p\})\right)\,.
\end{aligned}$$

We evaluate the expectations of such observables by means of the Feynman graph expansion discussed in Appendix B. The crucial step in the Feynman expansion is the commutation of the perturbative sum with the Gaussian integral. While the resulting sum over Feynman graphs is divergent and the expansion is only formal, much can be learned from it.

7.2.1 The Feynman expansion

Before performing the Feynman expansion it is convenient to rework the nonGaussian part of the probability measure in Eq. (7.1):

$$\sum_{q\geq 2}\frac{1}{q}t_q\,\mathrm{Tr}[(\mathbb{M}\mathbb{M}^\dagger)^q]\,.$$

Each of the terms in this sum is a bicolored cycle, hence a connected bicolored graph encoded into two permutations $\tau^{(1)}$ and $\tau^{(2)}$. Taking into account that there are $q!(q-1)!$ couples of permutations $\tau^{(1)},\tau^{(2)}$ such that $\mathrm{Graph}(\tau^{(1)},\tau^{(2)})$ is connected, we can rewrite the nonGaussian part as:

$$\sum_{q\geq 2}\frac{t_q}{q!q!}\sum_{\substack{\tau^{(1)},\tau^{(2)}\in\mathfrak{S}(q)\\ \mathrm{Graph}(\tau^{(1)},\tau^{(2)})\ \mathrm{connected}}}\ \sum_{a,\bar a}\prod_{j=1}^{q}\delta_{a^1_j\bar a^1_{\tau^{(1)}(j)}}\,\delta_{a^2_j\bar a^2_{\tau^{(2)}(j)}}\prod_{j=1}^{q}\mathbb{M}_{a^1_j a^2_j}\bar{\mathbb{M}}_{\bar a^1_j\bar a^2_j}\,.$$

Each term in this sum is a connected bicolored graph with labeled vertices. Using Chapter 2, Corollary 2.1, the Taylor expansion in the coupling constants t_q yields a sum over all bicolored graphs with labeled vertices having C_q connected components with colors 1, 2 with $q \geq 2$ white (and q black) vertices:

$$Z(\lambda;\{t_p\}) = \int [d\mathbb{M}d\mathbb{M}^\dagger]\, e^{-\frac{N}{\lambda}\mathrm{Tr}[MM^\dagger]}$$
$$\times \sum_{C_2,C_3\cdots\geq 0} \frac{1}{\left[(\sum qC_q)!\right]^2} N^{\sum C_q}\left(\prod_q (-t_q)^{C_q}\right)$$
$$\times \sum_{\substack{\tau^{(1)},\tau^{(2)}\in\mathfrak{S}(\sum qC_q)\\ \tau^{(1)}(j)\neq\tau^{(2)}(j),\,\forall j}} \sum_{a,\bar a} \prod_{j=1}^{\sum qC_q} \delta_{a^1_j \bar a^1_{\tau^{(1)}(j)}} \delta_{a^2_j \bar a^2_{\tau^{(2)}(j)}} \prod_{j=1}^{\sum qC_q} \mathbb{M}_{a^1_j a^2_j} \bar{\mathbb{M}}_{\bar a^1_j \bar a^2_j}\,,$$

where the restriction to permutations such that $\tau^{(1)}(j) \neq \tau^{(2)}(j)$, $\forall j$ comes from the fact that in the exponent all the connected cycles have at least two white vertices.

Commuting the sums over C_q with the Gaussian integral (ignoring for now the problems of convergence), and computing the latter as in Chapter 6, Section 6.1.1, we obtain:

$$Z(\lambda;\{t_p\}) = \sum_{C_2,C_3\cdots\geq 0} \frac{1}{\left[(\sum qC_q)!\right]^2} N^{\sum C_q}\left(\prod_q (-t_q)^{C_q}\right)$$
$$\times \sum_{\substack{\tau^{(1)},\tau^{(2)}\in\mathfrak{S}(\sum qC_q)\\ \tau^{(1)}(j)\neq\tau^{(2)}(j),\,\forall j}} \sum_{a,\bar a} \prod_{j=1}^{\sum qC_q} \delta_{a^1_j \bar a^1_{\tau^{(1)}(j)}} \delta_{a^2_j \bar a^2_{\tau^{(2)}(j)}}$$
$$\times \sum_{\tau^{(0)}\in\mathfrak{S}(\sum qC_q)} \prod_{j=1}^{\sum qC_q} \frac{\lambda}{N} \delta_{a^1_j \bar a^1_{\tau^{(0)}(j)}} \delta_{a^2_j \bar a^2_{\tau^{(0)}(j)}}\,. \tag{7.2}$$

As in Section 6.1.1, the sum over $\tau^{(0)}$ is a sum over the pairings of an $\mathbb{M}$ and an $\bar{\mathbb{M}}$. We represent each such pairing as a dashed edge to which we assign the color 0 (see Figure 7.1 for an example). The sums over $\tau^{(1)}, \tau^{(2)}$ and $\tau^{(0)}$ yield a sum over graphs $\mathcal{G}$ with three colors:

$$\mathcal{G} = \mathrm{Gr}(\tau^{(0)},\tau^{(1)},\tau^{(2)}) = \Big\{\{v_j, v_j | j = 1,\dots k\}\,,\ \bigcup_{c=0}^{2}\{(v_j, \bar v_{\tau^{(c)}(j)}) | j - 1,\dots k\}\Big\}$$

We call these three colored graphs *Feynman graphs*. Except for the constraint that they do not have faces of color 12 with only one white vertex, the Feynman graphs are arbitrary three colored graphs.

The difference between this expansion and the computation of the Gaussian expectation of an observable we performed in Section 6.1.1 is that this time the permutations $\tau^{(1)}$ and $\tau^{(2)}$ are *summed*, that is the partition function is a sum over *all* the three colored graphs. This is the source of the divergence of the perturbative expansion: the number of terms in Eq. (7.2) is $\left[(\sum qC_q)!\right]^3$ and despite the overall factor $\frac{1}{\left[(\sum qC_q)!\right]^2}$ (which takes care of the relabeling of the black and white vertices), the sum diverges.

$$\tau^{(1)} = (1)(2)(3)(4)(5)(6)(7)$$
$$\tau^{(2)} = (12)(34)(567)$$
$$\tau^{(0)} = (1542763)$$

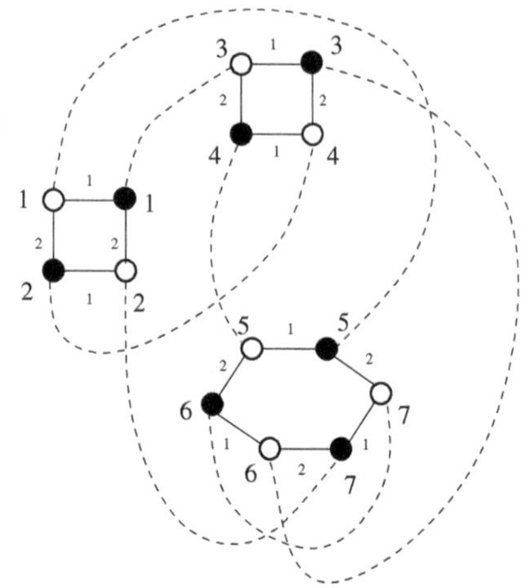

Figure 7.1: Feynman graph of a matrix model.

The subgraphs with colors $\{1,2\}$ of a Feynman graph $\mathcal{G}$ represent the terms in the perturbation potential, $\mathrm{Tr}[(\mathbb{M}\mathbb{M}^\dagger)^q]$, and are denoted $\mathcal{B} \subset \mathcal{G}$. As expected, the edges of color 0 play a distinguished role in a Feynman graph: besides bringing an explicit scaling with N, each edge of color 0 identifies all the indices of its two end vertices, whereas the edges of colors 1 and 2 identify only one index. The δ functions compose along the faces of colors $(0,1)$ and $(0,2)$ and yield a free sum, hence a factor N for each of them. The sum $\sum C_q$ simply counts the number of faces of color $(1,2)$ while the sum $\sum qC_q$ counts the number of white vertices of $\mathcal{G}$. Putting everything together we obtain:

$$Z(\lambda;\{t_p\}) = \sum_{\mathcal{G}} \frac{\lambda^{k(\mathcal{G})}}{k(\mathcal{G})!k(\mathcal{G})!} \left(\prod_{\mathcal{B}\subset\mathcal{G}} (-t_{|k(\mathcal{B})|}) \right) N^{-E^0(\mathcal{G})+F^{(0,1)}(\mathcal{G})+F^{(0,2)}(\mathcal{G})+F^{(1,2)}(\mathcal{G})} ,$$

where $E^0(\mathcal{G})$ denotes the number of edges of color 0 of $\mathcal{G}$ (which in particular equals $k(\mathcal{G})$, the number of white vertices of $\mathcal{G}$) and $F^{(c_1,c_2)}$ denotes the number of faces of colors (c_1,c_2) of $\mathcal{G}$.

The Feynman graph $\mathcal{G}$, having only three colors, is a combinatorial map with $2k(\mathcal{G})$ vertices and $3k(\mathcal{G})$ edges, thus:

$$-E^0(\mathcal{G}) + F^{(0,1)}(\mathcal{G}) + F^{(0,2)}(\mathcal{G}) + F^{(1,2)}(\mathcal{G}) = \sum_{\rho} \left(2 - 2g(\mathcal{G}_\rho)\right) ,$$

where $\mathcal{G}_\rho$ denotes the connected components of $\mathcal{G}$.

Using again Corollary 2.1 from Chapter 2 we conclude that the logarithm of Z is a sum over connected 3-colored graphs (again having no face $(1,2)$ with only one white vertex):

$$\boxed{\frac{1}{N^2} \ln Z(\lambda;\{t_p\}) = \sum_{\mathcal{G}\text{ connected}} \frac{\lambda^{k(\mathcal{G})}}{k(\mathcal{G})!k(\mathcal{G})!} \left(\prod_{\mathcal{B}\subset\mathcal{G}} (-t_{|k(\mathcal{B})|}) \right) N^{-2g(\mathcal{G})}} . \tag{7.3}$$

This is the perturbative expansion of matrix models. The expectations of invariants are computed as perturbative expansions by deriving term by term Eq. (7.3):

$$\frac{1}{N}\left\langle \mathrm{Tr}[\mathbb{M}\mathbb{M}^\dagger]\right\rangle = \lambda + \lambda^2 \frac{\partial}{\partial \lambda}\left(\frac{1}{N^2}\ln Z(\lambda;\{t_p\})\right),$$
$$\frac{1}{N}\left\langle \mathrm{Tr}[(\mathbb{M}\mathbb{M}^\dagger)^p]\right\rangle = -p\frac{\partial}{\partial t_p}\left(\frac{1}{N^2}\ln Z(t_p)\right). \tag{7.4}$$

Remark 7.1. Two remarks are in order:

Average number of vertices. As the number of white vertices is the exponent of λ in Eq. (7.3), the average number of vertices of a graph can be computed as a derivative:

$$\langle k\rangle = \lambda\partial_\lambda \ln\left[\frac{1}{N^2}\ln Z(\lambda;\{t_p\})\right].$$

Expectations of invariants. Deriving term by term Eq. (7.3) with respect to t_p, the derivative acts on one of the t_ps in the product and marks the corresponding subgraph with colors $(1,2)$ (and the overall factor p accounts for a choice of root white vertex on this subgraph). Denoting $\mathcal{B}^{(2p)}$ the rooted bicolored cycle with p white vertices, the expectations of invariants write as sums over graphs $\mathcal{G}$ having a marked subgraph $\mathcal{B}^{(2p)}$:

$$\frac{1}{N}\left\langle \mathrm{Tr}[(\mathbb{M}\mathbb{M}^\dagger)^p]\right\rangle$$
$$= \sum_{\substack{\mathcal{G}\text{ connected}\\ \mathcal{B}^{(2p)}\subset\mathcal{G}\text{ marked}}} \frac{\lambda^{k(\mathcal{G})}}{k(\mathcal{G})!k(\mathcal{G})!}\left(\prod_{\substack{\mathcal{B}\subset\mathcal{G}\\ \mathcal{B}\neq\mathcal{B}^{(2p)}}}(-t_{|k(\mathcal{B})|})\right)N^{-2g(\mathcal{G})}. \tag{7.5}$$

7.2.2 The $1/N$ expansion

Truncating the perturbative series in t_p to some finite order we express the partition function as a sum over a finite number of graphs. This can be done in any quantum field theory and, while leading to good estimations for weakly coupled theories (like quantum electrodynamics), is of limited use for strongly coupled theories (like quantum chromodynamics). In this second case one needs to deal with the perturbative series more carefully.

As observed by 't Hooft in the context of quantum chromodynamics [108], what is special about matrix models is that the size of the matrices, N, provides a new small parameter, $1/N$. This allows one to analyze the perturbative series in this case *order by order in* $1/N$. This is apparent in Eq. (7.3) and (7.4): as the genus of a map is nonnegative, the right hand side in (7.3) is a series in $1/N$:

$$\frac{1}{N^2}\ln Z(\lambda;\{t_p\}) \equiv W(\lambda;\{t_p\}) = \sum_{g\geq 0}\frac{1}{N^{2g}}W_g(\lambda;\{t_p\}),$$
$$W_g(\lambda;\{t_p\}) = \sum_{\substack{\mathcal{G}\text{ connected}\\ g(\mathcal{G})=g}}\frac{\lambda^{k(\mathcal{G})}}{k(\mathcal{G})!k(\mathcal{G})!}\left(\prod_{\mathcal{B}\subset\mathcal{G}}(-t_{|k(\mathcal{B})|})\right).$$

The planar graphs with $g(\mathcal{G}) = 0$ dominate this series, and the graphs with increasingly larger genus are increasingly suppressed. The $1/N$ expansion is markedly different from the perturbative expansion in the coupling constants: at any fixed genus, one sums an infinite family of graphs. Moreover, the family of graphs of fixed genus obtained at a fixed order in $1/N$ is exponentially bounded, that is $W_g(\lambda; \{t_p\})$ is an *absolutely convergent series* in λ. While of course the $1/N$ expansion is only formal (as the sum over genera does not converge), order by order it is controlled and at any fixed order it captures many features of the full model.

7.2.3 The continuum limit

It can be shown [47] that the number of rooted, bipartite quadrangulations of genus g with n quadrangles (which equals the number of rooted maps with arbitrary vertices and n edges) has the asymptotic behavior at large n:

$$W_{g;n}^{\text{rooted}} \sim W_g \, n^{\frac{5}{2}g-\frac{5}{2}} \, 12^n \Rightarrow q_{g;n}^{\text{unrooted}} \sim W_g \, n^{\frac{5}{2}g-\frac{7}{2}} \, 12^n \, .$$

Two things are remarkable about this formula:

- 12 does not depend on g, that is the exponential growth with n is *the same* for all the genera.
- the sub exponential growth $n^{\frac{5}{2}g-\frac{3}{2}}$ has a *linear behavior* with the genus.

Due to these two features one can take the continuum and the double scaling limits of the model. Let us for the moment focus on the case $t_2 = 1$ and $t_p = 0$ for $p > 2$. The graphs contributing to $W_g(\lambda)$ are 3-colored graphs such that all their faces with color 12 have four vertices (two white and two black), as in Figure 7.2.

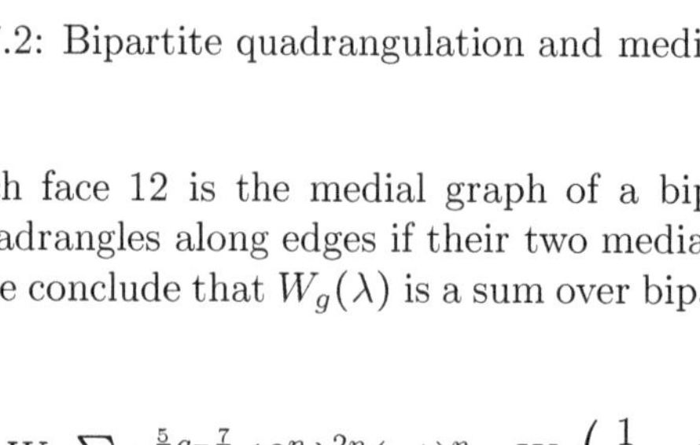

Figure 7.2: Bipartite quadrangulation and medial graph.

Observing that each face 12 is the medial graph of a bipartite quadrangle, and gluing pairs of such quadrangles along edges if their two medial vertices are connected by an edge of color 0, we conclude that $W_g(\lambda)$ is a sum over bipartite quadrangulations, therefore:

$$W_g(\lambda) \sim W_g \sum_{n \geq 0} n^{\frac{5}{2}g-\frac{7}{2}} \, 12^n \lambda^{2n} (-1)^n \sim W_g \left(\frac{1}{12} + \lambda^2 \right)^{\frac{5}{2}-\frac{5}{2}g} ,$$

$$W(\lambda) \sim \left(\frac{1}{12} + \lambda^2\right)^{\frac{5}{2}} \left[W_0 + \sum_{g \geq 1} W_g N^{-2g} \left(\frac{1}{12} + \lambda^2\right)^{-\frac{5}{2}g} \right],$$

and we ignore any analytic piece in λ as such pieces cancel after taking derivatives with λ.

Single scaling. Let us restrict to the planar sector (i.e. set $N = \infty$). The average number of vertices of the triangulations dual to the 3-colored graphs behaves like:

$$\langle k \rangle = \lambda \partial_\lambda \ln W(\lambda) \sim \frac{1}{\frac{1}{12} + \lambda^2},$$

and diverges as we send $\lambda^2 \to -\frac{1}{12}$. This is unsurprising: when approaching the boundary of the analyticity domain of some series, the higher order terms start to bring large contributions and ultimately spoil the convergence of the series.

In the critical regime graphs with infinitely many vertices dominate. As every graph is dual to a triangulation and the number of vertices of the graph is the number of triangles of the triangulation, it follows that in the critical regime infinitely refined planar surfaces dominate. In this critical regime matrix models reach a continuum phase that corresponds to two dimensional Liouville gravity [59]. A similar behavior is encountered at every fixed order in $1/N$.

Double scaling. The fact that all the W_gs diverge at the same critical value $\lambda_c^2 = -\frac{1}{12}$ allows one to explore in more detail the behavior of W at criticality. Introducing the rescaled variable $x = N(\lambda_c - \lambda)^{\frac{5}{4}}$, we obtain, up to an overall factor:

$$W\left(\lambda_c - \left(\frac{x}{N}\right)^{4/5}\right) \sim W_0' + \sum_{g \geq 1} W_g' \frac{1}{x^{2g}}.$$

In the double limit $N \to \infty$, $\lambda \to \lambda_c$, x fixed all the W_g's contribute on an equal footing to the partition function. In this regime one takes into account topology change, and the coupling constant x measures the genus of the dual surface. As in two dimensions the Einstein Hilbert action equals the Euler character; this is the limit in which random matrices describe quantum two-dimensional gravity at fixed Newton's constant [59].

Multicritical points. For typical values of the coupling constants t_p (for instance when $t_p \leq 0$ for all p) random matrix models exhibit the same kind of critical behavior (i.e. with the same critical exponents). In particular all these models fall in the universality class of pure gravity. Gravity with a matter content is described by multicritical points (which in particular exhibit different critical exponents). Such points can be obtained for specific choices of t_p, which lead to a dominant singular behavior [59]:

$$W_g \sim (\lambda_c - \lambda)^{\frac{2m+1}{m}(1-g)}.$$

A double scaling limit can be established close to a multicritical point by sending $N \to \infty$ while keeping $N(\lambda_c - \lambda)^{\frac{2m+1}{2m}}$ fixed.

7.3 The Schwinger–Dyson equations

One of the most useful tools of investigation of matrix models are the Schwinger–Dyson equations (SDE) [6, 125]. As the integral of a total derivative is 0, we have the trivial identity:

$$0 = \int [d\mathbb{M} d\mathbb{M}^\dagger] \sum_{a^1 a^2, \bar{a}^1 \bar{a}^2} \delta_{a^1 \bar{a}^1} \delta_{a^2 \bar{a}^2} \frac{\delta}{\delta \mathbb{M}_{a^1 a^2}} \times \left[\left(\frac{\delta}{\delta \bar{\mathbb{M}}_{\bar{a}_1 \bar{a}_2}} \frac{\mathrm{Tr}[(\mathbb{M}\mathbb{M}^\dagger)^{n+1}]}{n+1} \right) e^{-N\left\{ \frac{1}{\lambda} \mathrm{Tr}[\mathbb{M}\mathbb{M}^\dagger] + \sum_{p\geq 2} \frac{1}{p} t_p \mathrm{Tr}[(\mathbb{M}\mathbb{M}^\dagger)^p] \right\}} \right] .$$

Treating separately the case $n = 0$ we get:

$$\int d\mu_V(\mathbb{M}, \mathbb{M}^\dagger) \left[N^2 - \frac{N}{\lambda} \mathrm{Tr}[\mathbb{M}\mathbb{M}^\dagger] - N \sum_{p\geq 2} t_p \mathrm{Tr}[(\mathbb{M}\mathbb{M}^\dagger)^p] \right] = 0 ,$$

$$\int d\mu_V(\mathbb{M}, \mathbb{M}^\dagger) \left[\sum_{k=0}^{n} \mathrm{Tr}[(\mathbb{M}\mathbb{M}^\dagger)^k] \mathrm{Tr}[(\mathbb{M}\mathbb{M}^\dagger)^{n-k}] - \frac{N}{\lambda} \mathrm{Tr}[(\mathbb{M}\mathbb{M}^\dagger)^{n+1}] - N \sum_{p\geq 2} t_p \mathrm{Tr}[(\mathbb{M}\mathbb{M}^\dagger)^{n+p}] \right] = 0 . \qquad (7.6)$$

7.3.1 Graphical interpretation

The SDEs have a very transparent graphical interpretation (represented in Figure 7.3). It is convenient to multiply Eq. (7.6) by $\frac{\lambda}{N^2 Z(\lambda;\{t_p\})}$ and to write:

$$\frac{1}{N} \langle \mathrm{Tr}[\mathbb{M}\mathbb{M}^\dagger] \rangle = \lambda - \sum_{p\geq 2} t_p \frac{\lambda}{N} \langle \mathrm{Tr}[(\mathbb{M}\mathbb{M}^\dagger)^p] \rangle$$

$$\frac{1}{N} \langle \mathrm{Tr}[(\mathbb{M}\mathbb{M}^\dagger)^{n+1}] \rangle = \frac{\lambda}{N^2} \left\langle \sum_{k=0}^{n} \mathrm{Tr}[(\mathbb{M}\mathbb{M}^\dagger)^k] \mathrm{Tr}[(\mathbb{M}\mathbb{M}^\dagger)^{n-k}] \right\rangle - \sum_{p\geq 2} t_p \frac{\lambda}{N} \langle \mathrm{Tr}[(\mathbb{M}\mathbb{M}^\dagger)^{n+p}] \rangle .$$

Consider a graph contributing to the expectation of the observable:

$$\frac{1}{N} \mathrm{Tr}[(\mathbb{M}\mathbb{M}^\dagger)^{n+1}] ,$$

and the edge of color 0 originating at one of the black vertices of the observable. Then this edge either is connected to a white vertex on the observable, and one obtains the double trace term, or it is connected to another subgraph with colors (1, 2), and one obtains the t_p terms. The extra term λ for the SDE at $n = 0$ represents the graph formed by a unique edge of color 0.

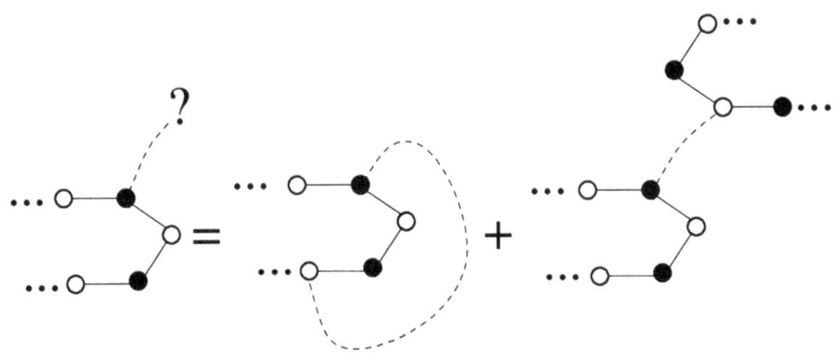

Figure 7.3: Graphical interpretation of the Schwinger–Dyson equations.

7.3.2 Algebra of constraints

Going back to Eq. (7.6), and denoting $t_1 = \frac{1}{\lambda}$ (and dropping the arguments of $Z(\lambda;\{t_p\})$) we observe that:

$$\int d\mu_V(\mathbb{M},\mathbb{M}^\dagger)\ \ \mathrm{Tr}[(\mathbb{M}\mathbb{M}^\dagger)^p] = -\frac{p}{N}\partial_{t_p} Z\ , \quad \forall p \geq 2\ ,$$
$$\int d\mu_V(\mathbb{M},\mathbb{M}^\dagger)\ \ \mathrm{Tr}[\mathbb{M}\mathbb{M}] = -\frac{1}{N} t_1^{N^2}\partial_{t_1}\left[t_1^{-N^2} Z\right],$$

which can be written in the compact form:

$$\int d\mu_V(\mathbb{M},\mathbb{M}^\dagger)\ \ \mathrm{Tr}[(\mathbb{M}\mathbb{M}^\dagger)^p] = -\frac{p}{N}[\partial_{t_p}]Z\ ,$$

where we denoted $[\partial_{t_p}]f \equiv t_1^{N^2}\partial_{t_p}\big(t_1^{-N^2} f\big)$.

With this notations we have:

$$\int d\mu_V(\mathbb{M},\mathbb{M}^\dagger)\ \ \mathrm{Tr}[(\mathbb{M}\mathbb{M}^\dagger)^p]\,\mathrm{Tr}[(\mathbb{M}\mathbb{M}^\dagger)^q] = \frac{pq}{N^2}[\partial_{t_p}][\partial_{t_q}]Z\ ,$$

and the SDEs take the simple form:

$$\left[N^2 + \sum_{p\geq 1} p t_p [\partial_{t_p}]\right] Z = 0$$
$$\left[-2n[\partial_{t_n}] + \sum_{k=1}^{n-1}\left(\frac{k(n-k)}{N^2}[\partial_{t_k}][\partial_{t_{n-k}}]\right) + \sum_{p\geq 1}(n+p)t_p[\partial_{t_{n+p}}]\right] Z = 0\ .$$

The SDEs can be encoded into a set of constraints obeyed by the partition function. Indeed, multiplying by $t_1^{-N^2}$ the SDEs become:

$$L_n(t_1^{-N^2} Z) = 0\ , \ \text{ for } n \geq 0\ ,$$
$$L_n = N^2\delta_{0,n} - 2n\frac{\partial}{\partial t_n} + \frac{1}{N^2}\sum_k k(n-k)\frac{\partial^2}{\partial t_k \partial t_{n-k}} + \sum_{p\geq 1}(n+p)\, t_p \frac{\partial}{\partial t_{n+p}}\ ,$$

where by convention $t_p = 0$ for $p \leq 0$ and the derivatives with respect to t_p, $p \leq 0$ are understood to be omitted.

A direct computation (involving some relabeling of discrete sums) shows [72] that the L_ns respect the commutation relations of the positive operators of the Witt algebra:

$$[L_m, L_n] = (m-n)\, L_{m+n} \quad \text{for } m,n \geq 0\,.$$

Note that, although *a priori* we only deal with L_m, $m \geq 0$, the definition of the operators L_n extends to negative n to yield a realization of the full Witt algebra.

Remark 7.2. A remark is in order.

Freeness. The double trace term in the SDEs factors at leading order in $1/N$. Indeed:

$$\begin{aligned}&\frac{pq}{N^4}\partial_{t_p}\left(\partial_{t_q}\ln Z\right)\\ &= \frac{1}{N^2}\left\langle \mathrm{Tr}[(\mathbb{M}\mathbb{M}^\dagger)^p]\,\mathrm{Tr}[(\mathbb{M}\mathbb{M}^\dagger)^q]\right\rangle - \frac{1}{N}\left\langle \mathrm{Tr}[(\mathbb{M}\mathbb{M}^\dagger)^p]\right\rangle \frac{1}{N}\left\langle \mathrm{Tr}[(\mathbb{M}\mathbb{M}^\dagger)^q]\right\rangle\,,\end{aligned}$$

which, from Eq. (7.5), scales like:

$$\frac{pq}{N^4}\partial_{t_p}\left(\partial_{t_q}\ln Z\right) = \frac{p}{N^2}\partial_{t_p}\left(\frac{1}{N}\left\langle \mathrm{Tr}[(\mathbb{M}\mathbb{M}^\dagger)^q]\right\rangle\right) \sim N^{-2}\,.$$

This property of factorization of the expectation of a product of observables is known as *freeness* [160, 161, 159].

Chapter 8

The perturbative expansion of tensor models

In this chapter we study the perturbative expansion of invariant tensor measures. We will show that, assuming that the perturbation and the Gaussian part scale at the same rate with N, the moments of such measures admit (as formal power series in the coupling constants) a $1/N$ expansion and that (still in the perturbative sense) all such measures are properly uniformly bounded. We then discuss the continuum limit of random tensor models and their Schwinger–Dyson equations.

8.1 Invariant probability measures revisited

The most general random tensor model is the probability measure:

$$\frac{1}{Z(\lambda;\{t_{\mathcal{B}}\})} d\mu_V(\mathbb{T},\bar{\mathbb{T}}) = \frac{1}{Z(\lambda;\{t_{\mathcal{B}}\})} e^{-N^{D-1}V(\mathbb{T},\bar{\mathbb{T}})}[d\mathbb{T}d\bar{\mathbb{T}}] ,$$
$$[d\mathbb{T}d\bar{\mathbb{T}}] = \prod_{a^{\mathcal{D}}} \frac{N^{D-1}}{\lambda} \frac{d\mathbb{T}_{a^{\mathcal{D}}} d\bar{\mathbb{T}}_{a^{\mathcal{D}}}}{2\pi\imath} ,$$
$$V(\mathbb{T},\bar{\mathbb{T}}) = \frac{1}{\lambda} \sum_{a^{\mathcal{D}},\bar{a}^{\mathcal{D}}} \mathbb{T}_{a^{\mathcal{D}}} \delta_{a^{\mathcal{D}}\bar{a}^{\mathcal{D}}} \bar{\mathbb{T}}_{\bar{a}^{\mathcal{D}}} + \sum_{\mathcal{B},\, k(\mathcal{B})\ge 2} \frac{t_{\mathcal{B}}(N)}{k(\mathcal{B})!k(\mathcal{B})!} \mathrm{Tr}_{\mathcal{B}}(\mathbb{T},\bar{\mathbb{T}}) , \tag{8.1}$$

where $\mathcal{B}$ runs over all the D-colored graphs with at least two white vertices, and the normalization constant $Z(\lambda;\{t_{\mathcal{B}}\})$ is called the partition function.

From now on we will make two simplifying assumptions. We will first assume that only connected invariants are included in the sum over $\mathcal{B}$ in Eq. (8.1). This is the equivalent of the single trace matrix models of Chapter 7. Second, we will assume that the couplings $t_{\mathcal{B}}(N)$ *do not depend* on N. In particular this comes to assuming that the perturbation and the Gaussian part of the measure both scale with the same global factor N^{D-1}. As we will see, if one considers that all the connected trace invariants in $V(\mathbb{T},\bar{\mathbb{T}})$ scale with the same power of N, then this scaling is the unique one leading to a $1/N$ expansion and a large N limit.

With these assumptions (recall that $\mathcal{B}^{(2)}$ denotes the fundamental melon with D colors) we have:

$$V(\mathbb{T},\bar{\mathbb{T}}) = \frac{1}{\lambda}\operatorname{Tr}_{\mathcal{B}^{(2)}}(\mathbb{T},\bar{\mathbb{T}}) + \sum_{\mathcal{B}\text{ connected, } k(\mathcal{B})\geq 2} \frac{t_{\mathcal{B}}}{k(\mathcal{B})!k(\mathcal{B})!}\operatorname{Tr}_{\mathcal{B}}(\mathbb{T},\bar{\mathbb{T}})$$
$$= \frac{1}{\lambda}\sum_{a,\bar{a}} \delta_{a^{\mathcal{D}}\bar{a}^{\mathcal{D}}}\mathbb{T}_{a^{\mathcal{D}}}\bar{\mathbb{T}}_{\bar{a}^{\mathcal{D}}} + \sum_{k\geq 2}\frac{1}{k!k!}$$
$$\times \sum_{\substack{\tau^{\mathcal{D}}\in[\mathfrak{S}(k)]^D \\ \mathrm{Gr}(\tau^{\mathcal{D}})\text{ connected}}} t_{\mathrm{Gr}(\tau^{\mathcal{D}})}\sum_{a,\bar{a}}\left(\prod_{j=1}^{k}\prod_{c=1}^{D}\delta_{a^c_j\bar{a}^c_{\tau^{(c)}(j)}}\right)\left(\prod_{j=1}^{k}\mathbb{T}_{a_j^{\mathcal{D}}}\bar{\mathbb{T}}_{\bar{a}_j^{\mathcal{D}}}\right).$$

As we saw in Chapter 2, the trace invariants $\operatorname{Tr}_{\mathcal{B}}(\mathbb{T},\bar{\mathbb{T}})$ are in fact class functions, that is they are invariant under the vertex relabelings of $\mathcal{B}$. We use this to rewrite $V(\mathbb{T},\bar{\mathbb{T}})$ as:

$$V(\mathbb{T},\bar{\mathbb{T}}) = \frac{1}{\lambda}\operatorname{Tr}_{\mathcal{B}^{(2)}}(\mathbb{T},\bar{\mathbb{T}}) + \sum_{\mathcal{B}\text{ connected, } k(\mathcal{B})\geq 2} \frac{t_{[\mathcal{B}]}}{k(\mathcal{B})!k(\mathcal{B})!}\operatorname{Tr}_{\mathcal{B}}(\mathbb{T},\bar{\mathbb{T}}),$$
$$\text{with } t_{[\mathcal{B}]} = \frac{\sum_{\mathcal{B}'\in[\mathcal{B}]} t_{\mathcal{B}'}}{\sum_{\mathcal{B}'\in[\mathcal{B}]} 1} \quad\text{and}\quad \operatorname{Tr}_{\mathcal{B}_1}(\mathbb{T},\bar{\mathbb{T}}) = \operatorname{Tr}_{\mathcal{B}_2}(\mathbb{T},\bar{\mathbb{T}}) \text{ if } \mathcal{B}_1\sim\mathcal{B}_2 .$$

The connected invariants $\operatorname{Tr}_{\mathcal{B}}(\mathbb{T},\bar{\mathbb{T}})$ are also the observables of tensor models. Their expectations can be recovered as derivatives of the partition function $Z(\lambda;\{t_{\mathcal{B}}\})$:

$$Z(\lambda;\{t_{\mathcal{B}}\}) \equiv \int d\mu_V(\mathbb{T},\bar{\mathbb{T}}),$$
$$\frac{1}{N}\left\langle \operatorname{Tr}_{\mathcal{B}^{(2)}}(\mathbb{T},\bar{\mathbb{T}})\right\rangle = \lambda + \lambda^2\frac{\partial}{\partial\lambda}\left(\frac{1}{N^D}\ln Z(\lambda;\{t_{\mathcal{B}}\})\right),$$
$$\frac{1}{N}\left\langle \operatorname{Tr}_{\mathcal{B}}(\mathbb{T},\bar{\mathbb{T}})\right\rangle = -[k(\mathcal{B})!]^2\frac{\partial}{\partial t_{\mathcal{B}}}\left(\frac{1}{N^D}\ln Z(\lambda;\{t_{\mathcal{B}}\})\right)$$
$$= -\frac{[k(\mathcal{B})!]^2}{\left(\sum_{\mathcal{B}'\in[\mathcal{B}]}1\right)}\frac{\partial}{\partial t_{[\mathcal{B}]}}\left(\frac{1}{N^D}\ln Z(\lambda;\{t_{\mathcal{B}}\})\right).$$

Observe that the derivative of $\ln Z(\lambda;\{t_{\mathcal{B}}\})$ with respect to $t_{\mathcal{B}}$ depends only on the class $[\mathcal{B}]$ of $\mathcal{B}$.

8.2 The 1/N expansion of tensor models

We perform a Taylor expansion in the coupling constants $t_{[\mathcal{B}]}$. We have in fact already computed the Taylor expansion of the exponential of $V(\mathbb{T},\bar{\mathbb{T}})$ in Chapter 2, Eq. (2.11). Let us recall the result. Denoting $\mathcal{B}_{(\rho)}$ the connected components of the (disconnected) graph $\mathcal{B}$ we have:

$$
\begin{aligned}
&\exp\left\{\sum_{\mathcal{B}\text{ connected},\ k(\mathcal{B})\geq 2}\frac{1}{k(\mathcal{B})!k(\mathcal{B})!}(-N^{D-1}t_{[\mathcal{B}]})\operatorname{Tr}_{\mathcal{B}}(\mathbb{T},\bar{\mathbb{T}})\right\}\\
&=\sum_{\mathcal{B},\ \mathcal{B}_{(\rho)}\neq\mathcal{B}^{(2)}\ \forall\rho}\frac{1}{k(\mathcal{B})!k(\mathcal{B})!}\left(\prod_{\rho}(-N^{D-1}t_{[\mathcal{B}_{(\rho)}]})\right)\left(\prod_{\rho}\operatorname{Tr}_{\mathcal{B}_{(\rho)}}(\mathbb{T},\bar{\mathbb{T}})\right)\\
&=\sum_{k\geq 2}\frac{1}{k!k!}\sum_{\substack{\tau^{\mathcal{D}}\in[\mathfrak{S}(k)]^D\\ \operatorname{Gr}(\tau^{\mathcal{D}})_{(\rho)}\neq\mathcal{B}^{(2)}\ \forall\rho}}\left[\prod_{\rho}(-N^{D-1}t_{[\operatorname{Gr}(\tau^{\mathcal{D}})_{(\rho)}]})\right]\\
&\quad\times\sum_{a,\bar{a}}\left(\prod_{j=1}^{k}\prod_{c=1}^{D}\delta_{a_j^c\bar{a}^c_{\tau^{(c)}(j)}}\right)\left(\prod_{j=1}^{k}\mathbb{T}_{a_j^{\mathcal{D}}}\bar{\mathbb{T}}_{\bar{a}_j^{\mathcal{D}}}\right).
\end{aligned}
$$

The partition function is then given by the Gaussian integral:

$$
\begin{aligned}
Z(\lambda;\{t_{\mathcal{B}}\})&=\int[d\mathbb{T}d\bar{\mathbb{T}}]\,e^{-\frac{N^{D-1}}{\lambda}\sum_{a^{\mathcal{D}},\bar{a}^{\mathcal{D}}}\mathbb{T}_{a^{\mathcal{D}}}\delta_{a^{\mathcal{D}}\bar{a}^{\mathcal{D}}}\bar{\mathbb{T}}_{\bar{a}^{\mathcal{D}}}}\\
&\quad\times\sum_{k\geq 2}\frac{1}{k!k!}\sum_{\substack{\tau^{\mathcal{D}}\in[\mathfrak{S}(k)]^D\\ \operatorname{Gr}(\tau^{\mathcal{D}})_{(\rho)}\neq\mathcal{B}^{(2)}\ \forall\rho}}\left[\prod_{\rho}(-N^{D-1}t_{[\operatorname{Gr}(\tau^{\mathcal{D}})_{(\rho)}]})\right]\\
&\quad\times\sum_{a,\bar{a}}\left(\prod_{j=1}^{k}\prod_{c=1}^{D}\delta_{a_j^c\bar{a}^c_{\tau^{(c)}(j)}}\right)\left(\prod_{j=1}^{k}\mathbb{T}_{a_j^{\mathcal{D}}}\bar{\mathbb{T}}_{\bar{a}_j^{\mathcal{D}}}\right).
\end{aligned}
$$

We commute the sums over k and $\tau^{\mathcal{D}}$ with the Gaussian integral. The Gaussian integral of a monomial in $\mathbb{T},\bar{\mathbb{T}}$ is computed (see Appendix B) as a sum over pairings of tensors $\mathbb{T}$ and $\bar{\mathbb{T}}$ (encoded in permutations $\tau^{(0)}$) of products of covariances:

$$
\begin{aligned}
Z(\lambda;\{t_{\mathcal{B}}\})&=\sum_{k\geq 2}\frac{1}{k!k!}\sum_{\substack{\tau^{\mathcal{D}}\in[\mathfrak{S}(k)]^D\\ \operatorname{Gr}(\tau^{\mathcal{D}})_{(\rho)}\neq\mathcal{B}^{(2)}\ \forall\rho}}\left[\prod_{\rho}(-N^{D-1}t_{[\operatorname{Gr}(\tau^{\mathcal{D}})_{(\rho)}]})\right]\\
&\quad\times\sum_{a,\bar{a}}\left(\prod_{j=1}^{k}\prod_{c=1}^{D}\delta_{a_j^c\bar{a}^c_{\tau^{(c)}(j)}}\right)\sum_{\tau^{(0)}\in\mathfrak{S}(k)}\left(\prod_{j=1}^{k}\frac{\lambda}{N^{D-1}}\prod_{c=1}^{D}\delta_{a_j^c\bar{a}^c_{\tau^{(0)}(j)}}\right).
\end{aligned}
$$

The sums over $\tau^{\mathcal{D}}=(\tau^{(1)},\dots\tau^{(D)})$ and $\tau^{(0)}$ yields a sum over $(D+1)$-colored *Feynman graphs* $\mathcal{G}$ having no subgraphs with colors $\mathcal{D}=\{1,\dots D\}$ with only two vertices. The subgraphs $\operatorname{Gr}(\tau^{\mathcal{D}})_{(\rho)}$ coming from the perturbation potential play the role of *effective vertices* and are connected by the edges of color 0, which play the role of *effective propagators*. An example of a Feynman graph is presented in Figure 8.1, where the edges of color 0 are represented as dashed.

Let us denote the subgraphs with colors $\mathcal{D}=\{1,\dots D\}$ of $\mathcal{G}$ (which are D-bubbles) by $\mathcal{B}$, $\mathcal{B}\subset\mathcal{G}$, and $B(\mathcal{G})$ their number. The number $B(\mathcal{G})$ is the number of connected components of the graph obtained from $\mathcal{G}$ by deleting all the edges of color 0:

$$
B(\mathcal{G})=C\left(\mathcal{G}\smallsetminus\mathcal{E}^0(\mathcal{G})\right).
$$

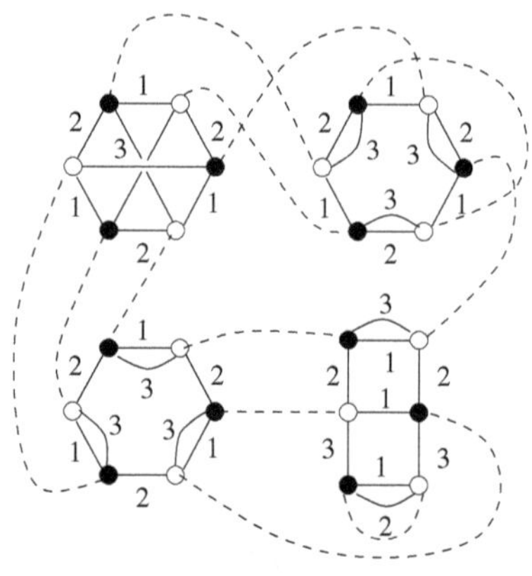

Figure 8.1: A Feynman graph.

A Feynman graph $\mathcal{G}$ has two kinds of faces: those with colors $(c_1, c_2), c_1, c_2 \in \mathcal{D}$, which also belong to some D-bubble $\mathcal{B}$ and those with colors $(0, c)$ for $c \in \mathcal{D}$, which involve the edges of color 0 in $\mathcal{G}$. The δ functions compose along the faces of colors $(0, c)$ and yield a free sum, hence a factor N for each such face:

$$Z(\lambda; \{t_{\mathcal{B}}\}) = \sum_{\mathcal{G}} \frac{\lambda^{k(\mathcal{G})}}{k(\mathcal{G})! k(\mathcal{G})!} \left(\prod_{\mathcal{B}, \mathcal{B} \subset \mathcal{G}} (-t_{[\mathcal{B}]}) \right) \\ \times N^{B(\mathcal{G})(D-1) - (D-1)E^0(\mathcal{G}) + \sum_{c \in \mathcal{D}} F^{(0,c)}(\mathcal{G})} , \tag{8.2}$$

where $E^0(\mathcal{G})$ denotes the number of edges of color 0 of $\mathcal{G}$ and in particular equals the number of white vertices, $E^0(\mathcal{G}) = k(\mathcal{G})$, and the sum over $\mathcal{G}$ runs over all the $(D+1)$-colored graphs having no bubble of colors $\{1, \ldots D\}$ with only two vertices.

The exponent of N can be evaluated further. The scaling with N factorizes over the connected components $\mathcal{G}_{(\rho)}$ of $\mathcal{G}$, and for each connected component we use:

$$\sum_{c \in \mathcal{D}} F^{(0,c)}(\mathcal{G}_{(\rho)}) = F(\mathcal{G}_{(\rho)}) - \sum_{\mathcal{B}, \mathcal{B} \subset \mathcal{G}_{(\rho)}} F(\mathcal{B}) ,$$

which, together with Eq. (3.1):

$$F(\mathcal{G}_{(\rho)}) = D + \frac{D(D-1)}{2} k(\mathcal{G}_{(\rho)}) - \frac{2}{(D-1)!} \omega(\mathcal{G}_{(\rho)})$$
$$F(\mathcal{B}) = (D-1) + \frac{(D-1)(D-2)}{2} k(\mathcal{B}) - \frac{2}{(D-2)!} \omega(\mathcal{B}) ,$$

leads to:

$$(D-1)B(\mathcal{G}_{(\rho)}) - (D-1)k(\mathcal{G}_{(\rho)}) + \sum_{c \in \mathcal{D}} F^{(0,c)}(\mathcal{G}_{(\rho)}) =$$

$$= D - \frac{2}{(D-1)!}\omega(\mathcal{G}_{(\rho)}) + \sum_{\mathcal{B},\, \mathcal{B}\subset\mathcal{G}_{(\rho)}} \frac{2}{(D-2)!}\omega(\mathcal{B}) .$$

Extracting the logarithm we finally obtain the $1/N$ expansion of tensor models:

$$\boxed{\begin{aligned}\frac{1}{N^D}\ln Z(\lambda;\{t_{\mathcal{B}}\}) = &\sum_{\substack{\mathcal{G}\text{ connected}\\ \mathcal{B}^{(2)}\not\subset\mathcal{G}}} \frac{\lambda^{k(\mathcal{G})}}{k(\mathcal{G})!k(\mathcal{G})!}\left(\prod_{\mathcal{B},\,\mathcal{B}\subset\mathcal{G}}(-t_{[\mathcal{B}]})\right)\\ &\times N^{-\frac{2}{(D-1)!}\omega(\mathcal{G})+\sum_{\mathcal{B},\,\mathcal{B}\subset\mathcal{G}}\frac{2}{(D-2)!}\omega(\mathcal{B})} .\end{aligned}} \tag{8.3}$$

The series in Eq. (8.3) is a series not only in the coupling constants $t_{[\mathcal{B}]}$ but also in $1/N$ because, for any connected $(D+1)$ colored graph $\mathcal{G}$, we have:

$$\begin{aligned}&-\frac{2}{(D-1)!}\omega(\mathcal{G}) + \sum_{\mathcal{B},\,\mathcal{B}\subset\mathcal{G}} \frac{2}{(D-2)!}\omega(\mathcal{B})\\ &\qquad = -\frac{2}{D!}\omega(\mathcal{G}) - \frac{2}{D(D-2)!}\left[\omega(\mathcal{G}) - D\sum_{\mathcal{B},\,\mathcal{B}\in\mathcal{G}}\omega(\mathcal{B})\right] \le 0 ,\end{aligned}$$

where for the last inequality we used Lemma 3.1.

8.2.1 The expectations of invariants

The expectations of invariants are computed as the derivatives of the logarithm of the partition function:

$$\begin{aligned}\frac{1}{N}\left\langle \mathrm{Tr}_{\mathcal{B}^{(2)}}(\mathbb{T},\bar{\mathbb{T}})\right\rangle &= \lambda + \lambda^2\frac{\partial}{\partial\lambda}\left(\frac{1}{N^D}\ln Z(\lambda;\{t_{\mathcal{B}}\})\right) ,\\ \frac{1}{N}\left\langle \mathrm{Tr}_{\mathcal{B}}(\mathbb{T},\bar{\mathbb{T}})\right\rangle &= -[k(\mathcal{B})!]^2\frac{\partial}{\partial t_{\mathcal{B}}}\left(\frac{1}{N^D}\ln Z(\lambda;\{t_{\mathcal{B}}\})\right) ,\end{aligned}$$

and the average number of vertices of a graph is again:

$$\langle k\rangle = \partial_\lambda \ln\left[\frac{1}{N^D}\ln Z(\lambda;\{t_{\mathcal{B}}\})\right] .$$

The expectation of an observable can computed explicitly: the derivative with respect to $t_{[\mathcal{B}]}$ acts on one of the $t_{[\mathcal{B}]}$s in the product and marks the corresponding subgraph with colors $\mathcal{D}$:

$$\begin{aligned}&\frac{1}{[k(\mathcal{B})!]^2}\frac{1}{N}\left\langle \mathrm{Tr}_{\mathcal{B}}(\mathbb{T},\bar{\mathbb{T}})\right\rangle\\ &\quad = \frac{1}{\left(\sum_{\mathcal{B}'\in[\mathcal{B}]}1\right)}\sum_{\mathcal{B}'\in[\mathcal{B}]}\sum_{\substack{\mathcal{G}\text{ connected}\\ \mathcal{B}'\subset\mathcal{G}}} \frac{\lambda^{k(\mathcal{G})}}{k(\mathcal{G})!k(\mathcal{G})!}\left(\prod_{\substack{\mathcal{B}'',\,\mathcal{B}''\subset\mathcal{G}\\ \mathcal{B}''\neq\mathcal{B}'}}(-t_{[\mathcal{B}'']})\right)\\ &\quad\times N^{-\frac{2}{(D-1)!}\omega(\mathcal{G})+\sum_{\mathcal{B}'',\,\mathcal{B}''\subset\mathcal{G}}\frac{2}{(D-2)!}\omega(\mathcal{B}'')}\end{aligned}$$

$$= \sum_{\substack{\mathcal{G} \text{ connected} \\ \mathcal{B} \subset \mathcal{G}}} \frac{\lambda^{k(\mathcal{G})}}{k(\mathcal{G})! k(\mathcal{G})!} \left(\prod_{\substack{\mathcal{B}'', \mathcal{B}'' \subset \mathcal{G} \\ \mathcal{B}'' \neq \mathcal{B}}} (-t_{[\mathcal{B}'']}) \right) \\ \times N^{-\frac{2}{(D-1)!}\omega(\mathcal{G}) + \sum_{\mathcal{B}'', \mathcal{B}'' \subset \mathcal{G}} \frac{2}{(D-2)!}\omega(\mathcal{B}'')} ,$$

where both $\mathcal{B}$ and $\mathcal{G}$ are graphs with labeled vertices.

8.3 Proper uniform boundedness

In this section we study in more detail the cumulants of perturbed Gaussian tensor measures. The generating function of the moments of such probability distributions is:

$$Z(J, \bar{J}) = \int d\mu_V(\mathbb{T}, \bar{\mathbb{T}}) \quad e^{\sum \bar{\mathbb{T}}_{\bar{a}^{\mathcal{D}}} \bar{J}_{\bar{a}^{\mathcal{D}}} + \sum J_{a^{\mathcal{D}}} \mathbb{T}_{a^{\mathcal{D}}}} ,$$

and the cumulants are given by:

$$\kappa_{2p}(\mathbb{T}_{a_1^{\mathcal{D}}}, \bar{\mathbb{T}}_{\bar{a}_1^{\mathcal{D}}}, \dots \mathbb{T}_{a_p^{\mathcal{D}}}, \bar{\mathbb{T}}_{\bar{a}_p^{\mathcal{D}}}) = \frac{\partial^{(2p)}}{\partial J_{a_1^{\mathcal{D}}} \partial \bar{J}_{\bar{a}_1^{\mathcal{D}}} \dots \partial J_{a_p^{\mathcal{D}}} \partial \bar{J}_{\bar{a}_p^{\mathcal{D}}}} \ln Z(J, \bar{J}) \Big|_{\substack{J=0 \\ \bar{J}=0}} .$$

Both the generating function of the moments and the cumulants can be computed in perturbation series as sums of Feynman graphs.

8.3.1 Feynman graphs for cumulants

The evaluation of the cumulants in perturbation series is similar to the one of the previous section, except that one now has to consider graphs containing new univalent vertices (coming from the terms $\sum \bar{\mathbb{T}}_{\bar{a}^{\mathcal{D}}} \bar{J}_{\bar{a}^{\mathcal{D}}} + \sum J_{a^{\mathcal{D}}} \mathbb{T}_{a^{\mathcal{D}}}$), that is open graphs in the sense of Definition 3.4. For each graph, the external vertices bring each a J or a $\bar{J}$. Again the δ functions compose along the faces of colors $(0, c)$, but this time they fall into two categories:

- the internal faces, as before, bring an N factor.
- the δ functions corresponding to the external faces contract pairwise the indices of J and $\bar{J}$, and reproduce the trace invariant operator associated to the boundary graph $\partial\mathcal{G}$ (see Definition 3.6).

Denoting the number of internal (resp. external) white vertices of a graph by $k_{\text{int}}(\mathcal{G})$ (resp. $k_{\text{ext}}(\mathcal{G})$), we note that the combinatorial factors coming from the $\mathcal{B}$ interactions yield an overall factor:

$$\frac{1}{k_{\text{int}}(\mathcal{G})! k_{\text{int}}(\mathcal{G})!} ,$$

while the external sources J and $\bar{J}$ contribute an overall factor:

$$\frac{1}{k_{\text{ext}}(\mathcal{G})! k_{\text{ext}}(\mathcal{G})!} .$$

Equation (8.2) is thus replaced by:

$$Z(J,\bar{J}) = \sum_{\mathcal{G}} \frac{\lambda^{k_{\text{int}}(\mathcal{G})+k_{\text{ext}}(\mathcal{G})}}{k_{\text{int}}(\mathcal{G})!k_{\text{int}}(\mathcal{G})!} \left(\prod_{\mathcal{B},\, \mathcal{B}\subset\mathcal{G}} (-t_{[\mathcal{B}]}) \right) \frac{\operatorname{Tr}_{\partial\mathcal{G}}(J,\bar{J})}{k_{\text{ext}}(\mathcal{G})!k_{\text{ext}}(\mathcal{G})!} \\ \times N^{(D-1)B(\mathcal{G})-(D-1)E^0(\mathcal{G})+\sum_{c\in\mathcal{D}} F^{(0,c)}_{\text{int}}(\mathcal{G})}, \tag{8.4}$$

where $E^0(\mathcal{G})$ denotes the total number of edges of color zero of $\mathcal{G}$ (both internal and external), $F^{(0,c)}_{\text{int}}(\mathcal{G})$ denotes the number of internal faces of colors $(0,c)$ of $\mathcal{G}$, and $B(\mathcal{G})$ denotes the number of bubbles (subgraphs) with colors $\mathcal{D} = \{1,\dots D\}$ of $\mathcal{G}$. The sum over $\mathcal{G}$ runs over open graphs whose internal and external vertices are labeled independently, that is the internal vertices of $\mathcal{G}$ are labeled $v_1, \bar{v}_1, \dots$ and its external vertices are labeled $w_1, \bar{w}_1, \dots$.

Again, the contribution of a graph factors over its connected components (including the boundary graph that is the union of the boundary graphs of the connected components). Counting the independent relabelings of the connected components of $\mathcal{G}$ and $\partial\mathcal{G}$ one obtains:

$$\ln Z(J,\bar{J}) = \sum_{\mathcal{G}\text{ connected}} \frac{\lambda^{k_{\text{int}}(\mathcal{G})+k_{\text{ext}}(\mathcal{G})}}{k_{\text{int}}(\mathcal{G})!k_{\text{int}}(\mathcal{G})!} \left(\prod_{\mathcal{B},\, \mathcal{B}\subset\mathcal{G}} (-t_{[\mathcal{B}]}) \right) \frac{\operatorname{Tr}_{\partial\mathcal{G}}(J,\bar{J})}{k_{\text{ext}}(\mathcal{G})!k_{\text{ext}}(\mathcal{G})!} \\ \times N^{(D-1)B(\mathcal{G})-(D-1)E^0(\mathcal{G})+\sum_{c\in\mathcal{D}} F^{(0,c)}_{\text{int}}(\mathcal{G})}. \tag{8.5}$$

Remark that $k_{\text{ext}}(\mathcal{G}) = k(\partial\mathcal{G})$. As already mentioned, in spite of the fact that $\mathcal{G}$ itself is connected, it may well be that the boundary graph $\partial\mathcal{G}$ is disconnected. Let us denote $\tau^{\mathcal{D}}_{\partial\mathcal{G}}$ the permutations encoding the boundary graph $\partial\mathcal{G}$. By taking derivatives with respect to the external sources, we conclude that any cumulant expands as a sum over trace invariant operators:

$$\kappa_{2p}\big(\mathbb{T}_{a_1^{\mathcal{D}}}, \bar{\mathbb{T}}_{\bar{a}_1^{\mathcal{D}}}, \dots \mathbb{T}_{a_p^{\mathcal{D}}}, \bar{\mathbb{T}}_{\bar{a}_p^{\mathcal{D}}}\big) = \sum_{\substack{\mathcal{G}\text{ connected}\\ k_{\text{ext}}(\mathcal{G})=p}} \frac{\lambda^{k_{\text{int}}(\mathcal{G})+k_{\text{ext}}(\mathcal{G})}}{k_{\text{int}}(\mathcal{G})!k_{\text{int}}(\mathcal{G})!} \left(\prod_{\mathcal{B}\, \mathcal{B}\subset\mathcal{G}} (-t_{[\mathcal{B}]}) \right) \\ \times N^{(D-1)B(\mathcal{G})-(D-1)E^0(\mathcal{G})+\sum_{c\in\mathcal{D}} F^{(0,c)}_{\text{int}}(\mathcal{G})} \frac{\sum_{\pi,\sigma\in\mathfrak{S}(p)} \prod_{c=1}^{D} \prod_{i=1}^{p} \delta_{a^c_{\pi(i)} \bar{a}^c_{\tau^c_{\partial\mathcal{G}}(\sigma(i))}}}{k(\partial\mathcal{G})!k(\partial\mathcal{G})!}.$$

We reorder the terms by the possible boundary graphs $\mathcal{B}$ (whose $C(\mathcal{B})$ connected components we label $\mathcal{B}_{(\rho)}$). The same graph $\mathcal{B}$ (with associated permutations $\tau^{\mathcal{D}}_{\mathcal{B}}$) is obtained exactly $k(\partial\mathcal{G})!k(\partial\mathcal{G})!$ times as, for every fixed π and σ, $\mathcal{B}$ is obtained from the graph $\mathcal{G}$ whose external vertices are labeled $w_{\pi^{-1}(1)}, \bar{w}_{\sigma^{-1}(1)}, \dots$, hence:

$$\boxed{\begin{aligned} &\kappa_{2p}\big(\mathbb{T}_{a_1^{\mathcal{D}}}, \bar{\mathbb{T}}_{\bar{a}_1^{\mathcal{D}}}, \dots \mathbb{T}_{a_p^{\mathcal{D}}}, \bar{\mathbb{T}}_{\bar{a}_p^{\mathcal{D}}}\big) \\ &= \sum_{\mathcal{B},\, k(\mathcal{B})=p} \left(\prod_{\rho=1}^{C(\mathcal{B})} \delta^{\mathcal{B}_{(\rho)}}_{a\bar{a}} \right) \sum_{\substack{\mathcal{G}\text{ connected}\\ \partial\mathcal{G}=\mathcal{B}}} \frac{\lambda^{k_{\text{int}}(\mathcal{G})+k_{\text{ext}}(\mathcal{G})}}{k_{\text{int}}(\mathcal{G})!k_{\text{int}}(\mathcal{G})!} \\ &\qquad \times \left(\prod_{\mathcal{B}',\, \mathcal{B}'\subset\mathcal{G}} (-t_{[\mathcal{B}']}) \right) N^{(D-1)B(\mathcal{G})-(D-1)E^0(\mathcal{G})+\sum_{c\in\mathcal{D}} F^{(0,c)}_{\text{int}}(\mathcal{G})}, \end{aligned}}$$

where the sum over $\mathcal{B}$ runs over D-colored graphs with vertices labeled:

$$w_1, \bar{w}_1 \dots w_p, \bar{w}_p \ ,$$

the sum over $\mathcal{G}$ runs over $D+1$-colored open graph with external vertices labeled:

$$w_1, \bar{w}_1 \dots w_p, \bar{w}_p \ ,$$

and internal vertices labeled:

$$v_1, \bar{v}_1, \dots v_{k_{\text{int}}(\mathcal{G})}, \bar{v}_{k_{\text{int}}(\mathcal{G})} \ ,$$

and

$$\left(\prod_{\rho=1}^{C(\mathcal{B})} \delta_{a\bar{a}}^{\mathcal{B}_{(\rho)}} \right) = \prod_{e=(v_i,\bar{v}_j)\in\mathcal{E}(\mathcal{B})} \delta_{a_i \bar{a}_j} \ .$$

8.3.2 Perturbative bounds

The measure μ_V is trace invariant. The rescaled cumulants of μ_V (recall Definition 6.5) are:

$$K(\mathcal{B},\mu_V) \equiv \frac{\mathfrak{K}(\mathcal{B},\mu_V)}{N^{D-2k(\mathcal{B})(D-1)-C(\mathcal{B})}} = \sum_{\substack{\mathcal{G} \text{ connected} \\ \partial\mathcal{G}=\mathcal{B}}} \frac{\lambda^{k_{\text{int}}(\mathcal{G})+k_{\text{ext}}(\mathcal{G})}}{k_{\text{int}}(\mathcal{G})! k_{\text{int}}(\mathcal{G})!} \left(\prod_{\mathcal{B}',\, \mathcal{B}'\subset\mathcal{G}} (-t_{[\mathcal{B}']}) \right)$$
$$\times N^{-D+2(D-1)k(\mathcal{B})+C(\mathcal{B})+(D-1)B(\mathcal{G})-(D-1)E^0(\mathcal{G})+\sum_{c\in\mathcal{D}} F_{\text{int}}^{(0,c)}(\mathcal{G})} \ .$$

Theorem 8.1 (Perturbative theorem). *The measure μ_V is properly uniformly bounded in the perturbative sense, that is:*

- *for any open connected $(D+1)$ colored graph $\mathcal{G}$:*

$$(D-1)B(\mathcal{G}) - (D-1)E^0(\mathcal{G}) + \sum_{c\in\mathcal{D}} F_{\text{int}}^{(0,c)}(\mathcal{G}) \le D - 2(D-1)k(\partial\mathcal{G}) - C(\partial\mathcal{G}) \ .$$

- *for $|t_{[\mathcal{B}]}| \le 1, \ \forall \mathcal{B}$ the series:*

$$\sum_{\substack{\mathcal{G} \text{ connected},\ \partial\mathcal{G}=\mathcal{B}^{(2)} \\ (D-1)B(\mathcal{G})-(D-1)E^0(\mathcal{G})+\sum_{c\in\mathcal{D}} F_{\text{int}}^{(0,c)}(\mathcal{G})=-D+1}} \frac{\lambda^{k_{\text{int}}(\mathcal{G})+1}}{(k_{\text{int}}(\mathcal{G})!)^2} \left(\prod_{\mathcal{B}',\, \mathcal{B}'\subset\mathcal{G}} (-t_{[\mathcal{B}']}) \right) , \tag{8.6}$$

is absolutely convergent for $|\lambda|$ small enough.

Proof. The first item is the statement of Theorem 3.2.

Concerning the second item, in the $N \to \infty$ limit only the graphs $\mathcal{G}$ with $\partial\mathcal{G} = \mathcal{B}^{(2)}$ and:

$$(D-1)B(\mathcal{G}) - (D-1)E^0(\mathcal{G}) + \sum_{c\in\mathcal{D}} F_{\text{int}}^{(0,c)}(\mathcal{G}) = -D+1 \ ,$$

contribute. Let us consider the closed graph $\bar{\mathcal{G}}$ obtained from $\mathcal{G}$ by reconnecting the two external edges into an edge of color 0 (and marking this new edge), hence $E^0(\bar{\mathcal{G}}) = E^0(\mathcal{G}) - 1$. The number of faces of color $(0,c)$ of $\bar{\mathcal{G}}$ is:

$$\sum_{c\in\mathcal{D}} F^{(0,c)}(\bar{\mathcal{G}}) = \sum_{c\in\mathcal{D}} F^{(0,c)}_{\text{int}}(\mathcal{G}) + D \,,$$

and the bubbles of colors $\mathcal{D}$ of $\mathcal{G}$ and $\bar{\mathcal{G}}$ are identical. The total number of faces of $\bar{\mathcal{G}}$ is:

$$\begin{aligned}
F(\bar{\mathcal{G}}) &= \sum_{c\in\mathcal{D}} F^{(0,c)}(\bar{\mathcal{G}}) + \sum_{\mathcal{B},\,\mathcal{B}\subset\mathcal{G}} F(\mathcal{B}) \\
&= D + (-D+1) - (D-1)B(\bar{\mathcal{G}}) + (D-1)[k(\bar{\mathcal{G}})+1] \\
&\quad + \sum_{\mathcal{B},\,\mathcal{B}\subset\bar{\mathcal{G}}} \left(\frac{(D-1)(D-2)}{2} k(\mathcal{B}) + (D-1) - \frac{2}{(D-2)!}\omega(\mathcal{B}) \right) \\
&= D + \frac{D(D-1)}{2} k(\bar{\mathcal{G}}) - \sum_{\mathcal{B},\,\mathcal{B}\subset\bar{\mathcal{G}}} \frac{2}{(D-2)!}\omega(\mathcal{B}) \,.
\end{aligned}$$

It follows that:

$$\frac{2}{(D-1)!}\omega(\bar{\mathcal{G}}) = \sum_{\mathcal{B},\,\mathcal{B}\subset\bar{\mathcal{G}}} \frac{2}{(D-2)!}\omega(\mathcal{B}) \Rightarrow$$

$$\frac{2}{D!}\omega(\bar{\mathcal{G}}) + \frac{2}{D(D-2)!}\left(\omega(\bar{\mathcal{G}}) - D \sum_{\mathcal{B},\,\mathcal{B}\subset\bar{\mathcal{G}}} \omega(\mathcal{B}) \right) = 0 \,.$$

By Lemma 3.1 both of these terms are nonnegative, hence $\omega(\bar{\mathcal{G}}) = 0$, which implies that $\bar{\mathcal{G}}$ is a closed melonic graph.

The series in Eq. (8.6) becomes:

$$\sum_{\mathcal{G}\text{ melonic}} \frac{\lambda^{k_{\text{int}}(\mathcal{G})+1}}{(k_{\text{int}}(\mathcal{G})!)^2} \left(\prod_{\mathcal{B}\,,\mathcal{B}\subset\mathcal{G}} (-t_{[\mathcal{B}]}) \right) .$$

Using Proposition 4.3 in Section 4.6.1, this series is convergent for $|t_{\mathcal{B}}| \le 1$ and $|\lambda| \le \frac{D^D}{(D+1)^{D+1}}$ and is bounded up to an overall factor $|\lambda|$ by the generating function of melonic graphs:

$$\left| \sum_{\mathcal{G}\text{ melonic}} \frac{\lambda^{k_{\text{int}}(\mathcal{G})+1}}{(k_{\text{int}}(\mathcal{G})!)^2} \left(\prod_{\mathcal{B},\,\mathcal{B}\subset\mathcal{G}} (-t_{[\mathcal{B}]}) \right) \right| \le |\lambda|\, T(|\lambda|) \,.$$

□

8.4 The large N limit

As the measure μ_V is properly uniformly bounded in the sense of perturbation theory, it follows that (still in the sense of perturbation theory) the universality Theorem 6.3

applies and in the large N limit μ_V becomes a Gaussian measure of covariance:

$$K(\mathcal{B}^{(2)}) \equiv K_2 = \sum_{\mathcal{G} \text{ melonic}} \frac{\lambda^{k_{\text{int}}(\mathcal{G})+1}}{k_{\text{int}}(\mathcal{G})! k_{\text{int}}(\mathcal{G})!} \left(\prod_{\mathcal{B},\, \mathcal{B} \subset \mathcal{G}} (-t_{[\mathcal{B}]}) \right) .$$

The moments of μ_V are:

$$\lim_{N \to \infty} N^{-1+\Omega(\mathcal{B})} \left\langle \text{Tr}_{\mathcal{B}}(\mathbb{T}, \bar{\mathbb{T}}) \right\rangle_{\mu_V} = K_2^{k(\mathcal{B})} R(\mathcal{B}) ,$$

with $R(\mathcal{B}) = 1$ if $\mathcal{B}$ is melonic and:

$$\begin{aligned} \Omega(\mathcal{B}) &= \frac{2}{(D-1)!} \omega(\mathcal{G}^{\text{min}}) - \frac{2}{(D-2)!} \omega(\mathcal{B}) \\ &= \frac{2}{D!} \omega(\mathcal{G}^{\text{min}}) + \frac{2}{D(D-2)!} \Big[\omega(\mathcal{G}) - D\omega(\mathcal{B}) \Big] \\ &\geq \frac{2}{D!} \omega(\mathcal{G}^{\text{min}}) \geq \frac{2}{(D-1)!} \omega(\mathcal{B}) \geq 0 . \end{aligned}$$

Observe that the expectation of $\text{Tr}_{\mathcal{B}^{(2)}}(\mathbb{T}, \bar{\mathbb{T}})$ equals in the large N limit the covariance:

$$\lim_{N \to \infty} N^{-1} \left\langle \text{Tr}_{\mathcal{B}^{(2)}}(\mathbb{T}, \bar{\mathbb{T}}) \right\rangle_{\mu_V} = K_2 .$$

The convergence order $\Omega(\mathcal{B})$ is 0 if and only if $\mathcal{B}$ is melonic. This is trivial for $D \geq 4$, while for $D = 3$ it comes from the fact that $\mathcal{B}$ must admit a melonic minimal covering graph $\mathcal{G}^{\text{min}}$. For melonic observables $\mathcal{B}$, only Feynman graphs consisting in the observable decorated by the full two point function connecting the pairs of external vertices of prime melonic subgraphs in $\mathcal{B}$ contribute, see Figure 8.2.

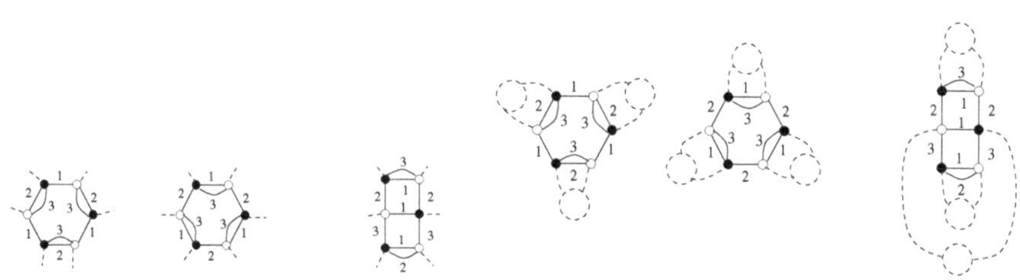

(a) Melonic observables.

(b) Graphs contributing to the expectations of melonic observables.

Figure 8.2: Feynman graphs contributing to melonic observables.

Let us denote $t_1 = \frac{1}{\lambda}$, and the free energy $W = \frac{1}{N^D} \ln Z(\lambda, \{t_{\mathcal{B}}\})$, such that:

$$\begin{aligned} \forall \mathcal{B} \neq \mathcal{B}^{(2)} , \quad \partial_{t_{\mathcal{B}}} W &= -\frac{1}{k(\mathcal{B})! k(\mathcal{B})!} \frac{1}{N} \left\langle \text{Tr}_{\mathcal{B}}(\mathbb{T}, \bar{\mathbb{T}}) \right\rangle_{\mu_V} , \\ \partial_{t_1} W &= \frac{1}{t_1} - \frac{1}{N} \left\langle \text{Tr}_{\mathcal{B}^{(2)}}(\mathbb{T}, \bar{\mathbb{T}}) \right\rangle_{\mu_V} , \end{aligned}$$

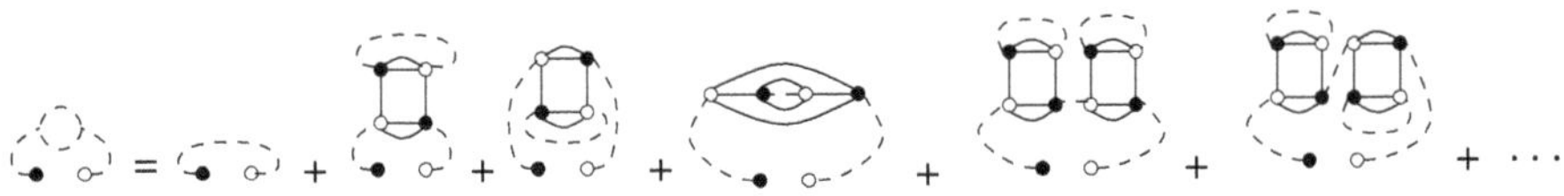

Figure 8.3: Graphs contributing to the K_2.

$$\langle k \rangle = \lambda \partial_\lambda \ln W = -t_1 \partial_{t_1} \ln W \,. \tag{8.7}$$

The covariance K_2 itself (the two point function in physics parlance) is a nontrivial sum over all the melonic graphs (see Figure 8.3).

The large N covariance. In order to analyze the possible behaviors of K_2 it is more convenient not to start from its explicit series expansion, but to derive a self consistency equation for it. This can be obtained from the simplest Schwinger–Dyson equation:

$$\frac{1}{N^D} \frac{1}{Z(\lambda;\{t_{\mathcal{B}}\})} \int [d\mathbb{T} d\bar{\mathbb{T}}] \sum_{a^{\mathcal{D}}} \frac{\partial}{\partial \mathbb{T}_{a^{\mathcal{D}}}} \Big[\mathbb{T}_{a^{\mathcal{D}}} \, e^{-N^{D-1} V(\mathbb{T},\bar{\mathbb{T}})} \Big] = 0 \,,$$

which computes to:

$$1 - \frac{1}{N} t_1 \left\langle \mathrm{Tr}_{\mathcal{B}^{(2)}}(\mathbb{T},\bar{\mathbb{T}}) \right\rangle_{\mu_V} - \sum_{\mathcal{B},\, k(\mathcal{B}) \ge 2} \frac{t_{\mathcal{B}}}{k(\mathcal{B})! k(\mathcal{B})!} k(\mathcal{B}) \frac{1}{N} \left\langle \mathrm{Tr}_{\mathcal{B}}(\mathbb{T},\bar{\mathbb{T}}) \right\rangle_{\mu_V} = 0 \,.$$

In the $N \to \infty$ limit only the melonic observables survive, and denoting the *melonic* part of the perturbation potential:

$$V_{\text{melo}}(x) = \sum_{\substack{\mathcal{B} \text{ melonic} \\ k(\mathcal{B}) \ge 2}} \frac{t_{\mathcal{B}}}{k(\mathcal{B})! k(\mathcal{B})!} \; x^{k(\mathcal{B})}$$

$$V'_{\text{melo}}(x) = \sum_{\substack{\mathcal{B} \text{ melonic} \\ k(\mathcal{B}) \ge 2}} \frac{t_{\mathcal{B}}}{k(\mathcal{B})! k(\mathcal{B})!} \; k(\mathcal{B}) x^{k(\mathcal{B})-1} \,,$$

the covariance respects the self consistency equation:

$$\boxed{1 - t_1 K_2 - K_2 V'_{\text{melo}}\Big(K_2\Big) = 0 \,.} \tag{8.8}$$

Remark that for polynomial potentials this is an algebraic equation, hence the covariance will typically have interesting critical behavior.

The large N free energy. Denoting the large N free energy $W_\infty = \lim_{N\to\infty} W$, in the large N limit Eq. (8.7) becomes:

$$\partial_{t_{\mathcal{B}}} W_\infty = -\frac{1}{k(\mathcal{B})! k(\mathcal{B})!} K_2^{k(\mathcal{B})} \,, \qquad \partial_{t_1} W_\infty = \frac{1}{t_1} - K_2 \,, \qquad \langle k \rangle = -t_1 \partial_{t_1} W_\infty \,,$$

and we recall that when computing the average number of vertices, the analytic parts of W_∞ should be ignored.

Proposition 8.1. *The large N free energy is:*

$$\boxed{W_\infty = 1 + \ln(t_1 K_2) - t_1 K_2 - V_{\text{melo}}\Big(K_2\Big) .}$$

Proof. Note that if $t_{\mathcal{B}} = 0$ for all $\mathcal{B}$ then, on the one hand, $W_\infty|_{t_{\mathcal{B}}=0} = 0$ and, on the other, $V_{\text{melo}} = 0$, thus Eq. (8.8) becomes:

$$1 - t_1 K_2 = 0 \Rightarrow 1 + \ln(t_1 K_2) - t_1 K_2 - V_{\text{melo}}\Big(K_2\Big) = 0 .$$

For generic values of $t_{\mathcal{B}}$ we have:

$$\begin{aligned} &\partial_{t_1}\Big[1 + \ln(t_1 K_2) - t_1 K_2 - V_{\text{melo}}\Big(K_2\Big)\Big] \\ &= \frac{1}{t_1} + \frac{\partial_{t_1} K_2}{K_2} - K_2 - t_1 \partial_{t_1} K_2 - V'_{\text{melo}}\Big(K_2\Big)\partial_{t_1} K_2 = \frac{1}{t_1} - K_2 , \end{aligned}$$

and:

$$\begin{aligned} &\partial_{t_{\mathcal{B}}}\Big[1 + \ln(t_1 K_2) - t_1 K_2 - V_{\text{melo}}\Big(K_2\Big)\Big] \\ &= \frac{\partial_{t_{\mathcal{B}}} K_2}{K_2} - t_1 \partial_{t_{\mathcal{B}}} K_2 - \frac{1}{k(\mathcal{B})!k(\mathcal{B}!)} K_2^{k(\mathcal{B})} - V'_{\text{melo}}\Big(K_2\Big)\partial_{t_{\mathcal{B}}} K_2 \\ &= -\frac{1}{k(\mathcal{B})!k(\mathcal{B}!)} K_2^{k(\mathcal{B})} , \end{aligned}$$

hence the differential:

$$d\Big[1 + \ln(t_1 K_2) - t_1 K_2 - V_{\text{melo}}\Big(K_2\Big) - W_\infty\Big] = 0 .$$

□

8.5 The continuum limit

The possible continuum limits are defined by the various possible critical behaviors of the covariance.

It is convenient to rewrite the self consistency equation (8.8) for the covariance as:

$$t_1 = \frac{1}{K_2} - V'_{\text{melo}}\Big(K_2\Big) , \qquad V_{\text{melo}}(x) = \sum_{\substack{\mathcal{B}\text{ melonic} \\ k(\mathcal{B})\geq 2}} \frac{t_{\mathcal{B}}}{k(\mathcal{B})!k(\mathcal{B})!} x^{k(\mathcal{B})} .$$

This equation, supplemented by the boundary condition:

$$\lim_{t_1 \to \infty} K_2(t_1) = 0 ,$$

defines implicitly the two point function K_2 as a function of t_1.

Criticality is reached for couples (t_1^c, K_2^c) such that the Jacobian:

$$\left(\frac{\partial t_1}{\partial K_2}\right)_{t_1^c, K_2^c} = \left[-\frac{1}{(K_2)^2} - V''_{\text{melo}}\big(K_2\big)\right]_{t_1^c, K_2^c} = 0 ,$$

and the implicit function K_2 ceases to be defined. The critical behavior of the two point function is obtained by Taylor expanding around the critical point.

Case 1. $t_{\mathcal{B}} \leq 0$ for all $\mathcal{B}$. This case is represented in Figure 8.4.

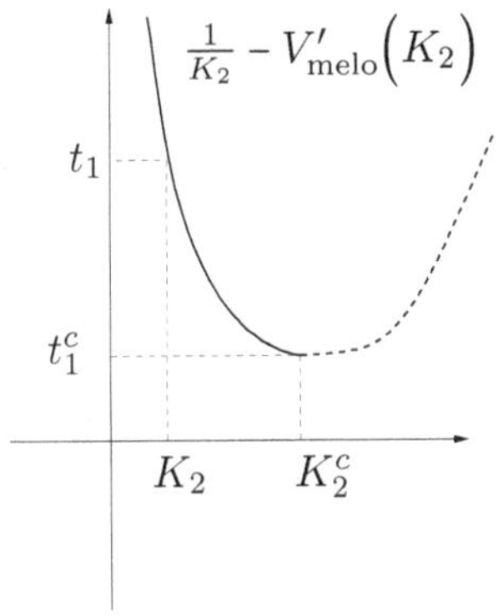

Figure 8.4: Critical point.

In this case the function $f(K_2) = \frac{1}{K_2} - V'_{\text{melo}}\big(K_2\big)$ is always positive, goes to infinity for $K_2 = 0$ and its derivatives are:

$$\frac{\partial f}{\partial_{K_2}} = -\frac{1}{(K_2)^2} - V''_{\text{melo}}\big(K_2\big) , \qquad \frac{\partial^2 f}{\partial K_2 \partial K_2} = \frac{2}{(K_2)^3} - V'''_{\text{melo}}\big(K_2\big) > 0 ,$$

hence f has a unique global minimum. This minimum is the critical point, and close to criticality we have:

$$t_1 \sim t_1^c + \frac{1}{\alpha^2}(K_2 - K_2^c)^2 \Rightarrow K_2 \sim K_2^c + \alpha(t_1 - t_1^c)^{\frac{1}{2}}$$

and, as $\partial_{t_1} W_\infty = \frac{1}{t_1} - K_2$, we obtain a leading critical behavior of the free energy:

$$W_\infty \sim (t_1 - t_1^c)^{1+\frac{1}{2}} .$$

Case 2. $t_{\mathcal{B}} > 0$ for some $\mathcal{B}$. In this case the situation depicted in Figure 8.4 can arise.

In this case, while typically the criticality equation has a solution, it might very well happen that:

$$\left(\frac{\partial^2 t_1}{\partial K_2 \partial K_2}\right)_{t_1^c, K_2^c} = \cdots = \left(\frac{\partial^{m-1} t_1}{\partial K_2^{m-1}}\right)_{t_1^c, K_2^c} = 0 , \qquad \left(\frac{\partial^m t_1}{\partial K_2^m}\right)_{t_1^c, K_2^c} \neq 0 .$$

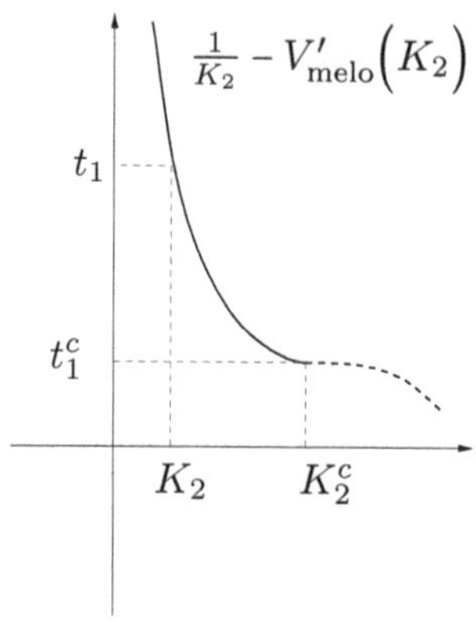

Figure 8.5: Multicritical point.

In such a situation, close to criticality we have:

$$t_1 \sim t_1^c + \frac{1}{\alpha^m}(K_2 - K_2^c)^m \;\Rightarrow\; K_2 \sim K_2^c + \alpha(t_1 - t_1^c)^{\frac{1}{m}}$$

and a leading critical behavior of the free energy:

$$W_\infty \sim (t_1 - t_1^c)^{1+\frac{1}{m}} \ .$$

Observe that in both cases (ignoring the analytic parts, which are washed out by taking derivatives) the average number of vertices diverges like $\frac{1}{t_1 - t_1^c}$, hence the critical points correspond to continuum limits of infinitely refined higher dimensional geometries. In order to keep the total volume constant, the volume of every simplex must go to zero. If one associates with a graph the equilateral triangulation with edge length a, then a must go to zero like $a \sim (t_1 - t_1^c)^{1/D}$.

8.6 The algebra of constraints

We will now derive the Schwinger–Dyson equations of a generic tensor model.

We first define a Lie algebra indexed by an observable $\mathcal{B}$ and a black vertex $\bar{v}$ of $\mathcal{B}$. In a second part we show that this is an algebra of constraints for the partition function of the tensor models, that is there exists a representation of this algebra as differential operators $L_{\mathcal{B}|\bar{v}}$ such that $L_{\mathcal{B}|\bar{v}}\left(t_1^{-N^D} Z(\lambda, \{t_{\mathcal{B}}\})\right) = 0$.

8.6.1 A Lie algebra indexed by observables

Let $L_{\mathcal{B}|\bar{v}}$ be a set of operators indexed by an observable $\mathcal{B}$ and a black vertex $\bar{v}$ of the observable, $\bar{v} \in \mathcal{V}^b(\mathcal{B})$. Recall the composition of graphs introduced in Definition 3.8, Section 3.3.2, Chapter 3.

Theorem 8.2. *The vector space* $\{\sum \alpha_{\mathcal{B}|\bar{v}} L_{\mathcal{B}|\bar{v}} | \alpha_{\mathcal{B}|\bar{v}} \in \mathbb{C}\})$ *is a Lie algebra with Lie bracket (the extension by bi linearity of):*

$$[L_{\mathcal{B}_1|\bar{v}_1}, L_{\mathcal{B}_2|\bar{v}_2}] = \sum_{v\in\mathcal{V}^w(\mathcal{B}_1)} L_{(\mathcal{B}_1\uplus\mathcal{B}_2)/\{(v,\bar{v}_2)\}|\bar{v}_1} - \sum_{v\in\mathcal{V}^w(\mathcal{B}_2)} L_{(\mathcal{B}_2\uplus\mathcal{B}_1)/\{(v,\bar{v}_1)\}|\bar{v}_2} ,$$

where $\mathcal{V}^w(\mathcal{B})$ *denotes the white vertices of* $\mathcal{B}$.

Proof. The bracket is by definition bilinear and trivially antisymmetric. It remains only to check that it respects the Jacobi identity:

$$\begin{aligned}\big[L_{\mathcal{B}_1|\bar{v}_1}, [L_{\mathcal{B}_2|\bar{v}_2}, L_{\mathcal{B}_3|\bar{v}_3}]\big] + \big[L_{\mathcal{B}_2|\bar{v}_2}, [L_{\mathcal{B}_3|\bar{v}_3}, L_{\mathcal{B}_1|\bar{v}_1}]\big]\\ + \big[L_{\mathcal{B}_3|\bar{v}_3}, [L_{\mathcal{B}_1|\bar{v}_1}, L_{\mathcal{B}_2|\bar{v}_2}]\big] = 0 .\end{aligned}$$

This is nontrivial as the composition of graphs *is not* associative. However, the Jacobi identity holds. Let us detail the first term:

$$\begin{aligned}&\big[L_{\mathcal{B}_1|\bar{v}_1}, [L_{\mathcal{B}_2|\bar{v}_2}, L_{\mathcal{B}_3|\bar{v}_3}]\big]\\ &= \Big[L_{\mathcal{B}_1|\bar{v}_1}, \sum_{v'\in\mathcal{V}^w(\mathcal{B}_2)} L_{(\mathcal{B}_2\uplus\mathcal{B}_3)/\{(v',\bar{v}_3)\}|\bar{v}_2} - \sum_{v'\in\mathcal{V}^w(\mathcal{B}_3)} L_{(\mathcal{B}_3\uplus\mathcal{B}_2)/\{(v',\bar{v}_2)\}|\bar{v}_3}\Big]\\ &= \sum_{v'\in\mathcal{V}^w(\mathcal{B}_2)}\Bigg[\sum_{v\in\mathcal{V}^w(\mathcal{B}_1)} L_{(\mathcal{B}_1\uplus\mathcal{B}_2\uplus\mathcal{B}_3)/\{(v',\bar{v}_3),(v,\bar{v}_2)\}|\bar{v}_1}\\ &\qquad - \sum_{v\in\mathcal{V}^w\left((\mathcal{B}_2\uplus\mathcal{B}_3)/\{(v',\bar{v}_3)\}\right)} L_{(\mathcal{B}_2\uplus\mathcal{B}_3\uplus\mathcal{B}_1)/\{(v',\bar{v}_3),(v,\bar{v}_1)\}|\bar{v}_2}\Bigg]\\ &\quad - \sum_{v'\in\mathcal{V}^w(\mathcal{B}_3)}\Bigg[\sum_{v\in\mathcal{V}^w(\mathcal{B}_1)} L_{(\mathcal{B}_1\uplus\mathcal{B}_3\uplus\mathcal{B}_2)/\{(v',\bar{v}_2),(v,\bar{v}_3)\}|\bar{v}_1}\\ &\qquad - \sum_{v\in\mathcal{V}^w\left((\mathcal{B}_3\uplus\mathcal{B}_2)/\{(v',\bar{v}_2)\}\right)} L_{(\mathcal{B}_3\uplus\mathcal{B}_2\uplus\mathcal{B}_1)/\{(v',\bar{v}_2),(v,\bar{v}_1)\}|\bar{v}_3}\Bigg] .\end{aligned}$$

Taking into account that a vertex $v \in \mathcal{V}^w\big((\mathcal{B}_2 \uplus \mathcal{B}_3)/\{(v',\bar{v}_3)\}\big)$ with $v' \in \mathcal{V}^w(\mathcal{B}_2)$ is either one of the white vertices of $\mathcal{B}_3$ or one of the white vertices of $\mathcal{B}_2$ different from v', we split this double commutator into six terms:

$$\begin{aligned}T_{1^{2^3}} &= \sum_{v'\in\mathcal{V}^w(\mathcal{B}_2),v\in\mathcal{V}^w(\mathcal{B}_1)} L_{(\mathcal{B}_1\uplus\mathcal{B}_2\uplus\mathcal{B}_3)/\{(v',\bar{v}_3),(v,\bar{v}_2)\}|\bar{v}_1}\\ T_{2^{13}} &= \sum_{v',v\in\mathcal{V}^w(\mathcal{B}_2),v\neq v'} L_{(\mathcal{B}_2\uplus\mathcal{B}_3\uplus\mathcal{B}_1)/\{(v',\bar{v}_3),(v,\bar{v}_1)\}|\bar{v}_2}\\ T_{2^{3^1}} &= \sum_{v'\in\mathcal{V}^w(\mathcal{B}_2),v\in\mathcal{V}^w(\mathcal{B}_3)} L_{(\mathcal{B}_2\uplus\mathcal{B}_3\uplus\mathcal{B}_1)/\{(v',\bar{v}_3),(v,\bar{v}_1)\}|\bar{v}_2}\\ T_{1^{3^2}} &= \sum_{v'\in\mathcal{V}^w(\mathcal{B}_3),v\in\mathcal{V}^w(\mathcal{B}_1)} L_{(\mathcal{B}_1\uplus\mathcal{B}_3\uplus\mathcal{B}_2)/\{(v',\bar{v}_2),(v,\bar{v}_3)\}|\bar{v}_1}\\ T_{3^{12}} &= \sum_{v',v\in\mathcal{V}^w(\mathcal{B}_3),v\neq v'} L_{(\mathcal{B}_3\uplus\mathcal{B}_2\uplus\mathcal{B}_1)/\{(v',\bar{v}_2),(v,\bar{v}_1)\}|\bar{v}_3}\end{aligned}$$

$$T_{32^1} = \sum_{v' \in \mathcal{V}^w(\mathcal{B}_3), v \in \mathcal{V}^w(\mathcal{B}_2)} L_{(\mathcal{B}_3 \uplus \mathcal{B}_2 \uplus \mathcal{B}_1)/\{(v',\bar{v}_2),(v,\bar{v}_1)\}|\bar{v}_3} \,.$$

It follows that the sum of the three double commutators is:

$$\begin{aligned} & T_{12^3} - T_{21^3} - T_{23^1} - T_{13^2} + T_{31^2} + T_{32^1} \\ & + T_{23^1} - T_{32^1} - T_{31^2} - T_{21^3} + T_{12^3} + T_{13^2} \\ & + T_{31^2} - T_{13^2} - T_{12^3} - T_{32^1} + T_{23^1} + T_{21^3} \,, \end{aligned}$$

and all the terms cancel.

□

8.6.2 Schwinger–Dyson equations

We now derive the SDE's of tensor models and prove that they can be recast as constraints satisfied by the partition function, which respect the Lie algebra introduced in Theorem 8.2.

For any D-colored graph $\mathcal{B}$ and vertex $\bar{v}_j \in \mathcal{B}$, we have the trivial identity:

$$\int [d\mathbb{T}d\bar{\mathbb{T}}] \sum_{b,\bar{b}} \frac{\partial}{\partial \mathbb{T}_{b_j^{\mathcal{D}}}} \left[\delta_{b_j^{\mathcal{D}} \bar{b}_j^{\mathcal{D}}} \left(\prod_{i=1}^{k(\mathcal{B})} \mathbb{T}_{b_i^{\mathcal{D}}} \right) \left(\prod_{\substack{i=1 \\ i \neq j}}^{k(\mathcal{B})} \bar{\mathbb{T}}_{\bar{b}_i^{\mathcal{D}}} \right) \delta^{\mathcal{B}}_{b,\bar{b}} \, e^{-N^{D-1} V(\mathbb{T},\bar{\mathbb{T}})} \right] = 0 \,,$$

which is rewritten as:

$$\begin{aligned} & \int [d\mathbb{T}d\bar{\mathbb{T}}] \, e^{-N^{D-1} V(\mathbb{T},\bar{\mathbb{T}})} \\ & \left(\sum_{v \in \mathcal{V}^w(\mathcal{B})} \mathrm{Tr}_{\mathcal{B}/\{(v,\bar{v}_j)\}}[\mathbb{T},\bar{\mathbb{T}}] - N^{D-1} \frac{1}{\lambda} \mathrm{Tr}_{\mathcal{B}}[\mathbb{T},\bar{\mathbb{T}}] \right. \\ & \left. - N^{D-1} \sum_{\mathcal{B}' \neq \mathcal{B}^{(2)}} \frac{t_{\mathcal{B}'}}{k(\mathcal{B}')! k(\mathcal{B}')!} \sum_{v \in \mathcal{V}^w(\mathcal{B}')} \mathrm{Tr}_{\mathcal{B}' \uplus \mathcal{B}/\{(v,\bar{v}_j)\}}[\mathbb{T},\bar{\mathbb{T}}] \right) = 0 \,. \end{aligned}$$

This equation admits a neat graphical interpretation. Consider the edge of color 0 originating in $\bar{v}_j$ for a graph contributing to the expectation of $\mathrm{Tr}_{\mathcal{B}}[\mathbb{T},\bar{\mathbb{T}}]$. This edge either connects back to a white vertex $v \in \mathcal{B}$, or it connects to the white vertex v of some other invariant $\mathcal{B}'$. The two terms have been discussed in Subsection 3.3.2 and are represented in Figure 3.12 and Figure 3.13.

The graph $\mathcal{B}/\{(v,\bar{v}_j)\}$ might be composed of several connected components (which are graphs with labeled vertices), which we denote $\left[\mathcal{B}/\{(v,\bar{v}_j)\}\right]_{(\rho)}$. Remark that some of these components can consist of a unique edge closed onto itself. We have:

$$\mathrm{Tr}_{\mathcal{B}/\{(v,\bar{v}_j)\}}[\mathbb{T},\bar{\mathbb{T}}] = \prod_{\rho} \mathrm{Tr}_{\left[\mathcal{B}/\{(v,\bar{v}_j)\}\right]_{(\rho)}}[\mathbb{T},\bar{\mathbb{T}}] \,,$$

with the convention that the trace associated to a connected component consisting of a unique edge closed onto itself is N. Denoting $\frac{1}{\lambda} = t_1 = t_{\mathcal{B}^{(2)}}$, the SDE's translate into:

$$\mathbf{L}_{\mathcal{B}|\bar{v}_j}(t_1^{-N^D} Z) = 0 \,;$$

$$\mathbf{L}_{\mathcal{B}|\bar{v}_j} = \sum_{v \in \mathcal{V}^w(\mathcal{B})} \prod_{\rho} \left(-\frac{\left[k\left(\left[\mathcal{B}/\{(v,\bar{v}_j)\}\right]_{(\rho)}\right)!\right]^2}{N^{D-1}} \frac{\partial}{\partial t_{\left[\mathcal{B}/\{(v,\bar{v}_j)\}\right]_{(\rho)}}} \right)$$
$$+ \sum_{\mathcal{B}'} \left(\frac{[k(\mathcal{B}) + k(\mathcal{B}') - 1]!}{k(\mathcal{B}')!} \right)^2 t_{\mathcal{B}'} \sum_{v \in \mathcal{V}^w(\mathcal{B}')} \frac{\partial}{\partial t_{\mathcal{B}' \uplus \mathcal{B}/\{(v,\bar{v}_j)\}}} ,$$

where:

$$\left(-\frac{\left[k\left(\left[\mathcal{B}/\{(v,\bar{v}_j)\}\right]_{(\rho)}\right)!\right]^2}{N^{D-1}} \frac{\partial}{\partial t_{\left[\mathcal{B}/\{(v,\bar{v}_j)\}\right]_{(\rho)}}} \right)$$

is replaced by N if $\left[\mathcal{B}/\{(v,\bar{v}_j)\}\right]_{(\rho)}$ consists of a unique edge closing onto itself.

Theorem 8.3 ([97]). *The operators* $\mathbf{L}_{\mathcal{B}|\bar{v}_j}$ *are a representation of the Lie algebra of Theorem 8.2.*

Proof. The proof is a straightforward computation. We evaluate the commutator of two differential operators:

$$[\mathbf{L}_{\mathcal{B}_1|\bar{v}_1}, \mathbf{L}_{\mathcal{B}_2|\bar{v}_2}] ,$$

term by term. We denote:

$$D_{\mathcal{B}_1|\bar{v}_1} = \sum_{v \in \mathcal{V}^w(\mathcal{B}_1)} \prod_{\rho} \left(-\frac{\left[k\left(\left[\mathcal{B}_1/\{(v,\bar{v}_1)\}\right]_{(\rho)}\right)!\right]^2}{N^{D-1}} \frac{\partial}{\partial t_{\left[\mathcal{B}_1/\{(v,\bar{v}_1)\}\right]_{(\rho)}}} \right) ,$$
$$M_{\mathcal{B}_1|\bar{v}_1} = \sum_{\mathcal{B}'} \left(\frac{[k(\mathcal{B}_1) + k(\mathcal{B}') - 1]!}{k(\mathcal{B}')!} \right)^2 t_{\mathcal{B}'} \sum_{v \in \mathcal{V}^w(\mathcal{B}')} \frac{\partial}{\partial t_{\mathcal{B}' \uplus \mathcal{B}_1/\{(v,\bar{v}_1)\}}} .$$

We have:

$$[D_{\mathcal{B}_1|\bar{v}_1}, D_{\mathcal{B}_2|\bar{v}_2}] = 0 ,$$

while, relabeling some dummy variables, we have:

$$[D_{\mathcal{B}_1|\bar{v}_1}, M_{\mathcal{B}_2|\bar{v}_2}] = \sum_{v_1 \in \mathcal{V}^w(\mathcal{B}_1)} \left(\prod_{\rho'} \frac{-\left[k\left(\left[\mathcal{B}_1/\{(v_1,\bar{v}_1)\}\right]_{(\rho')}\right)!\right]^2}{N^{D-1}} \right)$$
$$\times \sum_{\rho}^{C\left(\mathcal{B}_1/\{(v_1,\bar{v}_1)\}\right)} \left(\frac{\left[k(\mathcal{B}_2) + k\left(\left[\mathcal{B}_1/\{(v_1,\bar{v}_1)\}\right]_{(\rho)}\right) - 1\right]!}{k\left(\left[\mathcal{B}_1/\{(v_1,\bar{v}_1)\}\right]_{(\rho)}\right)!} \right)^2$$
$$\times \sum_{v \in \mathcal{V}^w\left(\left[\mathcal{B}_1/\{(v_1,\bar{v}_1)\}\right]_{(\rho)}\right)} \left(\prod_{\rho' \neq \rho} \frac{\partial}{\partial t_{\left[\mathcal{B}_1/\{(v_1,\bar{v}_1)\}\right]_{(\rho')}}} \right) \frac{\partial}{\partial t_{\left[\mathcal{B}_1/\{(v_1,\bar{v}_1)\}\right]_{(\rho)} \uplus \mathcal{B}_2/\{(v,\bar{v}_2)\}}} .$$

Consider the bubble $(\mathcal{B}_1 \uplus \mathcal{B}_2)/\{(v,\bar{v}_2)\}$. When reducing with respect to $(v_1,\bar{v}_1)$ with v_1 in $\mathcal{B}_1$ it will disconnect into several connected components:

$$
\begin{aligned}
&\big[(\mathcal{B}_1 \uplus \mathcal{B}_2)/\{(v,\bar{v}_2),(v_1,\bar{v}_1)\}\big]_{(\rho)} \\
&= \begin{cases} \big[\mathcal{B}_1/(v_1,\bar{v}_1)\big]_{(\rho)} & \text{if } v \notin \mathcal{V}^w\big([\mathcal{B}_1/(v_1,\bar{v}_1)]_{(\rho)}\big) \\ \big([\mathcal{B}_1/\{(v_1\bar{v}_1)\}]_{(\rho)} \uplus \mathcal{B}_2\big)/\{(v,\bar{v}_2)\} & \text{if } v \in \mathcal{V}^w\big([\mathcal{B}_1/(v_1,\bar{v}_1)]_{(\rho)}\big) \end{cases} .
\end{aligned}
$$

Also, $\sum_{\rho=1}^{C\left(\mathcal{B}_1/\{(v_1,\bar{v}_1)\}\right)} \sum_{v\in\mathcal{V}^w\left([\mathcal{B}_1/\{(v_1,\bar{v}_1)\}]_{(\rho)}\right)} = \sum_{v\in\mathcal{V}^w(\mathcal{B}_1),\, v\neq v_1}$, hence we get:

$$
\begin{aligned}
&[D_{\mathcal{B}_1|\bar{v}_1}, M_{\mathcal{B}_2|\bar{v}_2}] = \sum_{v,v_1\in\mathcal{V}^w(\mathcal{B}_1),\, v\neq v_1} \\
&\prod_{\rho}\left(-\frac{\big[k\big([(\mathcal{B}_1\uplus\mathcal{B}_2)/\{(v,\bar{v}_2),(v_1,\bar{v}_1)\}]_{(\rho)}\big)!\big]^2}{N^{D-1}} \frac{\partial}{\partial t_{\left[(\mathcal{B}_1\uplus\mathcal{B}_2)/\{(v,\bar{v}_2),(v_1,\bar{v}_1)\}\right]_{(\rho)}}} \right),
\end{aligned}
$$

therefore adding and subtracting the cross terms we get:

$$
\begin{aligned}
&[D_{\mathcal{B}_1|\bar{v}_1}, M_{\mathcal{B}_2|\bar{v}_2}] - [D_{\mathcal{B}_2|\bar{v}_2}, M_{\mathcal{B}_1|\bar{v}_1}] \\
&= \sum_{v\in\mathcal{V}^w(\mathcal{B}_1)} \Bigg[\sum_{v_1\in\mathcal{V}^w\left((\mathcal{B}_1\uplus\mathcal{B}_2)/\{(v,\bar{v}_2)\}\right)} \prod_{\rho} \\
&\quad \Bigg(-\frac{\Big[k\Big(\big[\big(\mathcal{B}_1\uplus\mathcal{B}_2/\{(v,\bar{v}_2)\}\big)/\{(v_1,\bar{v}_1)\}\big]_{(\rho)}\Big)!\Big]^2}{N^{D-1}} \\
&\qquad\qquad \times \frac{\partial}{\partial t_{\left[\left(\mathcal{B}_1\uplus\mathcal{B}_2/\{(v,\bar{v}_2)\}\right)/\{(v_1,\bar{v}_1)\}\right]_{(\rho)}}} \Bigg) \Bigg] \\
&- \sum_{v\in\mathcal{V}^w(\mathcal{B}_2)} \Bigg[\sum_{v_2\in\mathcal{V}^w\left((\mathcal{B}_1\uplus\mathcal{B}_2)/\{(v,\bar{v}_1)\}\right)} \prod_{\rho} \\
&\quad \Bigg(-\frac{\Big[k\Big(\big[\big(\mathcal{B}_1\uplus\mathcal{B}_2/\{(v,\bar{v}_1)\}\big)/\{(v_2,\bar{v}_2)\}\big]_{(\rho)}\Big)!\Big]^2}{N^{D-1}} \\
&\qquad\qquad \times \frac{\partial}{\partial t_{\left[\left(\mathcal{B}_1\uplus\mathcal{B}_2/\{(v,\bar{v}_1)\}\right)/\{(v_2,\bar{v}_2)\}\right]_{(\rho)}}} \Bigg) \Bigg] \\
&= \sum_{v\in\mathcal{V}^w(\mathcal{B}_1)} D_{(\mathcal{B}_1\uplus\mathcal{B}_2)/\{(v,\bar{v}_2)\}|\bar{v}_1} - \sum_{v\in\mathcal{V}^w(\mathcal{B}_2)} D_{(\mathcal{B}_1\uplus\mathcal{B}_2)/\{(v,\bar{v}_1)\}|\bar{v}_2} .
\end{aligned}
$$

It remains to compute the final commutator:

$$
\begin{aligned}
&[M_{\mathcal{B}_1|\bar{v}_1}, M_{\mathcal{B}_2|\bar{v}_2}] \\
&= \sum_{\mathcal{B}'} \left(\frac{[k(\mathcal{B}_1) + k(\mathcal{B}') - 1]!}{k(\mathcal{B}')!} \right)^2 t_{\mathcal{B}'} \sum_{v \in \mathcal{V}^w(\mathcal{B}')} \left(\frac{[k(\mathcal{B}_2) + k(\mathcal{B}' \uplus \mathcal{B}_1/\{(v,\bar{v}_1)\}) - 1]!}{k(\mathcal{B}' \uplus \mathcal{B}_1/\{(v,\bar{v}_1)\})!} \right)^2 \\
&\quad \sum_{v' \in \mathcal{V}^w(\mathcal{B}' \uplus \mathcal{B}_1/\{(v,\bar{v}_1)\})} \frac{\partial}{\partial t_{\mathcal{B}' \uplus \mathcal{B}_1 \uplus \mathcal{B}_2/\{(v,\bar{v}_1)(v',\bar{v}_2)\}}} - \{1 \leftrightarrow 2\} .
\end{aligned}
$$

Taking into account that the terms with both $v, v' \in \mathcal{V}^w(\mathcal{B}')$ cancel we get:

$$
\begin{aligned}
&\sum_{v' \in \mathcal{V}^w(\mathcal{B}_1)} \sum_{\mathcal{B}'} \left(\frac{[k((\mathcal{B}_1 \uplus \mathcal{B}_2)/\{(v',\bar{v}_2)\}) + k(\mathcal{B}') - 1]!}{k(\mathcal{B}')!} \right)^2 t_{\mathcal{B}'} \\
&\qquad \times \sum_{v \in \mathcal{V}^w(\mathcal{B}')} \frac{\partial}{\partial t_{\mathcal{B}' \uplus [\mathcal{B}_1 \uplus \mathcal{B}_2/\{(v',\bar{v}_2)\}]/\{(v,\bar{v}_1)\}}} - \{1 \leftrightarrow 2\} ,
\end{aligned}
$$

and finally:

$$
[M_{\mathcal{B}_1|\bar{v}_1}, M_{\mathcal{B}_2|\bar{v}_2}] = \sum_{v \in \mathcal{V}^w(\mathcal{B}_1)} M_{(\mathcal{B}_1 \uplus \mathcal{B}_2)/\{(v,\bar{v}_2)\}|\bar{v}_1} - \sum_{v \in \mathcal{V}^w(\mathcal{B}_2)} M_{(\mathcal{B}_1 \uplus \mathcal{B}_2)/\{(v,\bar{v}_1)\}|\bar{v}_2} ,
$$

which concludes the proof.

□

This Lie algebra admits a closed Lie subalgebra. The leading order observables, the melons, are indexed by trees. It is easy to check that the gluing of the observables $\mathcal{B}_1 \uplus \mathcal{B}_2/\{(v_1, \bar{v}_2)\}$ reproduces the gluing of their associated trees as defined in [92]. The melonic observables are closed under this composition (as the gluing of trees leads to trees), hence the algebra indexed by D-ary trees identified in [92] is a Lie subalgebra of the full constraints algebra.

Chapter 9

The quartic tensor model

In this chapter we will study in more detail the tensor model with an arbitrary quartic interaction. This model has several distinguishing features which render it very interesting.

Using the Hubbard-Stratonovich intermediate field representation, the quartic model can be completely reformulated in terms of a model of edge colored maps that we will define precisely. One can obtain directly the $1/N$ expansion of the quartic model in this formulation. More importantly, one can prove that the $1/N$ expansion of the model holds in the *constructive sense* [76]. The perturbative $1/N$ expansion we have seen in the last chapter consists of showing that each term in the formal power series expansion of a cumulant in the coupling constants obeys a scaling bound in $1/N$. Establishing the $1/N$ expansion in the constructive sense consists of proving that any cumulant is an analytic function in a certain domain in the complex plane, and then showing that in this domain the cumulant writes as a sum of explicit terms in $1/N$ plus a rest term that is analytic and obeys an appropriate scaling bound in $1/N$ [98, 58].

9.1 The quartic models

As we saw in Chapter 2, the unique quadratic invariant is the (scalar) Hermitian pairing of a tensor $\mathbf{T}$ and its dual $\mathbf{T}^{\vee}$:

$$\sum_{n^{\mathcal{D}}\bar{n}^{\mathcal{D}}} \bar{\mathbb{T}}_{\bar{n}^{\mathcal{D}}}\delta_{\bar{n}^{\mathcal{D}}n^{\mathcal{D}}}\mathbb{T}_{n^{\mathcal{D}}} \ .$$

The quadratic invariant is represented by the fundamental melon $\mathcal{B}^{(2)}$ consisting of two vertices connected by D edges.

The connected quartic trace invariants [58] are the invariants represented by the D-colored graphs $\mathcal{B}^{(4)}_{\mathcal{C}}$, indexed by nonempty subsets of colors:

$$\varnothing \neq \mathcal{C} \subset \mathcal{D} = \{1, \dots D\} \ ,$$

with associated permutations (in cycle notation):

$$\tau^{(c)} = \begin{cases} (1)(2)\,, & c \notin \mathcal{C} \\ (12)\,, & c \in \mathcal{C} \end{cases} .$$

Note that $\mathcal{B}^{(4)}_{\mathcal{C}}$ and $\mathcal{B}^{(4)}_{\mathcal{D}\setminus\mathcal{C}}$ are the same up to a relabeling of the vertices. The *quartic melonic invariants* are associated to the graphs $\mathcal{B}^{(4)}_{\{c\}}$ that we encountered in Chapter 5, Theorem 5.1. They are examples of quartic invariants with $|\mathcal{C}| = 1$. The quartic melonic invariants are the unique quartic invariants for $D = 3$, but for $D = 4$ other connected quartic invariants exist.

In order to count only the distinct invariants, from now on the subsets $\mathcal{C}$ always refers to subsets such that $\emptyset \neq \mathcal{C} \subset \mathcal{D}$, $|\mathcal{C}| \leq D/2$ and $1 \notin \mathcal{C}$ if $|\mathcal{C}| = D/2$, which we denote for short $\mathcal{C} \subset \mathcal{D}/2$.

For simplicity we will consider only models in which all the quartic invariants have the same coupling constant, λ. The *general quartic* model is the probability measure:

$$\begin{aligned} d\mu^{(4)} &= \frac{1}{Z^{(4)}(\lambda)} \left(\prod_{n^{\mathcal{D}}} N^{D-1} \frac{d\bar{\mathbb{T}}_{n^{\mathcal{D}}} d\mathbb{T}_{n^{\mathcal{D}}}}{2\imath\pi} \right) \\ &\qquad \times e^{-N^{D-1}\left(\sum_{n^{\mathcal{D}}\bar{n}^{\mathcal{D}}} \bar{\mathbb{T}}_{\bar{n}^{\mathcal{D}}} \delta_{\bar{n}^{\mathcal{D}} n^{\mathcal{D}}} \mathbb{T}_{n^{\mathcal{D}}} + \frac{\lambda}{2} \sum_{\mathcal{C}\subset\mathcal{D}/2} V^{(4)}_{\mathcal{C}}(\bar{\mathbb{T}},\mathbb{T})\right)} , \\ V^{(4)}_{\mathcal{C}}(\bar{\mathbb{T}},\mathbb{T}) &= \sum_{\bar{n}^{\mathcal{D}} n^{\mathcal{D}} m^{\mathcal{D}} \bar{m}^{\mathcal{D}}} \left(\bar{\mathbb{T}}_{\bar{n}^{\mathcal{D}}} \delta_{\bar{n}^{\mathcal{D}\setminus\mathcal{C}} n^{\mathcal{D}\setminus\mathcal{C}}} \mathbb{T}_{n^{\mathcal{D}}} \right) \\ &\qquad \times \delta_{\bar{n}^{\mathcal{C}} m^{\mathcal{C}}} \delta_{\bar{m}^{\mathcal{C}} n^{\mathcal{C}}} \left(\bar{\mathbb{T}}_{\bar{m}^{\mathcal{D}}} \delta_{\bar{m}^{\mathcal{D}\setminus\mathcal{C}} m^{\mathcal{D}\setminus\mathcal{C}}} \mathbb{T}_{m^{\mathcal{D}}} \right) . \end{aligned} \tag{9.1}$$

Of particular interest in the sequel is the so-called *quartic melonic* model in which the subsets $\mathcal{C}$ are restricted to consist of exactly one color $|\mathcal{C}| = 1$ (and in that case the measure is then denoted $d\mu^{(4)}_m$). Besides the general quartic model and the quartic melonic model, one can consider more exotic quartic models where only some of the $\mathcal{C}$s are present, or the coupling constants are different.

The generating function of the cumulants and the cumulants themselves of the measure $d\mu^{(4)}$ are:

$$Z^{(4)}(J,\bar{J}) = \int d\mu^{(4)} \; e^{\sum \bar{\mathbb{T}}_{\bar{a}^{\mathcal{D}}} \bar{J}_{\bar{a}^{\mathcal{D}}} + \sum J_{a^{\mathcal{D}}} \mathbb{T}_{a^{\mathcal{D}}}} ,$$

$$\kappa_{2k}\left(\mathbb{T}_{a_1^{\mathcal{D}}}, \bar{\mathbb{T}}_{\bar{a}_1^{\mathcal{D}}}, \dots \mathbb{T}_{a_k^{\mathcal{D}}}, \bar{\mathbb{T}}_{\bar{a}_k^{\mathcal{D}}}\right) = \frac{\partial^{(2k)}}{\partial J_{a_1^{\mathcal{D}}} \partial \bar{J}_{\bar{a}_1^{\mathcal{D}}} \dots \partial J_{a_k^{\mathcal{D}}} \partial \bar{J}_{\bar{a}_k^{\mathcal{D}}}} \ln Z^{(4)}(J,\bar{J}) \Big|_{\substack{J=0 \\ \bar{J}=0}} .$$

9.1.1 Feynman graphs

The quartic model falls in the category of perturbed Gaussian models discussed in the previous chapter. The generating function of the moments of $\mu^{(4)}$ writes:

$$Z^{(4)}(J,\bar{J}) = \Big[e^{\frac{1}{N^{D-1}} \sum_{\bar{n}^{\mathcal{D}} n^{\mathcal{D}}} \frac{\partial}{\partial \bar{\mathbb{T}}_{\bar{n}^{\mathcal{D}}}} \delta_{\bar{n}^{\mathcal{D}} n^{\mathcal{D}}} \frac{\partial}{\partial \mathbb{T}_{n^{\mathcal{D}}}}} \\ e^{-N^{D-1} \frac{\lambda}{2} \sum_{\substack{\mathcal{C}\subset\mathcal{D} \\ |\mathcal{C}|\leq D/2}} V^{(4)}_{\mathcal{C}}(\bar{\mathbb{T}},\mathbb{T}) + \sum \bar{\mathbb{T}}_{\bar{a}^{\mathcal{D}}} \bar{J}_{\bar{a}^{\mathcal{D}}} + \sum J_{a^{\mathcal{D}}} \mathbb{T}_{a^{\mathcal{D}}}} \Big]_{\mathbb{T},\bar{\mathbb{T}}=0} ,$$

where we have used the expression of the Gaussian integral as a differential operator as detailed in Appendix B.

The generating function and its logarithm $W^{(4)}(J,\bar{J})$ expand in sums over open graphs (respectively open connected graphs) $\mathcal{G}$ having $D+1$ colors, such that all their subgraphs with colors $\{1,\dots D\}$ are one of the quartic bubbles $\mathcal{B}^{(4)}_{\mathcal{C}}$. Such a graph is represented in Figure 9.1. The graphs $\mathcal{G}$ are such that each connected component

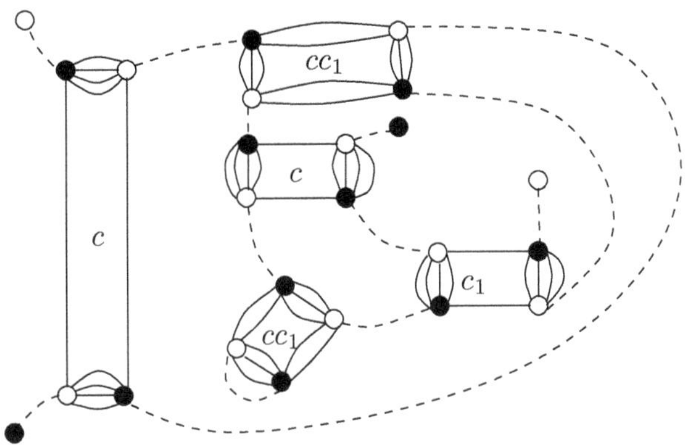

Figure 9.1: A graph of the general quartic model.

$[\mathcal{G}\setminus\mathcal{E}^0(\mathcal{G})]_{(\rho)}$ of the graph obtained from $\mathcal{G}$ by deleting the edges of color zero is one of the $\mathcal{B}^{(4)}_{\mathcal{C}}$, which we denote by:

$$[\mathcal{G}\setminus\mathcal{E}^0(\mathcal{G})]_{(\rho)}=\mathcal{B}^{(4)}_{\mathcal{C}}\ .$$

The equations (8.5) and (8.4) become:

$$\begin{aligned}
Z^{(4)}(J,\bar{J}) = & \sum_{\substack{\mathcal{G}\\ [\mathcal{G}\setminus\mathcal{E}^0(\mathcal{G})]_{(\rho)}=\mathcal{B}^{(4)}_{\mathcal{C}}}} \frac{(-\lambda)^{B(\mathcal{G})}}{k_{\text{int}}(\mathcal{G})!k_{\text{int}}(\mathcal{G})!} \\
& \times N^{(D-1)B(\mathcal{G})-(D-1)[k_{\text{int}}(\mathcal{G})+k_{\text{ext}}(\mathcal{G})]+\sum_{c\in\mathcal{D}}F^{(0,c)}_{\text{int}}(\mathcal{G})}\frac{\text{Tr}_{\partial\mathcal{G}}(J,\bar{J})}{k_{\text{ext}}(\mathcal{G})!k_{\text{ext}}(\mathcal{G})!}\ , \\
\frac{1}{N^D}\ln Z^{(4)}(J,\bar{J}) = & \sum_{\substack{\mathcal{G}\text{ connected}\\ [\mathcal{G}\setminus\mathcal{E}^0(\mathcal{G})]_{(\rho)}=\mathcal{B}^{(4)}_{\mathcal{C}}}} \frac{(-\lambda)^{B(\mathcal{G})}}{k_{\text{int}}(\mathcal{G})!k_{\text{int}}(\mathcal{G})!} \\
& \times N^{-D+(D-1)B(\mathcal{G})-(D-1)[k_{\text{int}}(\mathcal{G})+k_{\text{ext}}(\mathcal{G})]+\sum_{c\in\mathcal{D}}F^{(0,c)}_{\text{int}}(\mathcal{G})}\frac{\text{Tr}_{\partial\mathcal{G}}(J,\bar{J})}{k_{\text{ext}}(\mathcal{G})!k_{\text{ext}}(\mathcal{G})!}\ .
\end{aligned} \tag{9.2}$$

These expressions follow by observing that every bubble $\mathcal{B}^{(4)}_{\mathcal{C}}$ brings a $(-\lambda)N^{D-1}$ factor, every edge of color 0 brings a factor $1/N^{D-1}$, every internal face of colors $(0,c)$ brings an N and the external faces of $\mathcal{G}$ reconstitute the trace invariant corresponding to $\partial\mathcal{G}$. Remark that, as all the bubbles with colors $\{1,\dots D\}$ have two white vertices, $k_{\text{int}}(\mathcal{G}) = 2B(\mathcal{G})$, with $B(\mathcal{G})$ the number of these bubbles.

The graphs of the quartic model are one to one to edge multicolored maps with cilia, which we define now.

9.2 Edge multicolored maps

The notions we introduce here generalize the combinatorial maps of Chapter 3, Section 3.4.

Definition 9.1. *A* combinatorial map with external cilia *is:*

- *a finite set of halfedges $\mathcal{S}$, which is the disjoint union of the set of* internal halfedges $\mathcal{S}_{\text{int}}$ *and the set of external* cilia $\mathcal{S}_{\text{ext}}$. *We denote the elements of $\mathcal{S}$ by $h \in \mathcal{S}$.*
- *a permutation σ on $\mathcal{S}$.*
- *an involution α on the set of internal halfedges $\mathcal{S}_{\text{int}}$* having no fixed points. *The involution α on $\mathcal{S}_{\text{int}}$ is extended to an involution on $\mathcal{S}$ (which we denote also by α)* with fixed points *by imposing that $\alpha(h) = h$, $\forall h \in \mathcal{S}_{\text{ext}}$, that is by imposing that all the external cilia are fixed points of α.*

The presence of the cilia alters the notion of faces of the map.

Definition 9.2. *The cycles of the permutation $\sigma\alpha$ fall in two categories:*

- *the cycles of $\sigma\alpha$ which* do not contain any cilium, *that is such that:*
$$\alpha(h) \notin \mathcal{S}_{\text{ext}} \ ,$$
for any h in the cycle are called the internal faces *of the map.*
- *the cycles of $\sigma\alpha$ which contain cilia. They are further subdivided into sequences of halfedges separating two consecutive cilia:*
$$h, \sigma\alpha(h), \dots, (\sigma\alpha)^r(h) \quad \text{such that} \begin{cases} \alpha(h) \in \mathcal{S}_{\text{ext}} \\ \forall 0 < s < r, \ \alpha(\sigma\alpha)^s(h) \notin \mathcal{S}_{\text{ext}} \\ \alpha(\sigma\alpha)^r(h) \in \mathcal{S}_{\text{ext}} \end{cases} ,$$
called external strands.

Definition 9.3. *The* corners *of a combinatorial map are the couples:*
$$\big(h, \sigma(h)\big), \ \forall h \in \mathcal{S} \ .$$

The corners can be pictured as the pieces of vertices comprised between two consecutive halfedges or between a halfedge and a cilium.

An edge multicolored map with cilia is a map with cilia whose edges are furthermore colored by a subset $\mathcal{C}$ of colors.

Definition 9.4. *An* edge multicolored combinatorial map with external cilia $\mathcal{M}$ *is:*

- *a finite set $\mathcal{S}$ which is the disjoint union of the sets $\mathcal{S}_{\text{int}}^{(\mathcal{C})}$ of internal halfedges of the colors $\mathcal{C}$ and the set $\mathcal{S}_{\text{ext}}$ of external* cilia*:*

$$\mathcal{S} = \left(\bigsqcup_{\mathcal{C} \subset \mathcal{D}/2} \mathcal{S}_{\text{int}}^{(\mathcal{C})} \right) \bigsqcup \mathcal{S}_{\text{ext}} \,.$$

- *a permutation σ on $\mathcal{S}$.*
- *for every $\mathcal{C} \subset \mathcal{D}/2$, an involution $\alpha^{(\mathcal{C})}$ on the set of internal halfedges of colors $\mathcal{C}$, $\mathcal{S}_{\text{int}}^{(\mathcal{C})}$, having no fixed points. We extend the involutions $\alpha^{(\mathcal{C})}$ to the whole of $\mathcal{S}$ by setting $\alpha^{(\mathcal{C})}(h) = h$, $\forall h \in \mathcal{S} \setminus \mathcal{S}^{(\mathcal{C})}$.*

An example of an edge multicolored map with external cilia is presented in Figure 9.2.

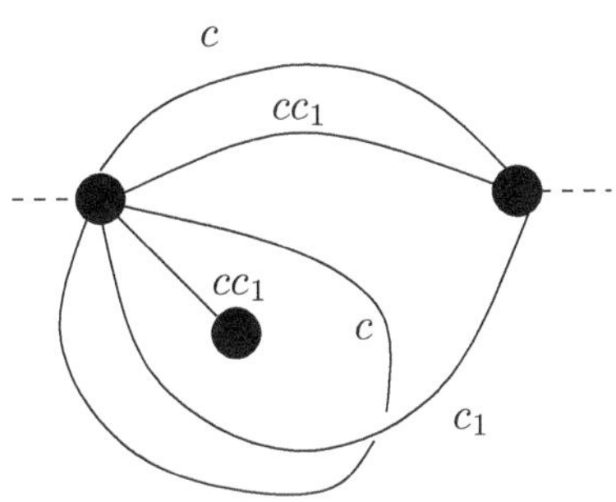

Figure 9.2: An edge multicolored map with external cilia (the cilia are the dashed halfedges in the figure).

Let us consider an edge multicolored map $\mathcal{M}$. For any color $c \in \{1, \dots D\}$, by erasing the edges of colors $\mathcal{C}$ with $c \notin \mathcal{C}$, one obtains the *submap* $\mathcal{M}^c$ with *external cilia and scars* of $\mathcal{M}$.

Definition 9.5. *The submap* $\mathcal{M}^c$ with external cilia and scars *of the edge multicolored map with cilia:*

$$\mathcal{M} = \left(\left(\bigsqcup_{\mathcal{C} \subset \mathcal{D}/2} \mathcal{S}_{\text{int}}^{(\mathcal{C})} \right) \bigsqcup \mathcal{S}_{\text{ext}}, \sigma, \left\{ \alpha^{(\mathcal{C})} \middle| \mathcal{C} \subset \mathcal{D}/2 \right\} \right) ,$$

is (see Figure 9.3 for an example):

- *the finite set $\mathcal{S}^c$ of halfedges of color c, which is the disjoint union of:*
  - *the* internal halfedges *of color c:*

$$\mathcal{S}_{\text{int}}^c = \bigsqcup_{\mathcal{C},\, c \in \mathcal{C} \subset \mathcal{D}/2} \mathcal{S}_{\text{int}}^{(\mathcal{C})}$$

– *the* external cilia*:*

$$\mathcal{S}^c_{\text{ext}} = \mathcal{S}_{\text{ext}}$$

– *the* scars*:*

$$\mathcal{S}^c_{\text{scars}} = \bigsqcup_{\mathcal{C},\ c\notin\mathcal{C}\subset\mathcal{D}/2} \mathcal{S}^{(\mathcal{C})}_{\text{int}}$$

that is:

$$\mathcal{S}^c = \mathcal{S}^c_{\text{int}} \bigsqcup \mathcal{S}^c_{\text{scars}} \bigsqcup \mathcal{S}^c_{\text{ext}} \,.$$

- *the permutation σ on $\mathcal{S}$.*
- *the involution α^c on $\mathcal{S}^c_{\text{int}}$ having no fixed points defined by:*

$$\alpha^c(h) = \alpha^{(\mathcal{C})}(h) \ \text{if}\ h \in \mathcal{S}^{(\mathcal{C})}_{\text{int}} \ \text{and}\ c \in \mathcal{C}\,.$$

The involution α^c is extended to an involution on $\mathcal{S}$ with fixed points by imposing that each scar and each cilium is a fixed point of α^c, i.e. $\alpha^c(h) = h$ for $h \in \mathcal{S}^c \setminus \mathcal{S}^c_{\text{int}}$.

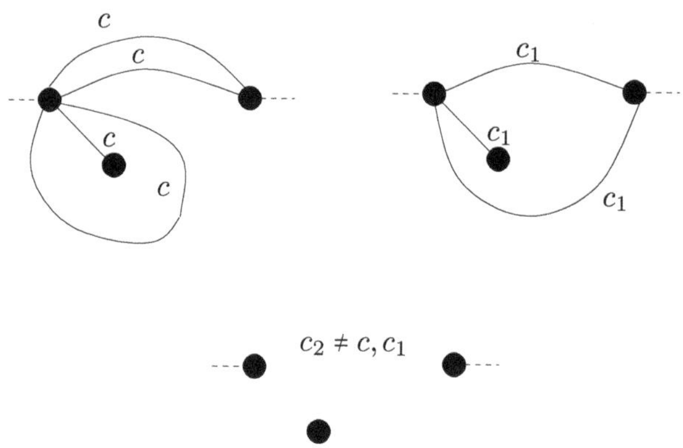

Figure 9.3: Submaps of color c, c_1 and respectively $c_2 \neq c, c_1$ of the map in Figure 9.2. The cilia are the dashed halfedges, and we did not represent the scars in the various submaps.

The map $\mathcal{M}^c$ can be disconnected, and some of its connected components can consist of an isolated vertex. The corners of an edge multicolored map $\mathcal{M}$ and its submaps $\mathcal{M}^c$ are still defined as the couples $\big(h, \sigma(h)\big)$, for all the halfedges h.

Definition 9.6. *The* faces *of an edge multicolored map with cilia and scars $\mathcal{M}^c$ are the cycles of the permutation $\sigma\alpha^c$. They are divided into:*

- internal faces. *They are the cycles of $\sigma\alpha^c$ that do not contain any external cilium, that is, for every h in the cycle:*

$$\alpha^c(h) \notin \mathcal{S}^c_{\text{ext}}\,.$$

The internal faces can, however, contain any number of scars,

$$\alpha^c(h) = h \in \mathcal{S}^c_{\text{scars}} .$$

- external faces. *They are the cycles of* $\sigma\alpha^c$ *containing cilia. As before, they are further subdivided in sequences of halfedges separating two consecutive cilia:*

$$h, \sigma\alpha^c(h), \dots, (\sigma\alpha^c)^r(h) \text{ such that } \begin{cases} \alpha^c(h) \in \mathcal{S}^c_{\text{ext}} \\ \forall s,\ 0 < s < r,\ \ \alpha^c(\sigma\alpha^c)^s(h) \notin \mathcal{S}^c_{\text{ext}} \\ \alpha^c(\sigma\alpha^c)^r(h) \in \mathcal{S}^c_{\text{ext}} \end{cases},$$

  called external strands.

If the map $\mathcal{M}^c$ *has a connected component consisting of an isolated vertex, it counts as a face.*

The internal faces and the external strands of color c *of an edge multicolored map* $\mathcal{M}$ *are the internal faces and the external strands of the submap* $\mathcal{M}^c$*. We denote by* $F^c_{\text{int}}(\mathcal{M})$ *the number of internal faces of color* c *of* $\mathcal{M}$ *and by* $F_{\text{int}}(\mathcal{M})$ *the total number of internal faces of* $\mathcal{M}$.

Remark that the faces of an edge multicolored map are colored by a *unique* color. For example, the faces of colors c, c_1 and $c_2 \neq c, c_1$ of the map in Figure 9.2 are the faces of its submaps represented in Figure 9.3. Also, observe that the scars play no role in the definition of the faces. Their only relevance is to keep track of the corners of the map, and in particular to ensure that an edge multicolored map $\mathcal{M}$ and its submaps $\mathcal{M}^c$ have the same corners.

There exists a bijection between, on the one hand, the $(D+1)$ colored graphs $\mathcal{G}$ such that all their subgraphs with colors $\{1, \dots D\}$ are one of the $\mathcal{B}^{(4)}_{\mathcal{C}}$ for some $\mathcal{C} \subset \mathcal{D}/2$ and, on the other, the edge multicolored maps with cilia $\mathcal{M}$.

Obtaining $\mathcal{M}$ *from* $\mathcal{G}$. The internal and boundary vertices of $\mathcal{G}$ are partitioned between the various $\mathcal{B}^{(4)}_{\mathcal{C}}$s. The four vertices of a bubble $\mathcal{B}^{(4)}_{\mathcal{C}}$ are further partitioned in two pairs, such that the vertices in a pair are connected by the edges in $\mathcal{D} \setminus \mathcal{C}$, and the pairs are connected in-between them by the edges in $\mathcal{C}$. We associate with each pair of vertices of $\mathcal{B}^{(4)}_{\mathcal{C}}$ connected by the edges in $\mathcal{D} \setminus \mathcal{C}$ an internal halfedge with colors $\mathcal{C}$, $h \in \mathcal{S}^{(\mathcal{C})}_{\text{int}}$ of the map.

The external vertices of $\mathcal{G}$ are partitioned into pairs connected by (simple) paths such that any two consecutive vertices in the path are either connected by an edge of color 0 or by the edges of colors $\mathcal{D} \setminus \mathcal{C}$ in some $\mathcal{B}^{(4)}_{\mathcal{C}}$. We associate an external cilium $h \in \mathcal{S}_{\text{ext}}$ with every such pair.

For any halfedge (internal or cilium) $h \in \mathcal{S}$, the white vertex of $\mathcal{G}$ in the pair corresponding to h is connected by an edge of color 0 to the black vertex of $\mathcal{G}$ belonging to the pair corresponding to some halfedge (internal or cilium) h'. We set $\sigma(h) = h'$.

Any halfedge $h \in \mathcal{S}_{\text{int}}^{(\mathcal{C})}$ with colors $\mathcal{C}$ represents a pair of vertices belonging to some $\mathcal{B}_{\mathcal{C}}^{(4)}$. The second pair of vertices of the same $\mathcal{B}_{\mathcal{C}}^{(4)}$ corresponds to some other halfedge of colors $\mathcal{C}$, h'. We set $\alpha^{(\mathcal{C})}(h) = h'$.

Obtaining $\mathcal{G}$ from $\mathcal{M}$. To every internal halfedge $h \in \mathcal{S}_{\text{int}}^{(\mathcal{C})}$ of colors $\mathcal{C}$ of the map $\mathcal{M}$ we associate a pair of black and white vertices connected by edges of colors $\mathcal{D} \setminus \mathcal{C}$, and we connect the pairs corresponding to h and $\alpha^{(\mathcal{C})}(h)$ by edges of colors $\mathcal{C}$. To any cilium of $\mathcal{C}$ we associate a pair of external black and white vertices. For every halfedge $h \in \mathcal{S}$, we connect the white vertex corresponding to h to the black vertex corresponding to $\sigma(h)$ by an edge of color 0.

The reader can convince themselves that the graph in Figure 9.1 and the edge multicolored map with cilia in Figure 9.2 are related by this construction.

Definition 9.7. *The* boundary graph $\partial\mathcal{M}$ *of an edge multicolored map with cilia $\mathcal{M}$ is the boundary graph $\partial\mathcal{G}$ of its associated edge colored graph.*

The boundary graph $\partial\mathcal{M}$ can be obtained directly from $\mathcal{M}$ by associating a black and a white external vertex to each of its cilia, and connecting the white vertex associated to the cilium h with the black vertex associated to the cilium h' by an edge of color c if there exists an external strand of color c of the map $\mathcal{M}$ (i.e. of its submap $\mathcal{M}^c$) going from h to h'.

An edge multicolored map can be represented either, as we have done so far, as a map with colored edges or as a *multiribbon graph.* This representation is obtained as

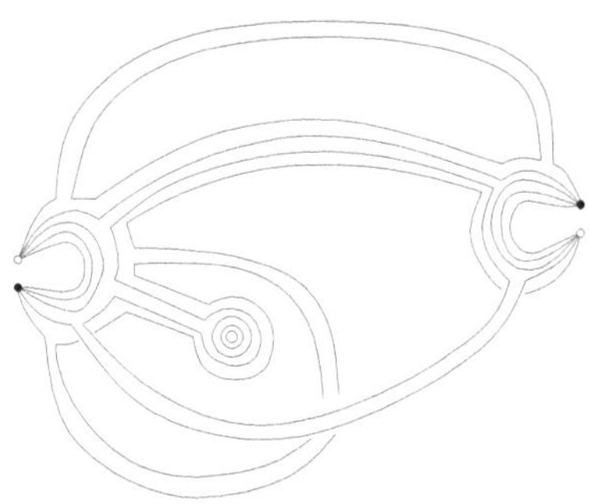

Figure 9.4: Multiribbon graph representation of an edge multicolored map with external cilia.

follows (see Figure 9.4 for the representation of the map in Figure 9.2):

- we represent every internal vertex by D concentric strands colored 1 to D. By convention we orient the strands clockwise.
- we represent every ciliated vertex as D strands connecting a white and a black external vertex. By convention we orient the strands from the white to the black vertex.

- for any edge of colors $\mathcal{C}$ we connect the strands with colors in the set $\mathcal{C}$ on its two end vertices by a ribbon. The edges of the graph are thus made of $|\mathcal{C}|$ parallel ribbons.

The advantage of this second representation is that the faces and the external strands are easily read off: the internal faces are the closed strands, while the external strands are the strands connecting a white and a black external vertex. The boundary graph of $\mathcal{M}$ is the graph obtained by erasing the internal faces of $\mathcal{M}$ in this representation.

All the relevant elements of the $D+1$ colored graph $\mathcal{G}$ are encoded in its associated edge multicolored map: the bubbles $\mathcal{B}^{(4)}_{\mathcal{C}}$ become the edges of the multicolored map, the internal faces of colors $(0,c)$ become the faces of color c of the map, while the external faces of colors $(0,c)$ become the external strands of color c of the map. Also, $k(\partial\mathcal{M})$, the number of white vertices of the boundary graph of $\mathcal{M}$, is nothing but the number of cilia of the map $\mathcal{M}$. Denoting $E(\mathcal{M})$ the number of edges of $\mathcal{M}$ and $F^c_{\text{int}}(\mathcal{M})$ the number of internal faces of color c of $\mathcal{M}$, the equations (9.2) become:

$$
\begin{aligned}
Z^{(4)}(J,\bar{J}) &= \sum_{\mathcal{M}} (-\lambda)^{E(\mathcal{M})}\,\mathrm{Tr}_{\partial\mathcal{M}}(J,\bar{J})\,, \\
&\qquad \times N^{-(D-1)E(\mathcal{M})-(D-1)k(\partial\mathcal{M})+\sum_{c=1}^{D}F^c_{\text{int}}(\mathcal{M})} \\
\frac{1}{N^D}\ln Z^{(4)}(J,\bar{J}) &= \sum_{\mathcal{M}\text{ connected}} (-\lambda)^{E(\mathcal{M})}\,\mathrm{Tr}_{\partial\mathcal{M}}(J,\bar{J}) \\
&\qquad \times N^{-D-(D-1)E(\mathcal{M})-(D-1)k(\partial\mathcal{M})+\sum_{c=1}^{D}F^c_{\text{int}}(\mathcal{M})}\,,
\end{aligned} \tag{9.3}
$$

where the sum runs over edge multicolored maps with *unlabeled* vertices. Each edge of the multicolored map contributes $N^{-(D-1)}$, each internal face contributes N and each cilium contributes an $N^{-(D-1)}$. These Feynman rules can be obtained directly as Feynman rules of a perturbed Gaussian measure.

It is important to note that the maps we obtain *never* have more than one cilium per ciliated vertex.

9.2.1 Maps with ρ and τ edges

Our purpose is to establish a bound on the number of internal faces of a map. This is somewhat involved and requires us to use a more complicated object, which we call a *map with ρ and τ edges*.

To any edge colored map $\mathcal{M}$ with k cilia $h_1,\dots h_k$, and any two D-uples of permutations of k elements $\tau^{\mathcal{D}}$ and $\rho^{\mathcal{D}}$ we associate an edge colored map $\mathcal{M}_{\rho^{\mathcal{D}}\tau^{\mathcal{D}}}$ obtained by replacing each cilium by $2D$ halfedges, a cilium and $2D$ halfedges in the order:

$$
h_i \Rightarrow \bar{h}_i^{1,\tau}\bar{h}_i^{2\tau}\dots\bar{h}_i^{D\tau}\bar{h}_i^{1,\rho}\bar{h}_i^{2\rho}\dots\bar{h}_i^{D\rho}h_i h_i^{D\rho}\dots h_i^{1,\rho}h_i^{D\tau}\dots h_i^{1,\tau}\,,
$$

and connecting $h_i^{c,\rho}$ with $\bar{h}^{c,\rho}_{\rho^{(c)}(i)}$ by an edge of color c and respectively $h_i^{c,\tau}$ with $\bar{h}^{c,\tau}_{\tau^{(c)}(i)}$ by an edge of color c. We call the new edges connecting $h_i^{c,\rho}$ with $\bar{h}^{c,\rho}_{\rho^{(c)}(i)}$ *ρ-edges* and the new edges connecting $h_i^{c,\tau}$ with $\bar{h}^{c,\tau}_{\tau^{(c)}(i)}$ *τ-edges*. An example of a map with ρ-edges and τ-edges is presented in Figure 9.5.

The internal faces and external strands of $\mathcal{M}_{\rho^{\mathcal{D}}\tau^{\mathcal{D}}}$ are classified as follows:

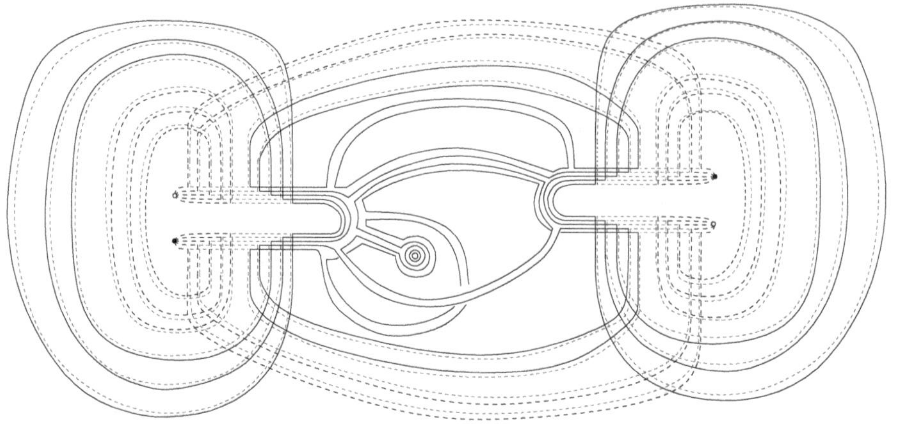

Figure 9.5: Map with ρ and τ edges.

- all the internal faces of $\mathcal{M}$ are internal faces of $\mathcal{M}_{\rho^{\mathcal{D}}\tau^{\mathcal{D}}}$. They are represented by solid closed strands in Figure 9.5. We denote their number by $F_{\text{int}}(\mathcal{M})$.
- the external strands of $\mathcal{M}$ are now reconnected among themselves by the τ-edges. They become internal faces of $\mathcal{M}_{\rho^{\mathcal{D}}\tau^{\mathcal{D}}}$ and are also represented by solid closed strands in Figure 9.5. We call such faces "intτ" faces of $\mathcal{M}_{\rho^{\mathcal{D}}\tau^{\mathcal{D}}}$ and denote their number by $F_{\text{int}\tau}(\mathcal{M}_{\rho^{\mathcal{D}}\tau^{\mathcal{D}}})$.
- there are new faces formed by alternating cycles of τ-edges and ρ-edges. They are internal faces of $\mathcal{M}_{\rho^{\mathcal{D}}\tau^{\mathcal{D}}}$ and are represented as dotted closed strands in Figure 9.5. We call them "$\rho\tau^{-1}$ faces" of $\mathcal{M}_{\rho^{\mathcal{D}}\tau^{\mathcal{D}}}$. Observe that the number of $\rho\tau^{-1}$ faces of color c is the number of cycles of the permutation $\rho^{(c)}(\tau^{(c)})^{-1}$. We denote the total number of such faces by $F_{\rho\tau^{-1}}(\mathcal{M}_{\rho^{\mathcal{D}}\tau^{\mathcal{D}}})$.
- the external strands of $\mathcal{M}_{\rho^{\mathcal{D}}\tau^{\mathcal{D}}}$ are formed by exactly one ρ-edge each. They are represented as dashed strands in Figure 9.5.

In this representation the ρ-edges are ribbon edges separating a dashed external strand and a dotted face, and the τ-edges are ribbon edges separating a dotted face and a solid face. The boundary graph $\partial\mathcal{M}_{\rho^{\mathcal{D}}\tau^{\mathcal{D}}}$ is exactly the edge colored graph associated to the D-uple of permutations $\rho^{\mathcal{D}}$. From the classification of faces of $\mathcal{M}_{\rho^{\mathcal{D}}\tau^{\mathcal{D}}}$ we have:

$$F_{\text{int}}(\mathcal{M}_{\rho^{\mathcal{D}}\tau^{\mathcal{D}}}) = F_{\text{int}}(\mathcal{M}) + F_{\text{int}\tau}(\mathcal{M}_{\rho^{\mathcal{D}}\tau^{\mathcal{D}}}) + F_{\rho\tau^{-1}}(\mathcal{M}_{\rho^{\mathcal{D}}\tau^{\mathcal{D}}}) \ .$$

Remark 9.1. A remark is in order.

Tracing the external strands. Among the $[k(\partial\mathcal{M})!]^2$ maps $\mathcal{M}_{\rho^{\mathcal{D}}\tau^{\mathcal{D}}}$ associated to $\mathcal{M}$ one stands out: the one in which the permutations $\rho^{\mathcal{D}}$ and $\tau^{\mathcal{D}}$ track the external strands of $\mathcal{M}$. This map is obtained by setting $\tau^{(c)}(i) = \rho^{(c)}(i) = j$ if h_j is the end cilium of the external strand of color c starting at the cilium h_i (the reader can convince himself that it is this map we represented in Figure 9.5). In this case:

- the boundary graphs of $\mathcal{M}_{\rho^{\mathcal{D}}\tau^{\mathcal{D}}}$ and $\mathcal{M}$ coincide:

$$\partial\mathcal{M}_{\rho^{\mathcal{D}}\tau^{\mathcal{D}}} = \partial\mathcal{M}\,.$$

- the intτ faces of $\mathcal{M}_{\rho^{\mathcal{D}}\tau^{\mathcal{D}}}$ are one to one to the external strands of $\mathcal{M}$ and in particular:

$$F_{\mathrm{int}\tau}(\mathcal{M}_{\rho^{\mathcal{D}}\tau^{\mathcal{D}}}) = Dk(\partial\mathcal{M})\,.$$

- all the $\rho\tau^{-1}$ faces of $\mathcal{M}_{\rho^{\mathcal{D}}\tau^{\mathcal{D}}}$ have length 2 and in particular:

$$F_{\rho\tau^{-1}}(\mathcal{M}_{\rho^{\mathcal{D}}\tau^{\mathcal{D}}}) = Dk(\partial\mathcal{M})\,.$$

The main bound on the number of faces of a map is encoded in the following theorem.

Theorem 9.1 (Main bound). *Let $\mathcal{M}$ be a connected edge multicolored map with $V(\mathcal{M})$ vertices (out of which $k(\partial\mathcal{M})$ are ciliated and each ciliated vertex has only one cilium) and $E(\mathcal{M})$ edges. Let furthermore $\rho^{\mathcal{D}}$ and $\tau^{\mathcal{D}}$ be two D-uple of permutations, and denote $C(\rho^{\mathcal{D}})$ the number of connected components of the graph of the permutations $\rho^{\mathcal{D}}$. We have:*

$$\begin{aligned} &F_{\mathrm{int}}(\mathcal{M}) + F_{\mathrm{int}\tau}(\mathcal{M}_{\rho^{\mathcal{D}}\tau^{\mathcal{D}}}) + F_{\rho\tau^{-1}}(\mathcal{M}_{\rho^{\mathcal{D}}\tau^{\mathcal{D}}}) \\ &\qquad \le 1 + (D+1)k(\partial\mathcal{M}) + (D-1)V(\mathcal{M}) - C(\rho^{\mathcal{D}}) \\ &\qquad\quad + \frac{D}{2}[E(\mathcal{M}) - V(\mathcal{M}) + 1]\,. \end{aligned} \tag{9.4}$$

In particular, choosing $\rho^{\mathcal{D}}$ and $\tau^{\mathcal{D}}$ to track the external strands of $\mathcal{M}$ we get:

$$\begin{aligned} F_{\mathrm{int}}(\mathcal{M}) \le\, & 1 - (D-1)k(\partial\mathcal{M}) - C(\partial\mathcal{M}) + (D-1)V(\mathcal{M}) \\ & + \frac{D}{2}[E(\mathcal{M}) - V(\mathcal{M}) + 1]\,. \end{aligned} \tag{9.5}$$

Furthermore, the two inequalities are saturated if and only if:

- *there exists a tree in $\mathcal{M}$ such that all the edges in this tree have exactly one color,*
- *adding the* excess edges *(the remaining edges, not belonging to the tree, i.e. the loop edges in the physics literature) one by one in some order, each addition of an edge (which can have up to $D/2$ colors) increases the number of faces by exactly $D/2$.*

If all the edges of the map have only one color, $|\mathcal{C}| = 1$, then the bounds become:

$$\begin{aligned} &F_{\mathrm{int}}(\mathcal{M}) + F_{\mathrm{int}\tau}(\mathcal{M}_{\rho^{\mathcal{D}}\tau^{\mathcal{D}}}) + F_{\rho\tau^{-1}}(\mathcal{M}_{\rho^{\mathcal{D}}\tau^{\mathcal{D}}}) \\ &\qquad \le 1 + (D+1)k(\partial\mathcal{M}) + (D-1)V(\mathcal{M}) - C(\rho^{\mathcal{D}}) \\ &\qquad\quad + [E(\mathcal{M}) - V(\mathcal{M}) + 1]\,, \end{aligned}$$

$$F_{\text{int}}(\mathcal{M}) \le 1 - (D-1)k(\partial\mathcal{M}) - C(\partial\mathcal{M}) + (D-1)V(\mathcal{M}) \\ + [E(\mathcal{M}) - V(\mathcal{M}) + 1] \,.$$

In this case the inequalities are saturated if and only if there exists a tree in $\mathcal{M}$ such that by adding the excess edges one by one in some order, each addition of an edge increases the number of faces by exactly 1.

Proof. All the edges of $\mathcal{M}$ are also edges of $\mathcal{M}_{\rho^{\mathcal{D}}\tau^{\mathcal{D}}}$. In $\mathcal{M}_{\rho^{\mathcal{D}}\tau^{\mathcal{D}}}$ these edges only involve the solid faces (the internal faces of $\mathcal{M}$ or the intτ faces of $\mathcal{M}_{\rho^{\mathcal{D}}\tau^{\mathcal{D}}}$) because the edges of $\mathcal{M}$ are (multi) ribbons made of solid strands only.

Deleting an edge of $\mathcal{M}$ and reconnecting the faces of colors $\mathcal{C}$ on its end vertices (i.e. replacing the edge by two scars on its end vertices) will change the structure of the solid faces of $\mathcal{M}_{\rho^{\mathcal{D}}\tau^{\mathcal{D}}}$, but it will not affect the permutations $\rho^{\mathcal{D}}$ and $\tau^{\mathcal{D}}$. In particular, the boundary graph $\partial\mathcal{M}_{\rho^{\mathcal{D}}\tau^{\mathcal{D}}}$, being the graph of the permutations $\rho^{\mathcal{D}}$, does not change. The deletion of an edge of the map in Figure 9.5 is presented in Figure 9.6.

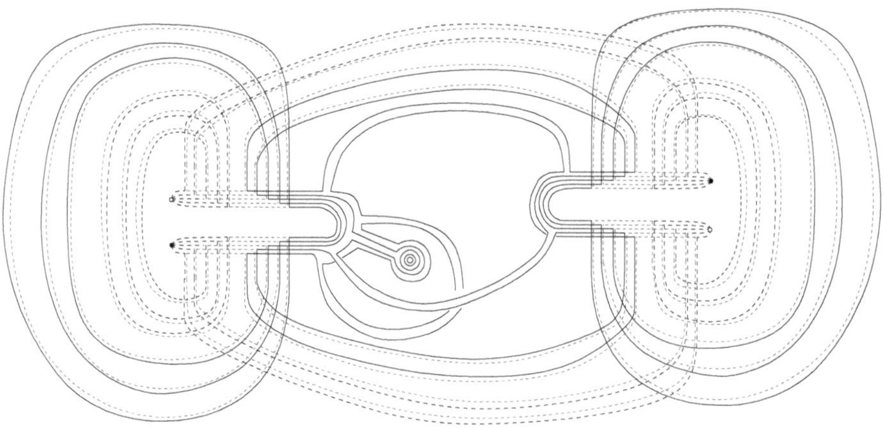

Figure 9.6: Deletion of an edge of $\mathcal{M}$.

Because there are at most $D/2$ faces going through an edge of $\mathcal{M}$, deleting it cannot increase the total number of solid faces of $\mathcal{M}_{\rho^{\mathcal{D}}\tau^{\mathcal{D}}}$ by more than $D/2$. Denoting $\mathfrak{T}$ a tree in $\mathcal{M}$, we obtain a first bound:

$$F_{\text{int}}(\mathcal{M}_{\rho^{\mathcal{D}}\tau^{\mathcal{D}}}) \le F_{\text{int}}(\mathfrak{T}_{\rho^{\mathcal{D}}\tau^{\mathcal{D}}}) + \frac{D}{2}[E(\mathcal{M}) - V(\mathcal{M}) + 1] \,.$$

If all the edges of the map have only one color then there is exactly one face going through any edge and the number of faces increases by at most one, hence:

$$F_{\text{int}}(\mathcal{M}_{\rho^{\mathcal{D}}\tau^{\mathcal{D}}}) \le F_{\text{int}}(\mathfrak{T}_{\rho^{\mathcal{D}}\tau^{\mathcal{D}}}) + [E(\mathcal{M}) - V(\mathcal{M}) + 1] \,.$$

Remark that, due to the ρ and τ-edges, $\mathfrak{T}_{\rho^{\mathcal{D}}\tau^{\mathcal{D}}}$ is not a tree (but of course it becomes a tree by deleting all the ρ and τ-edges). The theorem follows if one can show that for any edge multicolored tree with ρ and τ edges having V vertices out of which k are ciliated (and the ciliated vertices have only one cilium) the bound:

$$F_{\text{int}}(\mathfrak{T}_{\rho^{\mathcal{D}}\tau^{\mathcal{D}}}) \le 1 + (D+1)k - C(\rho^{\mathcal{D}}) + (D-1)V ,$$
$$\text{with} \quad F_{\text{int}}(\mathfrak{T}_{\rho^{\mathcal{D}}\tau^{\mathcal{D}}}) = F_{\text{int}}(\mathfrak{T}) + F_{\text{int}\tau}(\mathfrak{T}_{\rho^{\mathcal{D}}\tau^{\mathcal{D}}}) + F_{\rho\tau^{-1}}(\mathfrak{T}_{\rho^{\mathcal{D}}\tau^{\mathcal{D}}}) ,$$

holds.

The proof of this bound is significantly more involved: it consists of deleting one by one the leaves (univalent vertices) of the tree together with the tree edges they are hooked to and in tracking the evolution of the quantity:

$$F_{\text{int}}(\mathfrak{T}_{\rho^{\mathcal{D}}\tau^{\mathcal{D}}}) + C(\rho^{\mathcal{D}}) ,$$

under this move.

Deletion of a vertex. An edge multicolored tree with ρ and τ edges is presented in Figure 9.7.

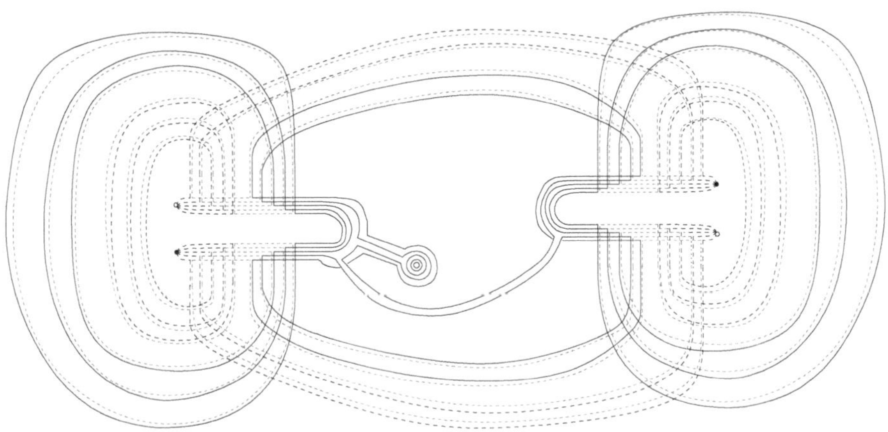

Figure 9.7: Tree with ρ and τ edges.

Choose a univalent vertex in $\mathfrak{T}$, labeled say v, connected to the rest of the tree by an edge with colors $\mathcal{C}$. By univalent we mean that v is hooked to a unique edge of $\mathfrak{T}$ (if v is ciliated then it is hooked furthermore to $2D$ ρ and τ-edges). There exists a unique vertex in the tree to which v is hooked, called its *ancestor*. The deletion of v is defined as follows:

- if v has no cilium the deletion consists of erasing all the faces with color in $\mathcal{D} \setminus \mathcal{C}$ containing v and reconnecting the faces with color in $\mathcal{C}$ passing through v directly on its ancestor (i.e. erasing v and the edge hooked to it and replacing the other end of the edge by a scar).
- if v has a cilium h_i, the deletion proceeds in two steps:

- If $\tau^{(c)}(i) \neq i$ (that is if the τ-edges of color c incident at the vertex v are distinct) then we replace it by the permutation $\tilde{\tau}^{(c)}$ defined as:

$$\begin{cases} \tilde{\tau}^{(c)}(i) = i \,, \\ \tilde{\tau}^{(c)} \left[\tau^{(c)}\right]^{-1}(i) = \tau^{(c)}(i) \,, \\ \tilde{\tau}^{(c)}(j) = \tau^{(c)}(j) \,, \quad \forall j \neq i, \left[\tau^{(c)}\right]^{-1}(i). \end{cases}$$

  Graphically this comes to cutting the τ-edges incident at v and reconnecting them the other way around.
  Similarly, if $\rho^{(c)}(i) \neq i$ then we replace it by the permutation $\tilde{\rho}^{(c)}$ defined as:

$$\begin{cases} \tilde{\rho}^{(c)}(i) = i \,, \\ \tilde{\rho}^{(c)} \left[\rho^{(c)}\right]^{-1}(i) = \rho^{(c)}(i) \,, \\ \tilde{\rho}^{(c)}(j) = \rho^{(c)}(j) \,, \quad \forall j \neq i, \left[\rho^{(c)}\right]^{-1}(i). \end{cases}$$

  Graphically this comes to cutting the ρ-edges incident at v and reconnecting them the other way around.
  Applying these flips to one of the vertices of the tree in Figure 9.7 leads to the tree in Figure 9.8.

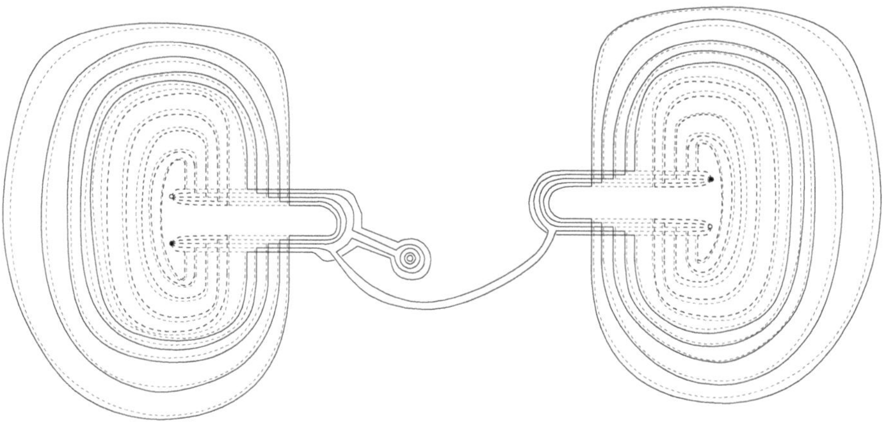

Figure 9.8: Flip of the ρ and τ-edges incident at a cilium.

- if $\tau^{(c)}(i) = i$ and $\rho^{(c)}(i) = i$ for all colors, the deletion consists of deleting v and all the faces incident only at v (solid, dotted or dashed), and reconnecting the solid faces of colors $\mathcal{C}$ passing through v directly on its ancestor (i.e. erasing v, the ρ and τ edges hooked to v and the edge of the multicolored tree hooked to v, and replacing the other end of this edge by a scar).

We denote $\tau^{\mathcal{D}}$, $\rho^{\mathcal{D}}$ and $\mathfrak{T}_{\rho^{\mathcal{D}}\tau^{\mathcal{D}}}$ the permutations and the tree before the deletion of v and $\hat{\tau}^{\mathcal{D}}$, $\hat{\rho}^{\mathcal{D}}$ and $\hat{\mathfrak{T}}_{\hat{\rho}^{\mathcal{D}}\hat{\tau}^{\mathcal{D}}}$ the permutations and the tree obtained after having erased v. If the vertex had a cilium and the deletion was done in two steps, at the intermediary step we have the same tree $\tilde{\mathfrak{T}} = \mathfrak{T}$, but the permutations are $\tilde{\tau}_{\mathcal{D}}$ and $\tilde{\rho}_{\mathcal{D}}$.

Evolution of $F_{\rm int}(\mathfrak{T}_{\rho^{\mathcal{D}}\tau^{\mathcal{D}}}) + C(\rho^{\mathcal{D}})$ under deletion. We distinguish the two cases:

v has no cilium. The permutations τ and ρ are unaffected by the deletion ($\hat\rho^{\mathcal{D}} = \rho^{\mathcal{D}}$ and $\hat\tau^{\mathcal{D}} = \tau^{\mathcal{D}}$) and furthermore the faces intτ cannot be deleted, thus:

$$
\begin{aligned}
& C(\rho^{\mathcal{D}}) = C(\hat\rho^{\mathcal{D}}) ,\\
& F_{\rho\tau^{-1}}(\mathfrak{T}_{\rho^{\mathcal{D}}\tau^{\mathcal{D}}}) = F_{\hat\rho\hat\tau^{-1}}(\hat{\mathfrak{T}}_{\hat\rho^{\mathcal{D}}\hat\tau^{\mathcal{D}}})\\
& F_{{\rm int}\tau}(\mathfrak{T}_{\rho^{\mathcal{D}}\tau^{\mathcal{D}}}) = F_{{\rm int}\tau}(\hat{\mathfrak{T}}_{\hat\rho^{\mathcal{D}}\hat\tau^{\mathcal{D}}}) .
\end{aligned}
$$

All the internal faces of $\mathfrak{T}$ containing the vertex and having colors in $\mathcal{D}\smallsetminus\mathcal{C}$ are deleted,

$$
F_{\rm int}(\mathfrak{T}) = F_{\rm int}(\hat{\mathfrak{T}}) + |\mathcal{D}\smallsetminus\mathcal{C}| ,
$$

and taking into account that $|\mathcal{D}\smallsetminus\mathcal{C}| \le D-1$ we have:

$$
F_{\rm int}(\mathfrak{T}_{\rho^{\mathcal{D}}\tau^{\mathcal{D}}}) + C(\rho^{\mathcal{D}}) \le F_{\rm int}(\hat{\mathfrak{T}}_{\hat\rho^{\mathcal{D}}\hat\tau^{\mathcal{D}}}) + C(\hat\rho^{\mathcal{D}}) + (D-1) ,
$$

and the graph obtained after deletion has one fewer vertex. Observe that equality holds only if we erase exactly $D-1$ faces, that is $|\mathcal{C}| = 1$.

v has a cilium. Suppose that we go through the intermediary graph $\tilde{\mathfrak{T}}_{\tilde\rho^{\mathcal{D}}\tilde\tau^{\mathcal{D}}}$:

- *The number of connected components $C(\rho^{\mathcal{D}})$.* By going to $\tilde\rho^{\mathcal{D}}$ we always create a connected component with a black and a white vertex (corresponding to the cilium h_i) connected by D edges $\tilde\rho^{(c)}(i) = i, \forall c$.
  The other connected components of the graph $\mathrm{Gr}(\rho^{\mathcal{D}})$ change as follows:
  - if $\rho^{(c)}(i) \neq i$ for all $c \in \mathcal{D}$ then there are two cases:
    - the black and white vertices corresponding to h_i belong to the same connected component of $\mathrm{Gr}(\rho^{\mathcal{D}})$. Then by going to $\tilde\rho_{\mathcal{D}}$ this component either survives or it is split into several connected components. Thus $C(\rho^{\mathcal{D}}) \le C(\tilde\rho^{\mathcal{D}}) - 1$.
    - the black and white vertices corresponding to h_i belong to two distinct connected components of $\mathrm{Gr}(\rho^{\mathcal{D}})$. Then the two are either merged into a unique component, or split into several. We always have $C(\rho^{\mathcal{D}}) \le C(\tilde\rho^{\mathcal{D}})$.
  - if there exists $c \in \mathcal{D}$ such that $\rho^{(c)}(i) = i$ then the black and white vertex corresponding to h_i belong to the same connected component of $\mathrm{Gr}(\rho^{\mathcal{D}})$. This component either survives or it is split into several by going to $\tilde\rho^{\mathcal{D}}$. Hence in this case we always have $C(\rho^{\mathcal{D}}) \le C(\tilde\rho^{\mathcal{D}}) - 1$.

  To summarize:

$$
\begin{cases}
C(\rho^{\mathcal{D}}) \le C(\tilde\rho^{\mathcal{D}}) - 1 \text{ if } \exists c\,,\ \rho^{(c)}(i) = i\,,\\
C(\rho^{\mathcal{D}}) \le C(\tilde\rho^{\mathcal{D}}) \text{ otherwise}\,.
\end{cases}
$$

- *The number of faces $F_{\rm int}(\mathfrak{T})$.* As $\tilde{\mathfrak{T}} = \mathfrak{T}$ we have:

$$
F_{\rm int}(\mathfrak{T}) = F_{\rm int}(\tilde{\mathfrak{T}}) .
$$

- *The number of faces* $F_{\text{int}\tau}(\mathfrak{T}_{\rho^{\mathcal{D}}\tau^{\mathcal{D}}})$. Fix a color c.
  - for $c \in \mathcal{C}$ the number of intτ faces of color c:
    - is constant if $\tau^{(c)}(i) = i$.
    - can at most decrease by 1 if $\tau^{(c)}(i) \neq i$.
  - for $c \in \mathcal{D} \setminus \mathcal{C}$ the number of intτ faces of color c is:
    - constant if $\tau^{(c)}(i) = i$.
    - increases by 1 if $\tau^{(c)}(i) \neq i$.

  We denote $s^{\mathcal{C}} \leq |\mathcal{C}|$ the number of colors in $\mathcal{C}$ such that $\tau^{(c)}(i) \neq i$, and $s^{\mathcal{D}\setminus\mathcal{C}}$ the number of colors in $\mathcal{D} \setminus \mathcal{C}$ such that $\tau^{(c)}(i) \neq i$ and we obtain:

$$F_{\text{int}\tau}(\mathfrak{T}_{\rho^{\mathcal{D}}\tau^{\mathcal{D}}}) \leq F_{\text{int}\tau}(\tilde{\mathfrak{T}}_{\tilde{\rho}^{\mathcal{D}}\tilde{\tau}^{\mathcal{D}}}) + s^{\mathcal{C}} - s^{\mathcal{D}\setminus\mathcal{C}} .$$

- *The number of faces* $F_{\rho\tau^{-1}}(\mathfrak{T}_{\rho^{\mathcal{D}}\tau^{\mathcal{D}}})$. Fix again a color c and denote the number of $\rho\tau^{-1}$ faces of color c of $\mathfrak{T}_{\rho^{\mathcal{D}}\tau^{\mathcal{D}}}$ by $F^{c}_{\rho\tau^{-1}}(\mathfrak{T}_{\rho^{\mathcal{D}}\tau^{\mathcal{D}}})$. We have:
  - if $\tau^{(c)}(i) = i$,
    - if $\rho^{(c)}(i) = i$,
$$F^{c}_{\rho\tau^{-1}}(\mathfrak{T}_{\rho^{\mathcal{D}}\tau^{\mathcal{D}}}) = F^{c}_{\rho\tau^{-1}}(\tilde{\mathfrak{T}}_{\tilde{\rho}^{\mathcal{D}}\tilde{\tau}^{\mathcal{D}}}) .$$
    - if $\rho^{(c)}(i) \neq i$,
$$F^{c}_{\rho\tau^{-1}}(\mathfrak{T}_{\rho^{\mathcal{D}}\tau^{\mathcal{D}}}) = F^{c}_{\rho\tau^{-1}}(\tilde{\mathfrak{T}}_{\tilde{\rho}^{\mathcal{D}}\tilde{\tau}^{\mathcal{D}}}) - 1 .$$
  - if $\tau^{(c)}(i) \neq i$,
    - if $\rho^{(c)}(i) = i$,
$$F^{c}_{\rho\tau^{-1}}(\mathfrak{T}_{\rho^{\mathcal{D}}\tau^{\mathcal{D}}}) = F^{c}_{\rho\tau^{-1}}(\tilde{\mathfrak{T}}_{\tilde{\rho}^{\mathcal{D}}\tilde{\tau}^{\mathcal{D}}}) - 1 .$$
    - if $\rho^{(c)}(i) \neq i$,
$$F^{c}_{\rho\tau^{-1}}(\mathfrak{T}_{\rho^{\mathcal{D}}\tau^{\mathcal{D}}}) \leq F^{c}_{\rho\tau^{-1}}(\tilde{\mathfrak{T}}_{\tilde{\rho}^{\mathcal{D}}\tilde{\tau}^{\mathcal{D}}}) .$$

  To summarize:

$$\begin{cases} F^{c}_{\rho\tau^{-1}}(\mathfrak{T}_{\rho^{\mathcal{D}}\tau^{\mathcal{D}}}) \leq F^{c}_{\rho\tau^{-1}}(\tilde{\mathfrak{T}}_{\tilde{\rho}^{\mathcal{D}}\tilde{\tau}^{\mathcal{D}}}) - 1 \text{ if } \tau^{(c)}(i) = i, \rho^{(c)}(i) \neq i , \\ F^{c}_{\rho\tau^{-1}}(\mathfrak{T}_{\rho^{\mathcal{D}}\tau^{\mathcal{D}}}) \leq F^{c}_{\rho\tau^{-1}}(\tilde{\mathfrak{T}}_{\tilde{\rho}^{\mathcal{D}}\tilde{\tau}^{\mathcal{D}}}) \text{ otherwise} . \end{cases}$$

As long as either $s^{\mathcal{C}} \leq |\mathcal{C}| - 1$ or $s^{\mathcal{C}} = |\mathcal{C}|$ and $s^{\mathcal{D}\setminus\mathcal{C}} \geq 1$, using the bound on the intτ faces and the worst bounds for all the other kinds of faces and for the number of connected components we get:

$$\begin{aligned} &F_{\text{int}}(\mathfrak{T}) + F_{\text{int}\tau}(\mathfrak{T}_{\rho^{\mathcal{D}}\tau^{\mathcal{D}}}) + F_{\rho\tau^{-1}}(\mathfrak{T}_{\rho^{\mathcal{D}}\tau^{\mathcal{D}}}) + C(\rho^{\mathcal{D}}) \\ &\quad \leq F_{\text{int}}(\tilde{\mathfrak{T}}) + F_{\text{int}\tau}(\tilde{\mathfrak{T}}_{\tilde{\rho}^{\mathcal{D}}\tilde{\tau}^{\mathcal{D}}}) + F_{\rho\tau^{-1}}(\tilde{\mathfrak{T}}_{\tilde{\rho}^{\mathcal{D}}\tilde{\tau}^{\mathcal{D}}}) + C(\tilde{\rho}^{\mathcal{D}}) + |\mathcal{C}| - 1 . \end{aligned}$$

If on the other hand both $s^{\mathcal{C}} = |\mathcal{C}|$ and $s^{\mathcal{D}\setminus\mathcal{C}} = 0$, then $\tau^{(c)}(i) = i$ for all $c \in \mathcal{D}\setminus\mathcal{C}$ and either $\rho^{(c)}(i) = i$ in which case the bound on the number of connected components is improved to:

$$C(\rho^{\mathcal{D}}) \leq C(\tilde{\rho}^{\mathcal{D}}) - 1,$$

or $\rho^{(c)}(i) \neq i$ and the bound on the $\rho\tau^{-1}$ faces is improved to:

$$F^{c}_{\rho\tau^{-1}}(\mathfrak{T}_{\rho^{\mathcal{D}}\tau^{\mathcal{D}}}) = F^{c}_{\rho\tau^{-1}}(\tilde{\mathfrak{T}}_{\tilde{\rho}^{\mathcal{D}}\tilde{\tau}^{\mathcal{D}}}) - 1 ,$$

therefore in this case we still get:

$$\begin{aligned} &F_{\text{int}}(\mathfrak{T}) + F_{\text{int}\tau}(\mathfrak{T}_{\rho^{\mathcal{D}}\tau^{\mathcal{D}}}) + F_{\rho\tau^{-1}}(\mathfrak{T}_{\rho^{\mathcal{D}}\tau^{\mathcal{D}}}) + C(\rho^{\mathcal{D}}) \\ &\quad \le F_{\text{int}}(\tilde{\mathfrak{T}}) + F_{\text{int}\tau}(\tilde{\mathfrak{T}}_{\tilde{\rho}^{\mathcal{D}}\tilde{\tau}^{\mathcal{D}}}) + F_{\rho\tau^{-1}}(\tilde{\mathfrak{T}}_{\tilde{\rho}^{\mathcal{D}}\tilde{\tau}^{\mathcal{D}}}) + C(\tilde{\rho}^{\mathcal{D}}) + |\mathcal{C}| - 1 . \end{aligned}$$

We emphasize that in order to obtain this bound we used the fact that in a tree with a cilium per vertex, if $s^{\mathcal{D}\setminus\mathcal{C}} = 0$, then the strands with colors $c \in \mathcal{D}\setminus\mathcal{C}$ are such that $\tau^{(c)}(i) = i$. This does not hold if we allow more than one cilium per vertex, as the τ edges with colors in $\mathcal{D}\setminus\mathcal{C}$ could connect the multiple cilia on the same vertex.

Finally, erasing a vertex such that $\tau^{(c)}(i) = i$ and $\rho^{(c)}(i) = i$ for all c we have

$$\begin{aligned} &F_{\text{int}}(\tilde{\mathfrak{T}}) = F_{\text{int}}(\hat{\mathfrak{T}}) , &&F_{\text{int}\tau}(\tilde{\mathfrak{T}}_{\tilde{\rho}^{\mathcal{D}}\tilde{\tau}^{\mathcal{D}}}) = F_{\text{int}\tau}(\hat{\mathfrak{T}}_{\hat{\rho}^{\mathcal{D}}\hat{\tau}^{\mathcal{D}}}) + |\mathcal{D}\setminus\mathcal{C}| , \\ &C(\tilde{\rho}^{\mathcal{D}}) = C(\hat{\rho}^{\mathcal{D}}) + 1 , &&F_{\rho\tau^{-1}}(\tilde{\mathfrak{T}}_{\tilde{\rho}^{\mathcal{D}}\tilde{\tau}^{\mathcal{D}}}) = F_{\rho\tau^{-1}}(\hat{\mathfrak{T}}_{\hat{\rho}^{\mathcal{D}}\hat{\tau}^{\mathcal{D}}}) + D , \end{aligned}$$

which, taking into account that $|\mathcal{C}| + |\mathcal{D}\setminus\mathcal{C}| = D$, leads to:

$$\begin{aligned} &F_{\text{int}}(\mathfrak{T}) + F_{\text{int}\tau}(\mathfrak{T}_{\rho^{\mathcal{D}}\tau^{\mathcal{D}}}) + F_{\rho\tau^{-1}}(\mathfrak{T}_{\rho^{\mathcal{D}}\tau^{\mathcal{D}}}) + C(\rho^{\mathcal{D}}) \\ &\quad \le F_{\text{int}}(\hat{\mathfrak{T}}) + F_{\text{int}\tau}(\hat{\mathfrak{T}}_{\hat{\rho}^{\mathcal{D}}\hat{\tau}^{\mathcal{D}}}) + F_{\rho\tau^{-1}}(\hat{\mathfrak{T}}_{\hat{\rho}^{\mathcal{D}}\hat{\tau}^{\mathcal{D}}}) + C(\hat{\rho}^{\mathcal{D}}) + 2D , \end{aligned}$$

and the numbers of vertices and of cilia both go down by 1.

Iteration. Iterating up to the last vertex, we either end up with a vertex with no cilium or with a vertex with a cilium and all $\rho^{(c)}(i) = \tau^{(c)}(i) = i$. Counting the faces and the number of connected components of $\rho^{\mathcal{D}}$ of the two possible end graphs we obtain:

$$\begin{aligned} F_{\text{int}}(\mathfrak{T}_{\rho^{\mathcal{D}}\tau^{\mathcal{D}}}) + C(\rho^{\mathcal{D}}) &= F_{\text{int}}(\mathfrak{T}) + F_{\text{int}\tau}(\mathfrak{T}_{\rho^{\mathcal{D}}\tau^{\mathcal{D}}}) + F_{\rho\tau^{-1}}(\mathfrak{T}_{\rho^{\mathcal{D}}\tau^{\mathcal{D}}}) + C(\rho^{\mathcal{D}}) \\ &\le \begin{cases} 2Dk + (D-1)(V-k-1) + D \\ 2D(k-1) + (D-1)(V-k) + 2D + 1 \end{cases} = 1 + (D+1)k + (D-1)V , \end{aligned}$$

which proves the theorem.

□

9.3 The intermediate field representation

The edge multicolored maps can be obtained as genuine Feynman graphs for *intermediate matrix fields*.

The intermediate field representation of the quartic tensor model (or the Hubbard Stratonovich representation) is obtained starting from the remark that, for any $\mathcal{C}$, the interaction $V_{\mathcal{C}}^{(4)}(\bar{\mathbb{T}},\mathbb{T})$ can be represented as a Gaussian integral over an intermediate $N^{|\mathcal{C}|} \times N^{|\mathcal{C}|}$ hermitian matrix $H^{\mathcal{C}}_{\bar{a}^{\mathcal{C}} a^{\mathcal{C}}} = \overline{H^{\mathcal{C}}_{a^{\mathcal{C}} \bar{a}^{\mathcal{C}}}}$:

$$e^{-N^{D-1}V_{\mathcal{C}}^{(4)}(\bar{\mathbb{T}},\mathbb{T})} = \int \Big(\prod_{\bar{a}^{\mathcal{C}} \neq a^{\mathcal{C}}} \frac{dH^{\mathcal{C}}_{\bar{a}^{\mathcal{C}} a^{\mathcal{C}}}}{\sqrt{2\pi\imath}} \Big) \Big(\prod_{\bar{a}^{\mathcal{C}} = a^{\mathcal{C}}} \frac{dH^{\mathcal{C}}_{\bar{a}^{\mathcal{C}} a^{\mathcal{C}}}}{\sqrt{2\pi}} \Big) \\ \times e^{-\frac{1}{2}\operatorname{Tr}_{\mathcal{C}}[H^{\mathcal{C}} H^{\mathcal{C}}] + \imath\sqrt{\lambda N^{D-1}} \sum_{n^{\mathcal{D}} \bar{n}^{\mathcal{D}}} H^{\mathcal{C}}_{\bar{n}^{\mathcal{C}} n^{\mathcal{C}}} \left(\bar{\mathbb{T}}_{\bar{n}^{\mathcal{D}}} \delta_{\bar{n}^{\mathcal{D}\setminus\mathcal{C}} n^{\mathcal{D}\setminus\mathcal{C}}} \mathbb{T}_{n^{\mathcal{D}}} \right)} ,$$

where $\operatorname{Tr}_{\mathcal{C}}$ denotes the trace over the indices of colors $c \in \mathcal{C}$. Let us denote $\mathbf{1}$ the $N \times N$ identity matrix, and $\mathbf{1}^{\mathcal{C}}$ the identity on the indices of colors $\mathcal{C}$. The generating function of the moments becomes:

$$Z^{(4)}(J,\bar{J}) = \int \left(\prod_{n^{\mathcal{D}}} N^{D-1} \frac{d\bar{\mathbb{T}}_{n^{\mathcal{D}}} d\mathbb{T}_{n^{\mathcal{D}}}}{2\imath\pi} \right) \\ \prod_{\mathcal{C} \subset \mathcal{D}/2} \left[\Big(\prod_{\bar{a}^{\mathcal{C}} \neq a^{\mathcal{C}}} \frac{dH^{\mathcal{C}}_{\bar{a}^{\mathcal{C}} a^{\mathcal{C}}}}{\sqrt{2\pi\imath}} \Big) \Big(\prod_{\bar{a}^{\mathcal{C}} = a^{\mathcal{C}}} \frac{dH^{\mathcal{C}}_{\bar{a}^{\mathcal{C}} a^{\mathcal{C}}}}{\sqrt{2\pi}} \Big) \right] e^{-\frac{1}{2}\sum_{\mathcal{C} \subset \mathcal{D}/2} \frac{1}{2} \operatorname{Tr}_{\mathcal{C}}[H^{\mathcal{C}} H^{\mathcal{C}}]} \\ e^{-N^{D-1} \left(\sum_{n^{\mathcal{D}} \bar{n}^{\mathcal{D}}} \bar{\mathbb{T}}_{\bar{n}^{\mathcal{D}}} \left[\mathbf{1}^{\mathcal{D}} - \imath \sqrt{\frac{\lambda}{N^{D-1}}} \sum_{\mathcal{C} \subset \mathcal{D}/2} H^{\mathcal{C}} \otimes \mathbf{1}^{\mathcal{D}\setminus\mathcal{C}} \right]_{\bar{n}^{\mathcal{D}} n^{\mathcal{D}}} \mathbb{T}_{n^{\mathcal{D}}} \right)} \\ e^{\sum \bar{\mathbb{T}}_{\bar{a}^{\mathcal{D}}} \bar{J}_{\bar{a}^{\mathcal{D}}} + \sum J_{a^{\mathcal{D}}} \mathbb{T}_{a^{\mathcal{D}}}} .$$

The integral over $\bar{\mathbb{T}}, \mathbb{T}$ is now Gaussian and can be computed explicitly. We obtain:

$$Z^{(4)}(J,\bar{J}) = \int \prod_{\mathcal{C} \subset \mathcal{D}/2} \left[\Big(\prod_{\bar{a}^{\mathcal{C}} \neq a^{\mathcal{C}}} \frac{dH^{\mathcal{C}}_{\bar{a}^{\mathcal{C}} a^{\mathcal{C}}}}{\sqrt{2\pi\imath}} \Big) \Big(\prod_{\bar{a}^{\mathcal{C}} = a^{\mathcal{C}}} \frac{dH^{\mathcal{C}}_{\bar{a}^{\mathcal{C}} a^{\mathcal{C}}}}{\sqrt{2\pi}} \Big) \right] \\ e^{-\frac{1}{2}\sum_{\mathcal{C} \subset \mathcal{D}/2} \operatorname{Tr}_{\mathcal{C}}[H^{\mathcal{C}} H^{\mathcal{C}}] - \operatorname{Tr}_{\mathcal{D}}\left[\ln\left(\mathbf{1}^{\mathcal{D}} - \imath\sqrt{\frac{\lambda}{N^{D-1}}} \sum_{\mathcal{C} \subset \mathcal{D}/2} H^{\mathcal{C}} \otimes \mathbf{1}^{\mathcal{D}\setminus\mathcal{C}} \right) \right]} \\ e^{\frac{1}{N^{D-1}} \sum_{\bar{n}^{\mathcal{D}} n^{\mathcal{D}}} J_{n^{\mathcal{D}}} \left[\left(\mathbf{1}^{\mathcal{D}} - \imath\sqrt{\frac{\lambda}{N^{D-1}}} \sum_{\mathcal{C} \subset \mathcal{D}/2} H^{\mathcal{C}} \otimes \mathbf{1}^{\mathcal{D}\setminus\mathcal{C}} \right)^{-1} \right]_{n^{\mathcal{D}} \bar{n}^{\mathcal{D}}} \bar{J}_{\bar{n}^{\mathcal{D}}}} .$$

We rescale the matrices H by $N^{-\frac{D-1}{2}}$ and we denote:

$$[dH^{\mathcal{C}}] \equiv \left[\Big(\prod_{\bar{a}^{\mathcal{C}} \neq a^{\mathcal{C}}} N^{\frac{D-1}{2}} \frac{dH^{\mathcal{C}}_{\bar{a}^{\mathcal{C}} a^{\mathcal{C}}}}{\sqrt{2\pi\imath}} \Big) \Big(\prod_{\bar{a}^{\mathcal{C}} = a^{\mathcal{C}}} N^{\frac{D-1}{2}} \frac{dH^{\mathcal{C}}_{\bar{a}^{\mathcal{C}} a^{\mathcal{C}}}}{\sqrt{2\pi}} \Big) \right] \\ JR(H)\bar{J} \equiv \sum_{\bar{n}^{\mathcal{D}} n^{\mathcal{D}}} J_{n^{\mathcal{D}}} R(H)_{n^{\mathcal{D}} \bar{n}^{\mathcal{D}}} \bar{J}_{\bar{n}^{\mathcal{D}}} ,$$

where $R(H)$ denotes the *resolvent* operator:

$$R(H) \equiv \frac{1}{\mathbf{1}^{\mathcal{D}} - \imath\sqrt{\lambda} \sum_{\mathcal{C} \subset \mathcal{D}/2} H^{\mathcal{C}} \otimes \mathbf{1}^{\mathcal{D}\setminus\mathcal{C}}} .$$

The intermediate field representation of the generating function of the moments is then:

$$Z^{(4)}(J,\bar{J}) = \int \left(\prod_{\mathcal{C} \subset \mathcal{D}/2} [dH^{\mathcal{C}}] \right)$$

$$e^{-\frac{1}{2}\sum_{\mathcal{C}\subset\mathcal{D}/2} N^{D-1}\,\mathrm{Tr}_{\mathcal{C}}[H^{\mathcal{C}}H^{\mathcal{C}}]+\mathrm{Tr}_{\mathcal{D}}[\ln R(H)]+\frac{1}{N^{D-1}}JR(H)\bar{J}}\,,$$

which is written alternatively, denoting:

$$\mathrm{Tr}_{\mathcal{C}}\left[\frac{\partial}{\partial H^{\mathcal{C}}}\frac{\partial}{\partial H^{\mathcal{C}}}\right] \equiv \sum_{\bar{a}^{\mathcal{C}},b^{\mathcal{C}}} \frac{\partial}{\partial H^{\mathcal{C}}_{\bar{a}^{\mathcal{C}}b^{\mathcal{C}}}}\frac{\partial}{\partial H^{\mathcal{C}}_{\bar{b}^{\mathcal{C}}\bar{a}^{\mathcal{C}}}}$$

as:

$$\boxed{\begin{aligned} Z^{(4)}(J,\bar{J}) =& \Big[e^{\frac{1}{2}\frac{1}{N^{D-1}}\sum_{\mathcal{C}\subset\mathcal{D}/2}\mathrm{Tr}_{\mathcal{C}}\left[\frac{\partial}{\partial H^{\mathcal{C}}}\frac{\partial}{\partial H^{\mathcal{C}}}\right]} \\ & \qquad \times e^{\mathrm{Tr}_{\mathcal{D}}[\ln R(H)]+\frac{1}{N^{D-1}}JR(H)\bar{J}}\Big]_{H^{\mathcal{C}}=0}\,. \end{aligned}} \tag{9.6}$$

9.3.1 Feynman graphs for the intermediate field

It is straightforward to derive the Feynman rules in the intermediate field representation starting from Eq. (9.6). We expand the exponential:

$$\begin{aligned} Z^{(4)}(J,\bar{J}) = \sum_{n_{\mathrm{int}},n_{\mathrm{ext}}\geq 0} \frac{1}{n_{\mathrm{int}}!n_{\mathrm{ext}}!} \Bigg[& e^{\frac{1}{2}\frac{1}{N^{D-1}}\sum_{\mathcal{C}\subset\mathcal{D}/2}\mathrm{Tr}_{\mathcal{C}}\left[\frac{\partial}{\partial H^{\mathcal{C}}}\frac{\partial}{\partial H^{\mathcal{C}}}\right]} \\ & \times \Big(\mathrm{Tr}_{\mathcal{D}}\left[\ln R(H)\right]\Big)^{n_{\mathrm{int}}} \left(\frac{1}{N^{D-1}}JR(H)\bar{J}\right)^{n_{\mathrm{ext}}} \Bigg]_{H^{\mathcal{C}}=0}\,, \end{aligned}$$

and we represent $\mathrm{Tr}_{\mathcal{D}}\left[\ln R(H)\right]$ as (internal) vertices and $\frac{1}{N^{D-1}}JR(H)\bar{J}$ as (external) vertices with a cilium signaling the presence of the external sources J and $\bar{J}$. The derivatives:

$$\begin{aligned} &\frac{\partial}{\partial H^{\mathcal{C}}_{\bar{a}^{\mathcal{C}}b^{\mathcal{C}}}}\mathrm{Tr}_{\mathcal{D}}\left[\ln R(H)\right] \\ &= \imath\sqrt{\lambda}\sum_{\bar{p}^{\mathcal{D}}q^{\mathcal{D}}} R(H)_{q^{\mathcal{D}}\bar{p}^{\mathcal{D}}}\left(\delta_{\bar{p}^{\mathcal{D}\setminus\mathcal{C}}q^{\mathcal{D}\setminus\mathcal{C}}}\delta_{\bar{p}^{\mathcal{C}}\bar{a}^{\mathcal{C}}}\delta_{q^{\mathcal{C}}b^{\mathcal{C}}}\right) \\ &\frac{\partial}{\partial H^{\mathcal{C}}_{\bar{a}^{\mathcal{C}}b^{\mathcal{C}}}}R(H)_{n^{\mathcal{D}}\bar{n}^{\mathcal{D}}} \\ &= \imath\sqrt{\lambda}\sum_{\bar{p}^{\mathcal{D}}q^{\mathcal{D}}} R(H)_{n^{\mathcal{D}}\bar{p}^{\mathcal{D}}}\left(\delta_{\bar{p}^{\mathcal{D}\setminus\mathcal{C}}q^{\mathcal{D}\setminus\mathcal{C}}}\delta_{\bar{p}^{\mathcal{C}}\bar{a}^{\mathcal{C}}}\delta_{q^{\mathcal{C}}b^{\mathcal{C}}}\right)R(H)_{q^{\mathcal{D}}\bar{n}^{\mathcal{D}}} \end{aligned}$$

create halfedges colored by the colors $\mathcal{C}$ of the corresponding field $H^{\mathcal{C}}$. The free indices of the resolvents at the end of the halfedges are then identified along the edges. The first derivative acting on a vertex creates a resolvent associated to the corner of the vertex. The second derivative acts on this resolvent and splits the corner of the vertex into two corners, each with its own resolvent. Subsequent derivatives act on each of the corners already present and split each corner into two (with a resolvent associated to each new corner). As one keeps track of the ordering of the corners around a vertex, we obtain a sum over edge multicolored combinatorial maps (see also the discussion in Appendix B).

Let us denote $\mathcal{S}_{\text{int}}(\mathcal{M}) = \sqcup_{\mathcal{C}\subset\mathcal{D}/2}\mathcal{S}^{(\mathcal{C})}_{\text{int}}(\mathcal{M})$, $\mathcal{S}_{\text{ext}}(\mathcal{M})$, $\sigma_{\mathcal{M}}$ and $\alpha^{(\mathcal{C})}_{\mathcal{M}}$, $\forall\mathcal{C}$ the internal halfedges, cilia, permutation and involutions of the edge multicolored map $\mathcal{M}$. Each edge of $\mathcal{M}$, $\big(h, h' = \alpha^{(\mathcal{C})}_{\mathcal{M}}(h)\big)$, is incident to four corners:

$$\big(\sigma^{-1}_{\mathcal{M}}(h), h\big)\,,\ \big(h, \sigma_{\mathcal{M}}(h)\big)\,,\ \big(\sigma^{-1}_{\mathcal{M}}(h'), h'\big)\,,\ \big(h', \sigma_{\mathcal{M}}(h')\big)\,,$$

and brings a scaling factor $N^{-(D-1)}$ and a pairwise identifications of the indices of the resolvents of these four corners. Each cilium brings a J and a $\bar{J}$ term. We therefore obtain a sum over edge multicolored combinatorial maps with cilia and labeled vertices. Denoting $\mathcal{E}(\mathcal{M})$ the set of edges of $\mathcal{M}$, $E(\mathcal{M}) = |\mathcal{E}(\mathcal{M})|$, $n_{\text{int}}(\mathcal{M})$ the number of internal (non ciliated) vertices and $n_{\text{ext}}(\mathcal{M})$ the number of external (ciliated vertices) of $\mathcal{M}$ we have:

$$\begin{aligned}
Z^{(4)}(J,\bar{J}) &= \sum_{\mathcal{M}} \frac{(-\lambda)^{E(\mathcal{M})}}{n_{\text{int}}(\mathcal{M})!\,n_{\text{ext}}(\mathcal{M})!} N^{-(D-1)E(\mathcal{M})} \\
&\times \sum_{a\bar{b}} \Bigg\{ \Bigg(\prod_{h\in\mathcal{S}(\mathcal{M})} R(H)_{a^{\mathcal{D}}_{(h,\sigma_{\mathcal{M}}(h))}\bar{b}^{\mathcal{D}}_{(h,\sigma_{\mathcal{M}}(h))}} \Bigg) \Bigg(\prod_{h\in\mathcal{S}_{\text{ext}}(\mathcal{M})} \frac{J_{a^{\mathcal{D}}_{(h,\sigma_{\mathcal{M}}(h))}} \bar{J}_{\bar{b}_{(\sigma^{-1}_{\mathcal{M}}(h),h)}}}{N^{D-1}} \Bigg) \\
&\times \Bigg[\prod_{e^{\mathcal{C}}=(h,h')\in\mathcal{E}(\mathcal{M})} \Bigg(\delta_{\bar{b}^{\mathcal{C}}_{(\sigma^{-1}_{\mathcal{M}}(h),h)} a^{\mathcal{C}}_{(h',\sigma_{\mathcal{M}}(h'))}} \delta_{\bar{b}^{\mathcal{C}}_{(\sigma^{-1}_{\mathcal{M}}(h'),h')} a^{\mathcal{C}}_{(h,\sigma_{\mathcal{M}}(h))}} \Bigg) \\
&\qquad \Bigg(\delta_{\bar{b}^{\mathcal{D}\smallsetminus\mathcal{C}}_{(\sigma^{-1}_{\mathcal{M}}(h),h)} a^{\mathcal{D}\smallsetminus\mathcal{C}}_{(h,\sigma_{\mathcal{M}}(h))}} \delta_{\bar{b}^{\mathcal{D}\smallsetminus\mathcal{C}}_{(\sigma^{-1}_{\mathcal{M}}(h'),h')} a^{\mathcal{D}\smallsetminus\mathcal{C}}_{(h',\sigma_{\mathcal{M}}(h'))}} \Bigg) \Bigg] \Bigg\}_{H^{\mathcal{C}}=0}\,,
\end{aligned}$$

where the sum runs over maps with *labeled* vertices.

If the map $\mathcal{M}$ has an isolated vertex with no cilium its contribution is slightly modified: such a vertex does not bring a resolvent but a factor $\text{Tr}_{\mathcal{D}}[\ln R(H)]$. The condition $H^{\mathcal{C}} = 0$ trivializes the resolvents: $R(0) = \mathbf{1}^{\mathcal{D}}$, and if the map had a vertex with no cilium then its contribution is zero because:

$$\text{Tr}_{\mathcal{D}}[\ln R(H)] = \text{Tr}_{\mathcal{D}}\big[\ln \mathbf{1}^{\mathcal{D}}\big] = 0\,.$$

The partition function becomes simply:

$$\begin{aligned}
Z^{(4)}(J,\bar{J}) &= \sum_{\mathcal{M}} \frac{(-\lambda)^{E(\mathcal{M})}}{n_{\text{int}}(\mathcal{M})!\,n_{\text{ext}}(\mathcal{M})!} N^{-(D-1)E(\mathcal{M})} \\
&\times \sum_{a\bar{b}} \Bigg(\prod_{h\in\mathcal{S}(\mathcal{M})} \delta_{a^{\mathcal{D}}_{(h,\sigma_{\mathcal{M}}(h))}\bar{b}^{\mathcal{D}}_{(h,\sigma_{\mathcal{M}}(h))}} \Bigg) \Bigg(\prod_{h\in\mathcal{S}_{\text{ext}}(\mathcal{M})} \frac{J_{a^{\mathcal{D}}_{(h,\sigma_{\mathcal{M}}(h))}} \bar{J}_{\bar{b}_{(\sigma^{-1}_{\mathcal{M}}(h),h)}}}{N^{D-1}} \Bigg) \\
&\times \Bigg[\prod_{e^{\mathcal{C}}=(h,h')\in\mathcal{E}(\mathcal{M})} \Bigg(\delta_{\bar{b}^{\mathcal{C}}_{(\sigma^{-1}_{\mathcal{M}}(h),h)} a^{\mathcal{C}}_{(h',\sigma_{\mathcal{M}}(h'))}} \delta_{\bar{b}^{\mathcal{C}}_{(\sigma^{-1}_{\mathcal{M}}(h'),h')} a^{\mathcal{C}}_{(h,\sigma_{\mathcal{M}}(h))}} \Bigg) \\
&\qquad \Bigg(\delta_{\bar{b}^{\mathcal{D}\smallsetminus\mathcal{C}}_{(\sigma^{-1}_{\mathcal{M}}(h),h)} a^{\mathcal{D}\smallsetminus\mathcal{C}}_{(h,\sigma_{\mathcal{M}}(h))}} \delta_{\bar{b}^{\mathcal{D}\smallsetminus\mathcal{C}}_{(\sigma^{-1}_{\mathcal{M}}(h'),h')} a^{\mathcal{D}\smallsetminus\mathcal{C}}_{(h',\sigma_{\mathcal{M}}(h'))}} \Bigg) \Bigg]\,.
\end{aligned}$$

Let us consider an index $a^c_{(h,\sigma_{\mathcal{M}}(h))}$. At the first step it is identified with the index $\bar{b}^c_{(h,\sigma_{\mathcal{M}}(h))}$. At the second step, there are two possibilities:

- either the edge hooked to $\sigma_{\mathcal{M}}(h)$ is colored by $\mathcal{C}$ with $c \notin \mathcal{C}$. Then the index is further identified with $a^c_{(\sigma_{\mathcal{M}}(h),\sigma^2_{\mathcal{M}}(h))}$ and subsequently identified with $\bar{b}^c_{(\sigma_{\mathcal{M}}(h),\sigma^2_{\mathcal{M}}(h))}$.
- or the edge hooked to $\sigma_{\mathcal{M}}(h)$ is colored by $\mathcal{C}$ with $c \in \mathcal{C}$. Then the index is further identified with $a^c_{(\alpha^{(\mathcal{C})}_{\mathcal{M}}\sigma_{\mathcal{M}}(h),\sigma_{\mathcal{M}}\alpha^{(\mathcal{C})}_{\mathcal{M}}\sigma_{\mathcal{M}}(h))}$ and subsequently identified with $b^c_{(\alpha^{(\mathcal{C})}_{\mathcal{M}}\sigma_{\mathcal{M}}(h),\sigma_{\mathcal{M}}\alpha^{(\mathcal{C})}_{\mathcal{M}}\sigma_{\mathcal{M}}(h))}$.

That is, as long as it encounters halfedges with colors $\mathcal{C}, c \notin \mathcal{C}$, the index follows the permutation $\sigma_{\mathcal{M}}$, and it follows the permutation $\alpha^{(\mathcal{C})}_{\mathcal{M}}$ when it encounters a halfedge with colors $\mathcal{C}, c \in \mathcal{C}$. The indices a^c are thus identified along the faces of colors c. If an index belongs to an internal face then we obtain a free sum and a factor N. If it belongs to an external strand, we obtain an identification of the indices of the external sources J and $\bar{J}$ corresponding to the two end cilia. Remark that the number of vertices bearing cilia of $\mathcal{M}$, $n_{\text{ext}}(\mathcal{M})$ is exactly the number of white vertices of $\partial\mathcal{M}$, $k(\partial\mathcal{M})$, hence we obtain:

$$Z^{(4)}(J,\bar{J}) = \sum_{\mathcal{M}} \frac{(-\lambda)^{E(\mathcal{M})}}{n_{\text{int}}(\mathcal{M})!} \frac{\text{Tr}_{\partial\mathcal{M}}(J,\bar{J})}{n_{\text{ext}}(\mathcal{M})!}, \\ \times N^{-(D-1)E(\mathcal{M})-(D-1)k(\partial\mathcal{M})+\sum_{c=1}^{D} F^c_{\text{int}}(\mathcal{M})}$$

$$\frac{1}{N^D}\ln Z^{(4)}(J,\bar{J}) = \sum_{\mathcal{M}\text{ connected}} \frac{(-\lambda)^{E(\mathcal{M})}}{n_{\text{int}}(\mathcal{M})!} \frac{\text{Tr}_{\partial\mathcal{M}}(J,\bar{J})}{n_{\text{ext}}(\mathcal{M})!} \\ \times N^{-D-(D-1)E(\mathcal{M})-(D-1)k(\partial\mathcal{M})+\sum_{c=1}^{D} F^c_{\text{int}}(\mathcal{M})}, \tag{9.7}$$

which, taking into account the relabelings of the vertices of $\mathcal{M}$, are exactly the equations (9.3).

It is important to note that, by construction, the ciliated vertices have only one cilium.

9.3.2 The perturbative $1/N$ expansion

The $1/N$ expansion of the partition function and its logarithm can be easily understood directly in the intermediate field representation. At $J = \bar{J} = 0$ the free energy of the model is:

$$\frac{1}{N^D}\ln Z^{(4)}(0,0) = \sum_{\mathcal{M}\text{ connected}} \frac{(-\lambda)^{E(\mathcal{M})}}{n_{\text{int}}(\mathcal{M})!} N^{-D-(D-1)E(\mathcal{M})+F_{\text{int}}(\mathcal{M})}, \tag{9.8}$$

where $F_{\text{int}}(\mathcal{M})$ is the number of internal faces of the map $\mathcal{M}$ and only the contribution from connected maps $\mathcal{M}$ with no cilia survive. Using Theorem 9.1, Eq. (9.5), the total number of faces of a map with no cilia is bounded by:

$$F_{\text{int}}(\mathcal{M}) \le 1 + (D-1)V(\mathcal{M}) + \frac{D}{2}[E(\mathcal{M}) - V(\mathcal{M}) + 1],$$

and equality holds only if there exists some tree $\mathfrak{T}$ in $\mathcal{M}$ such that all the edges of the tree have only one color, $|\mathcal{C}| = 1$, and all the subsequent edges one adds in order to obtain $\mathcal{M}$ always increase the number of faces by $D/2$.

The scaling in N of a connected map in Eq. (9.8) is then bounded by:

$$-\frac{D-2}{2}[E(\mathcal{M}) - V(\mathcal{M}) + 1], \tag{9.9}$$

and, as $\mathcal{M}$ is connected, $E(\mathcal{M}) - V(\mathcal{M}) + 1 \geq 0$. We thus conclude that:

- the global scaling with N of a map is negative or 0, that is the right hand side of Eq. (9.8) is a series in $1/N$.
- the global scaling with N of a map is exactly zero only if $E(\mathcal{M}) = V(\mathcal{M}) - 1$, that is $\mathcal{M}$ is a tree, and furthermore all the edges in this tree are such that $|\mathcal{C}| = 1$. The only surviving terms at leading order in $1/N$ are trees whose edges have a unique color.
- the larger the *excess* of edges of $\mathcal{M}$ (the number of loop edges of $\mathcal{M}$, that is $E(\mathcal{M}) - V(\mathcal{M}) + 1$, in the physics literature), the more suppressed the contribution of $\mathcal{M}$ is by the *a priori* bound (9.9).

While the first statement also holds in the case of matrices $D = 2$, the last two statements do not.

9.3.3 Perturbative uniform boundedness

The cumulants of the measure $\mu^{(4)}$:

$$\kappa_{2p}(\mathbb{T}_{a_1^{\mathcal{D}}}, \bar{\mathbb{T}}_{\bar{a}_1^{\mathcal{D}}}, \dots \mathbb{T}_{a_p^{\mathcal{D}}}, \bar{\mathbb{T}}_{\bar{a}_p^{\mathcal{D}}}) = \frac{\partial^{(2k)}}{\partial J_{a_1^{\mathcal{D}}} \partial \bar{J}_{\bar{a}_1^{\mathcal{D}}} \dots \partial J_{a_p^{\mathcal{D}}} \partial \bar{J}_{\bar{a}_p^{\mathcal{D}}}} \ln Z^{(4)}(J, \bar{J})\Big|_{\substack{J=0\\ \bar{J}=0}},$$

$$\ln Z^{(4)}(J, \bar{J}) = \sum_{\mathcal{M} \text{ connected}} \frac{(-\lambda)^{E(\mathcal{M})}}{n_{\text{int}}(\mathcal{M})!} \frac{\mathrm{Tr}_{\partial\mathcal{M}}(J, \bar{J})}{n_{\text{ext}}(\mathcal{M})!} \times N^{-(D-1)E(\mathcal{M}) - (D-1)k(\partial\mathcal{M}) + \sum_{c=1}^{D} F_{\text{int}}^{c}(\mathcal{M})},$$

expand in trace invariants:

$$\kappa_{2p}(\mathbb{T}_{a_1^{\mathcal{D}}}, \bar{\mathbb{T}}_{\bar{a}_1^{\mathcal{D}}}, \dots \mathbb{T}_{a_p^{\mathcal{D}}}, \bar{\mathbb{T}}_{\bar{a}_p^{\mathcal{D}}}) = \sum_{\substack{\mathcal{M} \text{ connected}\\ k(\partial\mathcal{M}) = p}} \frac{(-\lambda)^{E(\mathcal{M})}}{n_{\text{int}}(\mathcal{M})!} \frac{\sum_{\pi, \sigma \in \mathfrak{S}(k)} \prod_{c=1}^{D} \prod_{i=1}^{k} \delta_{a_{\pi(i)}^c \bar{a}^c_{\tau^c_{\partial\mathcal{M}}(\sigma(i))}}}{n_{\text{ext}}(\mathcal{M})!} \times N^{-(D-1)E(\mathcal{M}) - (D-1)k(\partial\mathcal{M}) + \sum_{c=1}^{D} F_{\text{int}}^{c}(\mathcal{M})},$$

and reordering the terms by the possible boundary graphs $\mathcal{B} = \partial\mathcal{M}$ (whose $C(\mathcal{B})$ connected components we label $\mathcal{B}_{(\rho)}$) we obtain:

$$\kappa_{2p}(\mathbb{T}_{a_1^{\mathcal{D}}}, \bar{\mathbb{T}}_{\bar{a}_1^{\mathcal{D}}}, \dots \mathbb{T}_{a_p^{\mathcal{D}}}, \bar{\mathbb{T}}_{\bar{a}_p^{\mathcal{D}}}) = \sum_{\mathcal{B},\, k(\mathcal{B}) = p} \left(\prod_{\rho=1}^{C(\mathcal{B})} \delta_{a\bar{a}}^{\mathcal{B}_{(\rho)}} \right) \sum_{\substack{\mathcal{M} \text{ connected}\\ \partial\mathcal{M} = \mathcal{B}}} \frac{(-\lambda)^{E(\mathcal{M})}}{n_{\text{int}}(\mathcal{M})!}$$

$$\times N^{-(D-1)E(\mathcal{M})-(D-1)k(\partial\mathcal{M})+\sum_{c=1}^{D} F_{\text{int}}^{c}(\mathcal{M})} ,$$

where $\mathcal{B}$ runs over all the D-colored graphs with $2p$ vertices labeled:

$$w_1, \bar{w}_1 \dots w_p, \bar{w}_p ,$$

and $\mathcal{M}$ runs over edge multicolored maps whose internal vertices bear a label v_1, ... $v_{n_{\text{int}}(\mathcal{M})}$ and whose external (ciliated) vertices bear two labels in the set:

$$w_1, \bar{w}_1 \dots w_p, \bar{w}_p ,$$

one for the white and one for the black vertex of $\partial\mathcal{M}$ associated to the ciliated vertex of $\mathcal{M}$. Finally, the rescaled cumulants defined in Eq. (6.5) are:

$$K(\mathcal{B}, \mu^{(4)}) \equiv \frac{\mathfrak{K}(\mathcal{B}, \mu^{(4)})}{N^{D-2k(\mathcal{B})(D-1)-C(\mathcal{B})}} = \sum_{\substack{\mathcal{M} \text{ connected} \\ \partial\mathcal{M}=\mathcal{B}}} \frac{(-\lambda)^{E(\mathcal{M})}}{n_{\text{int}}(\mathcal{M})!}$$
$$\times N^{-D+(D-1)k(\partial\mathcal{M})+C(\partial\mathcal{M})-(D-1)E(\mathcal{M})+F_{\text{int}}(\mathcal{M})} ,$$

with $F_{\text{int}}(\mathcal{M})$ the total number of internal faces of $\mathcal{M}$.

Theorem 9.2 (Intermediate field perturbative theorem). *The measure $\mu^{(4)}$ is properly uniformly bounded in the perturbative sense, that is:*

- *for any edge multicolored map $\mathcal{M}$ we have:*

$$-D + (D-1)k(\partial\mathcal{M}) + C(\partial\mathcal{M}) - (D-1)E(\mathcal{M}) + F_{\text{int}}(\mathcal{M}) \le 0 .$$

- *for $|\lambda|$ small enough the series:*

$$K(\mathcal{B}^{(2)}) = \sum_{\substack{\mathcal{M} \text{ connected, } \partial\mathcal{M}=\mathcal{B}^{(2)} \\ -D+(D-1)k(\partial\mathcal{M})+C(\partial\mathcal{M})-(D-1)E(\mathcal{M})+F_{\text{int}}(\mathcal{M})=0}} \frac{(-\lambda)^{E(\mathcal{M})}}{n_{\text{int}}(\mathcal{M})!} ,$$

is absolutely convergent.

Proof. From Theorem 9.1, Eq. (9.5) we obtain:

$$\begin{aligned} & -D + (D-1)k(\partial\mathcal{M}) + C(\partial\mathcal{M}) - (D-1)E(\mathcal{M}) + F_{\text{int}}(\mathcal{M}) \\ & \le -D + (D-1)k(\partial\mathcal{M}) + C(\partial\mathcal{M}) - (D-1)E(\mathcal{M}) \\ & \quad + 1 - (D-1)k(\partial\mathcal{M}) - C(\mathcal{M}) + (D-1)V(\mathcal{M}) \\ & \quad + \frac{D}{2}[E(\mathcal{M}) - V(\mathcal{M}) + 1] \le -\frac{D-2}{2}[E(\mathcal{M}) - V(\mathcal{M}) + 1] . \end{aligned}$$

This proves the first part of the theorem. Furthermore, this inequality shows that at leading order for any cumulant only trees contribute and that the contribution of a map is increasingly suppressed with the excess (number of loop edges), hence the cumulants admit a $1/N$ expansion.

For the second part of the theorem, we note that the maps $\mathcal{M}$ have only one cilium, and by the previous bound only trees with one cilium can contribute. Such trees can be rooted canonically at the cilium. Furthermore, the bound is saturated only if all the edges in the tree have exactly one color, $\mathcal{C} \subset \{1, \dots D\}$. Taking into account that there are $\frac{1}{n_{\text{int}}+1}\binom{2n_{\text{int}}}{n_{\text{int}}}$ trees with unlabeled vertices and uncolored edges we obtain:

$$K(\mathcal{B}^{(2)}) = \sum_{n \geq 0} (-D\lambda)^n \frac{1}{n+1}\binom{2n}{n},$$

which is absolutely convergent for $|\lambda| \leq \frac{1}{4D}$ and sums to:

$$K(\mathcal{B}^{(2)}) = \frac{-1 + \sqrt{1 + 4D\lambda}}{2D\lambda}.$$

□

9.4 The constructive expansions

We have so far established in the intermediate field representation the perturbative $1/N$ expansion of the cumulants of $\mu^{(4)}$: when expanded in λ, each term of the resulting formal power series can be bounded by the appropriate power of N. However, the series is not summable.

In this section we study in more depth the cumulants of $\mu^{(4)}$ as functions of λ and of $1/N$, without resorting to their divergent power series expansion. We show that the cumulants are Borel summable in λ uniformly in N (see Appendix C), and subsequently we establish the proper uniform boundedness as well as the existence of the $1/N$ expansion of the cumulants as functions of λ and $1/N$.

9.4.1 The loop vertex expansion

We first express the logarithm of the generating function of the moments of $\mu^{(4)}$ as an absolutely convergent series in a certain domain of the complex plane.

Our starting point is Eq. (9.6):

$$Z^{(4)}(J, \bar{J}) = \left[e^{\frac{1}{2}\frac{1}{N^{D-1}}\sum_{\mathcal{C}\subset\mathcal{D}/2}\operatorname{Tr}_{\mathcal{C}}\left[\frac{\partial}{\partial H^{\mathcal{C}}}\frac{\partial}{\partial H^{\mathcal{C}}}\right]} e^{\operatorname{Tr}_{\mathcal{D}}[\ln R(H)] + \frac{1}{N^{D-1}} J R(H) \bar{J}} \right]_{H^{\mathcal{C}}=0}.$$

Our first aim is to compute the logarithm of $Z^{(4)}(J, \bar{J})$ along the lines of Lemma D.1 in Appendix D. As for the perturbative treatment, we start by expanding the exponential, but instead of brutally computing the Gaussian integral we use the replica trick (see Appendix B) to transform the Gaussian integral into an integral over replicated fields $H^{(i)} = \{H^{\mathcal{C}(i)} | \mathcal{C} \subset \mathcal{D}/2\}$ with a degenerated covariance:

$$\sum_{n\geq 0}\frac{1}{n!}\left[e^{\frac{1}{2}\frac{1}{N^{D-1}}\sum_{\mathcal{C}}\operatorname{Tr}_{\mathcal{C}}\left[\frac{\partial}{\partial H^{\mathcal{C}}}\frac{\partial}{\partial H^{\mathcal{C}}}\right]} \left(\operatorname{Tr}_{\mathcal{D}}[\ln R(H)] + \frac{1}{N^{D-1}} J R(H) \bar{J} \right)^n \right]_{H^{\mathcal{C}}=0}$$

$$= \sum_{n\geq 0} \frac{1}{n!} \Bigg[e^{\frac{1}{2}\frac{1}{N^{D-1}}\sum_{i,j=1}^{n}\left(\sum_{\mathcal{C}} \mathrm{Tr}_{\mathcal{C}}\left[\frac{\partial}{\partial H^{\mathcal{C}(i)}}\frac{\partial}{\partial H^{\mathcal{C}(j)}}\right]\right)} \times \prod_{i=1}^{n} \left(\mathrm{Tr}_{\mathcal{D}}\left[\ln R(H^{(i)})\right] + \frac{1}{N^{D-1}} J R(H^{(i)}) \bar{J} \right) \Bigg]_{H^{\mathcal{C}(i)}=0} ,$$

where, from now on, the sums over $\mathcal{C}$ are always understood to run over $\mathcal{C} \subset \mathcal{D}/2$. The generating function can be written as a Gaussian integral with weakening parameters:

$$\sum_{n\geq 0} \frac{1}{n!} \Bigg[e^{\frac{1}{2}\frac{1}{N^{D-1}}\sum_{i,j=1}^{n} x_{ij}\left(\sum_{\mathcal{C}} \mathrm{Tr}_{\mathcal{C}}\left[\frac{\partial}{\partial H^{\mathcal{C}(i)}}\frac{\partial}{\partial H^{\mathcal{C}(j)}}\right]\right)} \times \prod_{i=1}^{n} \left(\mathrm{Tr}_{\mathcal{D}}\left[\ln R(H^{(i)})\right] + \frac{1}{N^{D-1}} J R(H^{(i)}) \bar{J} \right) \Bigg]_{H^{\mathcal{C}(i)}=0, x_{ij}=1} .$$

Treating this expression as a function of $x_{ij} = x_{ji}$, $i \neq j$ and applying the forest formula of Theorem D.1 in Appendix D yields:

$$Z^{(4)}(J,\bar{J}) = \sum_{n\geq 0} \frac{1}{n!} \sum_{\mathfrak{F}_n} \int_0^1 \left(\prod_{(i,j)\in\mathfrak{F}_n} du_{ij} \right) \times \Bigg\{ e^{\frac{1}{2}\frac{1}{N^{D-1}}\sum_{i,j=1}^{n} w_{ij}^{\mathfrak{F}_n}(u_{\mathfrak{F}_n})\left(\sum_{\mathcal{C}} \mathrm{Tr}_{\mathcal{C}}\left[\frac{\partial}{\partial H^{\mathcal{C}(i)}}\frac{\partial}{\partial H^{\mathcal{C}(j)}}\right]\right)} \times \left(\prod_{(i,j)\in\mathfrak{F}_n} \frac{1}{N^{D-1}} \mathrm{Tr}\left[\frac{\partial}{\partial H^{\mathcal{C}(i)}}\frac{\partial}{\partial H^{\mathcal{C}(j)}}\right] \right) \times \prod_{i=1}^{n} \left[\mathrm{Tr}_{\mathcal{D}}\left[\ln R(H^{(i)})\right] + \frac{1}{N^{D-1}} J R(H^{(i)}) \bar{J} \right] \Bigg\}_{H^{\mathcal{C}(i)}=0} ,$$

where $\mathfrak{F}_n$ runs over forests over n vertices labeled $1, \ldots n$ and:

$$w_{ij}^{\mathfrak{F}_n}(u_{\mathfrak{F}_n}) = \begin{cases} 1 , & \text{if } i = j \\ \inf_{(k,l)\in P_{i\to j}^{\mathfrak{F}_n}} (u_{kl}) , & \text{if } i \neq j \end{cases} ,$$

where $P_{i\to j}^{\mathfrak{F}_n}$ denotes the unique path in the forest $\mathfrak{F}_n$ joining the vertices i and j, and the infimum is set to zero if such a path does not exist.

As the Gaussian integral, the integrals over the parameters u, and the integrand factorize over the trees in the forest, it follows that the logarithm of $Z^{(4)}(J,\bar{J})$ is a sum over combinatorial trees over n labeled vertices:

$$\ln Z^{(4)}(J,\bar{J}) = \sum_{n\geq 1} \frac{1}{n!} \sum_{\mathfrak{T}_n} \int_0^1 \left(\prod_{(i,j)\in\mathfrak{T}_n} du_{ij} \right) \times \Bigg\{ e^{\frac{1}{2}\frac{1}{N^{D-1}}\sum_{i,j=1}^{n} w_{ij}^{\mathfrak{T}_n}(u_{\mathfrak{T}_n})\left(\sum_{\mathcal{C}} \mathrm{Tr}_{\mathcal{C}}\left[\frac{\partial}{\partial H^{\mathcal{C}(i)}}\frac{\partial}{\partial H^{\mathcal{C}(j)}}\right]\right)}$$

$$\times \left(\prod_{(i,j)\in\mathfrak{T}_n} \frac{1}{N^{D-1}} \operatorname{Tr}_{\mathcal{C}} \left[\frac{\partial}{\partial H^{\mathcal{C}(i)}} \frac{\partial}{\partial H^{\mathcal{C}(j)}} \right] \right)$$

$$\times \prod_{i=1}^{n} \left[\operatorname{Tr}_{\mathcal{D}} \left[\ln R(H^{(i)}) \right] + \frac{1}{N^{D-1}} J R(H^{(i)}) \bar{J} \right] \Bigg\}_{H^{\mathcal{C}(i)}=0} ,$$

$$w_{ij}^{\mathfrak{T}_n}(u_{\mathfrak{T}_n}) = \inf_{(k,l)\in P_{i\to j}^{\mathfrak{T}_n}} (u_{kl}) ,$$

where $P_{i\to j}^{\mathfrak{T}_n}$ denotes the unique path in the tree $\mathfrak{T}_n$ joining the vertices i and j, and the infimum is set to one if $i = j$.

We expand the product over n and obtain a sum over trees with n vertices out of which k are ciliated, corresponding to the $JR(H^{(i)})\bar{J}$ terms:

$$\sum_{\substack{i_1,\dots i_k=1 \\ i_1<i_2<\dots<i_k}}^{n} \left(\prod_{i\neq i_1,\dots i_k} \operatorname{Tr}_{\mathcal{D}} \left[\ln R(H^{(i)}) \right] \right) \left(\prod_{l=1}^{k} \frac{1}{N^{D-1}} J R(H^{(i_l)}) \bar{J} \right) .$$

We denote $\iota = \{i_1, \dots i_k\}$ the labels of the ciliated vertices. The evaluation of the derivatives is similar to the one performed for the perturbative expansion. The derivatives of the resolvent lead to embedded trees (with a prescribed order of the halfedges around a vertex) with edges colored by subsets $\mathcal{C}$ of colors, $\mathfrak{T}^{n,\iota}$. We denote $\mathcal{S}(\mathfrak{T}^{n,\iota})$ the set of halfedges and cilia of the tree $\mathfrak{T}^{n,\iota}$ and $\mathcal{E}(\mathfrak{T}^{n,\iota})$ the set of its edges.

We will use the shorthand notation σ instead of $\sigma_{\mathfrak{T}^{n,\iota}}$ to signify the permutation encoding the successor halfedge in the plane tree $\mathfrak{T}^{n,\iota}$. As corners are one to one with halfedges we will sometimes denote the corner $(h, \sigma(h))$ by the halfedge it originates from, $[h]$. Every corner $[h] \equiv (h, \sigma(h))$ of the embedded tree belongs to a vertex which we denote $i[h]$.

Every corner of the tree has an associated resolvent and two associated multi indices $a_{[h]}^{\mathcal{D}}$ and $\bar{b}_{[h]}^{\mathcal{D}}$. The edges identify pairs of indices $\bar{b}$ and a: the edge $e^{\mathcal{C}} = (h, h')$:

- identifies the indices with colors in $\mathcal{D}\setminus\mathcal{C}$ of the resolvents associated to the corners $[\sigma^{-1}(h)] \equiv (\sigma^{-1}(h), h)$ and $[h] \equiv (h, \sigma(h))$, respectively $[\sigma^{-1}(h')] \equiv (\sigma^{-1}(h'), h')$ and $[h'] = (h', \sigma(h'))$.

- identifies the indices with colors in $\mathcal{C}$ of the resolvents of $[\sigma^{-1}(h)] \equiv (\sigma^{-1}(h), h)$ and $[h'] \equiv (h', \sigma(h'))$, respectively $[\sigma^{-1}(h')] \equiv (\sigma^{-1}(h'), h')$ and $[h] \equiv (h, \sigma(h))$.

For the ciliated vertices, we will also use a second notation: for the cilium h_{i_l} decorating the vertex $i_l \in \iota$ we will sometimes denote $a_l^{\mathcal{D}} \equiv a_{[h_{i_l}]}^{\mathcal{D}}$ (that is $a_l^{\mathcal{D}}$ are the a indices of the resolvent of the corner that starts at the cilium) and $\bar{b}_l^{\mathcal{D}} \equiv \bar{b}_{[\sigma^{-1}(h_{i_l})]}$ (that is $\bar{b}_l^{\mathcal{D}}$ are the b indices of the resolvent of the corner that ends at the cilium). We thus obtain the following lemma.

Lemma 9.1. *The logarithm of the partition function of the quartic tensor model writes as a sum over plane trees with labeled vertices and cilia $\mathfrak{T}^{n,\iota}$ whose edges are colored by* *subsets of colors $\mathcal{C} \subset \mathcal{D}/2$:*

$$\ln Z^{(4)}(J,\bar{J}) = \sum_{n\geq 1}\sum_{k=0}^{n} \frac{1}{n!}\left(\frac{-\lambda}{N^{D-1}}\right)^{n-1}\left(\frac{1}{N^{D-1}}\right)^{k} \sum_{\substack{i_1,\dots i_k=1\\ i_1<i_2<\dots<i_k}}^{n} \sum_{\mathfrak{T}^{n,\iota}} \mathcal{A}(\mathfrak{T}^{n,\iota}) ,$$

where the amplitude of each tree is given by:

$$\mathcal{A}(\mathfrak{T}^{n,\iota}) = \int_0^1 \left(\prod_{(i,j)\in\mathfrak{T}^{n,\iota}} du_{ij}\right)\Bigg\{ e^{\frac{1}{2}\frac{1}{N^{D-1}}\sum_{i,j=1}^{n} w_{ij}^{\mathfrak{T}^{n,\iota}}(u_{\mathfrak{T}^{n,\iota}})\left(\sum_{\mathcal{C}} \mathrm{Tr}_{\mathcal{C}}\left[\frac{\partial}{\partial H^{\mathcal{C}(i)}}\frac{\partial}{\partial H^{\mathcal{C}(j)}}\right]\right)}$$
$$\times \sum_{a\bar{b}}\left(\prod_{h\in\mathcal{S}(\mathfrak{T}^{n,\iota})} R(H^{i[h]})_{a^{\mathcal{D}}_{[h]}\bar{b}^{\mathcal{D}}_{[h]}}\right)\left(\prod_{l=1}^{k} J_{a^{\mathcal{D}}_l}\bar{J}_{\bar{b}^{\mathcal{D}}_l}\right)$$
$$\times \Bigg[\prod_{\substack{e^{\mathcal{C}}\in\mathcal{E}(\mathfrak{T}^{n,\iota})\\ e^{\mathcal{C}}=(h,h')}} \left(\delta_{\bar{b}^{\mathcal{C}}_{[\sigma^{-1}(h)]}a^{\mathcal{C}}_{[h']}}\,\delta_{\bar{b}^{\mathcal{C}}_{[\sigma^{-1}(h')]}a^{\mathcal{C}}_{[h]}}\right)\left(\delta_{\bar{b}^{\mathcal{D}\setminus\mathcal{C}}_{[\sigma^{-1}(h)]}a^{\mathcal{D}\setminus\mathcal{C}}_{[h]}}\,\delta_{\bar{b}^{\mathcal{D}\setminus\mathcal{C}}_{[\sigma^{-1}(h')]}a^{\mathcal{D}\setminus\mathcal{C}}_{[h']}}\right)\Bigg]\Bigg\}_{H=0} ,$$

and σ is the permutation encoding the successor halfedge in the map $\mathfrak{T}^{n,\iota}$.

Remark 9.2. A remark is in order.

Counting trees. Denoting $\mathcal{N}$ the number of possible colorings of the edges, the number of terms at fixed n and k in the previous sum is:

$$\sum_{\substack{i_1,\dots i_k=1\\ i_1<i_2<\dots<i_k}}^{n} \sum_{\mathfrak{T}^{n,\iota}} 1$$
$$= \mathcal{N}^{n-1} \sum_{\substack{i_1,\dots i_k=1\\ i_1<i_2<\dots<i_k}}^{n} \sum_{d_1\dots d_n\geq 1}^{\sum d_i=2(n-1)} \left[\frac{(n-2)!}{\prod_i (d_i-1)!}\left(\prod_i (d_i-1)!\right) d_{i_1}\dots d_{i_k}\right] ,$$

where d_i is the degree of the vertex i, $\frac{(n-2)!}{\prod_i(d_i-1)!}$ is the number of combinatorial trees with fixed degrees, $\prod_i (d_i-1)!$ is the number of associated plane trees, and $\prod_{l=1}^{k} d_{i_l}$ is the number of ways to put cilia on the vertices i_l. For any fixed $i_1,\dots i_k$, the sums over d_i yield the coefficient of the monomial x^{2n-2} in the expansion of:

$$\left[x\left(\frac{1}{1-x}\right)'\right]^k \frac{x^{n-k}}{(1-x)^{n-k}} = \frac{x^n}{(1-x)^{n+k}} = x^n \sum_p \binom{n+k+p-1}{p} x^p ,$$

hence the total number of terms is:

$$\mathcal{N}^{n-1}\frac{(2n+k-3)!}{(n+k-1)!}\sum_{\substack{i_1,\dots i_k=1\\ i_1<i_2<\dots<i_k}}^{n} 1 = \mathcal{N}^{n-1}\frac{(2n+k-3)!}{(n+k-1)!}\frac{1}{k!}\sum_{\substack{i_1,\dots i_k=1\\ i_l\neq i_{l'}}}^{n} 1$$
$$= \mathcal{N}^{n-1}\frac{(2n+k-3)!}{(n+k-1)!}\binom{n}{k} .$$

The expression of the generating function in Lemma 9.1 suffers from one flaw: it is not apparent why the measure $\mu^{(4)}$ is trace invariant. The invariance can be made explicit by noting that the Gaussian integral is invariant under a simultaneous change of variable for all the copies i:

$$H^{\mathcal{C}(i)} \Rightarrow (\otimes_{c\in\mathcal{C}} U^{(c)})(H^{\mathcal{C}(i)})(\otimes_{c\in\mathcal{C}} [U^{(c)}]^{\dagger}) \,,$$

with a fixed D-uple of unitary operators $U^{(c)}$, and then averaging over the $U^{(c)}$s:

$$\begin{aligned} \mathcal{A}(\mathfrak{T}^{n,\iota}) = & \int_0^1 \Bigg(\prod_{(i,j)\in\mathfrak{T}^{n,\iota}} du_{ij} \Bigg) \Bigg\{ e^{\frac{1}{2}\frac{1}{N^{D-1}} \sum_{i,j=1}^{n} w_{ij}^{\mathfrak{T}^{n,\iota}}(u_{\mathfrak{T}^{n,\iota}}) \left(\sum_{\mathcal{C}} \mathrm{Tr}_{\mathcal{C}}\left[\frac{\partial}{\partial H^{\mathcal{C}(i)}} \frac{\partial}{\partial H^{\mathcal{C}(j)}}\right]\right)} \\ & \times \sum_{a\bar{b}} \Bigg(\prod_{h\in\mathcal{S}(\mathfrak{T}^{n,\iota})} R(H^{i[h]})_{a^{\mathcal{D}}_{[h]} \bar{b}^{\mathcal{D}}_{[h]}} \Bigg) \\ & \times \int \Bigg(\prod_{c=1}^{D} [dU^{(c)}] \Bigg) \prod_{l=1}^{k} \Bigg(\sum_{n_l^{\mathcal{D}} \bar{m}_l^{\mathcal{D}}} J_{n_l^{\mathcal{D}}} \Bigg(\prod_{c=1}^{D} U^{(c)}_{n_l^{\mathcal{D}} a_l^{\mathcal{D}}} \bar{U}^{(c)}_{\bar{m}_l^{\mathcal{D}} \bar{b}_l^{\mathcal{D}}} \Bigg) \bar{J}_{\bar{m}_l^{\mathcal{D}}} \Bigg) \\ & \times \Bigg[\prod_{\substack{e^{\mathcal{C}}\in\mathcal{E}(\mathfrak{T}^{n,\iota}) \\ e^{\mathcal{C}}=(h,h')}} \Big(\delta_{\bar{b}^{\mathcal{C}}_{[\sigma^{-1}(h)]} a^{\mathcal{C}}_{[h']}} \delta_{\bar{b}^{\mathcal{C}}_{[\sigma^{-1}(h')]} a^{\mathcal{C}}_{[h]}} \Big) \Big(\delta_{\bar{b}^{\mathcal{D}\setminus\mathcal{C}}_{[\sigma^{-1}(h)]} a^{\mathcal{D}\setminus\mathcal{C}}_{[h]}} \delta_{\bar{b}^{\mathcal{D}\setminus\mathcal{C}}_{[\sigma^{-1}(h')]} a^{\mathcal{D}\setminus\mathcal{C}}_{[h']}} \Big) \Bigg] \Bigg\}_{H=0} \,, \end{aligned}$$

where the bar over the U's denotes complex conjugation. The integral over the Us is computed in terms of Weingarten functions (see Appendix A),

$$\int [dU] \left(\prod_{l=1}^{k} U_{n_l a_l} \right) \left(\prod_{l=1}^{k} \bar{U}_{\bar{m}_l \bar{b}_l} \right) = \sum_{\rho,\tau\in\mathfrak{S}(k)} \mathrm{Wg}(\rho\tau^{-1}, N) \prod_{l=1}^{k} \delta_{n_l \bar{m}_{\rho(l)}} \delta_{a_l \bar{b}_{\tau(l)}} \,.$$

The logarithm of the partition function thus becomes a sum over trees with ρ and τ edges. The ρ-edges (representing the permutations ρ) bring identifications of the indices of J and $\bar{J}$ and reconstitute the trace invariant associated to $\mathrm{Gr}(\rho^{\mathcal{D}})$, which is the boundary graph of $\mathfrak{T}^{n,\iota}_{\rho^{\mathcal{D}}\tau^{\mathcal{D}}}$, while the τ-edges (representing the permutation τ) identify the indices of the resolvents along the intτ faces.

Theorem 9.3 (Loop vertex expansion of the generating function). *The logarithm of the partition function of the quartic tensor model writes as a sum over plane trees with labeled vertices, cilia, ρ-edges and τ-edges, $\mathfrak{T}^{n,\iota}_{\rho^{\mathcal{D}}\tau^{\mathcal{D}}}$ with edges colored by subsets of colors $\mathcal{C} \subset \mathcal{D}/2$:*

$$\begin{aligned} \ln Z^{(4)}(J,\bar{J}) = & \sum_{n\geq 1} \sum_{k=0}^{n} \sum_{\rho^{\mathcal{D}},\tau^{\mathcal{D}}} \frac{1}{n!} \left(\frac{-\lambda}{N^{D-1}} \right)^{n-1} \left(\frac{1}{N^{D-1}} \right)^{k} \\ & \times \sum_{\substack{i_1,\dots i_k=1 \\ i_1<i_2<\dots<i_k}}^{n} \sum_{\mathfrak{T}^{n,\iota}_{\rho^{\mathcal{D}}\tau^{\mathcal{D}}}} \mathrm{Tr}_{\mathrm{Gr}(\rho^{\mathcal{D}})}(J,\bar{J}) \; \mathcal{A}(\mathfrak{T}^{n,\iota}_{\rho^{\mathcal{D}}\tau^{\mathcal{D}}}) \,, \end{aligned}$$

where the amplitude of a tree with ρ and τ edges is:

$$\mathcal{A}(\mathfrak{T}^{n,\iota}_{\rho^{\mathcal{D}}\tau^{\mathcal{D}}}) = \left(\prod_{c=1}^{D} \mathrm{Wg}(\rho^{(c)}[\tau^{(c)}]^{-1}, N) \right)$$

$$
\times \int_0^1 \Bigg(\prod_{(i,j)\in\mathfrak{T}^{n,\iota}} du_{ij} \Bigg) \Bigg\{ e^{\frac{1}{2}\frac{1}{N^{D-1}} \sum_{i,j=1}^n w_{ij}^{\mathfrak{T}^{n,\iota}} (u_{\mathfrak{T}^{n,\iota}}) \left(\sum_{\mathcal{C}} \mathrm{Tr}_{\mathcal{C}} \left[\frac{\partial}{\partial H^{\mathcal{C}(i)}} \frac{\partial}{\partial H^{\mathcal{C}(j)}} \right] \right)}
$$

$$
\times \sum_{a\bar b} \Bigg(\prod_{h\in\mathcal{S}(\mathfrak{T}^{n,\iota})} R(H^{i[h]})_{a^{\mathcal{D}}_{[h]} \bar b^{\mathcal{D}}_{[h]}} \Bigg) \Bigg(\prod_{l=1}^k \delta_{a_l \bar b_{\tau(l)}} \Bigg)
$$

$$
\times \Bigg[\prod_{\substack{e^{\mathcal{C}}\in\mathcal{E}(\mathfrak{T}^{n,\iota}) \\ e^{\mathcal{C}}=(h,h')}} \Big(\delta_{\bar b^{\mathcal{C}}_{[\sigma^{-1}(h)]} a^{\mathcal{C}}_{[h']}} \delta_{\bar b^{\mathcal{C}}_{[\sigma^{-1}(h')]} a^{\mathcal{C}}_{[h]}} \Big) \Big(\delta_{\bar b^{\mathcal{D}\setminus\mathcal{C}}_{[\sigma^{-1}(h)]} a^{\mathcal{D}\setminus\mathcal{C}}_{[h]}} \delta_{\bar b^{\mathcal{D}\setminus\mathcal{C}}_{[\sigma^{-1}(h')]} a^{\mathcal{D}\setminus\mathcal{C}}_{[h']}} \Big) \Bigg] \Bigg\}_{H=0} ,
$$

and σ denotes the permutation encoding the successor halfedge on the tree $\mathfrak{T}^{n,\iota}$.

The cumulants of the measure $\mu^{(4)}$ are again obtained by taking derivatives:

$$
\kappa_{2k}\big(\mathbb{T}_{a_1^{\mathcal{D}}}, \bar{\mathbb{T}}_{\bar a_1^{\mathcal{D}}}, \dots \mathbb{T}_{a_k^{\mathcal{D}}}, \bar{\mathbb{T}}_{\bar a_k^{\mathcal{D}}}\big) = \frac{\partial^{(2k)}}{\partial J_{a_1^{\mathcal{D}}} \partial \bar J_{\bar a_1^{\mathcal{D}}} \dots \partial J_{a_k^{\mathcal{D}}} \partial \bar J_{\bar a_k^{\mathcal{D}}}} \ln Z^{(4)}(J,\bar J) \Big|_{\substack{J=0\\ \bar J=0}} ,
$$

and are trace invariant:

$$
\kappa_{2k}[\mathbb{T}_{a_1^{\mathcal{D}}}, \bar{\mathbb{T}}_{\bar a_1^{\mathcal{D}}} \dots \mathbb{T}_{a_k^{\mathcal{D}}}, \bar{\mathbb{T}}_{\bar a_k^{\mathcal{D}}}] = \sum_{\mathcal{B},\, k(\mathcal{B})=k} \mathfrak{K}(\mathcal{B},\mu^{(4)}) \prod_{\rho=1}^{C(\mathcal{B})} \delta_{a\bar a}^{\mathcal{B}_{(\rho)}} ,
$$

where $\mathcal{B}$ runs over all the D-colored graphs with $2k$ labeled vertices and $C(\mathcal{B})$ denotes the number of connected components (labeled $\mathcal{B}_{(\rho)}$) of $\mathcal{B}$. Observe that the derivatives with respect to J and $\bar J$ act only on the trace invariant $\mathrm{Tr}_{\mathrm{Gr}(\rho^{\mathcal{D}})}(J,\bar J)$:

$$
\frac{\partial^{(2k)}}{\partial J_{a_1^{\mathcal{D}}} \partial \bar J_{\bar a_1^{\mathcal{D}}} \dots \partial J_{a_k^{\mathcal{D}}} \partial \bar J_{\bar a_k^{\mathcal{D}}}} \mathrm{Tr}_{\mathrm{Gr}(\rho^{\mathcal{D}})}(J,\bar J)
$$

$$
= \sum_{\pi,\bar\pi\in\mathfrak{S}(k)} \prod_{i=1}^k \prod_{c=1}^D \delta_{a^c_{\pi(i)} \bar a^c_{\rho^{(c)}\bar\pi(i)}} = k! \sum_{\bar\pi\in\mathfrak{S}(k)} \prod_{i=1}^k \prod_{c=1}^D \delta_{a^c_i \bar a^c_{\rho^{(c)}\bar\pi(i)}} ,
$$

and each graph $\mathcal{B}$ associated to the D-uple of permutations $\rho^{\mathcal{D}}_{\mathcal{B}}$ is obtained $k!$ times: for $\bar\pi$, the graph $\mathcal{B}$ is obtained for $\rho^{\mathcal{D}} = \rho^{\mathcal{D}}_{\mathcal{B}} \bar\pi^{-1}$.

Lemma 9.2. *With the notations above the rescaled cumulants:*

$$
K(\mathcal{B},\mu^{(4)}) \equiv \frac{\mathfrak{K}(\mathcal{B},\mu^{(4)})}{N^{D-2k(\mathcal{B})(D-1)-C(\mathcal{B})}} ,
$$

are:

$$
K(\mathcal{B},\mu^{(4)}) = \frac{1}{N^{D-2k(\mathcal{B})(D-1)-C(\mathcal{B})}} [k(\mathcal{B})!]^2
$$

$$
\times \sum_{n\ge k(\mathcal{B})} \sum_{\tau^{\mathcal{D}}} \frac{1}{n!} \left(\frac{-\lambda}{N^{D-1}} \right)^{n-1} \left(\frac{1}{N^{D-1}} \right)^{k(\mathcal{B})} \sum_{\substack{i_1,\dots i_k=1\\ i_1<i_2<\dots<i_k}}^{n} \sum_{\mathfrak{T}^{n,\iota}_{\rho^{\mathcal{D}}_{\mathcal{B}}\tau^{\mathcal{D}}}} \mathcal{A}(\mathfrak{T}^{n,\iota}_{\rho^{\mathcal{D}}_{\mathcal{B}}\tau^{\mathcal{D}}}) ,
$$

where $\rho_{\mathcal{B}}^{\mathcal{D}}$ is the D-uple of permutations encoding the graph $\mathcal{B}$ and:

$$\mathcal{A}(\mathfrak{T}^{n,\iota}_{\rho_{\mathcal{B}}^{\mathcal{D}}\tau^{\mathcal{D}}}) = \left(\prod_{c=1}^{D} \mathrm{Wg}(\rho_{\mathcal{B}}^{(c)}[\tau^{(c)}]^{-1}, N)\right) \int_0^1 \left(\prod_{(i,j)\in\mathfrak{T}^{n,\iota}} du_{ij}\right)$$

$$\times \left\{ e^{\frac{1}{2}\frac{1}{N^{D-1}}\sum_{i,j=1}^{n} w_{ij}^{\mathfrak{T}^{n,\iota}}(u_{\mathfrak{T}^{n,\iota}})\left(\sum_{\mathcal{C}} \mathrm{Tr}_{\mathcal{C}}\left[\frac{\partial}{\partial H^{\mathcal{C}(i)}}\frac{\partial}{\partial H^{\mathcal{C}(j)}}\right]\right)} \right.$$

$$\times \sum_{a\bar{b}} \left(\prod_{h\in\mathcal{S}(\mathfrak{T}^{n,\iota})} R(H^{i[h]})_{a^{\mathcal{D}}_{[h]}\bar{b}^{\mathcal{D}}_{[h]}}\right)\left(\prod_{l=1}^{k}\delta_{a_l\bar{b}_{\tau(l)}}\right)$$

$$\left.\times \left[\prod_{\substack{e^{\mathcal{C}}\in\mathcal{E}(\mathfrak{T}^{n,\iota})\\ e^{\mathcal{C}}=(h,h')}} \left(\delta_{\bar{b}^{\mathcal{C}}_{[\sigma^{-1}(h)]}a^{\mathcal{C}}_{[h']}}\,\delta_{\bar{b}^{\mathcal{C}}_{[\sigma^{-1}(h')]}a^{\mathcal{C}}_{[h]}}\right)\left(\delta_{\bar{b}^{\mathcal{D}\setminus\mathcal{C}}_{[\sigma^{-1}(h)]}a^{\mathcal{D}\setminus\mathcal{C}}_{[h]}}\,\delta_{\bar{b}^{\mathcal{D}\setminus\mathcal{C}}_{[\sigma^{-1}(h')]}a^{\mathcal{D}\setminus\mathcal{C}}_{[h']}}\right)\right]\right\}_{H=0} .$$

9.4.2 The mixed expansion

One can refine the loop vertex expansion of the cumulants of $\mu^{(4)}$ by expanding a number of loop edges. This is done by introducing an additional Taylor variable s in the Gaussian measure:

$$e^{\frac{1}{2}\frac{1}{N^{D-1}}\sum_{i,j=1}^{n} w_{ij}^{\mathfrak{T}^{n,\iota}}(u_{\mathfrak{T}^{n,\iota}})\left(\sum_{\mathcal{C}} \mathrm{Tr}_{\mathcal{C}}\left[\frac{\partial}{\partial H^{\mathcal{C}(i)}}\frac{\partial}{\partial H^{\mathcal{C}(j)}}\right]\right)} =$$
$$= \left[e^{\frac{s}{2}\frac{1}{N^{D-1}}\sum_{i,j=1}^{n} w_{ij}^{\mathfrak{T}^{n,\iota}}(u_{\mathfrak{T}^{n,\iota}})\left(\sum_{\mathcal{C}} \mathrm{Tr}_{\mathcal{C}}\left[\frac{\partial}{\partial H^{\mathcal{C}(i)}}\frac{\partial}{\partial H^{\mathcal{C}(j)}}\right]\right)}\right]_{s=1} ,$$

and Taylor expanding up to order r with integral reminder:

$$f(1) = \sum_{q=0}^{r-1}\frac{1}{q!}f^{(q)}(0) + \frac{1}{(r-1)!}\int_0^1 dt\,(1-t)^{r-1}f^{(r)}(t) .$$

Every derivative with respect to s will bring down an operator:

$$\frac{1}{2}\frac{1}{N^{D-1}}\sum_{i,j=1}^{n} w_{ij}^{\mathfrak{T}_n}(u_{\mathfrak{T}_n})\left(\sum_{\mathcal{C}}\mathrm{Tr}_{\mathcal{C}}\left[\frac{\partial}{\partial H^{\mathcal{C}(i)}}\frac{\partial}{\partial H^{\mathcal{C}(j)}}\right]\right) ,$$

which will act on the integrand. This operator creates a loop edge (and the sum over i and j, the fact that the resolvents around a vertex are ordered plus the sum over $\mathcal{C}$ yields a sum over all the possible ways to add a colored loop edge to the tree $\mathfrak{T}^{n,\iota}_{\rho_{\mathcal{B}}^{\mathcal{D}}\tau^{\mathcal{D}}}$). Expanding up to order q we obtain q loop edges labeled 1 up to q.

We obtain a sum indexed by trees with ρ and τ edges $\mathfrak{T}^{n,\iota}_{\rho_{\mathcal{B}}^{\mathcal{D}}\tau^{\mathcal{D}}}$ and ordered loop edges $\mathcal{L} = \{e_1, \dots e_q\}$. The couple $(\mathfrak{T}^{n,\iota}_{\rho_{\mathcal{B}}^{\mathcal{D}}\tau^{\mathcal{D}}}, \mathcal{L})$ is a combinatorial map with ρ and τ edges whose halfedges we denote $\mathcal{S}(\mathfrak{T}^{n,\iota}, \mathcal{L})$ and whose set of edges we denote $\mathcal{E}(\mathfrak{T}^{n,\iota}, \mathcal{L})$ (and whose ρ and τ edges are of course given by $\rho_{\mathcal{B}}^{\mathcal{D}}$ and $\tau^{\mathcal{D}}$). The edges of this map consist of the edges of the tree plus the loop edges $\mathcal{L}$. The corners of this map are encoded in a permutation, $\sigma_{\mathfrak{T}^{n,\iota},\mathcal{L}}$.

Theorem 9.4 (The mixed expansion). *The rescaled cumulants of the measure $\mu^{(4)}$ are:*

$$
\begin{aligned}
K(\mathcal{B},\mu^{(4)}) &= \frac{1}{N^{D-2k(\mathcal{B})(D-1)-C(\mathcal{B})}}[k(\mathcal{B})!]^2 \\
&\times \sum_{n\geq k(\mathcal{B})}\sum_{\tau^{\mathcal{D}}}\frac{1}{n!}\left(\frac{-\lambda}{N^{D-1}}\right)^{n-1}\left(\frac{1}{N^{D-1}}\right)^{k(\mathcal{B})}\sum_{\substack{i_1,\dots,i_k=1\\ i_1<i_2<\dots<i_k}}^{n}\sum_{\mathfrak{T}^{n,\iota}_{\rho^{\mathcal{D}}_{\mathcal{B}}\tau^{\mathcal{D}}}} \\
&\times\Bigg[\sum_{\mathcal{L},|\mathcal{L}|<r}\frac{1}{|\mathcal{L}|!}\left(\frac{-\lambda}{N^{D-1}}\right)^{|\mathcal{L}|}\mathcal{A}(\mathfrak{T}^{n,\iota}_{\rho^{\mathcal{D}}_{\mathcal{B}}\tau^{\mathcal{D}}},\mathcal{L},0) \\
&\qquad + \sum_{\mathcal{L},|\mathcal{L}|=r}\frac{1}{(r-1)!}\left(\frac{-\lambda}{N^{D-1}}\right)^{r}\int_0^1 dt(1-t)^{r-1}\mathcal{A}(\mathfrak{T}^{n,\iota}_{\rho^{\mathcal{D}}_{\mathcal{B}}\tau^{\mathcal{D}}},\mathcal{L},t)\Bigg],
\end{aligned}
$$

and the amplitude $\mathcal{A}(\mathfrak{T}^{n,\iota}_{\rho^{\mathcal{D}}_{\mathcal{B}}\tau^{\mathcal{D}}},\mathcal{L},t)$ of a tree with loop edges $\mathfrak{T}^{n,\iota}_{\rho^{\mathcal{D}}_{\mathcal{B}}\tau^{\mathcal{D}}},\mathcal{L}$ and weakened Gaussian measure is:

$$
\begin{aligned}
&\left(\prod_{c=1}^{D}\mathrm{Wg}(\rho^{(c)}_{\mathcal{B}}[\tau^{(c)}]^{-1},N)\right)\int_0^1\left(\prod_{(i,j)\in\mathfrak{T}^{n,\iota}}du_{ij}\right)\left(\prod_{(i',j')\in\mathcal{L}}w^{\mathfrak{T}^{n,\iota}}_{i'j'}(u_{\mathfrak{T}^{n,\iota}})\right) \\
&\times\Bigg\{e^{\frac{t}{2}\frac{1}{N^{D-1}}\sum_{i,j=1}^{n}w^{\mathfrak{T}^{n,\iota}}_{ij}(u_{\mathfrak{T}^{n,\iota}})\left(\sum_{\mathcal{C}}\mathrm{Tr}_{\mathcal{C}}\left[\frac{\partial}{\partial H^{\mathcal{C}(i)}}\frac{\partial}{\partial H^{\mathcal{C}(j)}}\right]\right)} \\
&\qquad\times\sum_{a\bar{b}}\left(\prod_{h\in\mathcal{S}(\mathfrak{T}^{n,\iota},\mathcal{L})}R(H^{i[h]})_{a^{\mathcal{D}}_{[h]}\bar{b}^{\mathcal{D}}_{[h]}}\right)\left(\prod_{l=1}^{k}\delta_{a_l\bar{b}_{\tau(l)}}\right) \\
&\qquad\times\prod_{\substack{e^{\mathcal{C}}\in\mathcal{E}(\mathfrak{T}^{n,\iota},\mathcal{L})\\ e^{\mathcal{C}}=(h,h')}}\left(\delta_{\bar{b}^{\mathcal{C}}_{[\sigma^{-1}_{\mathfrak{T}^{n,\iota},\mathcal{L}}(h)]}a^{\mathcal{C}}_{[h']}}\delta_{\bar{b}^{\mathcal{C}}_{[\sigma^{-1}_{\mathfrak{T}^{n,\iota},\mathcal{L}}(h')]}a^{\mathcal{C}}_{[h]}}\right) \\
&\qquad\qquad\left(\delta_{\bar{b}^{\mathcal{D}\setminus\mathcal{C}}_{[\sigma^{-1}_{\mathfrak{T}^{n,\iota},\mathcal{L}}(h)]}a^{\mathcal{D}\setminus\mathcal{C}}_{[h]}}\delta_{\bar{b}^{\mathcal{D}\setminus\mathcal{C}}_{[\sigma^{-1}_{\mathfrak{T}^{n,\iota},\mathcal{L}}(h')]}a^{\mathcal{D}\setminus\mathcal{C}}_{[h']}}\right)\Bigg\}_{H=0}.
\end{aligned}
$$

where $\rho^{\mathcal{D}}_{\mathcal{B}}$ is the D-uple of permutations encoding the graph $\mathcal{B}$ and $\sigma_{\mathfrak{T}^{n,\iota},\mathcal{L}}$ is the permutation encoding the successor halfedge of the map $(\mathfrak{T}^{n,\iota},\mathcal{L})$.

Observe that the tree amplitude $\mathcal{A}(\mathfrak{T}^{n,\iota}_{\rho^{\mathcal{D}}_{\mathcal{B}}\tau^{\mathcal{D}}})$ in Lemma 9.2 is nothing but the amplitude $\mathcal{A}(\mathfrak{T}^{n,\iota}_{\rho^{\mathcal{D}}_{\mathcal{B}}\tau^{\mathcal{D}}},\varnothing,1)$, that is the expansion in Lemma 9.2 can be interpreted as the mixed expansion at the order $r = 0$.

Remark 9.3. A remark is in order.

Combinatorial trees with loop edges. A fixed tree $\mathfrak{T}^{n,\iota}_{\rho^{\mathcal{D}}_{\mathcal{B}}\tau^{\mathcal{D}}}$ with n vertices out of which k are ciliated has $2n-2+k$ corners. Each end of a loop edge hooks to an existing corner and splits it into two, that is the first loop edge has $\frac{(2n-2+k)(2n-2+k+1)}{2}$ possibilities (as the same loop edge is obtained irrespective of the order in which its ends are hooked). Counting $\mathcal{N}$ colorings for each loop edge we obtain:

$$\Big(\sum_{\mathcal{L},|\mathcal{L}|=q} 1\Big)\Big|_{\mathfrak{T}^{n,\iota}_{\rho^{\mathcal{D}}_{\mathcal{B}}\tau^{\mathcal{D}}}\text{ fixed}} = \mathcal{N}^q \frac{(2n-3+k+2q)!}{2^q(2n-3+k)!} \, .$$

Therefore, using the counting of the trees in Remark 9.2, we obtain:

$$\sum_{\substack{i_1,\dots i_k=1\\ i_1<i_2<\dots<i_k}}^{n} \sum_{\mathfrak{T}^{n,\iota}} \sum_{\mathcal{L},|\mathcal{L}|=q} 1 = \mathcal{N}^{q+n-1}\frac{(2n-3+k+2q)!}{2^q(n+k-1)!}\binom{n}{k} \, . \tag{9.10}$$

The first terms in the above expansion can be simplified further. Indeed, setting $t=0$ trivializes the Gaussian integral and $\mathcal{A}(\mathfrak{T}^{n,\iota}_{\rho^{\mathcal{D}}_{\mathcal{B}}\tau^{\mathcal{D}}},\mathcal{L},0)$ computes to:

$$\begin{aligned}
&\Big(\prod_{c=1}^{D}\mathrm{Wg}(\rho^{(c)}_{\mathcal{B}}[\tau^{(c)}]^{-1},N)\Big)\int_0^1\Big(\prod_{(i,j)\in\mathfrak{T}^{n,\iota}} du_{ij}\Big)\Big(\prod_{(i',j')\in\mathcal{L}} w^{\mathfrak{T}^{n,\iota}}_{i'j'}(u_{\mathfrak{T}^{n,\iota}})\Big)\\
&\times\sum_{a\bar b}\Big(\prod_{h\in\mathcal{S}(\mathfrak{T}^{n,\iota},\mathcal{L})}\delta_{a^{\mathcal{D}}_{[h]}\bar b^{\mathcal{D}}_{[h]}}\Big)\Big(\prod_{l=1}^{k}\delta_{a_l\bar b_{\tau(l)}}\Big)\\
&\prod_{\substack{e^{\mathcal{C}}\in\mathcal{E}(\mathfrak{T}^{n,\iota},\mathcal{L})\\ e^{\mathcal{C}}=(h,h')}}\Big(\delta_{\bar b^{\mathcal{C}}_{[\sigma^{-1}_{\mathfrak{T}^{n,\iota},\mathcal{L}}(h)]}a^{\mathcal{C}}_{[h']}}\delta_{\bar b^{\mathcal{C}}_{[\sigma^{-1}_{\mathfrak{T}^{n,\iota},\mathcal{L}}(h')]}a^{\mathcal{C}}_{[h]}}\Big)\Big(\delta_{\bar b^{\mathcal{D}\setminus\mathcal{C}}_{[\sigma^{-1}_{\mathfrak{T}^{n,\iota},\mathcal{L}}(h)]}a^{\mathcal{D}\setminus\mathcal{C}}_{[h]}}\delta_{\bar b^{\mathcal{D}\setminus\mathcal{C}}_{[\sigma^{-1}_{\mathfrak{T}^{n,\iota},\mathcal{L}}(h')]}a^{\mathcal{D}\setminus\mathcal{C}}_{[h']}}\Big)\\
&=\Big(\prod_{c=1}^{D}\mathrm{Wg}(\rho^{(c)}_{\mathcal{B}}[\tau^{(c)}]^{-1},N)\Big)\int_0^1\Big(\prod_{(i,j)\in\mathfrak{T}^{n,\iota}} du_{ij}\Big)\Big(\prod_{(i',j')\in\mathcal{L}} w^{\mathfrak{T}^{n,\iota}}_{i'j'}(u_{\mathfrak{T}^{n,\iota}})\Big)\\
&\quad\times N^{F_{\mathrm{int}}(\mathfrak{T}^{n,\iota},\mathcal{L})+F_{\mathrm{int}\tau}(\mathfrak{T}^{n,\iota}_{\rho^{\mathcal{D}}_{\mathcal{B}}\tau^{\mathcal{D}}},\mathcal{L})} \, ,
\end{aligned} \tag{9.11}$$

where the last equality follows by noting that the indices are identified along the corners (all the indices in $\mathcal{D}$ per corner), along the edges (for an edge of colors $\mathcal{C}$ the indices $\mathcal{C}$ are identified among the pairs of corners incident at the two halfedges while the indices $\mathcal{D}\setminus\mathcal{C}$ are identified among the pairs of corners incident at each halfedge) and along the τ edges, hence we get a free sum over an index for every internal face of the map $(\mathfrak{T}^{n,\iota},\mathcal{L})$ and one free sum for every intτ face of the map $(\mathfrak{T}^{n,\iota}_{\rho^{\mathcal{D}}_{\mathcal{B}}\tau^{\mathcal{D}}},\mathcal{L})$.

The integral over the u parameters is computed in Appendix D:

$$\int_0^1\Big(\prod_{(i,j)\in\mathfrak{T}^{n,\iota}} du_{ij}\Big)\Big(\prod_{(i',j')\in\mathcal{L}} w^{\mathfrak{T}^{n,\iota}}_{i'j'}(u_{\mathfrak{T}^{n,\iota}})\Big) = w\big((\mathfrak{T}^{n,\iota},\mathcal{L}),\mathfrak{T}^{n,\iota}\big) \, ,$$

where $w\big((\mathfrak{T}^{n,\iota},\mathcal{L}),\mathfrak{T}^{n,\iota}\big)$ is the percentage of Hepp sectors of the map $(\mathfrak{T}^{n,\iota},\mathcal{L})$ in which $\mathfrak{T}^{n,\iota}$ is the leading spanning tree. One can reorganize the explicit terms in the mixed expansion by edge multicolored maps with fixed excess using:

$$\sum_{\mathfrak{T}\subset\mathcal{G}} w(\mathcal{G},\mathfrak{T}) = 1 \, ,$$

and recovering identically the first terms in the perturbative expansion of the cumulants in terms of maps. This is of secondary importance for the analyticity results we establish in Section 9.5.

9.5 Non perturbative tensor models

The mixed expansion is the starting point for all the non perturbative results we will establish for the quartic tensor model. The idea is to bound uniformly in N the contribution of each tree with loop edges. It turns out that such bounds can be obtained for melonic models (with $|\mathcal{C}| = 1$) by a direct method, while for the generic quartic models the same bounds require the use of iterated Cauchy–Schwarz inequalities. As this second technique is somewhat more involved, we will first detail the results for melonic models and we will establish them in full generality only afterwards.

We are interested in the behavior of the cumulants as functions of $1/N$ and $\lambda = |\lambda|e^{\imath\varphi} \in \mathbb{C}$ with $-\pi < \varphi < \pi$.

9.5.1 The melonic models

Any resolvent can be represented using a Schwinger parameter α as:

$$R(H) = \frac{1}{e^{\imath\frac{\varphi}{2}}} \int_0^\infty d\alpha \, e^{-\alpha(\mathbf{1}^{\mathcal{D}} e^{-\imath\frac{\varphi}{2}} - \imath\sqrt{|\lambda|}\sum_{\mathcal{C}} H^{\mathcal{C}} \otimes \mathbf{1}^{\mathcal{D}\setminus\mathcal{C}})} \, .$$

The crucial difference between the melonic models (with $|\mathcal{C}| = 1$) and the general quartic models (for which $|\mathcal{C}|$ can be larger than 1) is that the terms in the sum $\sum_{c=1}^{D} H^c \otimes \mathbf{1}^{\mathcal{D}\setminus\{c\}}$ *commute* for melonic models, while the terms in the sum $\sum_{\mathcal{C}} H^{\mathcal{C}} \otimes \mathbf{1}^{\mathcal{D}\setminus\mathcal{C}}$ *do not commute* for generic quartic models. It follows that for melonic models the resolvent factors over the colors, that is:

$$R(H) = \int_0^\infty \left(e^{-\imath\frac{\varphi}{2}} d\alpha\right) e^{-\alpha e^{-\imath\frac{\varphi}{2}}} \bigotimes_{c=1}^{D} e^{\imath\alpha\sqrt{|\lambda|}H^c} \, .$$

This representation allows us to establish the following bound.

Lemma 9.3. *In the melonic quartic model the amplitude of a plane tree with loop edges, and ρ and τ edges, respects the bound:*

$$\left|\mathcal{A}(\mathfrak{T}^{n,\iota}_{\rho^{\mathcal{D}}_{\mathcal{B}}\tau^{\mathcal{D}}}, \mathcal{L}, t)\right| \le \frac{2^{2Dk(\mathcal{B})} N^{-2Dk(\mathcal{B}) + F_{\rm int}(\mathfrak{T}^{n,\iota},\mathcal{L}) + F_{{\rm int}\tau}(\mathfrak{T}^{n,\iota}_{\rho^{\mathcal{D}}_{\mathcal{B}}\tau^{\mathcal{D}}},\mathcal{L}) + F_{\rho\tau^{-1}}(\mathfrak{T}^{n,\iota}_{\rho^{\mathcal{D}}_{\mathcal{B}}\tau^{\mathcal{D}}},\mathcal{L})}}{\left(\cos\frac{\varphi}{2}\right)^{2n-2+2|\mathcal{L}|+k(\mathcal{B})}} \, .$$

Proof. Using a Schwinger parameter for each of the resolvents, and taking into account the fact that the edges have a unique color $\mathcal{C} = \{c\}, c = 1, \dots D$, the amplitude of a tree with loop edges, ρ edges and τ edges becomes:

$$\begin{aligned}
&\mathcal{A}(\mathfrak{T}^{n,\iota}_{\rho^{\mathcal{D}}_{\mathcal{B}}\tau^{\mathcal{D}}}, \mathcal{L}, t) \\
&= \left(\prod_{c=1}^{D} \mathrm{Wg}(\rho^{(c)}_{\mathcal{B}}[\tau^{(c)}]^{-1}, N)\right) \int_0^1 \left(\prod_{(i,j)\in\mathfrak{T}^{n,\iota}} du_{ij}\right)\left(\prod_{(i',j')\in\mathcal{L}} w^{\mathfrak{T}^{n,\iota}}_{i'j'}(u_{\mathfrak{T}^{n,\iota}})\right) \\
&\quad \times \int_0^\infty \left(\prod_{h\in\mathcal{S}(\mathfrak{T}^{n,\iota},\mathcal{L})} e^{-\imath\frac{\varphi}{2}} d\alpha_{[h]} \, e^{-\alpha_{[h]}e^{-\imath\frac{\varphi}{2}}}\right)
\end{aligned}$$

$$\times \left\{ e^{\frac{t}{2}\frac{1}{N^{D-1}}\sum_{i,j=1}^{n} w_{ij}^{\mathfrak{T}^n}(u_{\mathfrak{T}^n})\left(\sum_c \mathrm{Tr}_c\left[\frac{\partial}{\partial H^{c(i)}}\frac{\partial}{\partial H^{c(j)}}\right]\right)} \sum_{a\bar{b}} \left(\prod_{l=1}^{k} \delta_{a_l \bar{b}_{\tau(l)}} \right) \right.$$

$$\times \left(\prod_{h\in\mathcal{S}(\mathfrak{T}^{n,\iota},\mathcal{L})} \prod_{c=1}^{D} \left[e^{\imath\alpha_{[h]}\sqrt{|\lambda|}(H^c)^{i[h]}} \right]_{a^c_{[h]}\bar{b}^c_{[h]}} \right)$$

$$\times \prod_{\substack{e^{\mathcal{C}}\in\mathcal{E}(\mathfrak{T}^{n,\iota},\mathcal{L})\\ e^{\mathcal{C}}=(h,h')}} \left(\delta_{\bar{b}^{\mathcal{C}}_{[\sigma^{-1}_{\mathfrak{T}^{n,\iota},\mathcal{L}}(h)]} a^{\mathcal{C}}_{[h']}} \delta_{\bar{b}^{\mathcal{C}}_{[\sigma^{-1}_{\mathfrak{T}^{n,\iota},\mathcal{L}}(h')]} a^{\mathcal{C}}_{[h]}} \right)$$

$$\left. \left(\delta_{\bar{b}^{\mathcal{D}\setminus\mathcal{C}}_{[\sigma^{-1}_{\mathfrak{T}^{n,\iota},\mathcal{L}}(h)]} a^{\mathcal{D}\setminus\mathcal{C}}_{[h]}} \delta_{\bar{b}^{\mathcal{D}\setminus\mathcal{C}}_{[\sigma^{-1}_{\mathfrak{T}^{n,\iota},\mathcal{L}}(h')]} a^{\mathcal{D}\setminus\mathcal{C}}_{[h']}} \right) \right\} \Bigg|_{H=0} .$$

Tracking the indices we see that the amplitude factors over the internal and intτ faces, $\mathcal{F}_{\mathrm{int}}(\mathfrak{T}^{n,\iota},\mathcal{L})$ and $\mathcal{F}_{\mathrm{int}\tau}(\mathfrak{T}^{n,\iota}_{\rho^{\mathcal{D}}_{\mathcal{B}}\tau^{\mathcal{D}}},\mathcal{L})$, and that furthermore for each face we obtain an ordered product of exponentials of $\imath H^c$ corresponding to the corners that belong to the face:

$$\mathcal{A}(\mathfrak{T}^{n,\iota}_{\rho^{\mathcal{D}}_{\mathcal{B}}\tau^{\mathcal{D}}},\mathcal{L},t)$$

$$= \left(\prod_{c=1}^{D} \mathrm{Wg}(\rho^{(c)}_{\mathcal{B}}[\tau^{(c)}]^{-1},N) \right) \int_0^1 \left(\prod_{(i,j)\in\mathfrak{T}^{n,\iota}} du_{ij} \right) \left(\prod_{(i',j')\in\mathcal{L}} w_{i'j'}^{\mathfrak{T}^{n,\iota}}(u_{\mathfrak{T}^{n,\iota}}) \right)$$

$$\times \int_0^\infty \left(\prod_{h\in\mathcal{S}(\mathfrak{T}^{n,\iota},\mathcal{L})} e^{-\imath\frac{\varphi}{2}} d\alpha_{[h]}\, e^{-\alpha_{[h]}e^{-\imath\frac{\varphi}{2}}} \right)$$

$$\times \left\{ e^{\frac{t}{2}\frac{1}{N^{D-1}}\sum_{i,j=1}^{n} w_{ij}^{\mathfrak{T}^n}(u_{\mathfrak{T}^n})\left(\sum_c \mathrm{Tr}_c\left[\frac{\partial}{\partial H^{c(i)}}\frac{\partial}{\partial H^{c(j)}}\right]\right)} \right.$$

$$\left. \times \prod_{f\in\mathcal{F}_{\mathrm{int}}(\mathfrak{T}^{n,\iota},\mathcal{L})\cup\mathcal{F}_{\mathrm{int}\tau}(\mathfrak{T}^{n,\iota}_{\rho^{\mathcal{D}}_{\mathcal{B}}\tau^{\mathcal{D}}},\mathcal{L})} \mathrm{Tr}\left[\prod_{[h]\in f} e^{\imath\alpha_{[h]}\sqrt{|\lambda|}(H^c)^{i[h]}} \right] \right\} \Bigg|_{H=0} ,$$

where $[h]\in f$ runs over the corners belonging to the face f.

As the trace of a product of unitary operators is bounded by N, the Gaussian integral is bounded by:

$$N^{F_{\mathrm{int}}(\mathfrak{T}^{n,\iota},\mathcal{L})+F_{\mathrm{int}\tau}(\mathfrak{T}^{n,\iota}_{\rho^{\mathcal{D}}_{\mathcal{B}}\tau^{\mathcal{D}}},\mathcal{L})} .$$

Each integral over a parameter α is bounded by $\left(\cos\frac{\varphi}{2}\right)^{-1}$, and the integrals over the u parameters are bounded by 1.

Taking into account the large N behavior of the Weingarten functions (see Appendix A), these functions can be bounded for N large enough as:

$$\left|\mathrm{Wg}(\sigma,N)\right| \le 2^{2k} N^{C(\sigma)-2k} ,$$

where σ is a permutation of k elements and $C(\sigma)$ denotes the number of cycles of σ. As the number of cycles of the permutation $\rho^{(c)}_{\mathcal{B}}[\tau^{(c)}]^{-1}$ is exactly the number of $\rho\tau^{-1}$

faces of color c of $(\mathfrak{T}^{n,\iota}_{\rho^{\mathcal{D}}_{\mathcal{B}}\tau^{\mathcal{D}}}, \mathcal{L})$ we obtain an overall bound:

$$\frac{2^{2Dk(\mathcal{B})} N^{-2Dk(\mathcal{B})+F_{\text{int}}(\mathfrak{T}^{n,\iota},\mathcal{L})+F_{\text{int}\mathcal{T}}(\mathfrak{T}^{n,\iota}_{\rho^{\mathcal{D}}_{\mathcal{B}}\tau^{\mathcal{D}}},\mathcal{L})+F_{\rho_{\mathcal{T}}^{-1}}(\mathfrak{T}^{n,\iota}_{\rho^{\mathcal{D}}_{\mathcal{B}}\tau^{\mathcal{D}}},\mathcal{L})}}{\left(\cos\frac{\varphi}{2}\right)^{2n-2+2|\mathcal{L}|+k(\mathcal{B})}}.$$

□

We are now in the position to state our non perturbative results on the quartic tensor model. Let us define the cardioid domain in the complex plane (see Figure 9.9):

$$\text{Car} = \left\{\lambda = |\lambda| e^{\imath\varphi} \in \mathbb{C} \,\middle|\, 4\mathcal{N}|\lambda| < \left(\cos\frac{\varphi}{2}\right)^2\right\}, \tag{9.12}$$

where $1 \le \mathcal{N} \le D$ is the number of distinct melonic interactions one starts with.

Theorem 9.5 (Analyticity). *The rescaled cumulants of the quartic melonic tensor model:*

$$\begin{aligned} K(\mathcal{B},\mu^{(4)}) &= \frac{1}{N^{D-2k(\mathcal{B})(D-1)-C(\mathcal{B})}}[k(\mathcal{B})!]^2 \\ &\times \sum_{n\ge k(\mathcal{B})}\sum_{\mathcal{T}^{\mathcal{D}}}\frac{1}{n!}\left(\frac{-\lambda}{N^{D-1}}\right)^{n-1}\left(\frac{1}{N^{D-1}}\right)^{k(\mathcal{B})}\sum_{\substack{i_1,\dots,i_k=1\\ i_1<i_2<\dots<i_k}}^{n}\sum_{\mathfrak{T}^{n,\iota}_{\rho^{\mathcal{D}}_{\mathcal{B}}\tau^{\mathcal{D}}}} A(\mathfrak{T}^{n,\iota}_{\rho^{\mathcal{D}}_{\mathcal{B}}\tau^{\mathcal{D}}},\varnothing,1), \end{aligned}$$

are analytic functions in λ for $\lambda \in \text{Car}$ and obey the bounds:

$$\left|K(\mathcal{B},\mu^{(4)})\right| \le \frac{(k!)^3 2^{2Dk+k-1}}{\left(\cos\frac{\varphi}{2}\right)^k}\left(\frac{4\mathcal{N}|\lambda|}{\left(\cos\frac{\varphi}{2}\right)^2}\right)^{k-1}\frac{1}{\left(1-\frac{4\mathcal{N}|\lambda|}{(\cos\frac{\varphi}{2})^2}\right)^k},$$

where $k \equiv k(\mathcal{B})$.

Proof. We will show that the expansion of the cumulants over trees is convergent and bounded as stated in Car. For this we bound the amplitude of each tree using Lemma 9.3 to obtain:

$$\begin{aligned} \left|K(\mathcal{B},\mu^{(4)})\right| &\le \frac{1}{N^{D-2k(\mathcal{B})(D-1)-C(\mathcal{B})}}[k(\mathcal{B})!]^2 \\ &\times \sum_{n\ge k(\mathcal{B})}\sum_{\mathcal{T}^{\mathcal{D}}}\frac{1}{n!}\left(\frac{|\lambda|}{N^{D-1}}\right)^{n-1}\left(\frac{1}{N^{D-1}}\right)^{k(\mathcal{B})}\sum_{\substack{i_1,\dots,i_k=1\\ i_1<i_2<\dots<i_k}}^{n}\sum_{\mathfrak{T}^{n,\iota}_{\rho^{\mathcal{D}}_{\mathcal{B}}\tau^{\mathcal{D}}}} \\ &\times \frac{2^{2Dk(\mathcal{B})} N^{-2Dk(\mathcal{B})+F_{\text{int}}(\mathfrak{T}^{n,\iota})+F_{\text{int}\mathcal{T}}(\mathfrak{T}^{n,\iota}_{\rho^{\mathcal{D}}_{\mathcal{B}}\tau^{\mathcal{D}}})+F_{\rho_{\mathcal{T}}^{-1}}(\mathfrak{T}^{n,\iota}_{\rho^{\mathcal{D}}_{\mathcal{B}}\tau^{\mathcal{D}}})}}{\left(\cos\frac{\varphi}{2}\right)^{2n-2+k(\mathcal{B})}}. \end{aligned}$$

Theorem 9.1 states that for any map $\mathcal{M}$:

$$F_{\text{int}}(\mathcal{M}) + F_{\text{int}\mathcal{T}}(\mathcal{M}_{\rho^{\mathcal{D}}\tau^{\mathcal{D}}}) + F_{\rho_{\mathcal{T}}^{-1}}(\mathcal{M}_{\rho^{\mathcal{D}}\tau^{\mathcal{D}}})$$

$$\leq 1 + (D+1)k(\partial\mathcal{M}) + (D-1)V(\mathcal{M}) - C(\rho^{\mathcal{D}}) + \frac{D}{2}[E(\mathcal{M}) - V(\mathcal{M}) + 1] ,$$

hence, as $\mathfrak{T}^{n,\iota}$ is a plane tree we obtain:

$$\begin{aligned}\left|K(\mathcal{B},\mu^{(4)})\right| &\leq \frac{1}{N^{D-2k(\mathcal{B})(D-1)-C(\mathcal{B})}}[k(\mathcal{B})!]^2 \\ &\times \sum_{n\geq k(\mathcal{B})}\sum_{\mathcal{T}^{\mathcal{D}}}\frac{1}{n!}\left(\frac{|\lambda|}{N^{D-1}}\right)^{n-1}\left(\frac{1}{N^{D-1}}\right)^{k(\mathcal{B})}\sum_{\substack{i_1,\dots i_k=1\\ i_1<i_2<\dots<i_k}}^{n}\sum_{\mathfrak{T}^{n,\iota}_{\rho^{\mathcal{D}}_{\mathcal{B}}\mathcal{T}^{\mathcal{D}}}} \\ &\times \frac{2^{2Dk(\mathcal{B})}N^{-2Dk(\mathcal{B})+1+(D+1)k(\mathcal{B})+(D-1)n-C(\mathcal{B})}}{\left(\cos\frac{\varphi}{2}\right)^{2n-2+k(\mathcal{B})}} ,\end{aligned}$$

and all the factors N cancel:

$$\left|K(\mathcal{B},\mu^{(4)})\right| \leq \frac{[k(\mathcal{B})!]^2 2^{2Dk(\mathcal{B})}}{\left(\cos\frac{\varphi}{2}\right)^{k(\mathcal{B})}}\sum_{n\geq k(\mathcal{B})}\sum_{\mathcal{T}^{\mathcal{D}}}\frac{1}{n!}|\lambda|^{n-1}\sum_{\substack{i_1,\dots i_k=1\\ i_1<i_2<\dots<i_k}}^{n}\sum_{\mathfrak{T}^{n,\iota}_{\rho^{\mathcal{D}}_{\mathcal{B}}\mathcal{T}^{\mathcal{D}}}}\frac{1}{\left(\cos\frac{\varphi}{2}\right)^{2n-2}} .$$

We use the counting in Eq. (9.10) and, using the shorthand notation k for $k(\mathcal{B})$, we obtain:

$$\left|K(\mathcal{B},\mu^{(4)})\right| \leq \frac{(k!)^3 2^{2Dk}}{\left(\cos\frac{\varphi}{2}\right)^{k}}\sum_{n\geq k}\frac{1}{n!}\frac{(2n-3+k)!}{(n+k-1)!}\binom{n}{k}\left(\frac{\mathcal{N}|\lambda|}{\left(\cos\frac{\varphi}{2}\right)^2}\right)^{n-1} .$$

In order to bound the sum over n we first note that:

$$(2n-3+k)! \leq 2^{2n-3+k}(n+k-2)!(n-1)! \leq 2^{2n-3+k}(n+k-1)!(n-1)! ,$$

therefore:

$$\begin{aligned}\left|K(\mathcal{B},\mu^{(4)})\right| &\leq \frac{(k!)^2 2^{2Dk+k-1}}{\left(\cos\frac{\varphi}{2}\right)^{k}}\sum_{n\geq k}\frac{(n-1)!}{(n-k)!}\left(\frac{4\mathcal{N}|\lambda|}{\left(\cos\frac{\varphi}{2}\right)^2}\right)^{n-1} \\ &= \frac{(k!)^2 2^{2Dk+k-1}}{\left(\cos\frac{\varphi}{2}\right)^{k}}\left(\frac{4\mathcal{N}|\lambda|}{\left(\cos\frac{\varphi}{2}\right)^2}\right)^{k-1}\sum_{p\geq 0}\frac{(p+k-1)!}{p!}\left(\frac{4\mathcal{N}|\lambda|}{\left(\cos\frac{\varphi}{2}\right)^2}\right)^{p} \\ &= \frac{(k!)^2 2^{2Dk+k-1}}{\left(\cos\frac{\varphi}{2}\right)^{k}}\left(\frac{4\mathcal{N}|\lambda|}{\left(\cos\frac{\varphi}{2}\right)^2}\right)^{k-1}\frac{(k-1)!}{\left(1-\frac{4\mathcal{N}|\lambda|}{\left(\cos\frac{\varphi}{2}\right)^2}\right)^{k}}\end{aligned}$$

which concludes the proof.

□

One can use the mixed expansion to perform the $1/N$ expansion of the cumulants in the non perturbative sense, that is to write every cumulant as an explicit term plus a rest term which is analytic in some domain in the complex plane and suppressed in powers of $1/N$.

Theorem 9.6 (Non perturbative $1/N$ expansion). *For any $r \in \mathcal{N}$, $r \geq 1$, the rescaled cumulants of the quartic melonic tensor model are written as:*

$$K(\mathcal{B}, \mu^{(4)}) = K^{(r)}_{\text{explicit}}(\mathcal{B}, \mu^{(4)}) + K^{(r)}_{\text{rest}}(\mathcal{B}, \mu^{(4)}) \,,$$

with $K^{(r)}_{\text{explicit}}(\mathcal{B}, \mu^{(4)})$:

$$\begin{aligned} K^{(r)}_{\text{explicit}}(\mathcal{B}, \mu^{(4)}) &= \frac{1}{N^{D-2k(\mathcal{B})(D-1)-C(\mathcal{B})}} [k(\mathcal{B})!]^2 \\ &\times \sum_{n \geq k(\mathcal{B})} \sum_{\tau^{\mathcal{D}}} \sum_{\substack{i_1,\dots i_k=1 \\ i_1<i_2<\dots<i_k}}^{n} \sum_{\mathfrak{T}^{n,\iota}_{\rho^{\mathcal{D}}_{\mathcal{B}} \tau^{\mathcal{D}}}} \sum_{\mathcal{L}, |\mathcal{L}|<r} \\ &\times \frac{1}{n!|\mathcal{L}|!} \left(\frac{-\lambda}{N^{D-1}} \right)^{n+|\mathcal{L}|-1} \left(\frac{1}{N^{D-1}} \right)^{k(\mathcal{B})} \mathcal{A}(\mathfrak{T}^{n,\iota}_{\rho^{\mathcal{D}}_{\mathcal{B}} \tau^{\mathcal{D}}}, \mathcal{L}, 0) \,, \end{aligned}$$

an analytic function in the disk $4\mathcal{N}|\lambda| < 1$ which respects the bound:

$$\left| K^{(r)}_{\text{explicit}}(\mathcal{B}, \mu^{(4)}) \right| \leq (k!)^3 2^{2(D+1)k} \frac{(4\mathcal{N}|\lambda|)^{k-1}}{(1-4\mathcal{N}|\lambda|)^k} \sum_{q=0}^{r-1} q! \frac{2^{3q}}{N^{(D-2)q}} \frac{(4\mathcal{N}|\lambda|)^q}{(1-4\mathcal{N}|\lambda|)^{2q}} \,,$$

and $K^{(r)}_{\text{rest}}(\mathcal{B}, \mu^{(4)})$:

$$\begin{aligned} K^{(r)}_{\text{rest}}(\mathcal{B}, \mu^{(4)}) &= \frac{1}{N^{D-2k(\mathcal{B})(D-1)-C(\mathcal{B})}} [k(\mathcal{B})!]^2 \\ &\times \sum_{n \geq k(\mathcal{B})} \sum_{\tau^{\mathcal{D}}} \sum_{\substack{i_1,\dots i_k=1 \\ i_1<i_2<\dots<i_k}}^{n} \sum_{\mathfrak{T}^{n,\iota}_{\rho^{\mathcal{D}}_{\mathcal{B}} \tau^{\mathcal{D}}}} \sum_{\mathcal{L}, |\mathcal{L}|=r} \\ &\times \frac{1}{n!(r-1)!} \left(\frac{-\lambda}{N^{D-1}} \right)^{n+r-1} \left(\frac{1}{N^{D-1}} \right)^{k(\mathcal{B})} \\ &\times \int_0^1 dt (1-t)^{r-1} \mathcal{A}(\mathfrak{T}^{n,\iota}_{\rho^{\mathcal{D}}_{\mathcal{B}} \tau^{\mathcal{D}}}, \mathcal{L}, t) \,, \end{aligned}$$

an analytic function in Car *which respect the bounds:*

$$\left| K^{(r)}_{\text{rest}}(\mathcal{B}, \mu^{(4)}) \right| \leq \frac{(k!)^3 2^{2(D+1)k}}{\left(\cos \frac{\varphi}{2}\right)^k} r! \frac{2^{3r}}{N^{(D-2)r}} \frac{\left(\frac{4\mathcal{N}|\lambda|}{(\cos \frac{\varphi}{2})^2} \right)^{k+r-1}}{\left(1 - \frac{4\mathcal{N}|\lambda|}{(\cos \frac{\varphi}{2})^2} \right)^{k+2r}} \,.$$

Proof. Our starting point is the mixed expansion in Theorem 9.4 of the cumulants:

$$\begin{aligned} K(\mathcal{B}, \mu^{(4)}) &= \frac{1}{N^{D-2k(\mathcal{B})(D-1)-C(\mathcal{B})}} [k(\mathcal{B})!]^2 \\ &\times \sum_{n \geq k(\mathcal{B})} \sum_{\tau^{\mathcal{D}}} \frac{1}{n!} \left(\frac{-\lambda}{N^{D-1}} \right)^{n-1} \left(\frac{1}{N^{D-1}} \right)^{k(\mathcal{B})} \sum_{\substack{i_1,\dots i_k=1 \\ i_1<i_2<\dots<i_k}}^{n} \sum_{\mathfrak{T}^{n,\iota}_{\rho^{\mathcal{D}}_{\mathcal{B}} \tau^{\mathcal{D}}}} \end{aligned}$$

$$\times \Bigg[\sum_{\mathcal{L}, |\mathcal{L}|<r} \frac{1}{|\mathcal{L}|!} \left(\frac{-\lambda}{N^{D-1}} \right)^{|\mathcal{L}|} \mathcal{A}(\mathfrak{T}^{n,\iota}_{\rho^{\mathcal{D}}_{\mathcal{B}} \tau^{\mathcal{D}}}, \mathcal{L}, 0)$$

$$+ \sum_{\mathcal{L}, |\mathcal{L}|=r} \frac{1}{(r-1)!} \left(\frac{-\lambda}{N^{D-1}} \right)^{r} \int_0^1 dt (1-t)^{r-1} \mathcal{A}(\mathfrak{T}^{n,\iota}_{\rho^{\mathcal{D}}_{\mathcal{B}} \tau^{\mathcal{D}}}, \mathcal{L}, t) \Bigg] ,$$

which we split into explicit terms having up to r loop edges:

$$K^{(r)}_{\text{explicit}}(\mathcal{B}, \mu^{(4)}) = \frac{1}{N^{D-2k(\mathcal{B})(D-1)-C(\mathcal{B})}} [k(\mathcal{B})!]^2$$

$$\times \sum_{n \geq k(\mathcal{B})} \sum_{\tau^{\mathcal{D}}} \sum_{\substack{i_1, \dots i_k = 1 \\ i_1 < i_2 < \dots < i_k}}^{n} \sum_{\mathfrak{T}^{n,\iota}_{\rho^{\mathcal{D}}_{\mathcal{B}} \tau^{\mathcal{D}}}} \sum_{\mathcal{L}, |\mathcal{L}|<r}$$

$$\times \frac{1}{n! |\mathcal{L}|!} \left(\frac{-\lambda}{N^{D-1}} \right)^{n+|\mathcal{L}|-1} \left(\frac{1}{N^{D-1}} \right)^{k(\mathcal{B})} \mathcal{A}(\mathfrak{T}^{n,\iota}_{\rho^{\mathcal{D}}_{\mathcal{B}} \tau^{\mathcal{D}}}, \mathcal{L}, 0) ,$$

and the rest terms:

$$K^{(r)}_{\text{rest}}(\mathcal{B}, \mu^{(4)}) = \frac{1}{N^{D-2k(\mathcal{B})(D-1)-C(\mathcal{B})}} [k(\mathcal{B})!]^2$$

$$\times \sum_{n \geq k(\mathcal{B})} \sum_{\tau^{\mathcal{D}}} \sum_{\substack{i_1, \dots i_k = 1 \\ i_1 < i_2 < \dots < i_k}}^{n} \sum_{\mathfrak{T}^{n,\iota}_{\rho^{\mathcal{D}}_{\mathcal{B}} \tau^{\mathcal{D}}}} \sum_{\mathcal{L}, |\mathcal{L}|=r}$$

$$\times \frac{1}{n! (r-1)!} \left(\frac{-\lambda}{N^{D-1}} \right)^{n+r-1} \left(\frac{1}{N^{D-1}} \right)^{k(\mathcal{B})}$$

$$\times \int_0^1 dt (1-t)^{r-1} \mathcal{A}(\mathfrak{T}^{n,\iota}_{\rho^{\mathcal{D}}_{\mathcal{B}} \tau^{\mathcal{D}}}, \mathcal{L}, t) .$$

The amplitude $\mathcal{A}(\mathfrak{T}^{n,\iota}_{\rho^{\mathcal{D}}_{\mathcal{B}} \tau^{\mathcal{D}}}, \mathcal{L}, 0)$ arising in the explicit terms can be explicitly computed (hence the name) as in Eq. (9.11):

$$\mathcal{A}(\mathfrak{T}^{n,\iota}_{\rho^{\mathcal{D}}_{\mathcal{B}} \tau^{\mathcal{D}}}, \mathcal{L}, 0) = \left(\prod_{c=1}^{D} \mathrm{Wg}(\rho^{(c)}_{\mathcal{B}} [\tau^{(c)}]^{-1}, N) \right) N^{F_{\text{int}}(\mathfrak{T}^{n,\iota}, \mathcal{L}) + F_{\text{int}\tau}(\mathfrak{T}^{n,\iota}_{\rho^{\mathcal{D}}_{\mathcal{B}} \tau^{\mathcal{D}}}, \mathcal{L})}$$

$$\times \int_0^1 \left(\prod_{(i,j) \in \mathfrak{T}^{n,\iota}} du_{ij} \right) \left(\prod_{(i',j') \in \mathcal{L}} w^{\mathfrak{T}^{n,\iota}}_{i'j'}(u_{\mathfrak{T}^{n,\iota}}) \right) ,$$

and using the same bound on the Weingarten function as in Lemma 9.3 we get:

$$\left| \mathcal{A}(\mathfrak{T}^{n,\iota}_{\rho^{\mathcal{D}}_{\mathcal{B}} \tau^{\mathcal{D}}}, \mathcal{L}, 0) \right| \leq 2^{2Dk(\mathcal{B})} N^{-2Dk(\mathcal{B}) + F_{\text{int}}(\mathfrak{T}^{n,\iota}, \mathcal{L}) + F_{\text{int}\tau}(\mathfrak{T}^{n,\iota}_{\rho^{\mathcal{D}}_{\mathcal{B}} \tau^{\mathcal{D}}}, \mathcal{L}) + F_{\rho\tau^{-1}}(\mathfrak{T}^{n,\iota}_{\rho^{\mathcal{D}}_{\mathcal{B}} \tau^{\mathcal{D}}}, \mathcal{L})} ,$$

while the amplitude arising in the rest terms $\mathcal{A}(\mathfrak{T}^{n,\iota}_{\rho^{\mathcal{D}}_{\mathcal{B}} \tau^{\mathcal{D}}}, \mathcal{L}, t)$ is bounded by Lemma 9.3:

$$\left| \mathcal{A}(\mathfrak{T}^{n,\iota}_{\rho^{\mathcal{D}}_{\mathcal{B}} \tau^{\mathcal{D}}}, \mathcal{L}, t) \right| \leq \frac{2^{2Dk(\mathcal{B})} N^{-2Dk(\mathcal{B}) + F_{\text{int}}(\mathfrak{T}^{n,\iota}, \mathcal{L}) + F_{\text{int}\tau}(\mathfrak{T}^{n,\iota}_{\rho^{\mathcal{D}}_{\mathcal{B}} \tau^{\mathcal{D}}}, \mathcal{L}) + F_{\rho\tau^{-1}}(\mathfrak{T}^{n,\iota}_{\rho^{\mathcal{D}}_{\mathcal{B}} \tau^{\mathcal{D}}}, \mathcal{L})}}{\left(\cos \frac{\varphi}{2} \right)^{2n-2+2|\mathcal{L}|+k(\mathcal{B})}} .$$

Remark that the bound on the rest terms is weaker than the one on the explicit terms due to the presence of the denominator. This is the origin of the fact that the explicit terms are analytic in the disk $4\mathcal{N}|\lambda|$, while the rest terms are only analytic in Car. From here on the proof follows the same steps as in Theorem 9.5. Using Theorem 9.1, and taking into account the fact that for the melonic model all the edges of $\mathfrak{T}^{n,\iota},\mathcal{L}$ have only one color, we bound:

$$F_{\text{int}}(\mathfrak{T}^{n,\iota},\mathcal{L}) + F_{\text{int}\mathcal{T}}(\mathfrak{T}^{n,\iota}_{\rho^{\mathcal{D}}_{\mathcal{B}}\mathcal{T}^{\mathcal{D}}},\mathcal{L}) + F_{\rho\mathcal{T}^{-1}}(\mathfrak{T}^{n,\iota}_{\rho^{\mathcal{D}}_{\mathcal{B}}\mathcal{T}^{\mathcal{D}}},\mathcal{L}) \\ \leq 1 + (D+1)k(\mathcal{B}) + (D-1)n - C(\mathcal{B}) + |\mathcal{L}| \,,$$

hence the global scaling in N of a term in either the explicit or the rest terms is:

$$\frac{N^{-2Dk(\mathcal{B})+1+(D+1)k(\mathcal{B})+(D-1)n-C(\mathcal{B})+|\mathcal{L}|}}{N^{D-2k(\mathcal{B})(D-1)-C(\mathcal{B})}N^{(D-1)(n+|\mathcal{L}|-1)+(D-1)k(\mathcal{B})}} = \frac{1}{N^{(D-2)|\mathcal{L}|}} \,.$$

Using the counting in Eq. (9.10):

$$\sum_{\substack{i_1,\dots i_k=1\\ i_1<i_2<\dots<i_k}}^{n} \sum_{\mathfrak{T}^{n,\iota}} \sum_{\mathcal{L},|\mathcal{L}|=q} 1 = \mathcal{N}^{q+n-1}\frac{(2n-3+k+2q)!}{2^q(n+k-1)!}\binom{n}{k} \,,$$

and denoting $k(\mathcal{B}) = k$ we get:

$$\left|K^{(r)}_{\text{explicit}}(\mathcal{B},\mu^{(4)})\right| \leq (k!)^3 2^{2Dk} \\ \sum_{n\geq k}\sum_{q=0}^{r-1}\frac{1}{n!q!}\mathcal{N}^{q+n-1}\frac{(2n-3+k+2q)!}{2^q(n+k-1)!}\binom{n}{k}\frac{|\lambda|^{n+q-1}}{N^{(D-2)q}}$$

$$\left|K^{(r)}_{\text{rest}}(\mathcal{B},\mu^{(4)})\right| \leq (k!)^3 2^{2Dk}\frac{1}{N^{(D-2)r}} \\ \sum_{n\geq k}\frac{1}{n!(r-1)!}\mathcal{N}^{r+n-1}\frac{(2n-3+k+2r)!}{2^r(n+k-1)!}\binom{n}{k}\frac{|\lambda|^{n+r-1}}{\left(\cos\frac{\varphi}{2}\right)^{2n-2+2r+k}} \,.$$

We now use the binomial bound:

$$(2n-3+k+2q)! \leq 2^{2n-3+k+2q}(n+k-1)!(n-1+2q)! \,,$$

to get:

$$\left|K^{(r)}_{\text{explicit}}(\mathcal{B},\mu^{(4)})\right| \leq (k!)^2 2^{2Dk+k-1}\sum_{q=0}^{r-1}\frac{1}{q!2^q}\frac{1}{N^{(D-2)q}} \\ \times\sum_{n\geq k}\frac{(n-1+2q)!}{(n-k)!}(4\mathcal{N}|\lambda|)^{n+q-1}$$

$$\left|K^{(r)}_{\text{rest}}(\mathcal{B},\mu^{(4)})\right| \leq (k!)^2 2^{2Dk+k-1}\frac{1}{N^{(D-2)r}}\frac{1}{2^r(r-1)!}\frac{1}{\left(\cos\frac{\varphi}{2}\right)^k} \\ \times\sum_{n\geq k}\frac{(n-1+2r)!}{(n-k)!}\left(\frac{4\mathcal{N}|\lambda|}{\left(\cos\frac{\varphi}{2}\right)^2}\right)^{n+r-1} \,.$$

The sums over n can be explicitly evaluated to get:

$$\left|K^{(r)}_{\text{explicit}}(\mathcal{B},\mu^{(4)})\right| \le (k!)^2 2^{2Dk+k-1}\sum_{q=0}^{r-1}\frac{1}{q!2^q}\frac{1}{N^{(D-2)q}}\frac{(k+2q-1)!(4\mathcal{N}|\lambda|)^{k+q-1}}{(1-4\mathcal{N}|\lambda|)^{k+2q}}$$

$$\left|K^{(r)}_{\text{rest}}(\mathcal{B},\mu^{(4)})\right| \le (k!)^2 2^{2Dk+k-1}\frac{1}{N^{(D-2)r}}\frac{1}{2^r(r-1)!}\frac{1}{\left(\cos\frac{\varphi}{2}\right)^k}\frac{(k+2r-1)!\left(\frac{4\mathcal{N}|\lambda|}{(\cos\frac{\varphi}{2})^2}\right)^{k+r-1}}{\left(1-\frac{4\mathcal{N}|\lambda|}{(\cos\frac{\varphi}{2})^2}\right)^{k+2r}}.$$

Taking into account that $(k+2q-1)! \le 2^{k+2q}k!(2q-1)! \le 2^{k+2q}k!(2q)!$ we get:

$$\left|K^{(r)}_{\text{explicit}}(\mathcal{B},\mu^{(4)})\right| \le (k!)^3 2^{2(D+1)k}\sum_{q=0}^{r-1}\frac{(2q)!}{q!}\frac{2^q}{N^{(D-2)q}}\frac{(4\mathcal{N}|\lambda|)^{k+q-1}}{(1-4\mathcal{N}|\lambda|)^{k+2q}}$$

$$\left|K^{(r)}_{\text{rest}}(\mathcal{B},\mu^{(4)})\right| \le (k!)^3 2^{2(D+1)k}\frac{1}{N^{(D-2)r}}\frac{2^r(2r-1)!}{(r-1)!}\frac{1}{\left(\cos\frac{\varphi}{2}\right)^k}\frac{\left(\frac{4\mathcal{N}|\lambda|}{(\cos\frac{\varphi}{2})^2}\right)^{k+r-1}}{\left(1-\frac{4\mathcal{N}|\lambda|}{(\cos\frac{\varphi}{2})^2}\right)^{k+2r}},$$

and finally, as $(2q)! \le 2^{2q}q!q!$ and $(2r-1)! \le 2^{2r}r!(r-1)!$, we have:

$$\left|K^{(r)}_{\text{explicit}}(\mathcal{B},\mu^{(4)})\right| \le (k!)^3 2^{2(D+1)k}\frac{(4\mathcal{N}|\lambda|)^{k-1}}{(1-4\mathcal{N}|\lambda|)^k}\sum_{q=0}^{r-1}q!\frac{2^{3q}}{N^{(D-2)q}}\frac{(4\mathcal{N}|\lambda|)^q}{(1-4\mathcal{N}|\lambda|)^{2q}}$$

$$\left|K^{(r)}_{\text{rest}}(\mathcal{B},\mu^{(4)})\right| \le \frac{(k!)^3 2^{2(D+1)k}}{\left(\cos\frac{\varphi}{2}\right)^k}r!\frac{2^{3r}}{N^{(D-2)r}}\frac{\left(\frac{4\mathcal{N}|\lambda|}{(\cos\frac{\varphi}{2})^2}\right)^{k+r-1}}{\left(1-\frac{4\mathcal{N}|\lambda|}{(\cos\frac{\varphi}{2})^2}\right)^{k+2r}},$$

which concludes the proof.

□

The cumulant corresponding to $\mathcal{B}^{(2)}$ has a finite limit as $N \to \infty$ for $\lambda \in \text{Car}$.

Lemma 9.4. *We have:*

$$\lim_{N\to\infty} K(\mathcal{B}^{(2)},\mu^{(4)}) \equiv K(\mathcal{B}^{(2)}) = \frac{-1+\sqrt{1+4\mathcal{N}\lambda}}{2\mathcal{N}\lambda}.$$

Proof. Let us apply Theorem 9.6 for $K(\mathcal{B}^{(2)},\mu^{(4)})$, $r=1$ and $\lambda\in\text{Car}$:

$$K(\mathcal{B}^{(2)},\mu^{(4)}) = K^{(1)}_{\text{explicit}}(\mathcal{B}^{(2)},\mu^{(4)}) + K^{(1)}_{\text{rest}}(\mathcal{B}^{(2)},\mu^{(4)}) \,,$$

$$K^{(1)}_{\text{explicit}}(\mathcal{B}^{(2)},\mu^{(4)}) = \frac{1}{N^{D-2(D-1)-1}} \sum_{n\geq 1}\sum_{i_1=1}^{n}\sum_{\mathfrak{T}^{n,\iota}_{(1)^{\mathcal{D}}(1)^{\mathcal{D}}}} \times \frac{1}{n!}\left(\frac{-\lambda}{N^{D-1}}\right)^{n-1}\left(\frac{1}{N^{D-1}}\right)\mathcal{A}(\mathfrak{T}^{n,\iota}_{(1)^{\mathcal{D}}(1)^{\mathcal{D}}},\varnothing,0)\,,$$

$$\left|K^{(1)}_{\text{rest}}(\mathcal{B}^{(2)},\mu^{(4)})\right| \leq \frac{2^{2(D+1)}}{\cos\frac{\varphi}{2}}\,\frac{2^3}{N^{(D-2)}}\,\frac{\left(\frac{4\mathcal{N}|\lambda|}{(\cos\frac{\varphi}{2})^2}\right)}{\left(1-\frac{4\mathcal{N}|\lambda|}{(\cos\frac{\varphi}{2})^2}\right)^3}\,,$$

where:

$$\mathcal{A}(\mathfrak{T}^{n,\iota}_{(1)^{\mathcal{D}}(1)^{\mathcal{D}}},\varnothing,0) = \left(\prod_{c=1}^{D}\text{Wg}((1),N)\right)N^{F_{\text{int}}(\mathfrak{T}^{n,\iota})+F_{\text{int}\mathcal{T}}(\mathfrak{T}^{n,\iota}_{(1)^{\mathcal{D}}(1)^{\mathcal{D}}})}\,.$$

It follows that:

$$\lim_{N\to\infty} K^{(1)}_{\text{rest}}(\mathcal{B}^{(2)},\mu^{(4)}) = 0\,,$$

and, taking into account that $\text{Wg}((1),N)=\frac{1}{N}$ and that the bound in Theorem 9.1 is saturated for trees with a unique cilium whose edges have all only one color, we get:

$$\mathcal{A}(\mathfrak{T}^{n,\iota}_{(1)^{\mathcal{D}}(1)^{\mathcal{D}}},\varnothing,0) = N^{-2D+1+(D+1)+(D-1)n-1}\,,$$

hence $K^{(1)}_{\text{explicit}}(\mathcal{B}^{(2)},\mu^{(4)})$ computes explicitly to:

$$K^{(1)}_{\text{explicit}}(\mathcal{B}^{(2)},\mu^{(4)}) = \sum_{n\geq 1}\frac{1}{n!}\mathcal{N}^{n-1}\frac{(2n-2)!}{n!}\binom{n}{1}(-\lambda)^{n-1}$$
$$= \sum_{n'\geq 0}\frac{1}{n'+1}\binom{2n'}{n'}(-\mathcal{N}\lambda)^{n'} = \frac{-1+\sqrt{1+4\mathcal{N}\lambda}}{2\mathcal{N}\lambda}\,.$$

□

The following theorem makes the contact between the (convergent) constructive expansion on the one hand and the (divergent) perturbative expansion on the other. As the cumulants are Borel summable, they are the Borel sum of their perturbative expansions (which, while divergent, are Borel summable series).

Theorem 9.7. *The cumulants $K(\mathcal{B},\mu^{(4)})$ are Borel summable in λ uniformly in N.*

Proof. The Borel summability follows from Theorem C.1 in Appendix C provided one can prove a $s!|\lambda|^s$ bound on the rest term of order s in the Taylor series of the cumulants as functions of λ that is valid in some Borel disk in λ. For any cumulant we choose the order $s\geq k(\mathcal{B})$. We again start from the mixed expansion in Theorem 9.4:

$$K(\mathcal{B},\mu^{(4)}) = \frac{1}{N^{D-2k(\mathcal{B})(D-1)-C(\mathcal{B})}}[k(\mathcal{B})!]^2$$

$$\times \sum_{n \geq k(\mathcal{B})} \sum_{\mathcal{T}^{\mathcal{D}}} \frac{1}{n!} \left(\frac{-\lambda}{N^{D-1}} \right)^{n-1} \left(\frac{1}{N^{D-1}} \right)^{k(\mathcal{B})} \sum_{\substack{i_1,\dots i_k=1 \\ i_1<i_2<\dots<i_k}}^{n} \sum_{\mathfrak{T}^{n,\iota}_{\rho^{\mathcal{D}}_{\mathcal{B}} \mathcal{T}^{\mathcal{D}}}}$$

$$\times \left[\sum_{\mathcal{L},|\mathcal{L}|<r} \frac{1}{|\mathcal{L}|!} \left(\frac{-\lambda}{N^{D-1}} \right)^{|\mathcal{L}|} A(\mathfrak{T}^{n,\iota}_{\rho^{\mathcal{D}}_{\mathcal{B}} \mathcal{T}^{\mathcal{D}}}, \mathcal{L}, 0) \right.$$

$$\left. + \sum_{\mathcal{L},|\mathcal{L}|=r} \frac{1}{(r-1)!} \left(\frac{-\lambda}{N^{D-1}} \right)^{r} \int_0^1 dt (1-t)^{r-1} A(\mathfrak{T}^{n,\iota}_{\rho^{\mathcal{D}}_{\mathcal{B}} \mathcal{T}^{\mathcal{D}}}, \mathcal{L}, t) \right] .$$

In order to identify the rest term of order s in the expansion in λ of a cumulant we push the mixed expansion up to $|\mathcal{L}| = s+1-n$ edges for the trees with s vertices or less and up to order $|\mathcal{L}| = 0$ for the trees with more than s vertices. The rest term of order s in the expansion in λ is:

$$\mathcal{R}^{(s)}(\mathcal{B},\mu^{(4)}) = \frac{1}{N^{D-2k(\mathcal{B})(D-1)-C(\mathcal{B})}} [k(\mathcal{B})!]^2$$

$$\times \left\{ \sum_{n \geq k(\mathcal{B})}^{s} \sum_{\mathcal{T}^{\mathcal{D}}} \frac{1}{n!} \left(\frac{-\lambda}{N^{D-1}} \right)^{n-1} \left(\frac{1}{N^{D-1}} \right)^{k(\mathcal{B})} \sum_{\substack{i_1,\dots i_k=1 \\ i_1<i_2<\dots<i_k}}^{n} \sum_{\mathfrak{T}^{n,\iota}_{\rho^{\mathcal{D}}_{\mathcal{B}} \mathcal{T}^{\mathcal{D}}}} \right.$$

$$\times \sum_{\mathcal{L},|\mathcal{L}|=s+1-n} \frac{1}{(s-n)!} \left(\frac{-\lambda}{N^{D-1}} \right)^{s+1-n} \int_0^1 dt (1-t)^{s-n} A(\mathfrak{T}^{n,\iota}_{\rho^{\mathcal{D}}_{\mathcal{B}} \mathcal{T}^{\mathcal{D}}}, \mathcal{L}, t)$$

$$\left. + \sum_{n \geq s+1} \sum_{\mathcal{T}^{\mathcal{D}}} \frac{1}{n!} \left(\frac{-\lambda}{N^{D-1}} \right)^{n-1} \left(\frac{1}{N^{D-1}} \right)^{k(\mathcal{B})} \sum_{\substack{i_1,\dots i_k=1 \\ i_1<i_2<\dots<i_k}}^{n} \sum_{\mathfrak{T}^{n,\iota}_{\rho^{\mathcal{D}}_{\mathcal{B}} \mathcal{T}^{\mathcal{D}}}} A(\mathfrak{T}^{n,\iota}_{\rho^{\mathcal{D}}_{\mathcal{B}} \mathcal{T}^{\mathcal{D}}}, \emptyset, 1) \right\} .$$

Using the same bounds as in Theorem 9.6 (and denoting $k \equiv k(\mathcal{B})$) we obtain:

$$\left| \mathcal{R}^{(s)}(\mathcal{B},\mu^{(4)}) \right| \leq (k!)^3 2^{2Dk}$$

$$\left\{ \sum_{n \geq k}^{s} \frac{1}{n!(s-n)!} \mathcal{N}^s \frac{(2s+k-1)!}{2^{s-n+1}(n+k-1)!} \binom{n}{k} \frac{|\lambda|^s}{\left(\cos\frac{\varphi}{2}\right)^{2s+k}} \frac{1}{N^{(D-2)(s+1-n)}} \right.$$

$$\left. + \sum_{n \geq s+1} \frac{1}{n!} \mathcal{N}^{n-1} \frac{(2n-3+k)!}{(n+k-1)!} \binom{n}{k} \frac{|\lambda|^{n-1}}{\left(\cos\frac{\varphi}{2}\right)^{2n-2+k}} \right\} ,$$

and bounding $1/N, 1/2 \leq 1$, $(2s+k-1)! \leq 3^{2s+k-1} s!(s-n)!(n+k-1)!$ and $(2n-3+k)! \leq 2^{2n-3+k}(n+k-1)!(n-1)!$ we obtain:

$$\left| \mathcal{R}^{(s)}(\mathcal{B},\mu^{(4)}) \right| \leq (k!)^3 2^{2Dk} \left(\frac{4^s \mathcal{N}^s |\lambda|^s}{\left(\cos\frac{\varphi}{2}\right)^{2s+k}} \right)$$

$$\left\{ s! \frac{3^{2s+k-1}}{2^{2s}} \sum_{n \geq k}^{s} \frac{1}{k!(n-k)!} + \sum_{n \geq s+1} \frac{(n-1)!}{k!(n-k)!} \left(\frac{4\mathcal{N}|\lambda|}{\left(\cos\frac{\varphi}{2}\right)^2} \right)^{n-s-1} \right\} .$$

Taking into account that $\frac{1}{k!(n-k)!} \leq 1$ and $(n-k)! \geq (n-s-1)!$ we have:

$$\left|\mathcal{R}^{(s)}(\mathcal{B},\mu^{(4)})\right| \leq (k!)^3 2^{2Dk}\left(\frac{4^s\mathcal{N}^s|\lambda|^s}{\left(\cos\frac{\varphi}{2}\right)^{2s+k}}\right) \times\left\{(s+1)!\frac{3^{2s+k-1}}{2^{2s}} + \frac{(s+1)!}{k!\left(1-\frac{4\mathcal{N}|\lambda|}{\left(\cos\frac{\varphi}{2}\right)^2}\right)^{s+1}}\right\},$$

and, as $s+1 \leq e^s$, we finally obtain the bound:

$$\left|\mathcal{R}^{(s)}(\mathcal{B},\mu^{(4)})\right| \leq (k!)^3 2^{2Dk}\left(s!\frac{e^s 4^s\mathcal{N}^s|\lambda|^s}{\left(\cos\frac{\varphi}{2}\right)^{2s+k}}\right)\left\{\frac{3^{2s+k-1}}{2^{2s}} + \frac{1}{k!\left(1-\frac{4\mathcal{N}|\lambda|}{\left(\cos\frac{\varphi}{2}\right)^2}\right)^{s+1}}\right\}.$$

We apply this bound for λ in a Borel disk entirely contained in Car (such that the second term in the bound is finite). As $\Re\lambda \geq 0$ we have $2^{-1/2} \leq |\cos\frac{\varphi}{2}| \leq 1$ and there obviously exist some constants K and σ such that:

$$\left|\mathcal{R}^{(s)}(\mathcal{B},\mu^{(4)})\right| \leq K\sigma^s s!|\lambda|^s ,$$

which proves the theorem.

□

9.5.2 Summary of non perturbative results

Before presenting the generalization of these results to arbitrary quartic models, let us discuss their significance.

First, Theorem 9.6 together with Lemma 9.4 lead to the following immediate corollary.

Corollary 9.1. *The measure $\mu^{(4)}$ with only melonic quartic perturbation is properly uniformly bounded, hence it is universal in the sense of Theorem 6.3.*

The melonic quartic models are explicit examples of measures to which the universality Theorem 6.3 applies.

The following discussion is illustrated in Figure 9.9. The results of Theorem 9.6 and Lemma 9.4 are summarized as follows. The rescaled cumulants of the quartic melonic tensor model are written as:

$$K(\mathcal{B},\mu^{(4)}) = K^{(r)}_{\text{explicit}}(\mathcal{B},\mu^{(4)}) + K^{(r)}_{\text{rest}}(\mathcal{B},\mu^{(4)}) ,$$

where:

- the explicit part $K^{(r)}_{\text{explicit}}(\mathcal{B},\mu^{(4)})$ is analytic in the disk $4\mathcal{N}|\lambda| < 1$.

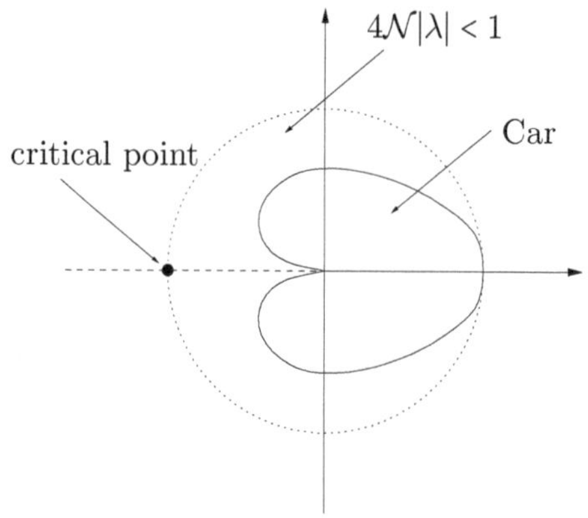

Figure 9.9: Domains of analyticity of the explicit and of the rest terms in the $1/N$ expansion.

- the rest term $K^{(r)}_{\text{rest}}(\mathcal{B}, \mu^{(4)})$ is analytic in the cardioid:

$$\text{Car} = \left\{ \lambda \in \mathbb{C} \middle| \; 4\mathcal{N}|\lambda| < \left(\cos\frac{\varphi}{2}\right)^2 \right\},$$

  and obeys the bound:

$$\left| K^{(r)}_{\text{rest}}(\mathcal{B}, \mu^{(4)}) \right| \le \frac{(k!)^3 2^{2(D+1)k}}{\left(\cos\frac{\varphi}{2}\right)^k} \, r! \frac{2^{3r}}{N^{(D-2)r}} \frac{\left(\frac{4\mathcal{N}|\lambda|}{(\cos\frac{\varphi}{2})^2}\right)^{k+r-1}}{\left(1 - \frac{4\mathcal{N}|\lambda|}{(\cos\frac{\varphi}{2})^2}\right)^{k+2r}} .$$

- furthermore:

$$\lim_{N\to\infty} K(\mathcal{B}^{(2)}, \mu^{(4)}) \equiv K(\mathcal{B}^{(2)}) = \frac{-1 + \sqrt{1 + 4\mathcal{N}\lambda}}{2\mathcal{N}\lambda} .$$

The domain of convergence of the explicit terms, the disk $4\mathcal{N}|\lambda| < 1$, is optimal. Indeed, $K(\mathcal{B}^{(2)})$ has a singularity at the critical value $\lambda_c = -\frac{1}{4\mathcal{N}}$. At this critical value the continuum phase of the model is reached. The constructive and perturbative evaluations of $K(\mathcal{B}^{(2)})$ computed in Lemma 9.4 and in Theorem 9.2 are identical. One can check that this is true for any cumulant and any fixed order r: the explicit part of the non perturbative $1/N$ expansion is identical with the perturbative evaluation. This is readily understood in light of Theorem 9.7: the perturbative expansion in λ of the cumulants is Borel summable. Reordering this Borel summable series in powers of $1/N$, while meaningless for the series itself, is in fact a reordering in powers of $1/N$ of its Borel sum, which is a well defined operation.

The main advantage of the non perturbative $1/N$ expansion of Theorem 9.6 is that it keeps the rest term of the $1/N$ expansion in a resummed form: $K^{(r)}_{\text{rest}}(\mathcal{B}, \mu^{(4)})$ is the *full*, *resummed contribution* of all the trees with more than r loop edges and it is analytic in the cardioid Car. This is a vast improvement with respect to the perturbative rest term which is itself a divergent series.

The resummed contribution in the rest term is indexed by trees with *exactly* r loop edges: each tree with r loop edges represents the resummation of the infinite family of graphs obtained by adding an arbitrary number of supplementary loop edges (encoded in the remaining Gaussian integral with its weakening parameters w). Instead of evaluating this Gaussian integral and obtaining a divergent series, we bound it. However, this comes at a price: bounding each resolvent brings a factor $\left(\cos\frac{\varphi}{2}\right)^{-2} \geq 1$ for each edge and the convergence domain deteriorates from the disk $4\mathcal{N}|\lambda| < 1$ to the cardioid $4\mathcal{N}|\lambda| \leq \left(\cos\frac{\varphi}{2}\right)^2$.

The divergence of the bound on the rest term at the boundary of Car is a remainder of the divergence of the perturbative series at fixed order in $1/N$ at criticality. One cannot expect the rest term to be analytic outside the disk $4\mathcal{N}|\lambda| < 1$. Furthermore, one cannot expect the rest term to be analytic for $\lambda \in \mathbb{R}_-$ as the perturbation is unstable in this case: the action is not bounded from below, which leads to a cut discontinuity in rest term. This motivates us to formulate the following conjecture.

Conjecture 9.1. *The rest term $K^{(r)}_{\text{rest}}(\mathcal{B},\mu^{(4)})$ (as a function of λ at N fixed) admits an analytic continuation in the domain:*

$$\left\{\lambda \in \mathbb{C}\setminus\mathbb{R}_- \middle| \quad 4\mathcal{N}|\lambda| < 1\right\}.$$

Assuming that this conjecture holds, two phenomena can be studied analytically:

The instanton effects. Consider the zero dimensional field theory:

$$\int d\bar{\phi}d\phi e^{-S(\bar{\phi},\phi)}\,, \qquad S(\bar{\phi},\phi) = \bar{\phi}\phi + \frac{\lambda}{2}(\bar{\phi}\phi)^2\,,$$

whose classical equations of motion $\partial_\phi S = \partial_{\bar{\phi}} S = 0$ have two solutions $\phi_1 = \bar{\phi}_1 = 0$ and $\bar{\phi}_2\phi_2 = -\frac{1}{\lambda}$. The second solution contributes to the integral only for $\lambda \in \mathbb{R}_-$, and in this case it leads to terms proportional to $e^{-S(\phi_2)} = e^{\frac{1}{2\lambda}}$. These are the instanton contributions and are identically zero in a naive perturbation series in λ.

The instanton effects are fully captured in the non perturbative expansion. Indeed, while the explicit terms are analytic in the disk $4\mathcal{N}|\lambda| < 1$, the rest term is not. The instanton effects are encoded in the discontinuity of $K^{(r)}_{\text{rest}}(\mathcal{B},\mu^{(4)})$ at the cut on the negative real axis.

The critical regime. The critical coupling of the phase transition to the continuum is $\lambda_c = -\frac{1}{4\mathcal{N}} \in \mathbb{R}_-$. The critical behavior of the rest term (as well as the instanton effects) are expected to play a crucial role in the behavior of the cumulants close to critical point.

A second remark is that, seen as a function of $1/N$, the bound on the rest term $|K^{(r)}_{\text{rest}}(\mathcal{B},\mu^{(4)})| \leq r!\left(\frac{1}{N^{D-2}}\right)^r$ strongly suggests that the cumulants are Borel summable (see Appendix C) in $\frac{1}{N^{D-2}}$ uniformly in λ. We conjecture that this is the case.

Conjecture 9.2. *As functions of* $1/N$ *the cumulants admit an analytic continuation in* $1/N$ *in some Borel disk:*

$$\Re N \geq R^{-1} .$$

9.5.3 The generic quartic model

The generalization of the non perturbative results to the generic quartic model follows the same lines as in the melonic case: the combinatorial counting as well as the exact computations of the explicit terms are unaffected by passing to the more general case.

The only thing that changes is that in the general case Lemma 9.3 does not hold, and one needs to work more in order to recover the appropriate bounds on the rest terms. A weaker result replaces Lemma 9.3 and in essence, with this new bound, one still obtains a suppression $N^{-\frac{D-2}{2}|\mathcal{L}|}$ of the contribution of a tree with loop edges to the mixed expansion of the cumulants. As the explicit terms are evaluated exactly as before, the strategy one uses consists in pushing the mixed expansion to higher orders, such that the weaker bound suffices to bound the integral rest term. When evaluating the explicit terms to this higher order, some of them are in fact at the same order as (or more suppressed than) the integral rest term. Including these explicit but highly suppressed terms in the rest allows one to generalize one by one all the non perturbative results.

Theorem 9.8. *For the generic quartic model the amplitude of a plane tree with loop edges, and* ρ *and* τ *edges respects the bound:*

$$\left|\mathcal{A}(\mathfrak{T}^{n,\iota}_{\rho^{\mathcal{D}}_{\mathcal{B}}\tau^{\mathcal{D}}}, \mathcal{L}, t)\right| \leq \frac{2^{2Dk(\mathcal{B})} N^{1+(D-1)n+\frac{D}{2}|\mathcal{L}|}}{\left(\cos\frac{\varphi}{2}\right)^{2n-2+2|\mathcal{L}|+k}} .$$

Proof. We go back to the definition of the amplitude $\mathcal{A}(\mathfrak{T}^{n,\iota}_{\rho^{\mathcal{D}}_{\mathcal{B}}\tau^{\mathcal{D}}}, \mathcal{L}, t)$ in Theorem 9.6:

$$\begin{aligned}
&\left(\prod_{c=1}^{D} \mathrm{Wg}(\rho^{(c)}_{\mathcal{B}}[\tau^{(c)}]^{-1}, N)\right) \int_0^1 \left(\prod_{(i,j)\in\mathfrak{T}^{n,\iota}} du_{ij}\right)\left(\prod_{(i',j')\in\mathcal{L}} w^{\mathfrak{T}^{n,\iota}}_{i'j'}(u_{\mathfrak{T}^{n,\iota}})\right) \\
&\times \Bigg\{ e^{\frac{t}{2}\frac{1}{N^{D-1}}\sum_{i,j=1}^{n} w^{\mathfrak{T}^{n,\iota}}_{ij}(u_{\mathfrak{T}^{n,\iota}})\left(\sum_{\mathcal{C}}\mathrm{Tr}\left[\frac{\partial}{\partial H^{\mathcal{C}(i)}}\frac{\partial}{\partial H^{\mathcal{C}(j)}}\right]\right)} \\
&\qquad \times \sum_{a\bar{b}} \left(\prod_{h\in\mathcal{S}(\mathfrak{T}^{n,\iota},\mathcal{L})} R(H^{i[h]})_{a^{\mathcal{D}}_{[h]}\bar{b}^{\mathcal{D}}_{[h]}}\right)\left(\prod_{l=1}^{k}\delta_{a_l\bar{b}_{\tau(l)}}\right) \\
&\qquad \times \prod_{\substack{e^{\mathcal{C}}\in\mathcal{E}(\mathfrak{T}^{n,\iota},\mathcal{L}) \\ e^{\mathcal{C}}=(h,h')}} \left(\delta_{\bar{b}^{\mathcal{C}}_{[\sigma^{-1}_{\mathfrak{T}^{n,\iota},\mathcal{L}}(h)]} a^{\mathcal{C}}_{[h']}}\delta_{\bar{b}^{\mathcal{C}}_{[\sigma^{-1}_{\mathfrak{T}^{n,\iota},\mathcal{L}}(h')]} a^{\mathcal{C}}_{[h]}}\right) \\
&\qquad\qquad \left(\delta_{\bar{b}^{\mathcal{D}\setminus\mathcal{C}}_{[\sigma^{-1}_{\mathfrak{T}^{n,\iota},\mathcal{L}}(h)]} a^{\mathcal{D}\setminus\mathcal{C}}_{[h]}}\delta_{\bar{b}^{\mathcal{D}\setminus\mathcal{C}}_{[\sigma^{-1}_{\mathfrak{T}^{n,\iota},\mathcal{L}}(h')]} a^{\mathcal{D}\setminus\mathcal{C}}_{[h']}}\right)\Bigg\}_{H=0} .
\end{aligned}$$

In order to bound the amplitude we start by bounding the integrand:

$$\sum_{a\bar{b}} \left(\prod_{h \in \mathcal{S}(\mathfrak{T}^{n,\iota},\mathcal{L})} R(H^{i[h]})_{a^{\mathcal{D}}_{[h]} \bar{b}^{\mathcal{D}}_{[h]}} \right) \left(\prod_{l=1}^{k} \delta_{a_l \bar{b}_{\tau(l)}} \right)$$
$$\times \prod_{\substack{e^{\mathcal{C}} \in \mathcal{E}(\mathfrak{T}^{n,\iota},\mathcal{L}) \\ e^{\mathcal{C}}=(h,h')}} \left(\delta_{\bar{b}^{\mathcal{C}}_{[\sigma^{-1}_{\mathfrak{T}^{n,\iota},\mathcal{L}}(h)]} a^{\mathcal{C}}_{[h']}} \delta_{\bar{b}^{\mathcal{C}}_{[\sigma^{-1}_{\mathfrak{T}^{n,\iota},\mathcal{L}}(h')]} a^{\mathcal{C}}_{[h]}} \right)$$
$$\left(\delta_{\bar{b}^{\mathcal{D}\setminus\mathcal{C}}_{[\sigma^{-1}_{\mathfrak{T}^{n,\iota},\mathcal{L}}(h)]} a^{\mathcal{D}\setminus\mathcal{C}}_{[h]}} \delta_{\bar{b}^{\mathcal{D}\setminus\mathcal{C}}_{[\sigma^{-1}_{\mathfrak{T}^{n,\iota},\mathcal{L}}(h')]} a^{\mathcal{D}\setminus\mathcal{C}}_{[h']}} \right)$$

using iterated Cauchy–Schwarz inequalities [123, 58]. As this factor does not depend at all on the ρ edges, we will from now on ignore them. The factor above is associated to a tree $\mathfrak{T}^{n,\iota}$ with loop edges $\mathcal{L}$ and τ-edges defined by the permutation τ.

Let us for now also forget about the τ edges and consider the map with cilia $\mathfrak{T}^{n,\iota}, \mathcal{L}$. All the $2n - 2 + 2|\mathcal{L}| + k$ corners $[h]$ of this map are decorated by a resolvent $R(H^{i[h]})$. For simplicity, let us assume for now that this number of corners is even. We call a Cauchy–Schwarz *cut* in $\mathfrak{T}^{n,\iota}, \mathcal{L}$ the choice of two corners $[h_1]$ and $[h_2]$ such that the number of resolvents encountered when going around the tree from $[h_1]$ to $[h_2]$ equals the number of resolvents encountered when going around the tree from $[h_2]$ to $[h_1]$. This has been represented in Figure 9.10 on the top, where the two corners in the cut are signaled by the white dot and the tree is represented in bold.

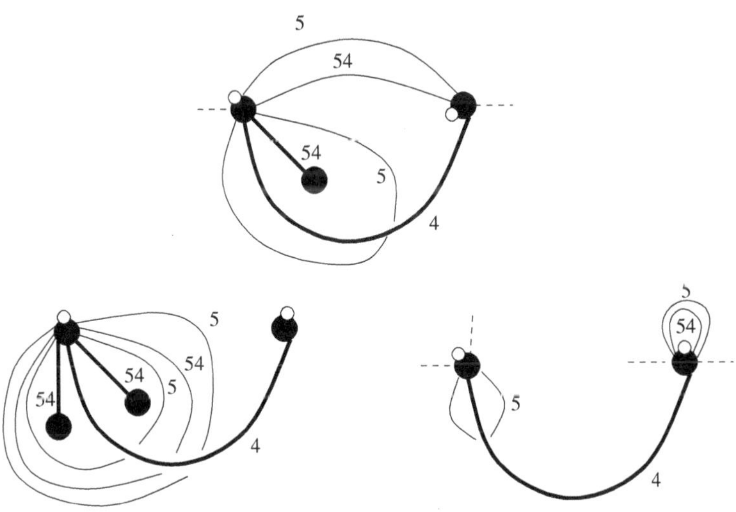

Figure 9.10: Cauchy–Schwarz cut and mirrored maps.

The technique we use here [58] was introduced in [123] in the context of tensor models. A Cauchy–Schwarz cut allows us to write the integrand as a scalar product in some Hilbert space:

$$\langle A| R(H^{i[h_1]}) \otimes R(H^{i[h_1]}) \otimes \mathbb{I} |B\rangle \ ,$$

where:

- A is the product of resolvents associated with the corners encountered when going from $[h_1]$ to $[h_2]$, whose indices are contracted along the edges whose both ends are incident to these corners.
- B is the product of resolvents associated with the corners encountered when going from $[h_2]$ to $[h_1]$, whose indices are contracted along the edges whose both ends are incident to these corners.
- $\mathbb{I}$ represent the contractions of the indices encoded in the edges that connect corners between $[h_1]$ and $[h_2]$ to corners between $[h_2]$ and $[h_1]$.

By the Cauchy–Schwarz inequality, taking into account that for any $\lambda \in \mathbb{C}$ the operator norm of any resolvent is bounded by $\|R(H^{i[h_1]})\| \leq \frac{1}{\cos\frac{\varphi}{2}}$, we obtain

$$\langle A|\, R(H^{i[h_1]}) \otimes R(H^{i[h_1]}) \otimes \mathbb{I} |B\rangle \leq \frac{1}{\left(\cos\frac{\varphi}{2}\right)^2} \sqrt{\langle A\,|\,A\rangle \langle B\,|\,B\rangle}\,.$$

The two factors $\langle A\,|\,A\rangle$ and $\langle B\,|\,B\rangle$ can each be represented as maps. Indeed, $\langle A\,|\,A\rangle$ is obtained by taking the mirror image with respect to the cut of all the corners and halfedges comprised between $[h_1]$ and $[h_2]$ and all the edges connecting them, and connecting a halfedge to its mirror image if it comes from an edge that connected A to B in the original map. This is represented in Figure 9.10 at the bottom. Remark that the original corners $[h_1]$ and $[h_2]$ appear in both maps, but now they just represent a contraction of indices (as if their resolvent have been replaced by the identity). We call $[h_1]$ and $[h_2]$ "cleaned" corners and we continue in Figure 9.10 to mark them with a white dot. In opposition, we call the corners bearing resolvents "not cleaned".

The exact same reasoning applies if one takes into account the τ edges, but, as the corners comprised between two τ edges on a vertex are an identification of indices and not a resolvent, they are from the beginning marked as cleaned with a white dot.

Let us denote $\mathcal{M}$ the original combinatorial map (and $\mathcal{M}_\tau$ represents the map decorated by some τ edges) and $\mathcal{M}_1$ and $\mathcal{M}_2$ (respectively $\mathcal{M}_{\tau 1}, \mathcal{M}_{\tau 2}$) the two maps obtained after applying this procedure. Any vertex of $\mathcal{M}$ is either copied into two vertices in $\mathcal{M}_1$ (or $\mathcal{M}_2$), or it is one of the vertices bearing the cut, hence it appears once in $\mathcal{M}_1$ and once in $\mathcal{M}_2$. A cilium of $\mathcal{M}$ will always be copied into two cilia either in $\mathcal{M}_1$ or in $\mathcal{M}_2$. Any edge of $\mathcal{M}$ is either copied into two edges of $\mathcal{M}_1$ (or $\mathcal{M}_2$) or it is split into two edges, one in $\mathcal{M}_1$ and one in $\mathcal{M}_2$:

$$V(\mathcal{M}) = \frac{V(\mathcal{M}_1) + V(\mathcal{M}_2)}{2}\,, \quad k(\partial\mathcal{M}) = \frac{k(\partial\mathcal{M}_1) + k(\partial\mathcal{M}_2)}{2}\,,$$
$$E(\mathcal{M}) = \frac{V(\mathcal{M}_1) + V(\mathcal{M}_2)}{2}\,.$$

Crucially, every corner bearing a resolvent different from $[h_1]$ and $[h_2]$ is always copied into two corners (both bearing resolvents), both belonging either to $\mathcal{M}_1$ or to $\mathcal{M}_2$. Thus if the original map had an even number $2n - 2 + 2|\mathcal{L}| + k$ of not cleaned corners, each of the two resulting maps has $2n-2+2|\mathcal{L}|+k-2$ not cleaned corners. We iterate the Cauchy–Schwarz bound for $\frac{2n-2+2|\mathcal{L}|+k}{2}$ cuts. Remark that we carefully defined a cut as two corners $[h_1]$ and $[h_2]$ separated by the same number of not cleaned corners, hence the number of not cleaned corners always goes down by two for every cut and after $\frac{2n-2+2|\mathcal{L}|+k}{2}$ iterations all the corners are cleaned. If one starts with an odd number of resolvents, at the first step we choose a cut such that the number of corners on the two sides differ by 1. All subsequent maps have an even number of corners.

It remains to estimate the integrand of a map in which all the corners have been cleaned:

$$\sum_{a\bar{b}}\left(\prod_{h\in\mathcal{S}(\mathcal{M})}\delta_{a^{\mathcal{D}}_{[h]}\bar{b}^{\mathcal{D}}_{[h]}}\right)\left(\prod_{l=1}^{k}\delta_{a_l\bar{b}_{\tau(l)}}\right)$$
$$\times\prod_{\substack{e^{\mathcal{C}}\in\mathcal{E}(\mathcal{M})\\ e^{\mathcal{C}}=(h,h')}}\left(\delta_{\bar{b}^{\mathcal{C}}_{[\sigma^{-1}_{\mathcal{M}}(h)]}a^{\mathcal{C}}_{[h']}}\delta_{\bar{b}^{\mathcal{C}}_{[\sigma^{-1}_{\mathcal{M}}(h')]}a^{\mathcal{C}}_{[h]}}\right)$$
$$\left(\delta_{\bar{b}^{\mathcal{D}\smallsetminus\mathcal{C}}_{[\sigma^{-1}_{\mathcal{M}}(h)]}a^{\mathcal{D}\smallsetminus\mathcal{C}}_{[h]}}\delta_{\bar{b}^{\mathcal{D}\smallsetminus\mathcal{C}}_{[\sigma^{-1}_{\mathcal{M}}(h')]}a^{\mathcal{D}\smallsetminus\mathcal{C}}_{[h']}}\right)=N^{F_{\text{int}}(\mathcal{M})+F_{\text{int}\tau}(\mathcal{M})}\,.$$

At this point a subtlety arises. The maps $\mathcal{M}$ obtained after Cauchy–Schwarz reflection are combinatorial maps with cilia like the original maps, with one crucial difference: as it is clear from Figure 9.10, $\mathcal{M}$ can have multiple cilia on the same vertex. In particular, for these maps Theorem 9.1 *does not apply*. However, a weaker result holds.

Lemma 9.5. *Let a combinatorial map with τ edges and cilia that can have multiple cilia on the same vertex. Then:*

$$F_{\text{int}}(\mathcal{M})+F_{\text{int}\tau}(\mathcal{M})\le$$
$$Dk(\partial\mathcal{M})+1+(D-1)V(\mathcal{M})+\frac{D}{2}\big[E(\mathcal{M})-V(\mathcal{M})+1\big]\,.$$

Proof. Erasing the cilia and promoting the τ edges to normal edges, the map $\mathcal{M}$ becomes a combinatorial map with no cilia having $E(\mathcal{M})+Dk(\partial\mathcal{M})$ edges. This new map has all the faces of $\mathcal{M}$ (either internal or intτ faces) and some new "$\tau\tau$" faces. Erasing the $Dk(\partial\mathcal{M})$ edges coming from the τ edges cannot disconnect the map, and each of these edges has a unique color. It follows that by deleting a τ edge the number of faces can at most go down by 1. Denoting $\mathcal{M}'$ the map obtained by deleting all the τ edges, we have:

$$F_{\text{int}}(\mathcal{M})+F_{\text{int}\tau}(\mathcal{M})\le F_{\text{int}}(\mathcal{M}')+Dk(\partial\mathcal{M})\,,$$

and using the Theorem 9.1 the Lemma follows.

□

As the number of vertices, edges and cilia exactly doubles at each iteration, but the Cauchy–Schwarz inequality involves a square root, we obtain:

$$\left|\sum_{a\bar{b}}\left(\prod_{h\in\mathcal{S}(\mathfrak{T}^{n,\iota},\mathcal{L})}R(H^{i[h]})_{a^{\mathcal{D}}_{[h]}\bar{b}^{\mathcal{D}}_{[h]}}\right)\left(\prod_{l=1}^{k}\delta_{a_l\bar{b}_{\tau(l)}}\right)\right.$$

$$\times \prod_{\substack{e^{\mathcal{C}} \in \mathcal{E}(\mathfrak{T}^{n,\iota},\mathcal{L}) \\ e^{\mathcal{C}}=(h,h')}} \left(\delta_{\bar{b}^{\mathcal{C}}_{[\sigma^{-1}_{\mathfrak{T}^{n,\iota},\mathcal{L}}(h)]}} a^{\mathcal{C}}_{[h']} \delta_{\bar{b}^{\mathcal{C}}_{[\sigma^{-1}_{\mathfrak{T}^{n,\iota},\mathcal{L}}(h')]}} a^{\mathcal{C}}_{[h]} \right)$$

$$\left. \left(\delta_{\bar{b}^{\mathcal{D}\setminus\mathcal{C}}_{[\sigma^{-1}_{\mathfrak{T}^{n,\iota},\mathcal{L}}(h)]}} a^{\mathcal{D}\setminus\mathcal{C}}_{[h]} \delta_{\bar{b}^{\mathcal{D}\setminus\mathcal{C}}_{[\sigma^{-1}_{\mathfrak{T}^{n,\iota},\mathcal{L}}(h')]}} a^{\mathcal{D}\setminus\mathcal{C}}_{[h']} \right) \right| \le \frac{N^{Dk(\mathcal{B})+1+(D-1)n+\frac{D}{2}|\mathcal{L}|}}{\left(\cos\frac{\varphi}{2}\right)^{2n-2+2|\mathcal{L}|+k}} .$$

Using the bound:

$$\mathrm{Wg}(\sigma, N) \le 2^{2k} N^{-k} ,$$

for σ a permutation of k elements, the Theorem 9.8 follows.

□

Chapter 10

The double scaling limit

In this chapter we will present the double scaling limit of tensor models. While this limit can be established (with various degrees of rigor) for many invariant models, we will only present the quartic melonic models discussed in Chapter 9. We will use extensively the intermediate field representation.

The leading order of tensor models is given by the melonic phase discussed at length in Chapter 5. As shown in Chapter 5, if one associates with a graph an equilateral triangulation (the natural metric assignment for the invariant tensor models) the melonic phase is geometrically a branched polymer. In order to study more interesting phases one would like to be able to study perturbations with respect to this melonic phase.

In the first part of this chapter we will show that the melonic family in the quartic melonic tensor model can be analytically resummed and that this resummation is encoded in a translation of the intermediate matrix fields to a nontrivial vacuum [57]. After performing this translation one obtains explicitly the generating function for the schemes (see Chapter 4) as a path integral. The most striking consequence of this translation is that one can now show that the continuum limit of tensor models corresponds to a genuine phase transition (in the field theory sense) associated with a symmetry breaking. We discuss this point in Section 10.1.

In the second part of this chapter we use the intermediate field representation to discuss the double scaling limit [51, 33, 107] of the quartic melonic model.

10.1 The continuum limit as a phase transition

We showed in Chapter 9 that the tensor model with quartic melonic interactions is written in the intermediate field representation (see Eq. (9.6)) as a coupled multi matrix model for D hermitian matrices H^c of size $N \times N$. We will show that $H^c = 0$ is *not* a solution of the classical equations of motion, hence it is not a vacuum of this multimatrix model. For small coupling constant λ the (unique) stable vacuum state is:

$$H^c = a_0 \mathbf{1} , \qquad a_0 = \frac{1 - \sqrt{1 + 4D\lambda}}{2\imath\sqrt{\lambda}D} ,$$

which in particular is invariant under unitary transformations. We call this vacuum state *the melonic vacuum.* Observe that the translation to the melonic vacuum is not a phase transition as no symmetry gets broken (the melonic vacuum possesses the full invariance of the theory). One can derive the effective theory for the fluctuation fields M^c around the melonic vacuum and write the partition function of the model, Eq. 10.3, as:

$$Z^{(4)}(0,0) = e^{-N^D\left(\frac{Da_0^2}{2} + \ln(1-\imath\sqrt{\lambda}Da_0)\right)} (1 - a_0^2)^{\frac{D-1}{2}} \int [dM^c]\, e^{-S(M)} ,$$

$$S(M) = \frac{1}{2} N^{D-1}(1 - a_0^2) \sum_{c=1}^{D} \mathrm{Tr}_c [M^c M^c] - \frac{1}{2} N^{D-2} \frac{D-1}{D} a_0^2 \left(\sum_{c=1}^{D} \mathrm{Tr}_c[M^c] \right)^2 - Q(M) ,$$

with $Q(M)$ some interaction term starting at order $(M^c)^3$.

Starting from this formula one can establish two facts. First, the translation to the melonic vacuum completely resums the leading order in $1/N$: the effective theory for the perturbation fields around the melonic vacuum contributes only to lower orders[1] in $1/N$, namely (see Corollary 10.1):

$$W = \frac{1}{N^D} \ln Z^{(4)}(0,0) = -\frac{D}{2} a_0^2 - \ln(1 - \imath\sqrt{\lambda}Da_0) + O\left(\frac{1}{N}\right) .$$

Second, the mass matrix of the effective theory for the perturbation M^c has eigenvalues $N^{D-1}(1 - a_0^2)$ with degeneracy $DN^2 - 1$ and $N^{D-1}(1 - Da_0^2)$ with degeneracy 1.

Dynamical triangulations continuum limit. The dynamical triangulations continuum limit is obtained from the leading order free energy:

$$W_\infty = -\frac{D}{2} a_0^2 - \ln(1 - \imath\sqrt{\lambda}Da_0) = \frac{1}{2} + \frac{1 - \sqrt{1 + 4D\lambda}}{4D\lambda} + \ln\left(\frac{1 - \sqrt{1 + 4D\lambda}}{-2D\lambda} \right) .$$

At leading order the free energy becomes critical for the critical constant:

$$\boxed{\lambda_c = -\frac{1}{4D}} .$$

In the critical regime $\frac{\lambda}{\lambda_c} = -4D\lambda = 1 - \epsilon$ one obtains:

$$W_\infty = \frac{1}{2} - \frac{1 - \sqrt{\epsilon}}{1 - \epsilon} + \ln\left(2\frac{1 - \sqrt{\epsilon}}{1 - \epsilon} \right) = \frac{1}{2} + \ln 2 + \frac{1}{2}\epsilon - \frac{2}{3}\epsilon^{3/2} + O(\epsilon^2) ,$$

hence the critical exponent $\frac{3}{2}$.

[1] The situation is quite different for matrix models. A similar procedure cannot be used for matrices: so far at least, the action of no matrix model could be translated to an explicit "planar vacuum state" such that the effective theory around it contributes only to lower orders in $1/N$.

Field theory phase transition. The effective field theory for the perturbation field M^c (which is subleading in $1/N$) undergoes a phase transition when the mass matrix has a zero eigenvalue, that is for $1 - Da_0^2 = 0$. At this point the invariant vacuum is no longer stable and the symmetry of the theory is broken. This occurs for the coupling constant $\lambda_{p.t.}$ solution of:

$$\frac{1}{\sqrt{D}} = \frac{1 - \sqrt{1 + 4D\lambda_{p.t.}}}{2\imath D\sqrt{\lambda_{p.t.}}} \Rightarrow \sqrt{1 + 4D\lambda_{p.t.}} = 1 - 2\imath\sqrt{D}\sqrt{\lambda_{p.t.}} \Rightarrow \boxed{\lambda_{p.t.} = -\frac{1}{4D}} .$$

We conclude that for the quartic melonic tensor model the continuum limit is a phase transition associated to a symmetry breaking. This phase transition is of an entirely different nature from the usual phase transitions studied in the context of dynamical triangulations and matrix models:

- *Dynamical triangulations.* In dynamical triangulations [4, 5] one takes the continuum limit and obtains various continuum phases. In Euclidean dynamical triangulations [4] one typically obtains two continuum phases (crumpled and branched polymer), while in causal dynamical triangulations [5] one obtains three continuum phases. One is then interested in the various transitions between continuum phases. These transitions have nothing to do with the phase transition from discrete to continuum that we discuss here.

- *Matrix models.* In matrix models [127] one studies the transition from a one cut resolvent to a multicut resolvent. The phase transition we identify here is very different, and occurs in a completely different range of parameters. In matrix models the continuum limit is a change of behavior of the spectral density of a one cut solution (one passes from a square root law to a 3/2 power law for the distribution of eigenvalues). It is the equivalent of this change of behavior in the case of tensors that we show here is a genuine phase transition.

This phase transition corresponds to the spontaneous breaking of the symmetry of the theory (color permutation and unitary invariance). However, in order to perform a complete study of this phase transition one needs to identify the appropriate noninvariant vacuum state and study perturbations around it. This vacuum state has not been found so far in the case of the melonic model with all the interactions.

10.2 The melonic phase revisited

Let us go back to the generating function of the moments of the quartic melonic model in the intermediate field representation in Eq. (9.6):

$$S(H) = \frac{1}{2}N^{D-1}\sum_{c=1}^{D}\mathrm{Tr}_c[H^cH^c] - \mathrm{Tr}_{\mathcal{D}}\left[\ln R(H)\right] ,$$
$$Z^{(4)}(J,\bar{J}) = \int\left(\prod_{c=1}^{D}[dH^c]\right)e^{-S(H)+\frac{1}{N^{D-1}}JR(H)\bar{J}}$$

$$= \left[e^{\frac{1}{2}\frac{1}{N^{D-1}}\sum_{c=1}^{D}\mathrm{Tr}_c\left[\frac{\partial}{\partial H^c}\frac{\partial}{\partial H^c}\right]} e^{\mathrm{Tr}_{\mathcal{D}}[\ln R(H)] + \frac{1}{N^{D-1}} J R(H) \bar{J}} \right]_{H^c=0} ,$$

where Tr_c is the trace over the index of color c, $\mathrm{Tr}_{\mathcal{D}}$ is the trace over all the indices and the resolvent is:

$$R(H) = \frac{1}{\mathbf{1}^{\mathcal{D}} - \imath\sqrt{\lambda}\sum_{c=1}^{D} H^c \otimes \mathbf{1}^{\mathcal{D}\setminus\{c\}}} ,$$

$$\frac{\partial}{\partial H^c_{\bar{a}^c b^c}} \mathrm{Tr}_{\mathcal{D}}\left[\ln R(H)\right] = \imath\sqrt{\lambda} \sum_{\bar{p}^{\mathcal{D}} q^{\mathcal{D}}} R(H)_{q^{\mathcal{D}}\bar{p}^{\mathcal{D}}} \left(\delta_{\bar{p}^{\mathcal{D}\setminus\{c\}} q^{\mathcal{D}\setminus\{c\}}} \delta_{\bar{p}^c \bar{a}^c} \delta_{q^c b^c}\right) .$$

In the previous chapter we dealt with this Gaussian integral over H^c. In physics language we have studied the model around the field configuration $H^c = 0$. However, seen as a field theory, this model is defined by the action $S(H)$ and $H^c = 0$ is *not* a solution of the classical equations of motion:

$$\frac{\partial S(H)}{\partial H^c_{b^c a^c}} = N^{D-1} H^c_{a^c b^c} - \imath\sqrt{\lambda} \sum_{p^{\mathcal{D}} q^{\mathcal{D}}} R(H)_{q^{\mathcal{D}} p^{\mathcal{D}}} \left(\delta_{p^{\mathcal{D}\setminus\{c\}} q^{\mathcal{D}\setminus\{c\}}} \delta_{p^c b^c} \delta_{q^c a^c}\right) = 0 ,$$

that is $H^c = 0$ is not a *stable vacuum* of the theory.

A full analysis of the action $S(H)$ consists of identifying the stable vacua and studying the perturbations around these vacua. Each of the vacua represents a phase of the theory and the perturbations around these vacua are the different effective theories in the various phases of the model.

The melonic vacuum and the instanton

In order to solve the classical equations of motion we express them in matrix form:

$$H^c - \imath\sqrt{\lambda}\frac{1}{N^{D-1}} \mathrm{Tr}_{\mathcal{D}\setminus\{c\}} \left[\frac{1}{\mathbf{1}^{\mathcal{D}} - \imath\sqrt{\lambda}\sum_{c'=1}^{D} H^{c'} \otimes \mathbf{1}^{\mathcal{D}\setminus\{c'\}}} \right] = 0 ,$$

where $\mathrm{Tr}_{\mathcal{D}\setminus\{c\}}$ denotes the trace over the indices with colors in $\mathcal{D}\setminus\{c\}$.

Assuming that H^c is a series in $\sqrt{\lambda}$ (that is H^c is perturbative in $\sqrt{\lambda}$), it is immediate to see order by order that $H^c = H^{c'}, \forall c, c'$ and that furthermore H^c is proportional to the identity. Considering an ansatz $H^c = a\mathbf{1}$, the classical equations of motion become:

$$a = \frac{\imath\sqrt{\lambda}}{1 - \imath\sqrt{\lambda}Da} .$$

This equation has two solutions:

The melonic vacuum. The perturbative solution, denoted a_0, is the power series in $\sqrt{\lambda}$ solution of the equation:

$$a_0 = \frac{1 - \sqrt{1+4D\lambda}}{2\imath\sqrt{\lambda}D} = \imath\sqrt{\lambda}\ \frac{-1 + \sqrt{1+4D\lambda}}{2D\lambda} .$$

The instanton. The second solution, denoted a_{inst}, is an instanton solution:

$$a_{\text{inst}} = \frac{1+\sqrt{1+4D\lambda}}{2\imath\sqrt{\lambda}D} = \imath\sqrt{\lambda}\,\frac{-1-\sqrt{1+4D\lambda}}{2D\lambda}\,.$$

Remark 10.1. Several remarks are in order.

Relation to the melonic graphs. The vacuum expectation value a_0 is strongly reminiscent of the two point function of the quartic melonic model at large N, Lemma 9.4 and Theorem 9.2, which is a sum of melonic graphs. It is therefore natural that the translation of the intermediate field by a_0 should be related to the resummation of the melonic family. This will be made more precise below.

Contour deformation. For λ real, positive and small enough, both the vacuum expectation value a_0 and the instanton a_{inst} are imaginary while the intermediate fields H^c are hermitian. The translation to either of the two solutions is a deformation in the complex plane of the contour of integration of H^c.

No symmetry breaking. The perturbative vacuum and the instanton are unitary invariant and color symmetric $H^c \to UH^cU^\dagger$. The translation to either of these solutions *does not lead* to a spontaneous symmetry breaking, and the effective theory will still be unitary invariant.

We choose the cut of the complex logarithm to be the negative real axis. With this determination, both the vacuum expectation value a_0 and the instanton a_{inst} are analytic functions of λ for $\lambda = |\lambda|e^{\imath\varphi}$, $\varphi \in (-\pi,\pi)$. We are interested in bounding a_0 and a_{inst} in the domain:

$$\left\{\lambda \in \mathbb{C}\smallsetminus\mathbb{R}_-\middle|\ 4D|\lambda|<1\right\}.$$

Inside this domain:

- a_0 is the sum of the absolutely convergent series:

$$a_0 = \imath\sqrt{\lambda}\sum_{n\geq 0}\frac{1}{(n+1)!}\binom{2n}{n}(-D\lambda)^n\,,$$

  and a bound for the absolute value of a_0 is then:

$$|a_0| \leq \sqrt{|\lambda|}\,\frac{1-\sqrt{1-4D|\lambda|}}{2D|\lambda|} = \frac{2\sqrt{|\lambda|}}{1+\sqrt{1-4D|\lambda|}} \leq \frac{1}{\sqrt{D}}\,,$$

  as $\frac{2\sqrt{|\lambda|}}{1+\sqrt{1-4D|\lambda|}}$ is increasing with $|\lambda|$ and reaches its maximum for $|\lambda| = (4D)^{-1}$.

- as $a_0 a_{\text{inst}} = \frac{1}{D}$, $|a_{\text{inst}}| \geq \frac{1}{\sqrt{D}}$.

The two solutions a_0 and a_{inst} collapse at criticality $\lambda_c = -\frac{1}{4D}$.

10.2.1 Translating the intermediate field

In order to study the effective theory around a solution of the classical equations of motion we must translate the field to $H^c = a\mathbf{1} + M^c$. This is a translation (in the complex plane) by a of the contour of integration of all the diagonal entries of H^c. The effective theory is obtained by Taylor expanding in M^c. The translation of the contour of integration is less trivial than it might appear at first sight because, before translating a contour of integration, one needs to make sure that no singularities of the integrand are traversed in the process.

We discuss first the more general case of the translation of H^c by some arbitrary constant a, that is we pose $H^c = a\mathbf{1} + M^c$, where a is not necessarily a solution of the classical equations of motion.

We denote $M = \sum_{c=1}^{D} M^c \otimes \mathbf{1}^{\mathcal{D}\setminus\{c\}}$ and we define:

$$b(a) \equiv \frac{\imath\sqrt{\lambda}}{1 - \imath\sqrt{\lambda}Da} \, .$$

Observe that the perturbative vacuum and the instanton are the two solutions of the self consistency equation $a = b(a)$. In order to simplify the notation we will use the shorthand notation b instead of $b(a)$ but the reader should keep in mind that b is a function of a, which equals a for the two solutions of the classical equations of motion.

In terms of the perturbation field M, the resolvent and the trace of its logarithm is written as:

$$R(a\mathbf{1}^{\mathcal{D}} + M) = \frac{1}{(1-\imath\sqrt{\lambda}Da)\mathbf{1}^{\mathcal{D}} - \imath\sqrt{\lambda}M} = \frac{1}{(1-\imath\sqrt{\lambda}Da)} \frac{1}{\mathbf{1}^{\mathcal{D}} - bM} \, ,$$

$$\mathrm{Tr}_{\mathcal{D}}\left[\ln R(a\mathbf{1}^{\mathcal{D}} + M)\right] = -N^D \ln(1-\imath\sqrt{\lambda}Da) + \mathrm{Tr}_{\mathcal{D}}\left[\ln \frac{1}{\mathbf{1}^{\mathcal{D}} - bM}\right] .$$

After translation by a the action becomes:

$$\frac{1}{2}N^D Da^2 + N^{D-1} a \sum_{c=1}^{D} \mathrm{Tr}_c[M^c] + \frac{1}{2}N^{D-1}\sum_{c=1}^{D}\mathrm{Tr}_c[M^c M^c]$$
$$+ N^D \ln(1-\imath\sqrt{\lambda}Da) - \mathrm{Tr}_{\mathcal{D}}\left[\ln\frac{1}{\mathbf{1}^{\mathcal{D}} - bM}\right] ,$$

and the generating function becomes, in terms of the perturbation field M:

$$Z^{(4)}(J,\bar{J}) = e^{-N^D\left(\frac{Da^2}{2} + \ln(1-\imath\sqrt{\lambda}Da)\right)}$$
$$\times \int \left(\prod_{c=1}^{D}[dM^c]\right) e^{-\frac{N^{D-1}}{2}\sum_{c=1}^{D}\mathrm{Tr}_c[M^cM^c] - a\,\mathrm{Tr}_{\mathcal{D}}[M]}$$
$$\times e^{\mathrm{Tr}_{\mathcal{D}}\left[\ln\frac{1}{\mathbf{1}^{\mathcal{D}}-bM}\right] + \frac{1}{N^{D-1}}\frac{1}{(1-\imath\sqrt{\lambda}Da)} J\left(\frac{1}{\mathbf{1}^{\mathcal{D}}-bM}\right)\bar{J}} \, .$$

One can use this form of the generating function, but it is possible to work some more and obtain a second form of the generating function, more adapted to the treatment we will be performing in this chapter.

The main remark is that the logarithm of the resolvent contains quadratic terms in M, which can be reabsorbed in the quadratic part of the action. Let us define the subtracted vertex function:

$$Q(M) = \mathrm{Tr}_{\mathcal{D}}\left[\ln \frac{1}{\mathbf{1}^{\mathcal{D}} - bM}\right] - \mathrm{Tr}_{\mathcal{D}}\left[aM\right] - \frac{1}{2}\mathrm{Tr}_{\mathcal{D}}\left[\left(bM\right)^2\right],$$

i.e. by definition the subtracted vertex function has no quadratic terms in M. The generating function is then:

$$\boxed{\begin{aligned} Z^{(4)}(J,\bar{J}) &= e^{-N^D\left(\frac{Da^2}{2}+\ln(1-\imath\sqrt{\lambda}Da)\right)} \\ &\times \int \left(\prod_{c=1}^{D}[dM^c]\right) e^{-\frac{N^{D-1}}{2}\sum_{c=1}^{D}\mathrm{Tr}_c[M^cM^c]+\frac{1}{2}\mathrm{Tr}_{\mathcal{D}}\left[\left(bM\right)^2\right]} \\ &\times e^{Q(M)+\frac{1}{N^{D-1}}\frac{1}{(1-\imath\sqrt{\lambda}Da)}J\left(\frac{1}{\mathbf{1}^{\mathcal{D}}-bM}\right)\bar{J}}. \end{aligned}} \tag{10.1}$$

Remark 10.2. Several remarks are in order.

Integral form of the subtracted vertex function. One can give a closed integral formula for the subtracted vertex function $Q(M)$. A Taylor expansion with integral rest yields:

$$\begin{aligned} \mathrm{Tr}_{\mathcal{D}}\left[\ln \frac{1}{\mathbf{1}^{\mathcal{D}} - bM}\right] &= \left\{\partial_t \mathrm{Tr}_{\mathcal{D}}\left[\ln \frac{1}{\mathbf{1}^{\mathcal{D}} - tbM}\right]\right\}_{t=0} \\ &+ \frac{1}{2}\left\{\partial_t^2 \mathrm{Tr}_{\mathcal{D}}\left[\ln \frac{1}{\mathbf{1}^{\mathcal{D}} - tbM}\right]\right\}_{t=0} \\ &+ \frac{1}{2}\int_0^1 dt\,(1-t)^2\partial_t^3 \mathrm{Tr}_{\mathcal{D}}\left[\ln \frac{1}{\mathbf{1}^{\mathcal{D}} - tbM}\right], \end{aligned}$$

and substituting the derivatives:

$$\partial_t^r \mathrm{Tr}_{\mathcal{D}}\left[\ln \frac{1}{\mathbf{1}^{\mathcal{D}} - tbM}\right] = (r-1)!\,\mathrm{Tr}_{\mathcal{D}}\left[\frac{1}{(\mathbf{1}^{\mathcal{D}} - tbM)^r}\left(bM\right)^r\right],$$

we obtain:

$$\begin{aligned} Q(M) &= \mathrm{Tr}_{\mathcal{D}}\left[\ln \frac{1}{\mathbf{1}^{\mathcal{D}} - bM}\right] - \mathrm{Tr}_{\mathcal{D}}\left[aM\right] - \frac{1}{2}\mathrm{Tr}_{\mathcal{D}}\left[\left(bM\right)^2\right] \\ &= (b-a)\,\mathrm{Tr}_{\mathcal{D}}\left[M\right] + \int_0^1 dt\,(1-t)^2\,\mathrm{Tr}_{\mathcal{D}}\left[\frac{1}{(\mathbf{1}^{\mathcal{D}} - tbM)^3}\left(bM\right)^3\right], \end{aligned}$$

that is $Q(M)$ is the sum of a linear piece and a piece starting at order M^3.

The linear term. The translation of H^c by a generic a should be interpreted as follows. At the price of modifying the covariance one introduces some univalent vertices in the theory. Choosing $a = a_0$ comes to choosing the weight of the univalent vertex as the counterterm of the melonic graphs, i.e. by translating to the vacuum one subtracts the contribution of the *entire melonic family.* Indeed, if one translates the field by a_0, as $a_0 = b(a_0)$, the linear term:

$$(b - a)\,\mathrm{Tr}_{\mathcal{D}}[M]\,,$$

which represents the univalent vertices, vanishes. As expected, in this case $M^c = 0$ is a solution of the classical equations of motion, due to the fact that the remaining piece of $Q(M)$ starts at order M^3.

However, no such interpretation exists for the translation to the instanton a_{inst}. While the linear term also cancels in this case, it is not true that the univalent vertices are the sum of the melonic family.

The free energy. We can compute the large N free energy of the model in the perturbative regime using Proposition 8.1 in Chapter 8. We have $V_{\text{melo}}(x) = \frac{D\lambda}{2}x^2$. The large N covariance K_2 and the large N free energy are:

$$1 - K_2 - D\lambda K_2^2 = 0 \Rightarrow K_2 = \frac{-1 + \sqrt{1 + 4D\lambda}}{2D\lambda}\,,$$
$$W_\infty = 1 + \ln(K_2) - K_2 - \frac{D\lambda}{2}K_2^2 = \frac{D\lambda}{2}K_2^2 + \ln(K_2)\,.$$

Noting that:

$$K_2 = -\frac{\imath}{\sqrt{\lambda}}a_0 = \frac{1}{1 - \imath\sqrt{\lambda}Da_0}\,,$$

the large N free energy writes in term of a_0 as:

$$W_\infty = -\frac{D}{2}a_0^2 - \ln(1 - \imath\sqrt{\lambda}Da_0)\,,$$

that is precisely the explicit prefactor in Eq. (10.1) if $a = a_0$.

The translation to the vacuum $H^c = a_0\mathbf{1}$ resums the contribution of the melonic sector, and the remaining integral over M (which, as we will see, only contributes at lower orders in $1/N$) is precisely the effective theory of a perturbation field around the melonic phase. We will show that the Feynman graphs of the remaining integral over M are exactly the *schemes* we defined in Chapter 4. Using the intermediate field representation we will give in particular an alternative proof of the fact that the number of schemes at a given order in $1/N$ is finite.

The case $D = 2$ The case of matrices, $D = 2$, is different. A careful inspection shows that for matrices, even after translating with $a = a_0$, the integral over the perturbation M contributes at leading order in $1/N$. This is natural, as for matrices the melonic graphs are only a subset of the planar graphs, and one cannot subtract the contribution of the planar sector by a single counterterm.

Contour translation

Before proceeding, let us identify in which regime of parameters one or the other of the two solutions of the classical equations of motion can be reached by a translation of the contour of integration. Let us denote $\lambda = |\lambda| e^{\imath\varphi}$, $\varphi \in (\pi, \pi)$.

Observe that the integrand involves the resolvent:

$$R(a\mathbf{1}^{\mathcal{D}} + M) = \frac{1}{(1 - \imath\sqrt{\lambda} D a)\mathbf{1}^{\mathcal{D}} - \imath\sqrt{\lambda} M} = \frac{1}{(1 - \imath\sqrt{\lambda} D a)} \frac{1}{\mathbf{1}^{\mathcal{D}} - bM},$$

and is singular whenever the resolvent can be zero. It follows that, before translating the contour of integration, one needs to make sure that the translated contour never reaches the singular value:

$$a_{\text{sing}} = \frac{1}{\imath\sqrt{\lambda} D} = -\imath e^{-\imath\frac{\varphi}{2}} \frac{1}{D\sqrt{|\lambda|}}, \qquad \Im a_{\text{sing}} = -\frac{\cos\frac{\varphi}{2}}{D\sqrt{|\lambda|}} \le 0 .$$

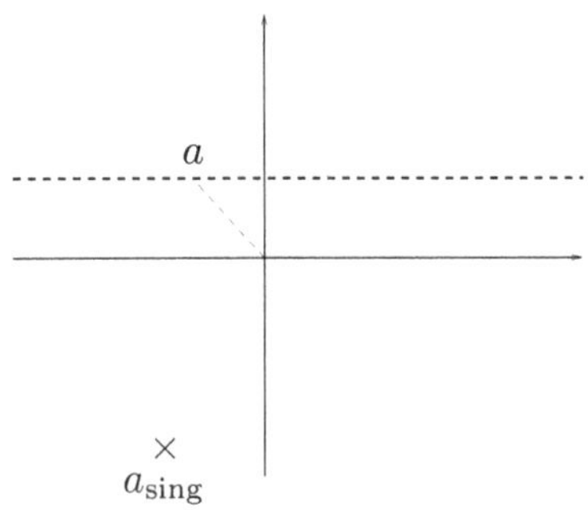

Figure 10.1: Translated integration contour.

The situation is presented in Figure 10.1. Only translations by a such that $\Im a > \Im a_{\text{sing}}$ are allowed. As $a_0 + a_{\text{inst}} = a_{\text{sing}}$ and $a_0 a_{\text{inst}} = \frac{1}{D}$ it follows that:

$$a_{\text{sing}} = |a_0| e^{\imath \arg a_0} + \frac{1}{D|a_0|} e^{-\imath \arg a_0} \Rightarrow \Im a_{\text{sing}} = \Im a_0 - \frac{1}{D|a_0|} \sin(\arg a_0) ,$$

hence the translation to a_0 is allowed as long as $\sin(\arg a_0) > 0$ and furthermore, as $\arg a_{\text{inst}} = -\arg a_0$, when the translation to one of the solutions is allowed, the translation to the other is forbidden. Observe that:

$$a_0 = \frac{1 - \sqrt{1 + 4D\lambda}}{2\imath\sqrt{\lambda} D} ,$$

is a continuous function of λ and behaves like $\imath\sqrt{\lambda}$ for small $|\lambda|$, thus for small $|\lambda|$ the contour can be translated to a_0. For small $|\lambda|$ the imaginary part of a_0 is positive. In fact the imaginary part of a_0 is positive for any $\lambda \in \mathbb{C} \setminus \mathbb{R}_-$, as:

$$a_0 = \bar{a}_0 \Leftrightarrow e^{-\imath\frac{\varphi}{2}} - \sqrt{e^{-\imath\varphi} + 4D|\lambda|} = -e^{\imath\frac{\varphi}{2}} + \sqrt{e^{\imath\varphi} + 4D|\lambda|}$$

$$\Leftrightarrow \left(e^{-\imath\frac{\varphi}{2}} + e^{\imath\frac{\varphi}{2}}\right)^2 = \left(\sqrt{e^{\imath\varphi} + 4D|\lambda|} + \sqrt{e^{-\imath\varphi} + 4D|\lambda|}\right)^2$$
$$\Leftrightarrow 1 - 4D|\lambda| = \sqrt{(e^{\imath\varphi} + 4D|\lambda|)\,(e^{-\imath\varphi} + 4D|\lambda|)}$$
$$\Leftrightarrow -8D|\lambda| = 4D|\lambda|(e^{\imath\varphi} + e^{-\imath\varphi}) \Rightarrow \varphi = \pm\pi \,.$$

It follows that for all $\lambda \in \mathbb{C} \setminus \mathbb{R}_-$ one can only translate the integration contour to the perturbative vacuum and not to the instanton.

10.2.2 Feynman rules

In order to derive the Feynman rules for the generating function written in terms of the perturbation field M in Eq. (10.1) we need to compute the effective covariance of the theory as well as the effective vertices.

The space of Hermitian $N \times N$ matrices is a N^2 dimensional real vector space $\mathcal{H}$ with inner product $\langle A|B\rangle \equiv \mathrm{Tr}[AB]$. The quadratic part of the action is a quadratic form on the direct sum of D such Hilbert spaces $\bigoplus_{c=1}^D \mathcal{H}^c$. Denoting a generic element in this direct sum space by:

$$\hat{M} = \begin{pmatrix} M^1 \\ \vdots \\ M^D \end{pmatrix} ,$$

the inner product in $\bigoplus_{c=1}^D \mathcal{H}^c$ is:

$$\langle \hat{P} | \hat{M} \rangle = \sum_{c=1}^{D} \mathrm{Tr}_c[P^c M^c] \,.$$

The Feynman rules of the effective theory are somewhat involved due to the fact that, while the quadratic part of the action is a quadratic form on the direct sum space $\bigoplus_{c=1}^D \mathcal{H}^c$, the interaction $Q(M)$ is a trace in the *tensor product* space $\bigotimes_{c=1}^D \mathcal{H}^c$. This is the root of all the subtleties we will need to deal with.

The effective covariance

The quadratic part of the action of the fluctuation field in Eq. (10.1) is:

$$N^{D-1}\sum_{c=1}^{D} \mathrm{Tr}_c[M^c M^c] - \mathrm{Tr}_{\mathcal{D}}[(bM)^2]$$
$$= N^{D-1}(1-b^2)\sum_{c=1}^{D} \mathrm{Tr}_c[(M^c)^2] + N^{D-2}b^2 \sum_{c=1}^{D}\left(\mathrm{Tr}_c[M^c]\right)^2 - N^{D-2}b^2\left(\sum_{c=1}^{D}\mathrm{Tr}_c[M^c]\right)^2 \,.$$

We denote by $\hat{I}^c$ the element of $\bigoplus_{c=1}^D \mathcal{H}^c$ consisting of the identity matrix in $\mathcal{H}^c$ and zero on the other components and by $\hat{I}$ the element consisting in the identity matrix in all the $\mathcal{H}^c$s:

$$\hat{I}_c = \begin{pmatrix} 0 \\ \vdots \\ 1 \\ \vdots \\ 0 \end{pmatrix} , \quad \hat{I} = \begin{pmatrix} 1 \\ \vdots \\ 1 \end{pmatrix} = \sum_{c=1}^{D} \hat{I}_c \,.$$

The orthogonal projectors on the one dimensional subspaces generated by $\hat{I}_c$ and $\hat{I}$ are:

$$\mathbf{P}_c\,\hat{M} = \frac{\langle \hat{I}_c \,|\, \hat{M}\rangle}{\langle \hat{I}_c \,|\, \hat{I}_c\rangle}\hat{I}_c = \frac{\mathrm{Tr}_c[M^c]}{N}\hat{I}_c \Rightarrow \langle \hat{M} \,|\, \mathbf{P}_c\,\hat{M}\rangle = \frac{\mathrm{Tr}_c[M^c]}{N}\mathrm{Tr}_c[M^c]$$

$$\mathbf{P}\hat{M} = \frac{\langle \hat{I} \,|\, \hat{M}\rangle}{\langle \hat{I} \,|\, \hat{I}\rangle}\hat{I} = \frac{\sum_{c=1}^D \mathrm{Tr}_c[M^c]}{DN}\hat{I}$$

$$\Rightarrow \langle \hat{M} \,|\, \mathbf{P}\hat{M}\rangle = \frac{\sum_{c=1}^D \mathrm{Tr}_c[M^c]}{DN}\left(\sum_{c'=1}^D \mathrm{Tr}_{c'}[M^{c'}]\right).$$

Let us denote $\mathbf{I}$ the identity operator on $\bigoplus_{c=1}^D \mathcal{H}^c$. The quadratic part is written then as $\langle \hat{M} \,|\, \mathbf{O}\,\hat{M}\rangle$ where the operator $\mathbf{O}$ is:

$$\mathbf{O} = N^{D-1}(1-b^2)\,\mathbf{I} + N^{D-1}b^2\sum_{c=1}^D \mathbf{P}_c - N^{D-1}Db^2\,\mathbf{P}\,.$$

The covariance of the effective theory is the operator $\mathbf{O}^{-1}$. In order to compute it, it is most convenient to diagonalize $\mathbf{O}$.

Lemma 10.1. *The operator $\mathbf{O}$ is diagonalized as:*

$$\mathbf{O} = N^{D-1}(1-b^2)\left[\mathbf{I} - \sum_{c=1}^D \mathbf{P}_c\right] + N^{D-1}\left[\sum_{c=1}^D \mathbf{P}_c - \mathbf{P}\right] + N^{D-1}(1-Db^2)\mathbf{P}\,,$$

where $\mathbf{I} - \sum_{c=1}^D \mathbf{P}_c$, $\sum_{c=1}^D \mathbf{P}_c - \mathbf{P}$ and $\mathbf{P}$ are the projectors on the eigenspaces of $\mathbf{O}$. The eigenvalues of $\mathbf{O}$ and their degeneracies are:

$$\Lambda_1 = N^{D-1}(1-b^2)\,,\quad \dim\left(\mathrm{Im}\left[\mathbf{I} - \sum_{c=1}^D \mathbf{P}_c\right]\right) = DN^2 - D\,,$$

$$\Lambda_2 = N^{D-1}\,,\quad \dim\left(\mathrm{Im}\left[\sum_{c=1}^D \mathbf{P}_c - \mathbf{P}\right]\right) = D-1\,,$$

$$\Lambda_3 = N^{D-1}(1-Db^2)\,,\quad \dim\left(\mathrm{Im}\left[\mathbf{P}\right]\right) = 1\,.$$

Proof. It is enough to check that the operators $\mathbf{I} - \sum_{c=1}^D \mathbf{P}_c$, $\sum_{c=1}^D \mathbf{P}_c - \mathbf{P}$ and $\mathbf{P}$ are three (mutually orthogonal) orthogonal projections. The three operators are obviously Hermitian and:

$$\mathbf{P}_{c'}\mathbf{P}_c\hat{M} = \frac{\mathrm{Tr}_c[M^c]}{N}\frac{\langle \hat{I}_{c'} \,|\, \hat{I}_c\rangle}{N}\hat{I}_{c'} = \delta_{cc'}\mathbf{P}_c\hat{M}\,,$$

$$\left(\sum_{c=1}^D \mathbf{P}_c\right)\mathbf{P}\hat{M} = \frac{\sum_{c'=1}^D \mathrm{Tr}_{c'}[M^{c'}]}{DN}\sum_{c=1}^D \frac{\langle \hat{I}_c \,|\, I\rangle}{N}\hat{I}_c = \mathbf{P}\hat{M}\,,$$

$$\mathbf{P}\left(\sum_{c=1}^D \mathbf{P}_c\right)\hat{M} = \sum_{c=1}^D \frac{\mathrm{Tr}_c[M^c]}{N}\frac{\langle \hat{I} \,|\, \hat{I}_c\rangle}{DN}\hat{I} = \mathbf{P}\hat{M}\,.$$

The rest of the proof is a straightforward computation:

$$\Big[\mathbf{I}-\sum_{c=1}^{D}\mathbf{P}_c\Big]\Big[\mathbf{I}-\sum_{c=1}^{D}\mathbf{P}_c\Big]=\mathbf{I}-2\sum_{c=1}^{D}\mathbf{P}_c+\sum_{c,c'=1}^{D}\mathbf{P}_c\mathbf{P}_{c'}=\mathbf{I}-\sum_{c=1}^{D}\mathbf{P}_c\,,$$

$$\begin{aligned}\Big[\sum_{c=1}^{D}\mathbf{P}_c-\mathbf{P}\Big]\Big[\sum_{c=1}^{D}\mathbf{P}_c-\mathbf{P}\Big]&=\sum_{c,c'=1}^{D}\mathbf{P}_c\mathbf{P}_{c'}-\mathbf{P}\Big(\sum_{c=1}^{D}\mathbf{P}_c\Big)-\Big(\sum_{c=1}^{D}\mathbf{P}_c\Big)\mathbf{P}+\mathbf{P}\\&=\sum_{c=1}^{D}\mathbf{P}_c-\mathbf{P}\,,\end{aligned}$$

$$\mathbf{P}^2=\mathbf{P}\,,$$

$$\Big[\mathbf{I}-\sum_{c=1}^{D}\mathbf{P}_c\Big]\Big[\sum_{c=1}^{D}\mathbf{P}_c-\mathbf{P}\Big]=\sum_{c=1}^{D}\mathbf{P}_c-\sum_{c,c'=1}^{D}\mathbf{P}_c\mathbf{P}_{c'}-\mathbf{P}+\Big(\sum_{c=1}^{D}\mathbf{P}_c\Big)\mathbf{P}=0\,,$$

$$\Big[\mathbf{I}-\sum_{c=1}^{D}\mathbf{P}_c\Big]\mathbf{P}=\mathbf{P}-\Big(\sum_{c=1}^{D}\mathbf{P}_c\Big)\mathbf{P}=0\,,$$

$$\Big[\sum_{c=1}^{D}\mathbf{P}_c-\mathbf{P}\Big]\mathbf{P}=\Big(\sum_{c=1}^{D}\mathbf{P}_c\Big)\mathbf{P}-\mathbf{P}^2=0\,.$$

Concerning the dimensions of the images of each of these operators, by definition $\mathbf{P}$ is one dimensional, $\mathbf{P}_c$ and $\mathbf{P}_{c'}$ are orthogonal if $c\neq c'$ hence $\big[\mathbf{I}-\sum_{c=1}^{D}\mathbf{P}_c\big]$ is DN^2-D dimensional and finally the sum of the three projectors is the identity.

□

Remark 10.3. We have the following remark.

Stability of the vacuum. In the domain $4D|\lambda|<1$ we have $|a_0|\leq\frac{1}{\sqrt{D}}$. It follows that, translating to the vacuum $a\equiv a_0=b(a_0)$, the three eigenvalues have a positive real part for λ in the disk of radius $(4D)^{-1}$, hence the perturbative vacuum is stable. In the critical regime $4D\lambda\to-1$, while the two eigenvalues Λ_1 and Λ_2 continue to have a finite positive real part, as:

$$\begin{aligned}a_0^2&=\frac{2+4D\lambda-2\sqrt{1+4D\lambda}}{-4\lambda D^2}=\frac{1}{D}-\frac{\sqrt{1+4D\lambda}\left(\sqrt{1+4D\lambda}-1\right)}{2\lambda D^2}\\\Rightarrow 1-Da_0^2&=\frac{\sqrt{1+4D\lambda}\left(\sqrt{1+4D\lambda}-1\right)}{2\lambda D}\,,\end{aligned}$$

the eigenvalue Λ_3 goes to zero.

From the point of view of the fluctuation field M, the tuning to criticality of the model is similar to the tuning in scalar field theory to a renormalization group trajectory leading to the infrared massless scalar field. In field theory this corresponds to a phase transition, usually accompanied by a spontaneous symmetry breaking.

The covariance of the effective theory is the inverse of $\mathbf{O}$:

$$\mathbf{O}^{-1}=\frac{1}{N^{D-1}(1-b^2)}\Big[\mathbf{I}-\sum_{c=1}^{D}\mathbf{P}_c\Big]+\frac{1}{N^{D-1}}\Big[\sum_{c=1}^{D}\mathbf{P}_c-\mathbf{P}\Big]+\frac{1}{N^{D-1}(1-Db^2)}\mathbf{P}$$

$$= \frac{1}{N^{D-1}(1-b^2)}\mathbf{I} - \frac{b^2}{N^{D-1}(1-b^2)}\left(\sum_{c=1}^{D}\mathbf{P}_c\right) + \frac{Db^2}{N^{D-1}(1-Db^2)}\mathbf{P}$$

and, taking into account the normalization of the Gaussian integral, the generating function in Eq. (10.1) becomes:

$$Z^{(4)}(J,\bar{J}) = \frac{e^{-N^D\left(\frac{Da^2}{2}+\ln(1-\imath\sqrt{\lambda}Da)\right)}}{\sqrt{[(1-b^2)]^{DN^2-D}[(1-Db^2)]}}\Big[e^{\frac{1}{2}\left\langle\frac{\partial}{\partial\hat{M}}\middle|\mathbf{O}^{-1}\frac{\partial}{\partial\hat{M}}\right\rangle} \times e^{Q(M)+\frac{1}{N^{D-1}}\frac{1}{(1-\imath\sqrt{\lambda}Da)}J\left(\frac{1}{\mathbf{1}^{\mathcal{D}}-bM}\right)\bar{J}}\Big]_{M^c=0}, \tag{10.2}$$

where, using the explicit form of the projectors $\mathbf{P}_c$ and $\mathbf{P}$, the covariance is written in terms of derivatives with respect to M^c as:

$$\begin{aligned}
&\left\langle\frac{\partial}{\partial\hat{M}}\middle|\mathbf{O}^{-1}\frac{\partial}{\partial\hat{M}}\right\rangle \\
&= \frac{1}{N^{D-1}(1-b^2)}\sum_{c=1}^{D}\mathrm{Tr}_c\left[\frac{\partial}{\partial M^c}\frac{\partial}{\partial M^c}\right] - \frac{b^2}{N^D(1-b^2)}\sum_{c=1}^{D}\mathrm{Tr}_c\left[\frac{\partial}{\partial M^c}\right]\mathrm{Tr}_c\left[\frac{\partial}{\partial M^c}\right] \\
&\quad + \frac{b^2}{N^D(1-Db^2)}\sum_{c,c'=1}^{D}\mathrm{Tr}_c\left[\frac{\partial}{\partial M^c}\right]\mathrm{Tr}_{c'}\left[\frac{\partial}{\partial M^{c'}}\right] \\
&= \frac{1}{N^{D-1}(1-b^2)}\sum_{c=1}^{D}\mathrm{Tr}_c\left[\frac{\partial}{\partial M^c}\frac{\partial}{\partial M^c}\right] + \frac{b^4(D-1)}{N^D(1-b^2)(1-Db^2)}\sum_{c=1}^{D}\mathrm{Tr}_c\left[\frac{\partial}{\partial M^c}\right]\mathrm{Tr}_c\left[\frac{\partial}{\partial M^c}\right] \\
&\quad + \frac{b^2}{N^D(1-Db^2)}\sum_{\substack{c,c'=1\\ c\neq c'}}^{D}\mathrm{Tr}_c\left[\frac{\partial}{\partial M^c}\right]\mathrm{Tr}_{c'}\left[\frac{\partial}{\partial M^{c'}}\right].
\end{aligned}$$

A moment's reflexion reveals that for $a \equiv a_0 = b(a_0)$ the three terms above correspond to the chains of $(D-1)$-dipoles discussed in Chapter 4. Indeed, the first term is nothing but the unbroken chains, while the second and third terms represent the broken chains with equal (respectively different) external colors (the numerical factors are easily recovered, taking into account that only a subset of dipoles is present in the quartic melonic models).

The covariance of the effective theory can be significantly simplified. The crucial observation is that the operator $\left\langle\frac{\partial}{\partial\hat{M}}\middle|\mathbf{O}^{-1}\frac{\partial}{\partial\hat{M}}\right\rangle$ acts on functions that depend only on $M = \sum_{c=1}^{D} M^c \otimes \mathbf{1}^{\mathcal{D}\setminus\{c\}}$. We observe that the action of a derivative operator:

$$\begin{aligned}
&\mathrm{Tr}_c\left[\frac{\partial}{\partial M^c}\right]\left(\sum_{c=1}^{D} M^c\otimes\mathbf{1}^{\mathcal{D}\setminus\{c\}}\right)^n \\
&= \sum_{n_1,\dots n_D\geq 0}^{n_1+\dots n_D=n}\frac{n!}{n_1!\dots n_D!}n_c(M^1)^{n_1}\otimes\dots(M^c)^{n_c-1}\dots\otimes(M^D)^{n_D} \\
&= n\sum_{n_1,\dots n_D\geq 0}^{n_1+\dots n_D=n-1}\frac{(n-1)!}{n_1!\dots n_D!}(M^1)^{n_1}\otimes\dots(M^c)^{n_c}\dots\otimes(M^D)^{n_D}
\end{aligned}$$

$$= n\left(\sum_c M^c \otimes \mathbf{1}^{\mathcal{D}\setminus\{c\}}\right)^{n-1} ,$$

does not depend on the color c. It follows that all the terms $\mathrm{Tr}_c[\frac{\partial}{\partial M^c}]$ can be replaced by $\frac{1}{D}\sum_{c=1}^{D}\mathrm{Tr}_c[\frac{\partial}{\partial M^c}]$, that is the operator $\left\langle \frac{\partial}{\partial \hat{M}} \middle| \mathbf{O}^{-1}\frac{\partial}{\partial \hat{M}} \right\rangle$ can in fact be replaced with:

$$\begin{aligned}
&\left\langle \frac{\partial}{\partial \hat{M}} \middle| \tilde{\mathbf{O}}^{-1}\frac{\partial}{\partial \hat{M}} \right\rangle \\
&= \frac{1}{N^{D-1}(1-b^2)}\sum_{c=1}^{D}\mathrm{Tr}_c\left[\frac{\partial}{\partial M^c}\frac{\partial}{\partial M^c}\right] \\
&\quad + \frac{b^2 D(D-1)}{N^D(1-b^2)(1-Db^2)}\left(\frac{1}{D}\sum_{c=1}^{D}\mathrm{Tr}_c\left[\frac{\partial}{\partial M^c}\right]\right)\left(\frac{1}{D}\sum_{c=1}^{D}\mathrm{Tr}_c\left[\frac{\partial}{\partial M^c}\right]\right) .
\end{aligned}$$

Observe that the new differential operator $\tilde{\mathbf{O}}^{-1}$ can be written in terms of $\mathbf{I}$ and $\mathbf{P}$ only:

$$\tilde{\mathbf{O}}^{-1} = \frac{1}{N^{D-1}(1-b^2)}\mathbf{I} + \frac{b^2(D-1)}{N^{D-1}(1-b^2)(1-Db^2)}\mathbf{P} ,$$

and can be diagonalized simply as:

$$\tilde{\mathbf{O}}^{-1} = \frac{1}{N^{D-1}(1-b^2)}(\mathbf{I}-\mathbf{P}) + \frac{1}{N^{D-1}(1-Db^2)}\mathbf{P} .$$

This operator corresponds to a new quadratic part of the action:

$$\begin{aligned}
\langle \hat{M} | \tilde{\mathbf{O}}\hat{M}\rangle &= \left\langle \hat{M} \middle| \left(N^{D-1}(1-b^2)\mathbf{I} - N^{D-1}(D-1)b^2\mathbf{P}\right)\hat{M}\right\rangle \\
&= N^{D-1}(1-b^2)\sum_{c=1}^{D}\mathrm{Tr}_c\left[M^cM^c\right] - N^{D-2}\frac{D-1}{D}b^2\left(\sum_{c=1}^{D}\mathrm{Tr}_c[M^c]\right)^2 .
\end{aligned}$$

With this remark, the generating function can then be rewritten in the simpler form:

$$\boxed{\begin{aligned}
Z^{(4)}(J,\bar{J}) &= \frac{e^{-N^D\left(\frac{Da^2}{2}+\ln(1-\imath\sqrt{\lambda}Da)\right)}}{\sqrt{[(1-b^2)]^{DN^2-D}[(1-Db^2)]}} \\
&\quad\times\Bigg[e^{\frac{1}{2}\frac{1}{N^{D-1}(1-b^2)}\sum_{c=1}^{D}\mathrm{Tr}_c[\frac{\partial}{\partial M^c}\frac{\partial}{\partial M^c}]} \\
&\quad\quad e^{+\frac{1}{2}\frac{b^2D(D-1)}{N^D(1-b^2)(1-Db^2)}(\frac{1}{D}\sum_{c=1}^{D}\mathrm{Tr}_c[\frac{\partial}{\partial M^c}])(\frac{1}{D}\sum_{c=1}^{D}\mathrm{Tr}_c[\frac{\partial}{\partial M^c}])} \\
&\quad\quad\quad\times e^{Q(M)+\frac{1}{N^{D-1}}\frac{1}{(1-\imath\sqrt{\lambda}Da)}J\left(\frac{1}{\mathbf{1}^{\mathcal{D}}-bM}\right)\bar{J}}\Bigg]_{M^c=0} \\
&= e^{-N^D\left(\frac{Da^2}{2}+\ln(1-\imath\sqrt{\lambda}Da)\right)}(1-b^2)^{\frac{D-1}{2}} \\
&\quad\int[dM^c]e^{-\frac{1}{2}N^{D-1}(1-b^2)\sum_{c=1}^{D}\mathrm{Tr}_c[M^cM^c]+\frac{1}{2}N^{D-2}\frac{D-1}{D}b^2(\sum_{c=1}^{D}\mathrm{Tr}_c[M^c])^2} \\
&\quad\times e^{Q(M)+\frac{1}{N^{D-1}}\frac{1}{(1-\imath\sqrt{\lambda}Da)}J\left(\frac{1}{\mathbf{1}^{\mathcal{D}}-bM}\right)\bar{J}} ,
\end{aligned}} \tag{10.3}$$

where, in the case of the translation to the melonic vacuum, one needs to substitute both a and b above with a_0. Observe that the mass eigenvalues for the perturbation field are $N^{D-1}(1-b^2)$ with degeneracy DN^2-1 and $N^{D-1}(1-Db^2)$ with degeneracy 1 (and again in the case of the translation to the melonic vacuum b should be replaced by a_0).

The effective vertices

The generating function of the cumulants of the melonic model can be computed directly starting from the integral over the fluctuation field M. Expanding the exponential we obtain:

$$Z^{(4)}(J,\bar{J}) = \sum_{n_{\text{int}},n_{\text{ext}}\geq 0} \frac{1}{n_{\text{int}}!n_{\text{ext}}!}\Bigg\{ e^{\frac{1}{2}\left\langle \frac{\partial}{\partial \hat{M}} \middle| \tilde{\mathbf{O}}^{-1} \frac{\partial}{\partial \hat{M}}\right\rangle} \times \Big[Q(M)\Big]^{n_{\text{int}}} \times \left(\frac{1}{N^{D-1}} \frac{1}{(1-\imath\sqrt{\lambda}Da)} J \left(\frac{1}{\mathbf{1}^{\mathcal{D}} - bM}\right) \bar{J}\right)^{n_{\text{ext}}} \Bigg\}_{M^c=0} .$$

We represent the terms $Q(M)$ as internal vertices and the $J\bar{J}$ terms as external ciliated vertices. The derivative in the operator $\left\langle \frac{\partial}{\partial \hat{M}} \middle| \tilde{\mathbf{O}}^{-1} \frac{\partial}{\partial \hat{M}}\right\rangle$ create pairs of halfedges connected into edges. The derivatives corresponding to the ends of the edges act on the vertices. For the ciliated vertices we use:

$$\frac{\partial}{\partial M^c_{a^c b^c}} \left[\frac{1}{\mathbf{1}^{\mathcal{D}} - bM}\right]_{q^{\mathcal{D}}p^{\mathcal{D}}} = b \sum_{n^{\mathcal{D}}m^{\mathcal{D}}} \left(\left[\frac{1}{\mathbf{1}^{\mathcal{D}} - bM}\right]_{q^{\mathcal{D}}n^{\mathcal{D}}} \left[\frac{1}{\mathbf{1}^{\mathcal{D}} - bM}\right]_{m^{\mathcal{D}}p^{\mathcal{D}}} \right) \times \left(\delta_{n^{\mathcal{D}\setminus\{c\}} m^{\mathcal{D}\setminus\{c\}}} \delta_{n^c a^c} \delta_{m^c b^c} \right) .$$

For the internal, nonciliated vertices, we need to estimate the derivatives of the subtracted vertex function:

$$Q(M) = \text{Tr}_{\mathcal{D}}\left[\ln \frac{1}{\mathbf{1}^{\mathcal{D}} - bM}\right] - \text{Tr}_{\mathcal{D}}\left[aM\right] - \frac{1}{2}\text{Tr}_{\mathcal{D}}\left[(bM)^2\right] .$$

Clearly, starting with the third order, the last two terms do not contribute. For the lower orders we use:

$$\begin{aligned} \frac{\partial}{\partial M^c_{a^c b^c}} \text{Tr}_{\mathcal{D}}\left[aM\right] &= \frac{\partial}{\partial M^c_{a^c b^c}} \left(a \sum_{c=1}^{D} \text{Tr}_{\mathcal{D}}[M^c \otimes \mathbf{1}^{\mathcal{D}\setminus\{c\}}] \right) \\ &= a \sum_{p^{\mathcal{D}}q^{\mathcal{D}}} \left[\mathbf{1}^{\mathcal{D}}\right]_{q^{\mathcal{D}}p^{\mathcal{D}}} \left(\delta_{p^{\mathcal{D}\setminus\{c\}} q^{\mathcal{D}\setminus\{c\}}} \delta_{p^c a^c} \delta_{q^c b^c} \right) , \end{aligned}$$

$$\frac{\partial}{\partial M^c_{a^c b^c}} \frac{1}{2} \text{Tr}_{\mathcal{D}}\left[(bM)^2\right] = \frac{b^2}{2} \frac{\partial}{\partial M^c_{a^c b^c}} \left(\sum_{c=1}^{D} \text{Tr}_{\mathcal{D}}[(M^c)^2 \otimes \mathbf{1}^{\mathcal{D}\setminus\{c\}}] + \sum_{\substack{c,c'=1 \\ c\neq c'}}^{D} \text{Tr}_{\mathcal{D}}[M^c \otimes M^{c'} \otimes \mathbf{1}^{\mathcal{D}\setminus\{c\}\setminus\{c'\}}] \right)$$

$$= b^2 \sum_{p^{\mathcal{D}} q^{\mathcal{D}}} M_{q^{\mathcal{D}} p^{\mathcal{D}}} \left(\delta_{p^{\mathcal{D}\setminus\{c\}} q^{\mathcal{D}\setminus\{c\}}} \delta_{p^c a^c} \delta_{q^c b^c} \right) ,$$

$$\frac{\partial}{\partial M^c_{a^c b^c}} \operatorname{Tr}_{\mathcal{D}} \left[\ln \frac{1}{\mathbf{1}^{\mathcal{D}} - bM} \right] = b \sum_{p^{\mathcal{D}} q^{\mathcal{D}}} \left[\frac{1}{\mathbf{1}^{\mathcal{D}} - bM} \right]_{q^{\mathcal{D}} p^{\mathcal{D}}} \left(\delta_{p^{\mathcal{D}\setminus\{c\}} q^{\mathcal{D}\setminus\{c\}}} \delta_{p^c a^c} \delta_{q^c b^c} \right) ,$$

hence the first derivative of the vertex function is:

$$\frac{\partial}{\partial M^c_{a^c b^c}} \left[Q(M) \right]$$
$$= \sum_{p^{\mathcal{D}} q^{\mathcal{D}}} \left[\frac{b}{\mathbf{1}^{\mathcal{D}} - bM} - a\mathbf{1}^{\mathcal{D}} - b^2 M \right]_{q^{\mathcal{D}} p^{\mathcal{D}}} \left(\delta_{p^{\mathcal{D}\setminus\{c\}} q^{\mathcal{D}\setminus\{c\}}} \delta_{p^c a^c} \delta_{q^c b^c} \right) ,$$

and the second derivative estimates using:

$$\frac{\partial}{\partial M^c_{a^c b^c}} \left[\frac{b}{\mathbf{1}^{\mathcal{D}} - bM} - a\mathbf{1}^{\mathcal{D}} - b^2 M \right]_{q^{\mathcal{D}} p^{\mathcal{D}}}$$
$$= b^2 \sum_{n^{\mathcal{D}} m^{\mathcal{D}}} \left(\left[\frac{1}{\mathbf{1}^{\mathcal{D}} - bM} \right]_{q^{\mathcal{D}} n^{\mathcal{D}}} \left[\frac{1}{\mathbf{1}^{\mathcal{D}} - bM} \right]_{m^{\mathcal{D}} p^{\mathcal{D}}} - [\mathbf{1}^{\mathcal{D}}]_{q^{\mathcal{D}} n^{\mathcal{D}}} [\mathbf{1}^{\mathcal{D}}]_{m^{\mathcal{D}} p^{\mathcal{D}}} \right)$$
$$\times \left(\delta_{n^{\mathcal{D}\setminus\{c\}} m^{\mathcal{D}\setminus\{c\}}} \delta_{n^c a^c} \delta_{m^c b^c} \right) .$$

Broken and unbroken edges

When evaluating the Gaussian integral in Eq. (10.3), the two terms in the differential operator $\left\langle \frac{\partial}{\partial \hat{M}} \middle| \tilde{\mathbf{O}}^{-1} \frac{\partial}{\partial \hat{M}} \right\rangle$ correspond to two kinds of edges:

- *unbroken edges* colored by a color c coming from the term:

$$\operatorname{Tr}_c \left[\frac{\partial}{\partial M^c} \frac{\partial}{\partial M^c} \right] .$$

- *broken edges* coming from the term:

$$\left(\frac{1}{D} \sum_{c=1}^{D} \operatorname{Tr}_c \left[\frac{\partial}{\partial M^c} \right] \right) \left(\frac{1}{D} \sum_{c=1}^{D} \operatorname{Tr}_c \left[\frac{\partial}{\partial M^c} \right] \right) .$$

  The broken edges have no color.

We draw the broken edges as dashed edges and the unbroken edges as solid edges having a color. An example of a map with broken and unbroken edges is presented in Figure 10.2 (where the white dots represent the cilia).

In the perturbative expansion the generating function becomes a sum over maps $\mathcal{M}$ with ciliated vertices whose edges fall in two categories: the unbroken edges $\mathcal{E}_u(\mathcal{M})$, which have a color c, and the broken edges, $\mathcal{E}_b(\mathcal{M})$. For the corners $[h]$ of the map we obtain associated matrix elements of operators:

- for any corner $[h]$ of a ciliated vertex, or of a nonciliated vertex of coordination at least three, we obtain:

$$b \left[\frac{1}{\mathbf{1}^{\mathcal{D}} - bM} \right]_{a^{\mathcal{D}}_{[h]} b^{\mathcal{D}}_{[h]}} ,$$

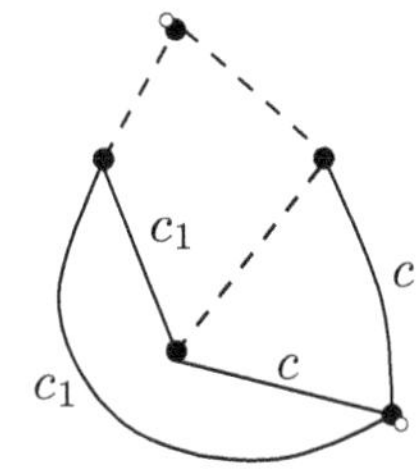

Figure 10.2: Map with cilia, broken and unbroken colored edges.

- for a corner $[h]$ of a nonciliated vertex of coordination one we obtain:

$$b\left[\frac{1}{\mathbf{1}^{\mathcal{D}}-bM}-\frac{a}{b}\mathbf{1}^{\mathcal{D}}-bM\right]_{a^{\mathcal{D}}_{[h]}b^{\mathcal{D}}_{[h]}},$$

- for the couple of corners $[h]$ and $[\sigma_{\mathcal{M}}(h)]$ belonging to a nonciliated vertex of coordination two we obtain:

$$b^2\Big(\left[\frac{1}{\mathbf{1}^{\mathcal{D}}-bM}\right]_{a^{\mathcal{D}}_{[h]}b^{\mathcal{D}}_{[h]}}\left[\frac{1}{\mathbf{1}^{\mathcal{D}}-bM}\right]_{a^{\mathcal{D}}_{[\sigma(h)]}b^{\mathcal{D}}_{[\sigma(h)]}}-[\mathbf{1}^{\mathcal{D}}]_{a^{\mathcal{D}}_{[h]}b^{\mathcal{D}}_{[h]}}[\mathbf{1}^{\mathcal{D}}]_{a^{\mathcal{D}}_{[\sigma(h)]}b^{\mathcal{D}}_{[\sigma(h)]}}\Big).$$

The edges contract the indices of the operators incident at their ends in a specific pattern, according to the nature of the edge (unbroken or broken).

Extracting the logarithm we obtain a sum over connected maps with edges broken or unbroken. Denoting $n_{\rm int}(\mathcal{M})$, $n_{\rm ext}(\mathcal{M})$, $E_u(\mathcal{M})$ and $E_b(\mathcal{M})$ the numbers of internal vertices, external vertices, unbroken edges and broken edges of $\mathcal{M}$ we obtain the generating function of the cumulants:

$$\begin{aligned}
\ln Z^{(4)}(J,\bar{J}) = & \sum_{\mathcal{M}\text{ connected}}\frac{b^{2E_u(\mathcal{M})+2E_b(\mathcal{M})}}{n_{\rm int}(\mathcal{M})!\,n_{\rm ext}(\mathcal{M})!}\\
&\times\sum_{ab}\Bigg[\Bigg(\prod_{h\in\mathcal{S}(\mathcal{M})}\mathfrak{R}(M)_{a^{\mathcal{D}}_{[h]}b^{\mathcal{D}}_{[h]}}\Bigg)\Bigg(\prod_{h\in\mathcal{S}_{\rm ext}(\mathcal{M})}\frac{J_{a^{\mathcal{D}}_{[h]}}\bar{J}_{b^{\mathcal{D}}_{[\sigma^{-1}_{\mathcal{M}}(h)]}}}{N^{D-1}(1-\imath\sqrt{\lambda}Da)}\Bigg)\\
&\times\Bigg[\prod_{e^c=(h,h')\in\mathcal{E}_u(\mathcal{M})}\frac{1}{N^{D-1}(1-b^2)}\Big(\delta_{b^c_{[\sigma^{-1}_{\mathcal{M}}(h')]}a^c_{[h']}}\delta_{b^c_{[\sigma^{-1}_{\mathcal{M}}(h')]}a^c_{[h]}}\Big)\\
&\qquad\Big(\delta_{b^{\mathcal{D}\setminus\{c\}}_{[\sigma^{-1}_{\mathcal{M}}(h)]}a^{\mathcal{D}\setminus\{c\}}_{[h]}}\delta_{b^{\mathcal{D}\setminus\{c\}}_{[\sigma^{-1}_{\mathcal{M}}(h')]}a^{\mathcal{D}\setminus\{c\}}_{[h']}}\Big)\Bigg]\\
&\times\Bigg[\prod_{e=(h,h')\in\mathcal{E}_b(\mathcal{M})}\frac{b^2D(D-1)}{N^D(1-b^2)(1-Db^2)}\\
&\qquad\Big(\delta_{b^{\mathcal{D}}_{[\sigma^{-1}_{\mathcal{M}}(h)]}a^{\mathcal{D}}_{[h]}}\delta_{b^{\mathcal{D}}_{[\sigma^{-1}_{\mathcal{M}}(h')]}a^{\mathcal{D}}_{[h']}}\Big)\Bigg]\Bigg]_{M=0},
\end{aligned}$$

where we have used the notation:

- for any corner $[h]$ of a ciliated vertex or of a nonciliated vertex of coordination at least three:

$$\mathfrak{R}(M)_{a^{\mathcal{D}}_{[h]} b^{\mathcal{D}}_{[h]}} \equiv \left[\frac{1}{\mathbf{1}^{\mathcal{D}} - bM} \right]_{a^{\mathcal{D}}_{[h]} b^{\mathcal{D}}_{[h]}} ,$$

- for a corner $[h]$ of a nonciliated vertex of coordination one:

$$\mathfrak{R}(M)_{a^{\mathcal{D}}_{[h]} b^{\mathcal{D}}_{[h]}} \equiv \left[\frac{1}{\mathbf{1}^{\mathcal{D}} - bM} - \frac{a}{b}\mathbf{1}^{\mathcal{D}} - bM \right]_{a^{\mathcal{D}}_{[h]} b^{\mathcal{D}}_{[h]}} ,$$

- for the couple of corners $[h]$ and $[\sigma_{\mathcal{M}}(h)]$ belonging to a nonciliated vertex of coordination two:

$$\begin{aligned} &\mathfrak{R}(M)_{a^{\mathcal{D}}_{[h]} b^{\mathcal{D}}_{[h]}} \mathfrak{R}(M)_{a^{\mathcal{D}}_{[\sigma(h)]} b^{\mathcal{D}}_{[\sigma(h)]}} \\ &\equiv \left[\frac{1}{\mathbf{1}^{\mathcal{D}} - bM} \right]_{a^{\mathcal{D}}_{[h]} b^{\mathcal{D}}_{[h]}} \left[\frac{1}{\mathbf{1}^{\mathcal{D}} - bM} \right]_{a^{\mathcal{D}}_{[\sigma(h)]} b^{\mathcal{D}}_{[\sigma(h)]}} - [\mathbf{1}^{\mathcal{D}}]_{a^{\mathcal{D}}_{[h]} b^{\mathcal{D}}_{[h]}} [\mathbf{1}^{\mathcal{D}}]_{a^{\mathcal{D}}_{[\sigma(h)]} b^{\mathcal{D}}_{[\sigma(h)]}} . \end{aligned}$$

10.2.3 Translating to the vacuum

If we translate to the perturbative vacuum $a \equiv a_0 = b(a_0)$ several significant simplifications can be made. Indeed, when setting $M = 0$ we obtain:

- for any corner $[h]$ of a ciliated vertex or of a nonciliated vertex of coordination at least three:

$$\mathfrak{R}(0)_{a^{\mathcal{D}}_{[h]} b^{\mathcal{D}}_{[h]}} = \mathbf{1}^{\mathcal{D}}_{a^{\mathcal{D}}_{[h]} b^{\mathcal{D}}_{[h]}} ,$$

- for a corner $[h]$ of a nonciliated vertex of coordination one:

$$\mathfrak{R}(0)_{a^{\mathcal{D}}_{[h]} b^{\mathcal{D}}_{[h]}} = \frac{b(a_0) - a_0}{b(a_0)} \mathbf{1}^{\mathcal{D}}_{a^{\mathcal{D}}_{[h]} b^{\mathcal{D}}_{[h]}} = 0 ,$$

- for the couple of corners $[h]$ and $[\sigma_{\mathcal{M}}(h)]$ belonging to a nonciliated vertex of coordination two:

$$\begin{aligned} &\mathfrak{R}(0)_{a^{\mathcal{D}}_{[h]} b^{\mathcal{D}}_{[h]}} \mathfrak{R}(0)_{a^{\mathcal{D}}_{[\sigma(h)]} b^{\mathcal{D}}_{[\sigma(h)]}} \\ &= [\mathbf{1}^{\mathcal{D}}]_{a^{\mathcal{D}}_{[h]} b^{\mathcal{D}}_{[h]}} [\mathbf{1}^{\mathcal{D}}]_{a^{\mathcal{D}}_{[\sigma(h)]} b^{\mathcal{D}}_{[\sigma(h)]}} - [\mathbf{1}^{\mathcal{D}}]_{a^{\mathcal{D}}_{[h]} b^{\mathcal{D}}_{[h]}} [\mathbf{1}^{\mathcal{D}}]_{a^{\mathcal{D}}_{[\sigma(h)]} b^{\mathcal{D}}_{[\sigma(h)]}} = 0 . \end{aligned}$$

Thus only maps whose internal vertices have coordination at least three (which we denote $\mathcal{M}^3$) survive.

Moreover, a map $\mathcal{M}^3$ has unbroken colored edges and broken edges. Erasing all the broken edges leads to a usual edge colored map, which we denote $\mathcal{M}^3_u$ (as it is formed only by the unbroken edges). Remark that $\mathcal{M}^3_u$ can be disconnected, even though $\mathcal{M}^3$ is connected (see Figure 10.3).

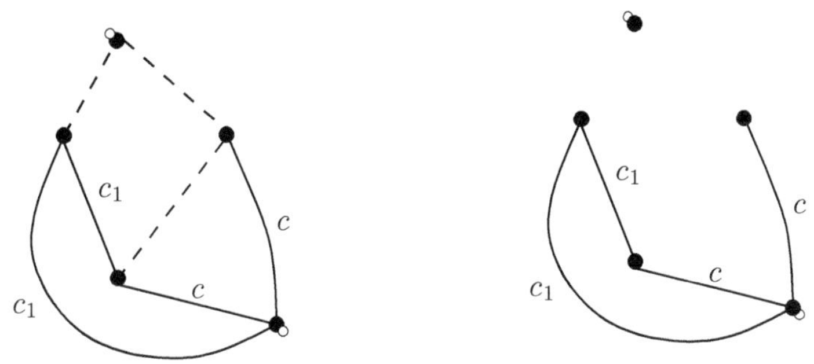

Figure 10.3: The maps $\mathcal{M}^3$ and $\mathcal{M}^3_u$.

As $\mathcal{M}^3_u$ is a usual edge colored map it has a well defined notion of boundary graph $\partial\mathcal{M}^3_u$ and (internal and external) faces. The indices are identified along the faces of $\mathcal{M}^3_u$. Taking into account the fact that the numbers of unbroken edges of $\mathcal{M}^3$ and $\mathcal{M}^3_u$ are equal, and that the number of ciliated vertices of $\mathcal{M}$ is the number of white vertices of $\partial\mathcal{M}^3_u$, we obtain:

$$\begin{aligned}\ln Z^{(4)}(J,\bar{J}) &= \sum_{\mathcal{M}^3 \text{ connected}} \frac{\mathrm{Tr}_{\partial\mathcal{M}^3_u}(J,\bar{J})}{k(\partial\mathcal{M}^3_u)!}\,\frac{1}{n_{\mathrm{int}}(\mathcal{M}^3)!}\\ &\times \frac{a_0^{2E_u(\mathcal{M}^3)+2E_b(\mathcal{M}^3)}}{(1-\imath\sqrt{\lambda}Da_0)^{k(\partial\mathcal{M}^3_u)}}\left[\frac{1}{1-a_0^2}\right]^{E_u(\mathcal{M}^3)}\left[\frac{a_0^2D(D-1)}{(1-a_0^2)(1-Da_0^2)}\right]^{E_b(\mathcal{M}^3)}\\ &\times N^{-DE_b(\mathcal{M}^3)-(D-1)E_u(\mathcal{M}^3)-(D-1)k(\partial\mathcal{M}^3_u)+F_{\mathrm{int}}(\mathcal{M}^3_u)}\ .\end{aligned}$$

The maps $\mathcal{M}^3$ in the above equation are the reduced schemes of Chapter 4. The constraint that the vertices of $\mathcal{M}^3$ have degree at least three means that the schemes one sums over are not only melon free but also free of chains, i.e. they are reduced. The amplitude associated to each scheme is exactly the counterpart for the melonic quartic model of the resummation of all the graphs corresponding to a reduced scheme, i.e. one term in Theorem 4.5.

The rescaled cumulants are computed by taking derivatives and dividing by the appropriate power of N:

$$\boxed{\begin{aligned}K(\mathcal{B},\mu^{(4)}) &= \sum_{\substack{\mathcal{M}^3 \text{ connected}\\ \partial\mathcal{M}^3_u=\mathcal{B}}} \frac{1}{n_{\mathrm{int}}(\mathcal{M}^3)!}\\ &\times \frac{a_0^{2E_u(\mathcal{M}^3)+4E_b(\mathcal{M}^3)}}{(1-\imath\sqrt{\lambda}Da_0)^{k(\partial\mathcal{M}^3_u)}}\,\frac{[D(D-1)]^{E_b(\mathcal{M}^3)}}{(1-a_0^2)^{E_u(\mathcal{M}^3)+E_b(\mathcal{M}^3)}}\\ &\times \frac{1}{(1-Da_0^2)^{E_b(\mathcal{M}^3)}}\\ &\times N^{-DE_b(\mathcal{M}^3)-(D-1)E_u(\mathcal{M}^3)-D+(D-1)k(\partial\mathcal{M}^3_u)+C(\partial\mathcal{M}^3_u)+F_{\mathrm{int}}(\mathcal{M}^3_u)}\ .\end{aligned}} \tag{10.4}$$

10.2.4 The $1/N$ expansion for the fluctuation field

The $1/N$ expansion of the cumulants can be derived directly starting from Eq. (10.4). We label $\mathcal{M}^3_{u;(\rho)}$ the connected components of $\mathcal{M}^3_u$ and we denote $C(\mathcal{M}^3_u)$ their number.

Definition 10.1. *For each connected component* $\mathcal{M}^3_{u;(\rho)}$ *we define its* deficit $\eta(\mathcal{M}^3_{u;(\rho)})$ *as:*

$$\eta(\mathcal{M}^3_{u;(\rho)}) = 2 + E_u(\mathcal{M}^3_{u;(\rho)}) + (D-2)V(\mathcal{M}^3_{u;(\rho)}) \\ - (D-1)k(\partial\mathcal{M}^3_{u;(\rho)}) - C(\partial\mathcal{M}^3_{u;(\rho)}) - F_{\text{int}}(\mathcal{M}^3_{u;(\rho)}) \,.$$

Furthermore, we define the excess of broken edges *of* $\mathcal{M}^3$, $L_b(\mathcal{M}^3)$, *and the* excess of unbroken edges *of* $\mathcal{M}^3_{u;(\rho)}$, $L_u(\mathcal{M}^3_{u;(\rho)})$, *as:*

$$L_b(\mathcal{M}^3) \equiv E_b(\mathcal{M}^3) - C(\mathcal{M}^3_u) + 1 \,, \quad L_u(\mathcal{M}^3_{u;(\rho)}) \equiv E_u(\mathcal{M}^3_{u;(\rho)}) - V(\mathcal{M}^3_{u;(\rho)}) + 1 \,.$$

The following result, encoding the $1/N$ expansion in the sense of perturbation theory for the fluctuation field, is the exact analogue of Theorem 4.4 and Theorem 4.5 in Chapter 4.

Theorem 10.1. *The rescaled cumulants of* $\mu^{(4)}$ *can be expressed as the formal series:*

$$K(\mathcal{B},\mu^{(4)}) \tag{10.5}$$
$$= \sum_{\substack{\mathcal{M}^3 \text{ connected} \\ \partial\mathcal{M}^3_u = \mathcal{B}}} \frac{1}{n_{\text{int}}(\mathcal{M}^3)!} \frac{a_0^{2E_u(\mathcal{M}^3)+4E_b(\mathcal{M}^3)}}{(1-\imath\sqrt{\lambda}Da_0)^{k(\partial\mathcal{M}^3_u)}} \frac{[D(D-1)]^{E_b(\mathcal{M}^3)}}{(1-a_0^2)^{E_u(\mathcal{M}^3)+E_b(\mathcal{M}^3)}} \\ \times \frac{1}{(1-Da_0^2)^{E_b(\mathcal{M}^3)}} N^{-DL_b(\mathcal{M}^3)-(D-2)\sum_{\rho=1}^{C(\mathcal{M}^3_u)} L_u(\mathcal{M}^3_{u;(\rho)}) - \sum_{\rho=1}^{C(\mathcal{M}^3_u)} \eta(\mathcal{M}^3_{u;(\rho)})} \,.$$

The right hand side of Eq. (10.5) *is a series in* $1/N$. *More importantly, there are only a finite number of maps* $\mathcal{M}^3$ *contributing to any fixed order in* $1/N$.

Proof. A straightforward computation leads from Eq. (10.4) to Eq. (10.5) taking into account that the boundary graph of $\mathcal{M}^3_u$ is distributed among the connected components $\mathcal{M}^3_{u;(\rho)}$ and that the numbers of vertices, ciliated vertices and unbroken edges of $\mathcal{M}^3$ and $\mathcal{M}^3_u$ are equal.

As $\mathcal{M}^3$ is connected, all the connected components of $\mathcal{M}^3_u$ must be connected in-between them by broken edges, hence $L_b(\mathcal{M}^3) \geq 0$. Similarly, as $\mathcal{M}^3_{u;(\rho)}$ is connected, $L_u(\mathcal{M}^3_{u;(\rho)}) \geq 0$. By Theorem 9.1 in Chapter 9, the number of internal faces of a connected edge colored map with cilia is bounded by:

$$F_{\text{int}}(\mathcal{M}) \leq 1 - (D-1)k(\partial\mathcal{M}) - C(\partial\mathcal{M}) + (D-1)V(\mathcal{M}) + \\ + [E(\mathcal{M}) - V(\mathcal{M}) + 1] \,,$$

where we have taken into account that we are dealing only with maps whose edges have a unique color. Thus for any connected component $\mathcal{M}^3_{u;(\rho)}$ its defect $\eta(\mathcal{M}^3_{u;(\rho)})$ is nonnegative.

We conclude that the right hand side of Eq. (10.5) is a series in $1/N$.

In order to prove the second part of the theorem, we note that, as $D > D-2$, the scaling in N is bounded by:

$$-(D-2)\Big[L_b(\mathcal{M}^3) + \sum_{\rho=1}^{C(\mathcal{M}^3_u)} L_u(\mathcal{M}^3_{u;(\rho)})\Big] = -(D-2)\big[E_b(\mathcal{M}^3) + E_u(\mathcal{M}^3) - V(\mathcal{M}^3) + 1\big] .$$

As the ciliated vertices in the map $\mathcal{M}^3$ have coordination at least one, while the nonciliated vertices have coordination at least three, we have:

$$2\big[E_b(\mathcal{M}^3) + E_u(\mathcal{M}^3)\big] \geq 3\big[V(\mathcal{M}^3) - k(\partial\mathcal{M}^3_u)\big] + k(\partial\mathcal{M}^3_u) \Rightarrow$$
$$\Rightarrow V(\mathcal{M}^3) \leq \frac{2}{3}\big[E_b(\mathcal{M}^3) + E_u(\mathcal{M}^3) + k(\partial\mathcal{M}^3_u)\big] ,$$

hence the scaling with N of a term on the right hand side of Eq. (10.4) is bounded by:

$$-(D-2)\left[\frac{E_b(\mathcal{M}^3) + E_u(\mathcal{M}^3)}{3} - \frac{2k(\partial\mathcal{M}^3_u)}{3} + 1\right] .$$

The number of external vertices of $\mathcal{M}^3_u$ is fixed, as it equals the number of the white vertices of the invariant $\mathcal{B}$ whose cumulant we evaluate. It follows that at any order in $1/N$ only connected maps with at most a finite number of edges contribute, and the second part of the theorem follows.

□

The shift to the nontrivial vacuum has thus analytically resummed *all* the series in λ at fixed order in $1/N$.

Corollary 10.1. *The free energy is written as:*

$$W = \frac{1}{N^D}\ln Z^{(4)}(0,0) = -\frac{D}{2}a_0^2 - \ln(1 - \imath\sqrt{\lambda}Da_0) + O\left(\frac{1}{N}\right) .$$

Proof. Setting the external sources to zero, and taking into account the prefactors in Eq. (10.3), the logarithm of the partition function of the quartic melonic model is written:

$$\ln Z^{(4)}(0,0) = -N^D\left(\frac{Da_0^2}{2} + \ln(1 - \imath\sqrt{\lambda}Da_0)\right) - \frac{D(N^2-1)}{2}\ln(1-a_0^2) - \frac{1}{2}\ln(1-Da_0^2)$$
$$+ N^D \sum_{\substack{\mathcal{M}^3 \text{ connected} \\ \partial\mathcal{M}^3=\emptyset}} \frac{1}{n_{\text{int}}(\mathcal{M}^3)!}$$
$$\times a_0^{2E_u(\mathcal{M}^3)+2E_b(\mathcal{M}^3)}\left[\frac{1}{1-a_0^2}\right]^{E_u(\mathcal{M}^3)}\left[\frac{a_0^2D(D-1)}{(1-a_0^2)(1-Da_0^2)}\right]^{E_b(\mathcal{M}^3)}$$
$$\times N^{-DL_b(\mathcal{M}^3)-(D-2)\sum_{\rho=1}^{C(\mathcal{M}^3_u)} L_u(\mathcal{M}^3_{u;(\rho)}) - \sum_{\rho=1}^{C(\mathcal{M}^3_u)} \eta(\mathcal{M}^3_{u;(\rho)})} .$$

The sum over $\mathcal{M}^3$ does not contribute at leading order in N. Indeed, a term in the sum with $L_b(\mathcal{M}^3) = L_u(\mathcal{M}^3_{u;(\rho)}) = 0$ would be a tree, hence would have univalent vertices, which is impossible as all the vertices $\mathcal{M}^3$ are at least three valent.

□

10.2.5 The first nontrivial order

Any colored graph $\mathcal{B}$ can be obtained as the boundary of some edge colored map $\mathcal{M}^3_u$ (this is a consequence of Theorem 5.1 in Chapter 5 and of the bijection between graphs built on quartic interactions and edge multicolored maps in Chapter 9).

In order to identify the first nontrivial order in the $1/N$ expansion of the cumulant of $\mathcal{B}$ one needs to identify the maps $\mathcal{M}^3$ with $\partial\mathcal{M}^3_u = \mathcal{B}$ and whose scaling in Eq. (10.5) is minimal. This is an open problem for general $\mathcal{B}$, but can be answered in some particular cases.

Lemma 10.2. *Let $\mathcal{B}$ be a* connected melonic graph *with $2k(\mathcal{B})$ vertices. Then there exist connected maps $\mathcal{M}^3_u$ with only colored unbroken edges such that:*

- $\mathcal{B} = \partial\mathcal{M}^3_u$
- $L_u(\mathcal{M}^3_u) = 0\,,\ \ \eta(\mathcal{M}^3_u) = 0$

hence which contribute to the first possible order in Eq. (10.5).

Proof. The construction we present here is somewhat reminiscent of the construction of the D-ary tree associated to the melon $\mathcal{B}$ in Chapter 4, but it is more general and leads to arbitrary trees.

Pick an edge of $\mathcal{B}$, say of color c, as root. We label the prime melonic subgraphs in the cut set of the root of $\mathcal{B}$ by 1, 2 up to q in the order they are encountered starting from the halfedge of the root incident at the black vertex. Remark that this labeling is flipped with respect to the one of Chapter 4. This flip is due to our conventions for drawing the boundary graph of a map.

We draw a ciliated vertex for every prime melonic subgraph in the cut set of the root. We connect these ciliated vertices into a branch of color c such that, when going around the branch, the cilia are encountered in the order 1 to q. Observe that, except for the constraint on the order we encounter the cilia, the branch can be chosen arbitrarily (it can even have vertices of coordination larger than two).

Consider now one of the prime melonic subgraphs and its corresponding ciliated vertex i. For each of its nontrivial melonic subgraphs we build the corresponding branch and attach it on the vertex i respecting the order of the cilia (considering the cilium on i as starting point of the contour walk of the branch).

Observe that, as all the vertices are ciliated, their degree is not constrained, hence this tree is a valid choice of a map with broken and unbroken colored edges. Furthermore, $\mathcal{M}^3$ has no broken edge, $\mathcal{M}^3 = \mathcal{M}^3_u$, and it has exactly $k(\mathcal{B})$ ciliated vertices. An example is presented in Figure 10.4.

□

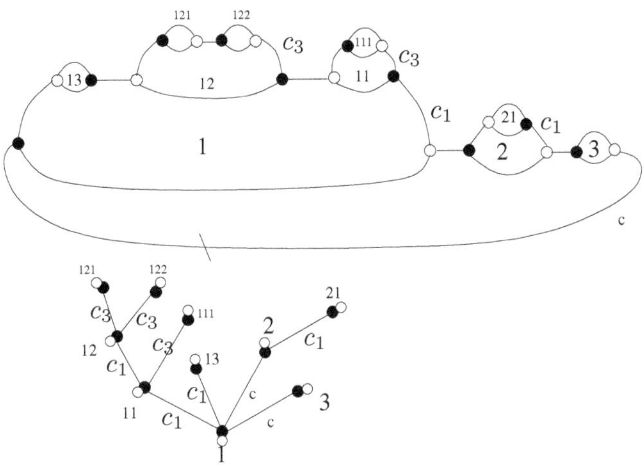

Figure 10.4: A tree whose boundary is a given melon.

Corollary 10.2. *If $\mathcal{B}$ is disconnected but each one of its connected components is a melonic graph then there exist maps $\mathcal{M}^3$ with:*

- $\partial \mathcal{M}^3_u = \mathcal{B}$,
- $L_b(\mathcal{M}^3) = 0$,
- $L_u(\mathcal{M}^3_u) = 0$ *and* $\eta(\mathcal{M}^3_u) = 0$.

Proof. An example can be built by choosing for each connected component of $\mathcal{B}$ a tree with ciliated vertices as in Lemma 10.2 and then joining these trees into a connected map by $C(\mathcal{B}) - 1$ broken edges.

□

One can also build other examples at the same order in $1/N$, by choosing for instance trees with some nonciliated trivalent internal vertices and cilia on the leaves.

Definition 10.2. *We call the* order *of a connected map having only unbroken colored edges the nonnegative integer:*

$$(D-2)L_u(\mathcal{M}^3_u) + \eta(\mathcal{M}^3_u) .$$

Let $\mathcal{B}$ be a connected edge colored graph. We call a connected map $\mathcal{M}^3_u$ having only unbroken edges such that $\partial\mathcal{M}^3_u = \mathcal{B}$ and having minimal order a minimal map *of $\mathcal{B}$. The order of the minimal maps of $\mathcal{B}$, denoted $\Omega(\mathcal{B})$, is called the order of $\mathcal{B}$. As, by Theorem 10.1, the number of maps at any fixed order is finite, the number of minimal maps of $\mathcal{B}$ is finite.*

In Lemma 10.2 we have identified a family of minimal maps for melonic boundary $\mathcal{B}$. In particular, we conclude that the order $\Omega(\mathcal{B})$ is zero in that case.

Lemma 10.3. *For any connected graph $\mathcal{B}$, the order of $\mathcal{B}$ obeys the bound:*

$$\Omega(\mathcal{B}) \le (D-1)(D-2)k(\mathcal{B}) + D - 1 \, .$$

Proof. We build a map whose boundary is $\mathcal{B}$ by the following algorithm. We associate to every pair $(v,\bar{v})$ of black and white vertices connected by an edge of color 1 in $\mathcal{B}$ a ciliated vertex. We then connect the two vertices bearing the cilia h and h' corresponding to the pairs $(v,\bar{v})$ and $(v',\bar{v}')$ by an edge of color c if the white vertex v is connected by an edge of color c in $\mathcal{B}$ to the black vertex $\bar{v}'$, such that the halfedge incident at h (resp. h') is the first successor (resp. last predecessor) of color c of the cilium. This map has $k(\mathcal{B})$ vertices, $(D-1)k(\mathcal{B})$ edges and no internal faces.

□

10.3 Double scaling

The double scaling limit is a refinement of the $N \to \infty$ limit, which allows one to probe lower orders in the $1/N$ series.

Let us take a closer look at a term in Eq. (10.5):

$$\begin{aligned} & K(\mathcal{B},\mu^{(4)}) \\ & = \sum_{\substack{\mathcal{M}^3 \text{ connected} \\ \partial\mathcal{M}^3_u = \mathcal{B}}} \frac{1}{n_{\text{int}}(\mathcal{M}^3)!} \frac{a_0^{2E_u(\mathcal{M}^3)+4E_b(\mathcal{M}^3)}}{(1-\imath\sqrt{\lambda}Da_0)^{k(\partial\mathcal{M}^3_u)}} \frac{[D(D-1)]^{E_b(\mathcal{M}^3)}}{(1-a_0^2)^{E_u(\mathcal{M}^3)+E_b(\mathcal{M}^3)}} \\ & \times \frac{1}{(1-Da_0^2)^{E_b(\mathcal{M}^3)}} N^{-DL_b(\mathcal{M}^3)-(D-2)\sum_{\rho=1}^{C(\mathcal{M}^3_u)} L_u(\mathcal{M}^3_{u;(\rho)}) - \sum_{\rho=1}^{C(\mathcal{M}^3_u)} \eta(\mathcal{M}^3_{u;(\rho)})} \, . \end{aligned}$$

where:

$$\begin{aligned} \eta(\mathcal{M}^3_{u;(\rho)}) &= 2 + E_u(\mathcal{M}^3_{u;(\rho)}) + (D-2)V(\mathcal{M}^3_{u;(\rho)}) \\ &\quad - (D-1)k(\partial\mathcal{M}^3_{u;(\rho)}) - C(\partial\mathcal{M}^3_{u;(\rho)}) - F_{\text{int}}(\mathcal{M}^3_{u;(\rho)}) \ge 0 \, , \\ L_b(\mathcal{M}^3) &= E_b(\mathcal{M}^3) - C(\mathcal{M}^3_u) + 1 \ge 0 \, , \\ L_u(\mathcal{M}^3_{u;(\rho)}) &= E_u(\mathcal{M}^3_{u;(\rho)}) - V(\mathcal{M}^3_{u;(\rho)}) + 1 \ge 0 \, . \end{aligned}$$

10.3.1 Enhancement at criticality

The crucial observation is that the term corresponding to the map $\mathcal{M}^3$ in the expansion of a cumulant $K(\mathcal{B},\mu^{(4)})$ comprises a factor:

$$\frac{1}{(1-Da_0^2)^{E_b(\mathcal{M}^3)}} \propto \frac{1}{(1+4D\lambda)^{\frac{E_b(\mathcal{M}^3)}{2}}} \, .$$

These factors diverge in the critical regime $\lambda \to \lambda_c = -\frac{1}{4D}$. The more broken edges $\mathcal{M}^3$ has (which automatically entails the more suppression in $1/N$), the more its contribution is boosted in the critical regime: the lower order terms in the $1/N$ series are enhanced at criticality.

This behavior allows one to build the *double scaling limit* of the model. In this limit one attempts to exactly balance the suppression in $1/N$ of a term with its enhancement at criticality such that in the double limit:

$$N \to \infty \ , \ \lambda \to \lambda_c \ , \ N^{\alpha}(\lambda - \lambda_c) = \text{constant} \ ,$$

(for some exponent α to be determined) we obtain:

- an infinite family of lower order terms in the $1/N$ series contributing at the same order for any cumulant.
- the cumulants are not arbitrarily divergent.

In the double scaling regime the coupling constant λ is traded for the new double scaling parameter x, the rescaled displacement, defined as:

$$\lambda = -\frac{1}{4D} + \frac{x}{4DN^{\alpha}} \Rightarrow a_0 = -\frac{1-\sqrt{\frac{x}{N^{\alpha}}}}{\sqrt{D}\sqrt{1-\frac{x}{N^{\alpha}}}} \ ,$$

and one must analyze the cumulants as functions of x. In the double scaling limit a term in Eq. (10.5) has a global scaling in N:

$$N^{\alpha \frac{E_b(\mathcal{M}^3)}{2} - DL_b(\mathcal{M}^3) - (D-2)\sum_{\rho=1}^{C(\mathcal{M}^3_u)} L_u(\mathcal{M}^3_{u;(\rho)}) - \sum_{\rho=1}^{C(\mathcal{M}^3_u)} \eta(\mathcal{M}^3_{u;(\rho)})} \ ,$$

that is the scaling of a term is boosted by a factor $N^{\alpha \frac{E_b(\mathcal{M}^3)}{2}}$. Our task is therefore to identify the (infinite) family of maps $\mathcal{M}^3$ with $\partial \mathcal{M}^3_u = \mathcal{B}$, which maximizes the number of broken edges $E_b(\mathcal{M}^3)$ at constant *a priori* scaling in $1/N$:

$$DL_b(\mathcal{M}^3) + (D-2) \sum_{\rho=1}^{C(\mathcal{M}^3_u)} L_u(\mathcal{M}^3_{u;(\rho)}) + \sum_{\rho=1}^{C(\mathcal{M}^3_u)} \eta(\mathcal{M}^3_{u;(\rho)}) \ ,$$

as such maps will dominate in the double scaling limit. Solving this combinatorial question will also fix uniquely the double scaling exponent α.

10.3.2 Maximal number of broken edges

Due to the combinatorial constraints on the map $\mathcal{M}^3$, one can derive an upper bound on the maximal number of broken edges of $\mathcal{M}^3$.

Definition 10.3. *We call a connected component $\mathcal{M}^3_{u;(\rho)}$ containing no cilia* trivial *if it is a tree. In this case both $L_u(\mathcal{M}^3_{u;(\rho)}) = 0$ and $\eta(\mathcal{M}^3_{u;(\rho)}) = 0$ (as the number of internal faces of a tree with no cilium is $1 + (D-1)V(\mathcal{M}^3_{u;(\rho)})$).*

Lemma 10.4. *Any connected component $\mathcal{M}^3_{u;(\rho)}$ is incident to* at least one *broken edge. A trivial component is incident to* at least three *broken edges.*

Proof. The first statement is trivial, as $\mathcal{M}^3$ is connected. Concerning the second statement, consider a connected component $\mathcal{M}^3_{u;(\rho)}$ with no cilium incident to exactly one or exactly two broken edges. As all the nonciliated vertices of $\mathcal{M}^3$ are at least three valent we have in both cases:

$$2E_u(\mathcal{M}^3_{u;(\rho)}) \geq 3V(\mathcal{M}^3_{u;(\rho)}) - 2 \Rightarrow$$
$$L_u(\mathcal{M}^3_{u;(\rho)}) = E_u(\mathcal{M}^3_{u;(\rho)}) - V(\mathcal{M}^3_{u;(\rho)}) + 1 \geq \frac{1}{2}V(\mathcal{M}^3_{u;(\rho)}) > 0 \,.$$

□

Let us denote $C_0(\mathcal{M}^3_u)$ the number of trivial connected components of $\mathcal{M}^3_u$ and $C_+(\mathcal{M}^3_u)$ the number of nontrivial ones. That is, $C_+(\mathcal{M}^3_u)$ counts the components containing cilia *irrespective* whether they are trees or not, together with the nonciliated components that *are not* trees. The previous lemma implies:

$$2E_b(\mathcal{M}^3) \geq C_+(\mathcal{M}^3_u) + 3C_0(\mathcal{M}^3_u) \,.$$

On the other hand:

$$E_b(\mathcal{M}^3) = L_b(\mathcal{M}^3) + C_+(\mathcal{M}^3_u) + C_0(\mathcal{M}^3_u) - 1 \,,$$

therefore:

$$2L_b(\mathcal{M}^3) + 2C_+(\mathcal{M}^3_u) + 2C_0(\mathcal{M}^3_u) - 2 \geq C_+(\mathcal{M}^3_u) + 3C_0(\mathcal{M}^3_u) \,,$$

and we obtain the following lemma.

Lemma 10.5. *The number of trivial connected components of $\mathcal{M}^3$ is bounded by the excess of broken edges and the number of nontrivial connected components as:*

$$C_0(\mathcal{M}^3_u) \leq 2L_b(\mathcal{M}^3) + C_+(\mathcal{M}^3_u) - 2 \,.$$

The total number of broken edges of $\mathcal{M}^3$ is bounded by

$$E_b(\mathcal{M}^3) \leq 3L_b(\mathcal{M}^3) + 2C_+(\mathcal{M}^3_u) - 3 \,.$$

10.3.3 Cumulants

We will treat separately the case where all the connected components of $\mathcal{B}$ are melonic graphs from the general case. We do this as, while the results are the same in the two cases, the order and minimal maps are more explicit in the first case.

Definition 10.4. *Let $\mathcal{B} = \bigcup_{\mu=1}^{C(\mathcal{B})} \mathcal{B}_{(\mu)}$ and each $\mathcal{B}_{(\mu)}$ is a connected melonic graph. We call a map $\mathcal{M}^3$* a candidate double scaling map *for $\mathcal{B}$ if:*

- *for any μ there exists exactly one ciliated component $\mathcal{M}^3_{u;(\rho)}$ such that $\partial\mathcal{M}^3_{u;(\rho)} = \mathcal{B}_{(\mu)}$. This component furthermore satisfies:*

$$L_u(\mathcal{M}^3_{u;(\rho)}) = 0 \,, \quad \eta(\mathcal{M}^3_{u;(\rho)}) = 0 \,.$$

Observe that by Lemma 10.2 such components always exist.

- *the nontrivial nonciliated components of $\mathcal{M}_u^3$ consist of a tree decorated by a loop edge.*
- *the trivial connected components of $\mathcal{M}_u^3$ are incident to exactly three broken edges and the nontrivial ones to exactly one broken edge.*

For general $\mathcal{B}$ one generalizes this definition as follows.

Definition 10.5. *Let $\mathcal{B} = \bigcup_{\mu=1}^{C(\mathcal{B})} \mathcal{B}_{(\mu)}$ with $\mathcal{B}_{(\mu)}$ connected. We call a map $\mathcal{M}^3$ a* candidate double scaling map *for $\mathcal{B}$ if:*

- *for any μ there exists exactly one ciliated component $\mathcal{M}^3_{u;(\rho)}$ such that $\partial\mathcal{M}^3_{u;(\rho)} = \mathcal{B}_{(\mu)}$ and $\mathcal{M}^3_{u;(\rho)}$ is a minimal map for $\mathcal{B}_{(\mu)}$ in the sense of Definition 10.2.*
- *the nontrivial nonciliated components of $\mathcal{M}_u^3$ consist of a tree decorated by a loop edge.*
- *the trivial connected components of $\mathcal{M}_u^3$ are incident to exactly three broken edges and the nontrivial ones to exactly one broken edge.*

Observe that candidate double scaling maps exist for any $\mathcal{B}$. They satisfy the following properties.

Theorem 10.2. *Let $\mathcal{M}^3$ be a candidate double scaling map for $\mathcal{B}$. Then:*

i) *$\mathcal{M}^3$ contributes to the cumulant $K(\mathcal{B},\mu^{(4)})$, that is $\partial\mathcal{M}^3_u = \mathcal{B}$.*

ii) *each connect component $\mathcal{M}^3_{u;(\rho)}$ either has no cilium or has a connected boundary:*

$$\forall \rho: \quad C(\partial\mathcal{M}^3_{u;(\rho)}) = 0 \text{ or } 1 \,.$$

iii) *the nontrivial nonciliated components of $\mathcal{M}^3_u$ consist of exactly one vertex decorated by an unbroken edge forming a self loop (a tadpole in physics literature) and connected to the rest of $\mathcal{M}^3$ by a broken edge.*

iv) *for any connected component $\mathcal{B}_{(\mu)}$ of $\mathcal{B}$, the number of possible nontrivial ciliated components such that $\partial\mathcal{M}^3_{u;(\rho)} = \mathcal{B}_{(\mu)}$ is finite.*

v) *the trivial components of $\mathcal{M}^3_u$ are three valent vertices connected to the rest of $\mathcal{M}^3$ by three broken edges.*

vi) *the scaling in $1/N$ of $\mathcal{M}^3$ is:*

$$DL_b(\mathcal{M}^3) + (D-2)\big[C_+(\mathcal{M}^3_u) - C(\mathcal{B})\big] + \sum_{\mu=1}^{C(\mathcal{B})} \Omega(\mathcal{B}_{(\mu)}) \,.$$

vii) *$\mathcal{M}^3$ has exactly:*

$$E_b(\mathcal{M}^3) = 3L_b(\mathcal{M}^3) + 2C_+(\mathcal{M}^3_u) - 3 \,,$$

broken edges.

viii) *for any $\mathcal{B}$ and any two integers $c_+ \geq C(\mathcal{B})$ and $l_b \geq 0$ one can always build candidate double scaling maps $\mathcal{M}^3$ having exactly $C_+(\mathcal{M}^3_u) = c_+$ nontrivial components and exactly $L_b(\mathcal{M}^3) = l_b$ excess of broken edges.*

Proof. Items i) and ii) are trivial.

For item iii), we observe that the nontrivial components of $\mathcal{M}^3_u$ having no cilia are trees decorated by a loop edge, that is such that $L_u(\mathcal{M}^3_{u;(\rho)}) = 1$. As on the other hand they are incident to exactly one broken edge and all the vertices have coordination at least three, we have $2E_u(\mathcal{M}^3_{u;(\rho)}) \geq 3V(\mathcal{M}^3_{u;(\rho)}) - 1$ which implies that $V(\mathcal{M}^3_{u;(\rho)}) = 1$.

Item iv) is a consequence of the fact that the order of $\mathcal{B}_{(\mu)}$ is fixed and from Theorem 10.1 that there are only a finite number of maps at any order.

It is instructive to give an alternative proof in the case where $\mathcal{B}_{(\mu)}$ is a melon. The nontrivial ciliated components $\mathcal{M}^3_{u;(\rho)}$ are incident to only one broken edge, hence all of their nonciliated vertices are three valent except at most one that is two valent. The $k(\mathcal{B}_{(\mu)})$ ciliated vertices of $\mathcal{M}^3_{u;(\rho)}$ are at least one valent. Furthermore, by Definition 10.4, $\mathcal{M}^3_{u;(\rho)}$ is a tree, thus:

$$
\begin{aligned}
2E_u(\mathcal{M}^3_{u;(\rho)}) &\geq k(\mathcal{B}_{(\mu)}) + 2 + 3\big[V(\mathcal{M}^3_{u;(\rho)}) - 1 - k(\mathcal{B}_{(\mu)})\big] \\
&= 3V(\mathcal{M}^3_{u;(\rho)}) - 2k(\mathcal{B}_{(\mu)}) - 1 , \\
E_u(\mathcal{M}^3_{u;(\rho)}) &= V(\mathcal{M}^3_{u;(\rho)}) - 1 ,
\end{aligned}
$$

hence $V(\mathcal{M}^3_{u;(\rho)}) \leq 2k(\mathcal{B}_{(\mu)}) - 1$ and there are only a finite number of trees with a finite number of vertices.

Concerning item v), we note that the trivial components are trees having only three valent vertices and incident to exactly three external edges. Thus:

$$
3V(\mathcal{M}^3_{u;(\rho)}) - 3 = 2E_u(\mathcal{M}^3_{u;(\rho)}) = 2\big[V(\mathcal{M}^3_{u;(\rho)}) - 1\big] \Rightarrow V(\mathcal{M}^3_{u;(\rho)}) = 1 .
$$

Items vi) and vii) are immediate consequences of the previous items and of Definition 10.4 and the Definition 10.5.

Finally, for item viii), one first builds a binary tree such that:

- the leaves of the tree are the nontrivial components. The $C(\mathcal{B})$ ciliated ones are chosen to be minimal maps for each connected component of $\mathcal{B}$, while the $c_+ - C(\mathcal{B})$ nonciliated ones are chosen as in item iii).
- the internal three valent vertices of the tree are trivial components (that is just simple nonciliated trivalent vertices).
- the edges of the tree are broken edges.

Subsequently one adds l_b loop edges by adding two intermediate vertices on two broken edges and connecting them by a new broken edge.

□

It follows that all the candidate double scaling maps have the structure presented in Figure 10.5.

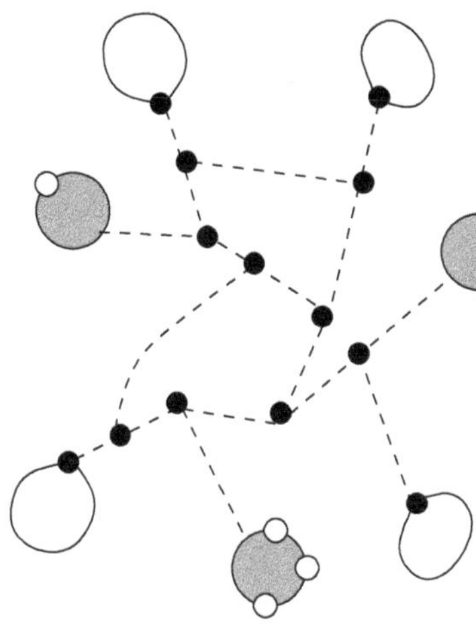

Figure 10.5: Candidate double scaling maps. We represented in gray the ciliated components of $\mathcal{M}^3_u$.

Melonic cumulants

We will first consider the case where all the connected components of $\mathcal{B}$ are melonic graphs.

Lemma 10.6. *Let $\mathcal{B} = \cup_{\mu=1}^{C(\mathcal{B})} \mathcal{B}_{(\mu)}$ with $\mathcal{B}_{(\mu)}$ connected melonic graphs. If a map $\mathcal{M}^3$ with $\partial\mathcal{M}^3_u = \mathcal{B}$ has a priori scaling in $1/N$:*

$$DL_b(\mathcal{M}^3) + (D-2)\big[C_+(\mathcal{M}^3_u) - C(\mathcal{B})\big]$$

and exactly:

$$E_b(\mathcal{M}^3) = 3L_b(\mathcal{M}^3) + 2C_+(\mathcal{M}^3_u) - 3$$

broken edges then it is a candidate double scaling map for $\mathcal{B}$.

Proof. The equality concerning the number of broken edges implies that all the nontrivial components of $\mathcal{M}^3_u$ are incident to one unbroken edge and all the trivial ones to three broken edges.

Observe that $C_+(\mathcal{M}^3_u) \geq C(\mathcal{B})$, otherwise the *a priori* order of the map would be strictly smaller that $DL_b(\mathcal{M}^3)$, which is impossible. The map $\mathcal{M}^3_u$ has at least $C_+(\mathcal{M}^3_u) - C(\mathcal{B})$ nonciliated nontrivial components. The *a priori* scaling in $1/N$ implies that the map has exactly:

- $C_+(\mathcal{M}^3_u) - C(\mathcal{B})$ nontrivial components with no cilia that have exactly one loop edge (as they are nontrivial they have at least one loop edge, and as they contribute only $D-2$ to the scaling they must have exactly one loop edge).

- $C(\mathcal{B})$ nontrivial components with cilia that are minimal maps for the connect components $\mathcal{B}_{(\mu)}$, that is such that:

$$
\begin{aligned}
&\partial \mathcal{M}^3_{u;(\rho)} = \mathcal{B}_{(\mu)} \\
&L_u(\mathcal{M}^3_{u;(\rho)}) = 0\,, \quad \eta(\mathcal{M}^3_{u;(\rho)}) = 0\,.
\end{aligned}
$$

□

The importance of the candidate double scaling maps comes from the following result.

Theorem 10.3. *All connected maps $\mathcal{M}^3$ with $\partial\mathcal{M}^3_u = \mathcal{B}$ that are not candidate double scaling maps are suppressed in a double scaling limit.*

Proof. Consider a map $\mathcal{M}^3$ with $\partial\mathcal{M}^3_u = \mathcal{B}$. We denote the number of connect components of $\mathcal{M}^3_u$ containing cilia by $c_+^{\text{cil}} \leq C(\mathcal{B})$, the number of nontrivial components of $\mathcal{M}^3_u$ not containing cilia by $c_+^{\text{n.cil}}$ and the total excess of broken edges of $\mathcal{M}^3$ by $l_b \equiv L_b(\mathcal{M}^3)$. The order in $1/N$ and the number of broken edges of $\mathcal{M}^3$ are bounded by:

$$
\begin{aligned}
DL_b(\mathcal{M}^3) + (D-2)\sum_{\rho=1}^{C(\mathcal{M}^3_u)} L_u(\mathcal{M}^3_{u;(\rho)}) + \sum_{\rho=1}^{C(\mathcal{M}^3_u)} \eta(\mathcal{M}^3_{u;(\rho)}) \\
\geq Dl_b + (D-2)c_+^{\text{n.cil}} \\
E_b(\mathcal{M}^3) \leq 3l_b + 2\big[c_+^{\text{cil}} + c_+^{\text{n.cil}}\big] - 3\,.
\end{aligned}
$$

We compare $\mathcal{M}^3$ with a candidate double scaling map having the same number $c_+^{\text{n.cil}}$ of nonciliated nontrivial connected components, but having $C(\mathcal{B})$ ciliated non trivial connected components and l_b excess of broken edges. This map has an *a priori* order in $1/N$:

$$
Dl_b + (D-2)c_+^{\text{n.cil}}\,,
$$

and exactly:

$$
3l_b + 2\big[C(\mathcal{B}) + c_+^{\text{n.cil}}\big] - 3\,,
$$

broken edges. Thus for any map one can always find a candidate double scaling map which is either *a priori* less suppressed in $1/N$ or has more broken edges.

□

Arbitrary cumulants

One would like to be able to generalize these results for arbitrary $\mathcal{B}$. However, a subtlety arises. This subtlety comes from the fact that in general the same connected component of $\mathcal{M}^3_u$ might contribute several connected components of the boundary graph. In order for the previous reasoning to go through one needs to bound the *a priori* $1/N$ order of such a connected component by the sum of the orders of the components of its boundary.

Conjecture 10.1. *Let $\mathcal{M}_u^3$ be a connect map with unbroken edges and with a disconnected boundary $\partial\mathcal{M}_u^3 = \sum_{\mu=1}^{C(\partial\mathcal{M}_u^3)} B_{(\mu)}$. Then:*

$$(D-2)L_u(\mathcal{M}_u^3) + \eta(\mathcal{M}_u^3) \geq \sum_{\mu=1}^{C(\partial\mathcal{M}_u^3)} \Omega(\mathcal{B}_{(\mu)}) .$$

Of course this conjecture holds if all the $\mathcal{B}_{(\mu)}$ are melonic graphs (the right hand side of the inequality is zero). However, we do not have proof of this conjecture for the general case.

Assuming that this conjecture also holds for general graphs $\mathcal{B}$, we can adapt with minimal modification the previous section and show that the candidate double scaling maps dominate all the other maps in a double scaling limit for arbitrary $\mathcal{B}$. First, Lemma 10.6 would be replaced by the following, more general, result.

Lemma 10.7. *If a map $\mathcal{M}^3$ with $\partial\mathcal{M}_u^3 = \mathcal{B}$ has a priori order in $1/N$:*

$$DL_b(\mathcal{M}^3) + (D-2)\big[C_+(\mathcal{M}_u^3) - C(\mathcal{B})\big] + \sum_{\mu=1}^{C(\mathcal{B})} \Omega(\mathcal{B}_{(\mu)})$$

and exactly:

$$E_b(\mathcal{M}^3) = 3L_b(\mathcal{M}^3) + 2C_+(\mathcal{M}_u^3) - 3$$

broken edges then it is a candidate double scaling map for $\mathcal{B}$.

The Theorem 10.3 would then generalize for an arbitrary map as, if the conjecture holds, we have:

$$\begin{aligned} DL_b(\mathcal{M}^3) + (D-2)\sum_{\rho=1}^{C(\mathcal{M}_u^3)} L_u(\mathcal{M}_{u;(\rho)}^3) + \sum_{\rho=1}^{C(\mathcal{M}_u^3)} \eta(\mathcal{M}_{u;(\rho)}^3) \\ \geq Dl_b + (D-2)c_+^{\text{n.cil}} + \sum_{\mu=1}^{C(\partial\mathcal{M}_u^3)} \Omega(\mathcal{B}_{(\mu)}) \end{aligned}$$

$$E_b(\mathcal{M}^3) \leq 3l_b + 2\big[c_+^{\text{cil}} + c_+^{\text{n.cil}}\big] - 3 ,$$

while for a candidate map with the same number of nonciliated nontrivial connected components, but having $C(\mathcal{B})$ ciliated nontrivial connected components and l_b excess of broken edges, we would get exactly an *a priori* order in $1/N$:

$$Dl_b + (D-2)c_+^{\text{n.cil}} + \sum_{\mu=1}^{C(\partial\mathcal{M}_u^3)} \Omega(\mathcal{B}_{(\mu)})$$

and exactly:

$$3l_b + 2\big[C(\mathcal{B}) + c_+^{\text{n.cil}}\big] - 3$$

broken edges.

However, even without proving the conjecture, one thing is for certain. For any $\mathcal{B}$, any map $\mathcal{M}^3$ with $\partial\mathcal{M}_u^3 = \mathcal{B}$ is suppressed in double scaling limit with respect to the maps corresponding to a melonic observable $\mathcal{B}'$ having $k(\mathcal{B})$ white vertices and $C(\mathcal{B})$ connected components. We are thus ensured that in the double scaling limit the nonmelonic cumulants are suppressed with respect to the melonic ones, hence in particular they cannot be arbitrarily divergent.

10.3.4 Critical dimension

In order to compute a cumulant $K(\mathcal{B}, \mu^{(4)})$ in the double scaling regime, we recall that in this regime the scaling with N of a candidate double scaling map $\mathcal{M}^3$, $\partial\mathcal{M}^3_u = \mathcal{B}$ is enhanced to:

$$\frac{\alpha}{2}\Big[3L_b(\mathcal{M}^3) + 2C_+(\mathcal{M}^3_u) - 3\Big] \\ - DL_b(\mathcal{M}^3) - (D-2)\big[C_+(\mathcal{M}^3_u) - C(\mathcal{B})\big] - \sum_{\mu=1}^{C(\partial\mathcal{M}^3_u)} \Omega(\mathcal{B}_{(\mu)}) \,,$$

where we have assumed the Conjecture 10.1. As already mentioned, even if the conjecture does not hold in general it holds for melonic $\mathcal{B}_{(\mu)}$, and all the other cumulants are strictly suppressed with respect to the melonic ones.

Our task is to identify the optimal candidate double scaling maps, that is the candidate double scaling maps having the largest number of broken edges for fixed *a priori* order. This equates to solving the following optimization problem:

$$\left\{ \max\left(3l_b + 2c_+^{\text{n.cil}}\right) \;\middle|\; Dl_b + (D-2)c_+^{\text{n.cil}} = \delta \,, \quad \delta \text{ constant} \right\} ,$$

or equivalently:

$$c_+^{\text{n.cil}} = \frac{1}{D-2}\delta - \frac{D}{D-2}l_b \,, \qquad \max_{\substack{l_b \\ l_b \le \frac{\delta}{D}}} \left(\frac{2}{D-2}\delta + \frac{D-6}{D-2}l_b\right) .$$

As a function of D, we obtain three very different double scaling behaviors:

- $D > 6$. The optimal double scaling maps have $l_b = \frac{\delta}{D}$, $c_+^{\text{n.cil}} = 0$. The appropriate scaling exponent (leading to an infinite family of maps at the same order) is $\alpha = \frac{2}{3}D$. The optimal maps are arbitrary three valent maps made of broken edges connecting the $C(\mathcal{B})$ ciliated components among them.

- $D = 6$. The optimal double scaling maps have *any* l_b with $0 \le l_b \le D\delta$. The scaling exponent is $\alpha = 4$ and the optimal maps include as a subset the arbitrary three valent maps of the previous case.

- $D < 6$. The optimal double scaling maps have $l_b = 0$, $c_+^{\text{n.cil}} = \frac{1}{D-2}\delta$. The scaling exponent is $\alpha = D-2$. The optimal maps are *binary trees* made of broken edges whose leaves consist of the nontrivial connected components of $\mathcal{M}^3_u$ (both the ciliated and the nonciliated ones).

The dimension $D = 6$ is a critical dimension for the double scaling limit. Above the critical dimension ($D \geq 6$) the optimal double scaling maps are not a summable family of maps. Whether in this case the cumulants are Borel summable is not clear. The double scaling limit behaves very differently below the critical dimension. Indeed in $D < 6$ the family of optimal maps is summable and has a finite radius of convergence.

We will now compute analytically the two and four point cumulants in the double scaling limit below the critical dimension.

10.3.5 Explicit computations

In order to compute explicitly the double scaled cumulant for some boundary $\mathcal{B} = \bigcup_{\mu=1}^{C(\mathcal{B})} \mathcal{B}_{(\mu)}$ one proceeds in three steps:

- for every $\mathcal{B}_{(\mu)}$, identify the minimal maps $\mathcal{M}_{u;(\rho)}$ with $\partial\mathcal{M}_{u;(\rho)} = \mathcal{B}_{(\mu)}$.
- build the optimal double scaling maps. They are binary trees with $C(\mathcal{B})$ leaves consisting of the minimal maps. Observe that the optimal maps are slightly nontrivial, as one needs to take into account all the possible ways to hook a minimal map to an unbroken edge.
- sum the resulting series.

In the explicit examples we present here we will be dealing repeatedly with melonic boundary graphs $\mathcal{B}$. Let us recall (see the proof of item iv in Theorem 10.2) the following fact. Let $\mathcal{B}_{(\mu)}$ be a melon. The nontrivial ciliated components $\mathcal{M}^3_{u;(\rho)}$ are incident to only one broken edge, hence all of their nonciliated vertices are three valent except at most one, which is two valent. The $k(\mathcal{B}_{(\mu)})$ ciliated vertices of $\mathcal{M}^3_{u;(\rho)}$ are at least one valent and furthermore $\mathcal{M}^3_{u;(\rho)}$ is a tree, thus:

$$\begin{aligned} 2E_u(\mathcal{M}^3_{u;(\rho)}) \geq & k(\mathcal{B}_{(\mu)}) + 2 + 3\big[V(\mathcal{M}^3_{u;(\rho)}) - 1 - k(\mathcal{B}_{(\mu)})\big] \\ & = 3V(\mathcal{M}^3_{u;(\rho)}) - 2k(\mathcal{B}_{(\mu)}) - 1 \ , \\ E_u(\mathcal{M}^3_{u;(\rho)}) = & V(\mathcal{M}^3_{u;(\rho)}) - 1 \ , \end{aligned}$$

hence:

$$V(\mathcal{M}^3_{u;(\rho)}) \leq 2k(\mathcal{B}_{(\mu)}) - 1 \ . \tag{10.6}$$

Two point cumulant $K(\mathcal{B}^{(2)}, \mu^{(4)})$ for $2 < D < 6$

The simplest cumulant one can study analytically in the double scaling limit is the two point cumulant $K(\mathcal{B}^{(2)}, \mu^{(4)})$. Let us apply the recipe in this case.

Minimal maps. We start by identifying the minimal maps for $\mathcal{B}^{(2)}$. Observe that such maps have exactly one cilium, and from Eq. (10.6) they have only one vertex (which is necessarily ciliated).

Optimal double scaling maps. Considering the ciliated vertex as root, the optimal double scaling maps are rooted binary trees whose leaves are the nonciliated nontrivial components of $\mathcal{M}_u^{(3)}$ which, from Theorem 10.2, are formed by an unbroken edge forming a self loop. A typical optimal map is presented in Figure 10.6.

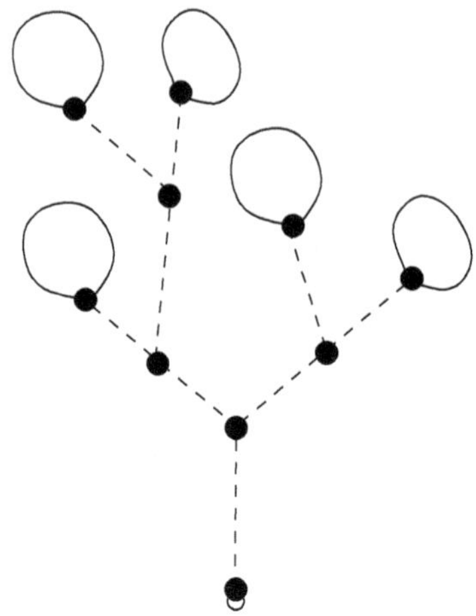

Figure 10.6: Optimal maps for $\mathcal{B}^{(2)}$ in $D < 6$.

Let us compute first as an intermediary the generating function $T(a_0)$ (and its double scaled leading order $T(x)_{\rm DS}$) of trees having an univalent vertex and at least one edge represented in Figure 10.7. There are $\frac{1}{n+1}\binom{2n}{n}$ rooted binary trees with $n+1$ leaves

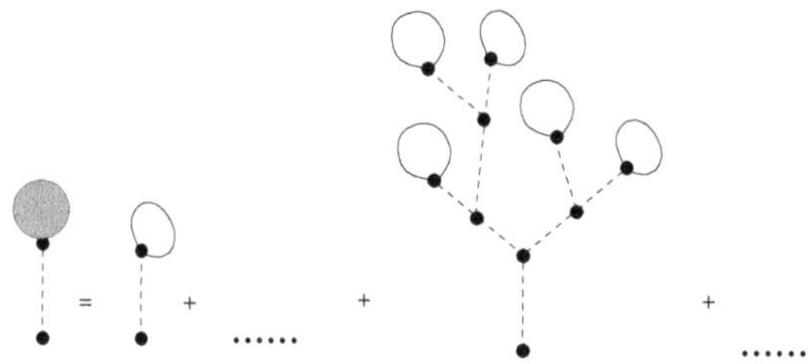

Figure 10.7: The generating function $T(a_0)$.

(nontrivial nonciliated components). The corresponding maps have $n+1$ unbroken edges (which can have D distinct colors) and $2n+1$ broken edges:

$$T(a_0) = \sum_{n\geq 0} \frac{1}{n+1}\binom{2n}{n} D^{n+1}[D(D-1)]^{2n+1} \frac{a_0^{10n+6}}{(1-a_0^2)^{3n+2}} \\ \times \frac{1}{(1-Da_0^2)^{2n+1}} \frac{1}{N^{(D-2)(n+1)}}$$

$$= \frac{D^2(D-1)a_0^6}{(1-a_0^2)^2(1-Da_0^2)N^{D-2}} \sum_{n\geq 0} \frac{1}{n+1}\binom{2n}{n}\left(\frac{D^3(D-1)^2a_0^{10}}{(1-a_0^2)^3(1-Da_0^2)^2N^{D-2}}\right)^n .$$

We perform a truncation at the first nontrivial order in the double scaling regime $\lambda = -\frac{1}{4D} + \frac{x}{4DN^{D-2}}$, $a_0 = -\frac{1-\sqrt{\frac{x}{N^{D-2}}}}{\sqrt{D}\sqrt{1-\frac{x}{N^{D-2}}}}$:

$$a_0^2 \approx \frac{1}{D}, \qquad 1 - Da_0^2 = 1 - \frac{\left(1-\sqrt{\frac{x}{N^{D-2}}}\right)^2}{1-\frac{x}{N^{D-2}}} \approx 2\sqrt{\frac{x}{N^{D-2}}} ,$$

hence the leading double scaling behavior of $T(a_0)$ is:

$$T(x)_{\text{DS}} \approx \frac{1}{2}\frac{D}{(D-1)\sqrt{x}N^{\frac{D-2}{2}}} \sum_{n\geq 0}\frac{1}{n+1}\binom{2n}{n}\left(\frac{D}{4(D-1)x}\right)^n$$
$$= N^{-\frac{D-2}{2}}\left(\sqrt{x}-\sqrt{x-\frac{D}{D-1}}\right) .$$

The contribution of the optimal maps to the two point cumulant is:

$$K(\mathcal{B}^{(2)},\mu^{(4)})_{\text{optimal}} = \frac{1}{(1-\imath\sqrt{\lambda}Da_0)} + \frac{1}{(1-\imath\sqrt{\lambda}Da_0)}T(a_0) ,$$

where we took into account that the map consisting of a unique ciliated vertex must be added separately, and one needs to multiply the contribution of $T(a_0)$ by a factor for the cilium. In the double scaling regime we need to evaluate the contribution of the cilium:

$$\frac{1}{(1-\imath\sqrt{\lambda}Da_0)} = \frac{a_0}{\imath\sqrt{\lambda}} = \frac{1-\sqrt{1+4D\lambda}}{-2\lambda D} \approx 2\left(1-\sqrt{\frac{x}{N^{D-2}}}\right) ,$$

hence the leading double scaling behavior of the two point cumulant is:

$$K(\mathcal{B}^{(2)},\mu^{(4)})_{\text{DS}} = 2 - 2N^{-\frac{D-2}{2}}\sqrt{x} + 2T(x)_{\text{DS}} = 2 - 2N^{-\frac{D-2}{2}}\sqrt{x-\frac{D}{D-1}} .$$

Remark 10.4. Two remarks are in order.

Domain of convergence. The family of optimal maps is summable for $x > \frac{D}{D-1}$, that is the series is summable provided that one is not too close to the critical point. When approaching the critical point the series looses summability. In this regime one can hope to identify a further scaling regime, a *triple scaling*:

$$\lambda = -\frac{1}{4D} + \frac{1}{4(D-1)N^{D-2}} + \frac{y}{N^{\beta}} , \qquad \beta > D-2 ,$$

in which a family of graphs larger than the candidate double scaling maps contributes.

D=2 . Unsurprisingly, the case $D = 2$ is special: in this case the trivalent maps are not suppressed in $1/N$ and before attempting to perform the double scaling limit one needs to rework the leading order contribution.

Four point cumulants for $2 < D < 6$

There are two melonic graphs with four vertices. One consists of two copies of $\mathcal{B}^{(2)}$, and we denoted it $\mathcal{B}^{(2)} \cup \mathcal{B}^{(2)}$, and the second one is one of the $\mathcal{B}^{(4)}_{\{c\}}$ for some color c.

The cumulant $K(\mathcal{B}^{(2)} \cup \mathcal{B}^{(2)}, \mu^{(4)})$

As we have already seen, the only minimal map of $\mathcal{B}^{(2)}$ consists of a ciliated vertex. The optimal double scaling maps for $\mathcal{B}^{(2)} \cup \mathcal{B}^{(2)}$ are binary trees having two univalent ciliated vertices, such that all the other leaves are the nontrivial nonciliated components of $\mathcal{M}^{(3)}_u$, as in Figure 10.8.

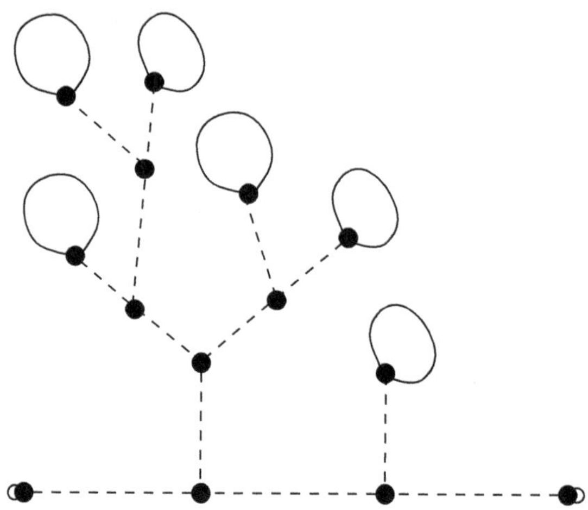

Figure 10.8: Optimal maps for $\mathcal{B}^{(2)} \cup \mathcal{B}^{(2)}$ in $D < 6$.

In order to compute the number of optimal maps, let us denote $p \geq 0$ the number of intermediate vertices on the path between the two ciliated vertices. Any of the intermediate vertices is a root for a binary tree. Taking into account that each binary tree can be attached on either of the two corners of the intermediate vertices, and taking into account the contribution of the broken edges in the path connecting the two ciliated vertices we obtain:

$$K(\mathcal{B}^{(2)} \cup \mathcal{B}^{(2)}, \mu^{(4)})_{\text{optimal}}$$
$$= \frac{1}{(1 - \imath\sqrt{\lambda}Da_0)^2} \sum_{p\geq 0} \frac{a_0^{4(p+1)}[D(D-1)]^{p+1}}{(1-a_0^2)^{p+1}(1-Da_0^2)^{p+1}} [2T(a_0)]^p ,$$

which becomes, at leading order in double scaling:

$$K(\mathcal{B}^{(2)} \cup \mathcal{B}^{(2)}, \mu^{(4)})_{\text{DS}} = 2\frac{N^{\frac{D-2}{2}}}{\sqrt{x}} \sum_{p\geq 0} \left[\frac{N^{\frac{D-2}{2}}}{\sqrt{x}} T(x)_{\text{DS}} \right]^p$$

$$= 2\frac{N^{\frac{D-2}{2}}}{\sqrt{x}} \sum_{p\geq 0} \left[1 - \sqrt{1 - \frac{D}{x(D-1)}}\right]^p = \frac{2N^{\frac{D-2}{2}}}{\sqrt{x - \frac{D}{D-1}}} \ .$$

Observe that this cumulant is *enhanced* in double scaling by the global factor $N^{\frac{D-2}{2}}$. This should come as no surprise, as the term of *a priori* order N^0 contributing to $K(\mathcal{B}^{(2)} \cup \mathcal{B}^{(2)}, \mu^{(4)})$ consists of a broken edge joining the two ciliated vertices, and in double scaling is boosted by a factor $N^{\frac{D-2}{2}}$.

The cumulants $K(\mathcal{B}^{(4)}_{\{c\}}, \mu^{(4)})$

The minimal maps for $\mathcal{B}^{(4)}_{\{c\}}$ are trees having, due to Eq. (10.6), at most three vertices. A rapid inspection reveals that the optimal maps must fall in one of the two categories represented in Figure 10.9. We thus obtain:

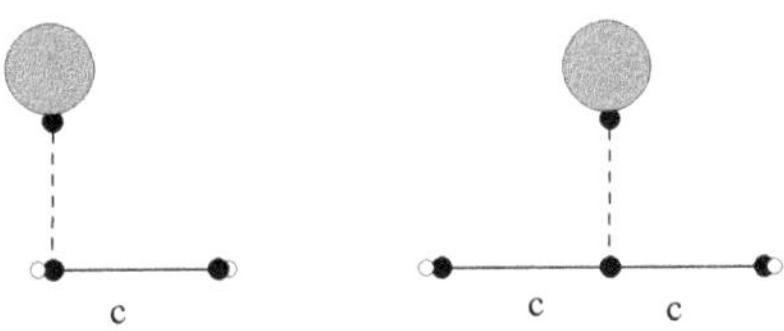

Figure 10.9: Optimal maps for $\mathcal{B}^{(4)}_{\{c\}}$.

$$K(\mathcal{B}^{(4)}_{\{c\}}, \mu^{(4)})_{\text{optimal}} = \frac{1}{(1 - \imath\sqrt{\lambda}Da_0)^2} \left[4\frac{a_0^2}{1-a_0^2}T(a_0) + 2\frac{a_0^4}{(1-a_0^2)^2}T(a_0)\right] ,$$

hence the double scaling value:

$$K(\mathcal{B}^{(4)}_{\{c\}}, \mu^{(4)})_{\text{DS}} = 4\frac{4D+1}{(D-1)^2} N^{-\frac{D-2}{2}} \left(\sqrt{x} - \sqrt{x - \frac{D}{D-1}}\right) .$$

The cumulants $K(\mathcal{B}^{(4)}_{\{c_1,c_2\}}, \mu^{(4)})$

As $D < 6$, nonmelonic four point cumulants $\mathcal{B}^{(4)}_{\{c_1,c_2\}}$ exist only for $D = 4, 5$. Observe that, as $\mathcal{B}^{(4)}_{\{c_1,c_2\}}$ is connected, the optimal double scaling maps dominate in the double scaling limit.

As the boundary of a tree is a union of melonic connected components, it follows that a minimal map for $\mathcal{B}^{(4)}_{\{c_1,c_2\}}$ must have at least one loop edge. It is possible to obtain $\mathcal{B}^{(4)}_{\{c_1,c_2\}}$ as the boundary of a map with unbroken edges and exactly one loop edge, as shown in Figure 10.10. Let us call this map $\mathcal{M}^3_{u;\{c_1,c_2\}}$. We have:

$$L_u(\mathcal{M}^3_{u;\{c_1,c_2\}}) = 1 \ , \quad \eta(\mathcal{M}^3_{u;\{c_1,c_2\}}) = 1 \ .$$

Figure 10.10: Minimal maps for $\mathcal{B}^{(4)}_{\{c_1,c_2\}}$.

Observe that in any minimal map of $\mathcal{B}^{(4)}_{\{c_1,c_2\}}$, each of the ciliated vertices is incident to at least two edges (otherwise the boundary graph of the map would have two vertices connected by D edges). Also, the minimal map has at most one two valent internal vertex. Hence, for a minimal map of $\mathcal{B}^{(4)}_{\{c_1,c_2\}}$ we have at the same time:

$$E_u(\mathcal{M}^3_{u;(\rho)}) \geq V(\mathcal{M}^3_{u;(\rho)}) \ , \ \ 2E_u(\mathcal{M}^3_{u;(\rho)}) \geq 3V(\mathcal{M}^3_{u;(\rho)}) - 3 \ ,$$

hence the map has either only the two ciliated vertices, or it has three vertices, but the third vertex is two valent. Maps with at least two loop edges arise at least at the order $2D - 4 > D - 1$, hence are suppressed. It follows that all the optimal maps for $\mathcal{B}^{(4)}_{\{c_1,c_2\}}$ consist of the map represented in Figure 10.10 decorated by $T(a_0)$ either on a corner of a vertex or on an edge and:

$$K(\mathcal{B}^{(4)}_{\{c_1,c_2\}}, \mu^{(4)})_{\text{DS}} \sim N^{-(D-1)-\frac{D-2}{2}} \left(\sqrt{x} - \sqrt{x - \frac{D}{D-1}} \right) \, .$$

Observe that, as expected, $K(\mathcal{B}^{(4)}_{\{c_1,c_2\}}, \mu^{(4)})$ is strictly suppressed in scaling with respect to $K(\mathcal{B}^{(4)}_{\{c\}}, \mu^{(4)})$.

Chapter 11

Symmetry breaking

Perhaps the most striking conclusion of Chapter 10 is that, at least for the quartic melonic model, the critical point at which the continuum limit is reached corresponds to a genuine phase transition accompanied by a symmetry breaking. What prevented us from performing a complete study of this phase transition is the fact that we were not able to identify an appropriate vacuum in the broken symmetry phase.

In this chapter we perform this complete study for the simplified melonic model possessing a unique interaction term. In the intermediate field representation this model becomes just a matrix model with an unusual scaling, and all the vacua can be obtained analytically [23]. We show here that for the model with only one interaction the phase transition corresponds to a breaking of the unitary symmetry. We analyze the model in the two phases and show that, in a double scaling limit, the symmetric phase corresponds to a theory of infinitely refined random surfaces, while the broken phase corresponds to a theory of infinitely refined random nodal surfaces. At leading order in the double scaling limit planar surfaces dominate in the symmetric phase, and planar nodal surfaces dominate in the broken phase.

The main result of this chapter is that, for the model with only one interaction, one obtains a genuinely new geometric phase in double scaling: instead of obtaining (at leading order in double scaling) a family of trees, one obtains (at leading order in double scaling) a family of planar graphs. The phase transition (in double scaling) is a phase transition between a phase of planar surfaces and a phase of planar nodal surfaces. Of course the emergent geometry obtained in double scaling is not yet genuinely D-dimensional, but (very likely) is a Brownian sphere. However, this is a first step in the quest for a genuinely new random geometry in higher dimensions: we succeeded for the first time in finding a nonbranched polymer state of higher dimensional random geometry in a scaling regime of tensor models.

11.1 The melonic model with only one interaction

In this chapter we will discuss a quartic melonic model with only one interaction. In this case one uses only one intermediate matrix field and (dropping the index c) the partition function of Eq. (9.6) simplifies to:

$$Z(g) = \int [dH]\, e^{-N^{D-1}S(H)} , \quad S(H) = \frac{1}{2}\operatorname{Tr}[H^2] - \operatorname{Tr}\left[\ln\left(\frac{1}{1-\imath\sqrt{\lambda}H}\right)\right] . \tag{11.1}$$

For convenience we relabel the coupling constant of the model $g \equiv \imath\sqrt{\lambda}$. In this form a tensor model in rank D looks very similar to a matrix model for a $N \times N$ matrix. The only difference comes from the modified scaling factor N^{D-1} instead of N for the action. This apparently innocuous modification leads to numerous consequences, which we will be exploring in detail.

Vacua of the theory

The equations of motion of the model defined in Eq. (11.1) are:

$$H - \frac{g}{1-gH} = 0 . \tag{11.2}$$

Due to the unitary invariance, Eq. (11.2) fixes only the eigenvalues λ_b of H: upon diagonalization, the equations of motion become:

$$\lambda_b - \frac{g}{1-g\lambda_b} = 0 \quad \text{for} \quad b = 1, \dots, N , \tag{11.3}$$

whose two solutions are:

$$a_\pm \equiv \frac{1 \pm \sqrt{1-4g^2}}{2g} .$$

We have a critical value of the coupling at $g_c = 1/2$, where we meet a square root singularity and the two solutions merge. The solution a_- has a regular Taylor expansion around $g = 0$ starting at linear order, $a_- = g + O(g^3)$, while the solution a_+ has a simple pole at the origin.

Up to conjugation by an arbitrary unitary matrix, any solution of the equations of motion (i.e. a *vacuum* of the theory) is of the form:

$$\bar{H}_{(N_+,N_-)} = \begin{pmatrix} a_+\mathbb{I}_+ & 0 \\ 0 & a_-\mathbb{I}_- \end{pmatrix} = a_+ \begin{pmatrix} \mathbb{I}_+ & 0 \\ 0 & 0 \end{pmatrix} + a_- \begin{pmatrix} 0 & 0 \\ 0 & \mathbb{I}_- \end{pmatrix} \equiv a_+ P_+ + a_- P_- , \tag{11.4}$$

where $\mathbb{I}_+$ and $\mathbb{I}_-$ are the identity matrices of size $N_+ \times N_+$ and respectively $N_- \times N_-$, with $N_+ + N_- = N$. There are $N+1$ possible solutions of this type.

Observe that from the onset we have neglected the contribution of the measure $[dH]$ to the equations of motion. In fact this contribution can profoundly alter the equations of motion and it is possible that the "vacua" we identified by just considering the action are nowhere near the true vacua of the theory. As we will see, this is in fact what happens in the case of matrices $D = 2$. However, for tensors ($D \geq 3$) the contribution of the measure is subleading in $1/N$ and the solutions in Eq. (11.4) are true vacuum states in the large N limit.

Each $\bar{H}_{(N_+,N_-)}$ is invariant under the residual symmetry group $\mathcal{U}(N_+) \times \mathcal{U}(N_-)$:

$$\bar{H}_{(N_+,N_-)} = \begin{pmatrix} U_+ & 0 \\ 0 & U_- \end{pmatrix} \bar{H}_{(N_+,N_-)} \begin{pmatrix} U_+^\dagger & 0 \\ 0 & U_-^\dagger \end{pmatrix}, \qquad \forall U_+ \in \mathcal{U}(N_+)\,, \forall U_- \in \mathcal{U}(N_-)\,,$$

and, being a representative in a conjugacy class of solutions of the classical equations of motions, is a saddle point of the action $S(H)$. The partition function is as a sum over the saddle points $\bar{H}_{(N_+,N_-)}$:

$$Z(g) = \sum_{N_+} C_{(N_+,N_-)} Z_{(N_+,N_-)} \qquad Z_{(N_+,N_-)} = \int_{H \text{ close to } \bar{H}_{(N_+,N_-)}} [dH]\, e^{-N^{D-1} S(H)}\,.$$

In the rest of this chapter we will discuss the path integral near a saddle point $\bar{H}$ and in order to simplify the notation we drop from now on the indices (N_+, N_-).

Evaluating the action at a generic solution $\bar{H}$, we obtain:

$$S(\bar{H}) = \frac{1}{2} \operatorname{Tr}\left[\bar{H}^2\right] - \operatorname{Tr}\left[\ln\left(\frac{1}{g}\bar{H}\right)\right] = N_+ \left(\frac{a_+^2}{2} - \ln\frac{a_+}{g}\right) + N_- \left(\frac{a_-^2}{2} - \ln\frac{a_-}{g}\right). \tag{11.5}$$

Near $g = 0$, regularity of the free energy demands $N_+ = 0$. However, at $g = g_c = 1/2$, the free energy develops a singularity, and beyond that, solutions with generic N_+ are possible. For $g > g_c$, it can be easily checked that the real part of (11.5) is independent of N_+ and N_-. Therefore, all these vacua contribute to the same leading exponential order, and in the broken phase we do not have at this level any reason to choose a particular value of N_+.

Close to a vacuum $\bar{H}$, we make the change of variables $H = \bar{H} + M$, splitting the field into a background $\bar{H}$ and a fluctuation field M. Observe that even if $\bar{H}$ is not hermitian (which happens if $a_\pm$ are not real), the perturbation M remains hermitian, and the translation by $\bar{H}$ is a translation of the diagonal entries of H in the complex plane. Using the equations of motion we obtain:

$$\begin{aligned}
S(\bar{H} + M) &= \frac{1}{2} \operatorname{Tr}[(\bar{H} + M)^2] - \operatorname{Tr}\left[\ln\left(\frac{1}{1 - g(\bar{H} + M)}\right)\right] \\
&= \frac{1}{2} \operatorname{Tr}[\bar{H}^2] + \operatorname{Tr}[\bar{H}M] + \frac{1}{2}\operatorname{Tr}[M^2] - \operatorname{Tr}\left[\ln\left(\frac{1}{1 - g\bar{H}}\right)\right] - \operatorname{Tr}\left[\ln\left(\frac{1}{1 - \frac{g}{1-g\bar{H}}M}\right)\right] \\
&= \frac{1}{2} \operatorname{Tr}[\bar{H}^2] - \frac{1}{g}\operatorname{Tr}[\bar{H}] + \operatorname{Tr}[\bar{H}M] + \frac{1}{2}\operatorname{Tr}[M^2] - \operatorname{Tr}\left[\ln\left(\frac{1}{1 - \bar{H}M}\right)\right] \\
&= \frac{1}{2} \operatorname{Tr}[\bar{H}^2] - \frac{1}{g}\operatorname{Tr}[\bar{H}] + \frac{1}{2}\left[\operatorname{Tr}[M^2] - \operatorname{Tr}\left[\bar{H}M\bar{H}M\right]\right] - \sum_{p\geq 3}\frac{1}{p}\operatorname{Tr}\left[(\bar{H}M)^p\right]\,,
\end{aligned}$$

therefore the effective quadratic term (defining the inverse covariance) for the fluctuation field is:

$$\frac{1}{2}\operatorname{Tr}\left[M^2\right] - \frac{1}{2}\operatorname{Tr}[\bar{H}M\bar{H}M].$$

Decomposing M into blocks:

$$M^{++} = P_+ M P_+ \quad M^{+-} = P_+ M P_-, \quad M^{--} = P_- M P_-, \qquad M = \begin{pmatrix} M^{++} & M^{+-} \\ M^{-+} & M^{--} \end{pmatrix},$$

and taking into account that $P_+ + P_- = \mathbb{I}$, the quadratic part for the fluctuation field is rewritten as:

$$\frac{1}{2}(1-a_+^2)\,\mathrm{Tr}[(M^{++})^2] + \frac{1}{2}(1-a_-^2)\,\mathrm{Tr}[(M^{--})^2] + (1-a_+a_-)\,\mathrm{Tr}[M^{+-}M^{-+}] .$$

The inverse covariance is an operator in the vectors space of hermitian matrices with inner product $\langle M_1 \,|\, M_2 \rangle = \mathrm{Tr}[M_1 M_2]$. Since:

$$(P_+ \otimes P_+)\,|M\rangle = |P_+ M P_+\rangle = |M^{++}\rangle ,$$

the inverse covariance is:

$$(1-a_+^2) P_+ \otimes P_+ + (1-a_-^2) P_- \otimes P_- + (1-a_+a_-)(P_+ \otimes P_- + P_- \otimes P_+) ,$$

and $P_+ \otimes P_+$, $P_- \otimes P_-$ and $P_+ \otimes P_- + P_- \otimes P_+$ are (mutually orthogonal) projectors. The mass eigenvalues for the fluctuation field M are then read off as $\Lambda_+ \equiv 1 - a_+^2$ with degeneracy N_+^2, $\Lambda_- \equiv 1 - a_-^2$ with degeneracy N_-^2, and $1 - a_+a_- = 0$ with degeneracy N_+N_-.

Defining $\epsilon = 1 - 4g^2$ we have $\Lambda_\pm = 2\sqrt{\epsilon}/(\sqrt{\epsilon} \mp 1)$, and we distinguish the following situations:

- For small $\epsilon > 0$, the leading behavior of the mass eigenvalues of the fluctuation field is $\Lambda_\pm \sim \mp 2\sqrt{\epsilon}$, and the only positive eigenvalue is Λ_-. It follows that for $g < g_c$ only the solution $N_- = N, N_+ = 0$ is stable, and therefore we are in the $\mathcal{U}(N)$ symmetric phase, in agreement with the small g analysis.
- At $\epsilon = 0$ (i.e. at $g = g_c$), all the mass eigenvalues are zero, signaling criticality.
- For small $\epsilon < 0$, the leading behavior is $\Lambda_\pm \sim 2|\epsilon| \mp i2\sqrt{|\epsilon|}$, and both Λ_+ and Λ_- have a positive real part. It follows that for negative ϵ all the vacua are stable and therefore we are in the broken phase, with residual symmetry $\mathcal{U}(N_+) \times \mathcal{U}(N_-)$. In this case the off-diagonal fluctuations M_{+-} and M_{-+}, lacking a quadratic term, have the interpretation of massless Goldstone modes.

Around a vacuum state the effective theory for the fluctuation field writes:

$$\begin{aligned} Z_{(N_+,N_-)} \simeq{}& e^{-N^{D-1}\left[N_+\left(\frac{a_+^2}{2} - \ln\frac{a_+}{g}\right) + N_-\left(\frac{a_-^2}{2} - \ln\frac{a_-}{g}\right)\right]} \\ & \int [dM^{+-}dM^{-+}] \int [dM^{++}][dM^{--}]\; e^{-N^{D-1}S^{\mathrm{eff}}} \\ S^{\mathrm{eff}} ={}& \frac{1}{2}(1-a_+^2)\,\mathrm{Tr}[(M^{++})^2] + \frac{1}{2}(1-a_-^2)\,\mathrm{Tr}[(M^{--})^2] \\ & - \sum_{p\geq 3} \frac{1}{p} \sum_{i_1,\dots i_p \in \{+,-\}} \prod_{q=1}^{p} a_{i_q} M^{i_q i_{q+1}} . \end{aligned} \tag{11.6}$$

The theory thus obtained looks quite complicated. We can considerably simplify it by getting rid of the off-diagonal blocks M^{+-} and M^{-+} (which only appear at cubic order in the above action), exploiting the unitary invariance. The procedure we detail here comes to explicitly integrating out the Goldstone modes in order to obtain the effective theory for the "radial" degrees of freedom.

11.2 Integrating out the massless modes

Knowing that at the saddle point the eigenvalues split in two sets of N_+ and N_- identical eigenvalues, it is convenient to consider the subgroup $\mathcal{H} = \mathcal{U}(N_+) \times \mathcal{U}(N_-) \subset \mathcal{U}(N)$ defined by:

$$\mathcal{H} = \left\{ W \in \mathcal{U}(N) \middle| W = e^{i\begin{pmatrix} W_+ & 0 \\ 0 & W_- \end{pmatrix}} = \begin{pmatrix} e^{iW_+} & 0 \\ 0 & e^{iW_-} \end{pmatrix} \right\},$$

for two arbitrary Hermitian matrices W_+ and W_- of sizes $N_+ \times N_+$ and respectively $N_- \times N_-$.

We first show that any hermitian matrix H can be brought into block diagonal form:

$$H = V \begin{pmatrix} H_+ & 0 \\ 0 & H_- \end{pmatrix} V^\dagger ,$$

for some $V \in \mathcal{U}(N)/\mathcal{H}$, i.e. for some V in the left coset of $\mathcal{H}$ in $\mathcal{U}(N)$. We subsequently perform the change of variables $H \to (H_+, H_-, V)$, and compute the Jacobian $[dH] = \det(J)[dH_+][dH_-][dV]$.

11.2.1 Block diagonalization

Let us write $H = UEU^\dagger$, where E is diagonal and $U \in \mathcal{U}(N)$ is a unitary transformation that diagonalizes H. Let us denote $V \in \mathcal{U}(N)/\mathcal{H}$ the representative in $\mathcal{U}(N)/\mathcal{H}$ of the equivalence class of U, $[U] = \{UW | W \in \mathcal{H}\}$. Then there exists a $W \in \mathcal{H}$ such that[1]:

$$H = UEU^\dagger = VW^\dagger EWV^\dagger = V \begin{pmatrix} e^{-iW_+} E_+ e^{iW_+} & 0 \\ 0 & e^{-iW_-} E_- e^{iW_-} \end{pmatrix} V^\dagger ,$$

where E_+ denotes the first N_+ eigenvalues of H and E_- denotes the subsequent N_- eigenvalues. Defining $H_+ = e^{iW_+} E_+ e^{-iW_+}$ and $H_- = e^{iW_-} E_- e^{-iW_-}$, shows that H can be brought into block diagonal form by conjugation by some $V \in \mathcal{U}(N)/\mathcal{H}$.

11.2.2 Jacobian

In order to compute the Jacobian J of the change of variables $H \to (H_+, H_-, V)$, we will first show that the Jacobian does not depend on V and then compute it for V close to the identity.

[1] For the sake of precision, $U \in \mathcal{U}'(N) \equiv \mathcal{U}(N)/\mathcal{U}(1)^{\otimes N}$, because a multiplication of U to the right by a diagonal matrix $\mathrm{diag}(e^{i\phi_1}, \ldots, e^{i\phi_N})$ leaves H unaffected. Similarly, we should also quotient $\mathcal{H} \to \mathcal{H}' \equiv \mathcal{H}/\mathcal{U}(1)^{\otimes N}$. However, $\mathcal{U}(N)/\mathcal{H} \simeq \mathcal{U}'(N)/\mathcal{H}'$.

Denoting W_+ and W_- two Hermitian matrices, and B a complex matrix (of sizes $N_+ \times N_+$, $N_- \times N_-$ and $N_+ \times N_-$, respectively), any $U \in \mathcal{U}(N)$ close to the identity is written as:

$$U = e^{i\begin{pmatrix} W_+ & B \\ B^\dagger & W_- \end{pmatrix}} = e^{i\begin{pmatrix} 0 & B \\ B^\dagger & 0 \end{pmatrix}} e^{i\begin{pmatrix} W_+ & 0 \\ 0 & W_- \end{pmatrix}} + O(BW) , \qquad e^{i\begin{pmatrix} W_+ & 0 \\ 0 & W_- \end{pmatrix}} \in \mathcal{H} ,$$

hence the class of U in $\mathcal{U}(N)/\mathcal{H}$ can be represented by:

$$V \approx \mathbb{I} + i\begin{pmatrix} 0 & B \\ B^\dagger & 0 \end{pmatrix} .$$

The infinitesimal variation of V is then:

$$dV = i\begin{pmatrix} 0 & dB \\ dB^\dagger & 0 \end{pmatrix} ,$$

and for any fixed $W \in \mathcal{U}(N)$ close to the identity, $W = \mathbb{I} + iT$ (with T a hermitian matrix), we have:

$$WV \approx \mathbb{I} + iT + i\begin{pmatrix} 0 & B \\ B^\dagger & 0 \end{pmatrix} + O(T^2, B^2, QB) \Rightarrow [d(WV)] \approx i\begin{pmatrix} 0 & dB \\ dB^\dagger & 0 \end{pmatrix} \approx [dV] ,$$

hence the Jacobian does not depend on V. In fact, the invariance of the measure induced on the quotient space by the Haar measure on the group can be proved under quite general conditions (which in particular apply to our case as $\mathcal{U}(N)$ is a compact group), see for example [129].

We now follow the standard strategy. First, we write an infinitesimal variation of the matrix H as:

$$dH = dV\begin{pmatrix} H_+ & 0 \\ 0 & H_- \end{pmatrix} V^\dagger + V\begin{pmatrix} H_+ & 0 \\ 0 & H_- \end{pmatrix} dV^\dagger + V\begin{pmatrix} dH_+ & 0 \\ 0 & dH_- \end{pmatrix} V^\dagger ,$$

and use $VV^\dagger = \mathbb{I} \Rightarrow dV\,V^\dagger + V\,dV^\dagger = 0$ to eliminate $dV^\dagger$. Then we use the invariance of the measure and evaluate the Jacobian at $V = \mathbb{I}$:

$$dH = i\begin{pmatrix} 0 & dBH_- \\ dB^\dagger H_+ & 0 \end{pmatrix} - i\begin{pmatrix} 0 & H_+ dB \\ H_- dB^\dagger & 0 \end{pmatrix} + \begin{pmatrix} dH_+ & 0 \\ 0 & dH_- \end{pmatrix} .$$

Finally, writing $dH_{ij} = J_{ij,kl} dX_{kl}$, for

$$X = (H_+, H_-, B, B^\dagger) , \qquad H = (H^{++}, H^{--}, H^{+-}, H^{-+}) ,$$

we read off the Jacobian:

$$J \equiv \frac{dH}{dX} = \begin{pmatrix} \mathbb{I}_+ \otimes \mathbb{I}_+ & 0 & 0 & 0 \\ 0 & \mathbb{I}_- \otimes \mathbb{I}_- & 0 & 0 \\ 0 & 0 & i(\mathbb{I}_+ \otimes H_- - H_+ \otimes \mathbb{I}_-) & 0 \\ 0 & 0 & 0 & -i(H_- \otimes \mathbb{I}_+ - \mathbb{I}_- \otimes H_+) \end{pmatrix} ,$$

with determinant $\det(J) = \det(H_+ \otimes \mathbb{I}_- - \mathbb{I}_+ \otimes H_-)^2$. We have thus shown that:

$$\boxed{\begin{aligned}\int [dH]\, e^{-N^{D-1}S(H)} = \int [dV] \int [dH_+][dH_-] \\ \Big[\det\left(H_+ \otimes \mathbb{I}_- - \mathbb{I}_+ \otimes H_-\right)\Big]^2 e^{-N^{D-1}S\begin{pmatrix} H_+ & 0 \\ 0 & H_- \end{pmatrix}}.\end{aligned}} \tag{11.7}$$

The integral over V is now completely decoupled and can be absorbed into a proper normalization.

This formula generalizes the usual reduction to eigenvalues, which can be recovered by taking $N_- = 1$ and then iteratively repeating the decomposition on the remaining $N_+ \times N_+$ block. This will iteratively generate the usual Vandermonde determinant $\Delta(\lambda)^2 = \prod_{i<j}(\lambda_i - \lambda_j)^2$. Keeping instead N_+ and N_- generic, but reducing to eigenvalues our formula (11.7), we obtain:

$$\begin{aligned}&\int [dH]\, e^{-N^{D-1}S(H)} \\ &= \int \left(\prod_{i=1}^{N_+} d\lambda_i^+\, \Delta(\lambda^+)^2\, e^{-N^{D-1}S(\lambda_i^+)}\right)\left(\prod_{j=1}^{N_-} d\lambda_j^-\, \Delta(\lambda^-)^2\, e^{-N^{D-1}S(\lambda_j^-)}\right) \prod_{i,j}(\lambda_i^+ - \lambda_j^-)^2\,,\end{aligned} \tag{11.8}$$

and of course the two sets of eigenvalues can be recombined as $\lambda = (\lambda^+, \lambda^-)$ to give the usual expression:

$$\int [dH]\, e^{-N^{D-1}S(H)} = \int \prod_{i=1}^{N} d\lambda_i\, \Delta(\lambda)^2\, e^{-N^{D-1}S(\lambda_i)}. \tag{11.9}$$

The effect of the scaling with N^{D-1} in Eq. (11.1) now comes into play. From Eq. (11.9) we conclude that we can evaluate the partition function by a saddle point method without having to care about the Vandermonde determinant if $D > 2$. In fact $\log \Delta(\lambda)$ is of order N^2, whereas $N^{D-1}\sum_i S(\lambda_i)$ is of order N^D. This is *de facto* what we have done when writing the saddle point Eq. (11.3) for the action alone (i.e. without measure contributions). Eq. (11.3) yields just two solutions for the eigenvalues at the saddle point. We can then directly use Eq. (11.8) and perform the saddle point approximation, taking in each block one of the solutions i.e. $\lambda^+ = a_+$ and $\lambda^- = a_-$. In order to study the fluctuations around the saddle point one would then write $\lambda_i^\pm = a_\pm + \mu_i^\pm$ and study the corrections coming from the diagonal fluctuations $\mu_i^\pm$.

11.3 The effective theory

If one computes the partition function using the eigenvalue decomposition, the connection between the Feynman graphs of the theory and random triangulations is lost. In order to keep this connection explicit and find an interpretation of the effective theory around a nontrivial vacuum in terms of random triangulations, one must start from Eq. (11.7), and decompose the matrices $H_\pm$ into a background and a fluctuation $H_\pm = a_\pm + M_\pm$. We thus obtain:

$$Z_{(N_+,N_-)} = e^{-N^{D-1}\left[N_+\left(\frac{a_+^2}{2}-\ln\frac{a_+}{g}\right)+N_-\left(\frac{a_-^2}{2}-\ln\frac{a_-}{g}\right)\right]}\int [dM_+][dM_-] \tag{11.10}$$

$$\left[\det\Big((a_+-a_-)\mathbb{I}_+\otimes\mathbb{I}_- + M_+\otimes\mathbb{I}_- - \mathbb{I}_+\otimes M_-\Big)\right]^2 e^{-N^{D-1}S^{\text{eff}}} ,$$

$$S^{\text{eff}} = \frac{1}{2}(1-a_+^2)\,\text{Tr}[M_+^2] + \frac{1}{2}(1-a_-^2)\,\text{Tr}[M_-^2] - \sum_{p\ge3}\frac{a_+^p}{p}\,\text{Tr}\,[M_+^p] - \sum_{p\ge3}\frac{a_-^p}{p}\,\text{Tr}\,[M_-^p] ,$$

achieving our goal of eliminating the off-diagonal blocks in Eq. (11.6), at the price of introducing a nontrivial determinant term. Note, however, that $M_\pm$ in Eq. (11.10) are not the same matrices as the M^{++} and M^{--} in Eq. (11.6), in particular they depend nontrivially on the off-diagonal blocks M^{+-} and M^{-+} in Eq. (11.6) just like the eigenvalues of a matrix depend on its off-diagonal elements.

11.3.1 Feynman graphs

The integral in Eq. (11.10) for the fluctuations around a saddle point:

$$(a_+-a_-)^{2N_+N_-}\int [dM_+][dM_-]\; e^{2\,\text{Tr}\left[\ln\left(\mathbb{I}_+\otimes\mathbb{I}_- + \frac{1}{a_+-a_+-}(M_+\otimes\mathbb{I}_- - \mathbb{I}_+\otimes M_-)\right)\right]}$$
$$\times e^{-N^{D-1}\left[\frac{1}{2}(1-a_+^2)\,\text{Tr}[M_+^2]+\frac{1}{2}(1-a_-^2)\,\text{Tr}[M_-^2]-\sum_{p\ge3}\frac{a_+^p}{p}\,\text{Tr}[M_+^p]-\sum_{p\ge3}\frac{a_-^p}{p}\,\text{Tr}[M_-^p]\right]} ,$$

can be written more explicitly by expanding the logarithm:

$$(a_+-a_-)^{2N_+N_-}\int [dM_+][dM_-]\; e^{2\sum_{q\ge1}\frac{(-1)^{q+1}}{q(a_+-a_-)^q}\sum_{q_-=0}^{q}\binom{q}{q_-}(-1)^{q_-}\,\text{Tr}[M_+^{q-q_-}]\,\text{Tr}[M_-^{q_-}]}$$
$$\times e^{-N^{D-1}\left[\frac{1}{2}(1-a_+^2)\,\text{Tr}[M_+^2]+\frac{1}{2}(1-a_-^2)\,\text{Tr}[M_-^2]-\sum_{p\ge3}\frac{a_+^p}{p}\,\text{Tr}[M_+^p]-\sum_{p\ge3}\frac{a_-^p}{p}\,\text{Tr}[M_-^p]\right]} ,$$

leading to the effective action:

$$\begin{aligned}\tilde{S}^{\text{eff}} &= N^{D-1}\frac{1}{2}(1-a_+^2)\,\text{Tr}[M_+^2] + N^{D-1}\frac{1}{2}(1-a_-^2)\,\text{Tr}[M_-^2] \\ &\quad - N^{D-1}\sum_{p_+\ge3}\frac{a_+^{p_+}}{p_+}\,\text{Tr}\,[M_+^{p_+}] - N^{D-1}\sum_{p_-\ge3}\frac{a_-^{p_-}}{p_-}\,\text{Tr}\,[M_-^{p_-}] \\ &\quad + \sum_{q_+,q_-\ge0}^{q_++q_-\ge1}\frac{2}{(a_+-a_-)^{q_++q_-}}\frac{(-1)^{q_+}}{q_++q_-}\binom{q_++q_-}{q_+}\text{Tr}\,[M_+^{q_+}]\,\text{Tr}\,[M_-^{q_-}] .\end{aligned} \tag{11.11}$$

Observe that the terms coming from the nontrivial determinant generate new quadratic terms. These terms could be included in the measure, but we prefer to treat them as bivalent vertices.

The logarithm of the partition function is a sum over connected graphs. The graphs have three kinds of vertices:

- vertices coming from $\sum_{p_+\geq 3} \frac{a_+^{p_+}}{p_+} \operatorname{Tr}[M_+^{p_+}]$. We denote v_{p_+} the number of such vertices of coordination p_+ and V_+ the total number of such vertices, $V_+ = \sum_{p_+\geq 3} v_{p_+}$. These vertices have coordination of at least 3. They represent polygons in the + sector, which will glue together to form + surfaces.

- vertices coming from $\sum_{p_-\geq 3} \frac{a_-^{p_-}}{p_-} \operatorname{Tr}[M_-^{p_-}]$. We denote v_{p_-} the number of such vertices of coordination p_- and V_- the total number of such vertices, $V_- = \sum_{p_-\geq 3} v_{p_-}$. These vertices have coordination of at least 3. They represent polygons in the − sector, which will glue together to form − surfaces.

- vertices coming from $\sum_{q_+,q_-\geq 0}^{q_++q_-\geq 1} \frac{2}{(a_+-a_-)^{q_++q_-}} \frac{(-1)^{q_+}}{q_++q_-} \binom{q_++q_-}{q_+} \operatorname{Tr}[M_+^{q_+}] \operatorname{Tr}[M_-^{q_-}]$. We denote $v_{q_+q_-}$ the number of such vertices of coordination q_+ and q_- and V_{+-} the total number of such vertices $V_{+-} = \sum_{q_+,q_-\geq 0}^{q_++q_-\geq 1} v_{q_+q_-}$. These vertices have a joint coordination $q_+ + q_-$ of at least one, but can have coordination of 0 in one of the sectors. These vertices represent a pair of a + and a − polygons connected at a nodal point. The + polygons (respectively the − polygons) will glue into + surfaces (respectively − surfaces), and the + and − surfaces will be connected at a mixed vertex by a nodal point.

We can represent the + and − vertices as solid vertices, and the +− vertices as two solid vertices (one for the + sector and one for the − sector) connected by a dashed edge; see Figure 11.1. Each graph represents a nodal surface (for definitions and properties of nodal surfaces see, for example, [137]). The E_+ edges connecting two M_+ halfedges

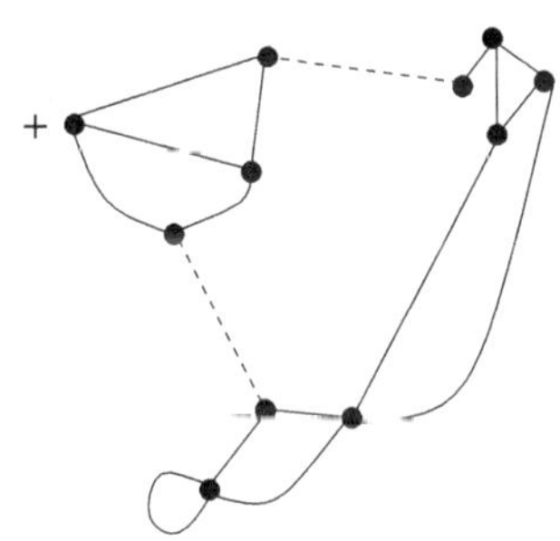

Figure 11.1: An example of a graph generated by the effective action in Eq. (11.11). Here the two dashed edges represent two +− vertices of the type $\operatorname{Tr}[M_+^2]\operatorname{Tr}[M_-^2]$ and $\operatorname{Tr}[M_+^2]\operatorname{Tr}[M_-]$.

(which we call the + edges) bring a factor $\frac{1}{N^{D-1}(1-a_+^2)}$. The E_- edges connecting two M_- halfedges (which we call the − edges) bring a factor $\frac{1}{N^{D-1}(1-a_-^2)}$.

The graph is connected but, by erasing the dashed edges, it might disconnect into several connected components. Let us call C_+ the number of the connected components of the graph made exclusively of the edges +, g_+ the sum of the genera of these components and F_+ the number of faces of these components (and similarly for the −

components). Up to irrelevant combinatorial factors the amplitude of a graph is:

$$\left(\frac{1}{N^{D-1}(1-a_+^2)}\right)^{E_+}\left(\frac{1}{N^{D-1}(1-a_-^2)}\right)^{E_-}$$
$$\times N^{(D-1)V_+}a_+^{\sum_{p_+\ge 3}p_+v_{p_+}}N^{(D-1)V_-}a_-^{\sum_{p_-\ge 3}p_-v_{p_-}}$$
$$\times \frac{(-1)^{\sum_{q_+,q_-\ge 0}^{q_++q_-\ge 1}(q_++1)v_{q_+q_-}}}{(a_+-a_-)^{\sum_{q_+,q_-\ge 0}^{q_++q_-\ge 1}(q_++q_-)v_{q_+q_-}}}N_+^{F_+}N_-^{F_-}\ .$$

Note that in the symmetric phase (i.e. $g < g_c$ and $N_+ = 0$) the amplitudes are positively definite, while in the broken phase (i.e. $g > g_c$ and $N_+ > 0$) they are complex. In the following we will not keep track of global minus signs for individual amplitudes.

The various numbers of vertices, edges and faces are related by the following topological relations:

$$2E_+ = \sum_{p_+\ge 3} p_+v_{p_+} + \sum_{q_+,q_-\ge 0}^{q_++q_-\ge 1} q_+v_{q_+q_-}\ ,\qquad 2E_- = \sum_{p_-\ge 3} p_-v_{p_-} + \sum_{q_+,q_-\ge 0}^{q_++q_-\ge 1} q_-v_{q_+q_-}\ ,\tag{11.12}$$

$$V_+ + V_{+-} - E_+ + F_+ = 2C_+ - 2g_+\ ,\qquad V_- + V_{+-} - E_- + F_- = 2C_- - 2g_-\ ,\tag{11.13}$$

$$\begin{aligned}&E_+ - V_+ - V_{+-} + C_+ \ge 0\ ,\qquad E_- - V_- - V_{+-} + C_- \ge 0\ ,\\ &E_+ + E_- - V_+ - V_- - V_{+-} + 1 \ge 0\ ,\\ &V_{+-} - C_+ - C_- + 1 \ge 0\ .\end{aligned}\tag{11.14}$$

Equations (11.12) and (11.13) are usual graph relations applied to the + and − subgraphs: the first simply states that summing the numbers of halfedges at every vertex amounts to counting twice the number of edges, while the second is simply the Euler characteristic formula. Equations (11.14) state that the first Betti number of a graph is always nonnegative, in particular it is zero for trees and positive for graphs with excess edges.

A connected graph, consisting of + and − components and dashed edges, is dual to a nodal surface with V_{+-} nodal points. The total genus of a nodal surface is:

$$g_+ + g_- + (V_{+-} - C_+ - C_- + 1)\ ,$$

i.e. the sum of the genera of the + components, the − components and the first Betti number of the abstract graph obtained by collapsing the + and − components to vertices connected by the dashed edges. Of course, $g_+ + g_- + (V_{+-} - C_+ - C_- + 1) \ge 0$, and the equality holds only for the case in which all the + and − components are planar, and the abstract graph is a tree.

Defining the filling fractions $n_+ = \frac{N_+}{N} \le 1$ and $n_- = \frac{N_-}{N} \le 1$, the amplitude of a graph is written:

$$\frac{a_+^{\sum_{p_+\ge 3}p_+v_{p_+}}a_-^{\sum_{p_-\ge 3}p_-v_{p_-}}}{(1-a_+^2)^{E_+}(1-a_-^2)^{E_-}(a_+-a_-)^{\sum_{q_+,q_-\ge 0}^{q_++q_-\ge 1}(q_++q_-)v_{q_+q_-}}}\,n_+^{F_+}n_-^{F_-}$$

$$\times N^{(D-1)V_+-(D-1)E_++F_+} N^{(D-1)V_--(D-1)E_-+F_-} ,$$

which, using:

$$\begin{aligned}(D-1)V_+ - (D-1)E_+ + F_+ &= (D-1)V_+ - (D-1)E_+ + E_+ - V_+ - V_{+-} + 2C_+ - 2g_+ \\ &= -(D-2)(E_+ - V_+) - V_{+-} + 2C_+ - 2g_+ ,\end{aligned}$$

becomes:

$$\frac{a_+^{\sum_{p_+\geq 3} p_+ v_{p_+}} a_-^{\sum_{p_-\geq 3} p_- v_{p_-}}}{(1-a_+^2)^{E_+}(1-a_-^2)^{E_-}(a_+ - a_-)^{\sum_{q_+,q_-\geq 0}^{q_++q_-\geq 1}(q_++q_-)v_{q_+q_-}}} n_+^{F_+} n_-^{F_-}$$
$$\times N^{-(D-2)(E_+-V_+)-(D-2)(E_--V_-)+2-2(V_{+-}-C_+-C_-+1)-2g_+-2g_-} . \quad (11.15)$$

11.4 The phases of the model and their geometry

The main difference between the $D = 2$ case of matrices and the $D \geq 3$ case of tensors comes from the following theorem.

Theorem 11.1. *The number of graphs at fixed order in* $1/N$ *is:*

- finite *for* $D \geq 3$,
- infinite *for* $D = 2$.

Proof. Let us consider the scaling with N of a term:

$$-(D-2)(E_+ - V_+) - (D-2)(E_- - V_-) + 2 - 2(V_{+-} - C_+ - C_- + 1) - 2g_+ - 2g_- .$$

As the graph is connected $V_{+-} - C_+ - C_- + 1 \geq 0$, hence this scaling is bounded from above by:

$$2 - (D-2)(E_+ - V_+) - (D-2)(E_- - V_-) .$$

The V_+ and V_- vertices are at least three valent, hence $2E_+ \geq 3V_+$ and $2E_- \geq 3V_-$, and the scaling with N of a term is bounded from above by:

$$2 - \frac{D-2}{3}(E_+ + E_-) .$$

Finally, the number of dashed edges equals V_{+-} and is itself bounded by $E_+ + E_- + 1$. Consequently, for $D \geq 3$, at each fixed scaling in $1/N$ only connected graphs with at most a finite number of (+, −, and dashed) edges contribute. As the number of such graphs is finite the first assertion of the theorem follows.

For $D = 2$ the above bound does not constrain the number of graphs (it just states that the amplitude of any graph is bounded by N^2). The scaling with N of a graph becomes for $D = 2$:

$$2 - 2(V_{+-} - C_+ - C_- + 1 + g_+ + g_-) ,$$

i.e. it is proportional with the total genus of the nodal surface dual to the graph. As the number of triangulated surfaces of fixed total genus is infinite, the second assertion of the theorem follows.

□

Having estimated a bound for the scaling in N of the graphs generated by the effective action Eq. (11.11), we can now address the question of whether our expansion around the "vacua" in Eq. (11.4) is justified.

The connected one point function $\langle \frac{1}{N}\operatorname{Tr}[M_-]\rangle_{\text{connected}}$, is a sum over graphs having an extra − univalent vertex representing the external field M_{--}. The scaling with N of such a graph is:

$$\begin{aligned}&-1-(D-1)\\&\quad -(D-2)(E_+-V_+)-(D-2)(E_--V_-)+2-2(V_{+-}-C_+-C_-+1)-2g_+-2g_- ,\end{aligned}$$

where V_- includes the extra univalent − vertex. As $2E_- \geq 1+3(V_--1)$. It follows that the scaling with N of a connected one point graph is bounded by:

$$-(D-2)(E_+-V_+)-(D-2)\big[E_- -(V_- -1)\big] \leq -\frac{D-2}{3}(E_+ + E_- +1) ,$$

hence:

$$\left\langle \frac{1}{N}\operatorname{Tr}[M_-]\right\rangle_{\text{connected}} \sim \begin{cases} O(1) , & D=2 \\ O(N^{-\frac{D-2}{3}}) , & D\geq 3 \end{cases} ,$$

that is, the connected one point function starts at order 1 for $D = 2$, but is strictly smaller in $1/N$ for $D \geq 3$.

Similarly, we can look at the free energy:

$$W = \frac{1}{N^D}\ln Z_{(N_+,N_-)} = -n_+\left(\frac{a_+^2}{2}-\ln\frac{a_+}{g}\right)-n_-\left(\frac{a_-^2}{2}-\ln\frac{a_-}{g}\right)+\frac{1}{N^{D-2}}\tilde{W} ,$$

where $\tilde{W}$ contains the graphs for which we proved that an upper bound for the scaling is $N^{-\frac{D-2}{3}(E_++E_-)}$. That is, the correction from the effective theory to the free energy computed at the saddle point is of the same order as the latter for $D = 2$, but it is subleading in $1/N$ for $D > 2$.

11.4.1 The matrix case $D = 2$

It is now apparent why the $D = 2$ case is very different from the $D \geq 3$. Indeed, for $D = 2$ the critical behavior of the amplitude (11.15) of an individual graph has no relevance: the free energy at fixed order in $1/N$ is a sum of an *infinite* family of graphs, and acquires a critical behavior due to the criticality of this infinite sum. In particular, it is well known that this sum becomes critical at the critical constant $g_c' = \frac{1}{\sqrt{12}}$, which is smaller than the critical value $g_c = \frac{1}{2}$ at which the amplitude of an individual graph becomes critical.

The fact that one still obtains an infinity of graphs at any fixed order in $1/N$ shows that the contribution coming from the measure (the Vandermonde determinant) cannot be neglected in this case. The "vacuum state" of a diagonal matrix with entries a_+ and a_-, which we obtained by neglecting the contributions of the measure, is far from a genuine vacuum state of the theory because, as we have shown above, $\langle \frac{1}{N}\operatorname{Tr}[M_-]\rangle_{\text{connected}}$ is of the same order in $1/N$ as $\langle \frac{1}{N}\operatorname{Tr}[\bar{H}]\rangle_{\text{connected}}$.

11.4.2 The tensor case $D \geq 3$

For $D \geq 3$ one can derive the critical behavior of the free energy just by looking at the critical behavior of individual graphs, as the free energy is just a finite sum over such contributions. The diagonal vacua we identified are close to the genuine vacua of the theory as the connected one point function is subleading in $1/N$.

The symmetric phase

We first consider the case $\epsilon = 1-4g^2 > 0$. The only positive mass eigenvalue is Λ_-, hence the unique stable vacuum corresponds to filling fractions $n_+ = 0, n_- = 1$. In this case we have no + components, and no nodal points. In this phase the graphs are made of a single connected − component, and their amplitude is:

$$\frac{a_-^{\sum_{p_-\geq 3} p_- v_{p_-}}}{(1-a_-^2)^{E_-}} \quad N^{2-(D-2)(E_--V_-)-2g_-} .$$

From the onset we observe that the nonplanar graphs are subleading. Indeed, if a graph is nonplanar $g_- > 0$, its amplitude is strictly smaller than the amplitude of a planar graph having the same number of edges E_- and the same number of loop edges $E_- - V_- + 1$, which is planar.

In the critical regime the amplitude of a planar graphs is:

$$\frac{1}{\sqrt{\epsilon}^{E_-}} N^{2-(D-2)(E_--V_-)} ,$$

and, as all the V vertices are at least three valent, we have $V_- \leq \frac{2}{3}E_-$, therefore this amplitude scales at most like:

$$N^2 \left(\sqrt{\epsilon} N^{\frac{D-2}{3}}\right)^{-E_-} .$$

The suppression in N of this amplitude can be offset by the enhancement at criticality. In the double scaling regime $N \to \infty$, $\alpha^{-1} \equiv \sqrt{\epsilon} N^{\frac{D-2}{3}}$ fixed, *arbitrary planar three-valent graphs* contribute at the same order and dominate the free energy.

This is to be contrasted with the double scaling limit in the symmetric phase for the tensor model that includes all the interactions, Eq. (9.1), discussed in Chapter 10. In that case, in the double scaling regime either only trees decorated with self loop edges (tadpole edges) on the leaves contribute for $D < 6$, or arbitrary three valent graphs contribute for $D \geq 6$. The family of arbitrary three valent graphs is not summable, hence for $D \geq 6$ the double scaling series is not summable, like it is for matrices. For $D < 6$, the family of trees decorated with self loop edges on the leaves is summable. However, it is very likely that it corresponds to a branched polymer phase.

For tensor models with only one interaction the dominant family in double scaling is summable, but it consists in all the planar three valent graphs. It is most likely that this double scaling regime corresponds to a Brownian sphere geometric phase of the model.

The free energy is constructed by summing all the planar three-valent graphs with (up to combinatorial factors) weight α^{E_-}, therefore we expect to find a finite critical value α_c at which graphs with an infinite number of edges (or vertices, i.e. triangles in the dual surface) dominate. This is the standard situation that one exploits to construct a continuum limit, but we will not push this further here.

The broken phase

When $\epsilon = 1 - 4g^2 < 0$ both mass eigenvalues Λ_+, Λ_- have a positive real part, hence we can have $n_+ > 0$. In this regime the amplitude of a graph is proportional to:

$$\frac{n_+^{F_+} n_-^{F_-}}{\sqrt{\epsilon}^{E_+ + E_- + \sum_{q_+, q_- \geq 0}^{q_+ + q_- \geq 1} (q_+ + q_-) v_{q_+ q_-}}} \times N^{-(D-2)(E_+ - V_+) - (D-2)(E_- - V_-) + 2 - 2(V_{+-} - C_+ - C_- + 1) - 2g_+ - 2g_-} .$$

As in the symmetric phase one can from the onset consider only the graphs such that g_+ and g_- are zero.

Using the topological relations, the amplitude of a graph with planar + and − components is:

$$\frac{n_+^{F_+} n_-^{F_-}}{\sqrt{\epsilon}^{3E_+ + 3E_-}} \sqrt{\epsilon}^{\sum_{p_+ \geq 3} p_+ v_{p_+} \sum_{p_- \geq 3} p_+ v_{p_-}} N^{-(D-2)(E_+ - V_+) - (D-2)(E_- - V_-) + 2 - 2(V_{+-} - C_+ - C_- + 1)} .$$

Let us consider a graph $\mathcal{G}$ and suppose that it possesses an internal V_+ vertex of coordination $p_+ > 3$. We compare the amplitude of this graph with the one of the graph $\mathcal{G}'$ that is identical to $\mathcal{G}$, except that the p_+ valent vertex is replaced by a three-valent vertex and a $p_+ - 1$ valent vertex connected by a + edge. As $\mathcal{G}'$ has one more + edge and one more + vertex with respect to $\mathcal{G}$, the scaling in N of the two amplitudes is the same. We have:

$$A(\mathcal{G}') = \frac{1}{\sqrt{\epsilon}^3} \sqrt{\epsilon}^{3 + p_+ - 1 - p_+} A(\mathcal{G}) \Rightarrow A(\mathcal{G}') > A(\mathcal{G}) .$$

It follows that the most singular graphs must have only three valent + and − vertices and their amplitude is:

$$\frac{n_+^{F_+} n_-^{F_-}}{\sqrt{\epsilon}^{3E_+ + 3E_-}} \sqrt{\epsilon}^{3V_+ + 3V_-} N^{-(D-2)(E_+ - V_+) - (D-2)(E_- - V_-) + 2 - 2(V_{+-} - C_+ - C_- + 1)}$$
$$= n_+^{2 + E_+/3} n_-^{2 + E_-/3} N^2 \left(\sqrt{\epsilon} N^{\frac{D-2}{3}} \right)^{-(E_+ + E_-)} N^{-2(V_{+-} - C_+ - C_- + 1)} .$$

In the double scaling regime $N \to \infty$, $\alpha^{-1} \equiv \sqrt{\epsilon} N^{\frac{D-2}{3}}$ fixed, graphs with:

- only three valent V_+ and V_- vertices,
- planar + and − components,

- dashed edges that form a tree connecting the planar + and − components,

contribute to the same order and dominate the free energy. Such graphs represents planar nodal surfaces, such that all the vertices which are not nodal points are three valent.

Leading graphs in the double scaling regime in the symmetric and broken phases are represented in Figure 11.2.

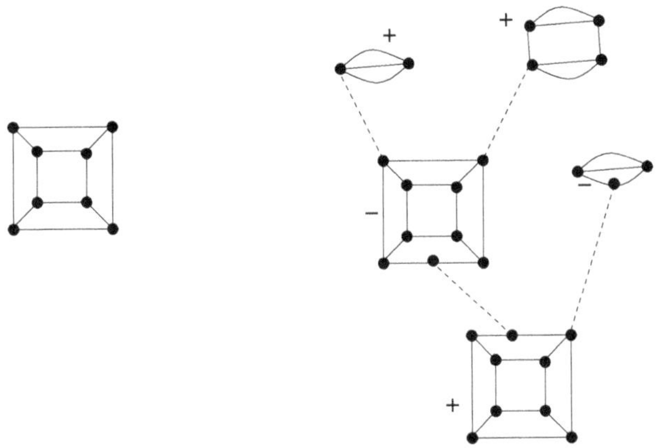

Figure 11.2: Examples of graphs dominating the double scaling regime. On the left a graph dominating the symmetric phase and on the right a graph dominating the broken phase.

In the double scaling regime, the free energy in the broken phase is constructed by summing all the graphs detailed above, with a weight proportional to:

$$(n_+^{1/3}\alpha)^{E_+}(n_-^{1/3}\alpha)^{E_-} .$$

Therefore, it is effectively a double series in the parameters $z_+ = n_+^{1/3}\alpha$ and $z_- = n_+^{1/3}\alpha$, which we expect to have a nontrivial domain of convergence in $\mathbb{C}^2$. For $n_+ = n_- = 1/2$ and hence $z_+ = z_- = z$, this series again becomes a simple series in one variable, z. In this case we expect to find a finite radius of convergence. Note, however, that the coefficients of the power series are now complex, therefore one might have to tune z to a complex value in order to attain a continuum limit with a standard geometrical interpretation.

Chapter 12

Conclusions

In this chapter we will briefly discuss some of the applications of tensor models as models of random geometry.

Tensor models generate weighted triangulations in any dimensions. The flexibility of tensor models comes from the freedom one has to choose the weight of the triangulation by choosing various vertex kernels and various covariances. These choices encode the physical interpretation of a model and there exist several choices that lead to different physical interpretations.

All the fundamental forces in nature (except gravity) are successfully described by perturbatively renormalizable quantum field theory [109, 80, 79, 81, 139] under the guise of the standard model of particle physics formulated by Weinberg, Salam and Glashow [162, 149, 75]. Classical gravity is a theory of the ambient space-time geometry, which is a background for all the other interactions. Crucially, this background is not fixed: the main lesson of general relativity is that the ambient geometry is dynamical and gravity must also be quantized.

While classical general relativity is a field theory, it cannot be quantized the same way the standard model is. As was demonstrated by 't Hooft [110] and Sagnotti [77], general relativity is *not* perturbatively renormalizable. Classical gravity thus predicts its own demise: a more fundamental theory, quantum gravity, must come into play at some high energy scale. Formulating a consistent theory of quantum gravity is one of the deepest outstanding puzzles in theoretical physics today. The lack of perturbative renormalizability suggests that this fundamental theory of nature is quite different from the classical theory of gravity: it would be somewhat surprising that just some small changes here and there can cure a wildly ill behaved theory.

Over the years, several compelling alternative frameworks have been developed, most notably string theory, but many questions still await an answer. While not much is known about the elusive fundamental theory of quantum gravity, one thing is certain:

whatever this theory is, it must make sense of an expression like:

$$\sum_{\text{topologies}} \int [\mathcal{D}g][dX]\, e^{-S_{EH}-S_{SM}-S_{\text{other}}} \,. \tag{12.1}$$

This is how quantum gravity should look [59]: a path integral over matter fields X and metrics g (including a sum over topologies) of the exponential of an action consisting in the Einstein–Hilbert action, the standard model action, and maybe other terms describing physics yet to be discovered.

From the beginning we included a sum over topologies in Eq. (12.1). This sum can be reduced to a single topology by fine tuning S_{other}, but *a priori* there is no reason to do this: the main lesson of quantum field theory is that whatever can happen will happen, and this includes topology change.

The main problem with Eq. (12.1) is that, as already hinted, its precise sense is unclear: the sum over topologies and metrics is ill defined and the naive attempts to use it as the starting point for a theory of quantum gravity fail. A strategy to make sense of this equation is to replace the sum over continuous geometries with a sum over triangulations (more generally discretizations):

$$\sum_{\text{topologies}} \int [\mathcal{D}g] \quad \to \sum_{\text{Triangulations}} ,$$

and to replace the continuum actions by their discretized versions.

However, the moment we replace the sum over continuous geometries by a sum over discretized geometries two questions arise. First, what is the weight (probability distribution) one should use to sum over triangulations? Second, how does one go back from discrete to continuum geometries?

In two dimensions the answer to both questions is provided by the theory of random matrices [59, 128, 88]. The perturbative evaluation of the partition function (and the cumulants) of a matrix model in terms of ribbon Feynman graphs (combinatorial maps) dual to surfaces yields a canonical theory of random discretized two dimensional surfaces and consequently provide an answer to the first question above. In the large N (planar) limit and in the critical regime, infinitely refined surfaces arise. This answers the second question above. The behavior of matter coupled to Liouville gravity in two dimensions is very well captured by matrix models [59, 114, 35, 36, 115, 6, 72, 125, 60, 54, 116, 62, 64] and the double scaling limit [38, 63, 78] corresponds to a continuum theory with finite renormalized Newton's constant.

The lesson one draws from this is that in two dimensions the minimal choices suffice. In order to relate matrix models to quantum gravity one can study the simplest invariant models and consider that the dual triangulation is made of equilateral building blocks. This strategy goes under the name of Euclidean dynamical triangulations [55, 4]. As random tensors generalize random matrices, it is only natural to try to apply tensor models in the same way in higher dimensions.

However, as gravity is quite special in two dimensions (and it is special in a different way in three dimensions, and it is special in yet another different way in four), one should keep an open mind for alternatives. A first natural extension of tensor models one can consider is to change the covariance of the theory and extend tensor models to group field theories [34, 133, 134, 9, 8] and tensor field theories [143, 144, 146]. The main merit of these alternatives is that, while combinatorially they are quite similar to tensor models, the interpretation in terms of metric triangulations is quite different. This extension of tensor models has also a second merit: while the actual physics is still far, one can try to understand the generic features and universal behaviors of such theories.

Euclidean dynamical triangulations

Let us discuss more precisely the interpretation of tensor models as generators of equilateral Euclidean dynamical triangulations. In particular, one would like to understand the relation between the parameters of a tensor model and the physical cosmological and Newton's constants.

Let us consider a D-dimensional triangulation and let us denote the D-simplices in the triangulation by σ_D, and the $(D-2)$-simplices by σ_{D-2}. The Euclidean Einstein–Hilbert action:

$$S_{\text{EH}} = \frac{\Lambda}{8\pi G}\int d^D x\sqrt{g} - \frac{1}{16\pi G}\int d^D x\sqrt{g}R\,,$$

can be discretized on the triangulation following Regge's prescription [140] as:

$$\frac{\Lambda}{8\pi G}\sum_{\sigma_D}\text{vol}(\sigma_D) - \frac{1}{16\pi G}\sum_{\sigma_{D-2}}\text{vol}(\sigma_{D-2})\,[2\pi - \theta(\sigma_{D-2})]\ ,$$

where $\text{vol}(\sigma_D)$ (resp. $\text{vol}(\sigma_{D-2})$) is the volume of the simplex σ_D (resp. σ_{D-2}), and $\theta(\sigma_{D-2})$ is the total dihedral angle spanned by the D-simplices meeting at σ_{D-2}. In the case of an equilateral triangulation with edge length a, the n-dimensional simplices have volume $V_n = a^n\frac{\sqrt{n+1}}{n!\sqrt{2^n}}$, and a D-simplex spans an angle $\arccos\frac{1}{D}$ at each of its $\binom{D+1}{2}$ subsimplices of dimension $(D-2)$. Denoting N_D and N_{D-2} the numbers of D and $(D-2)$-simplices the Regge action on an equilateral triangulation writes:

$$\frac{\Lambda}{8\pi G}V_D N_D - \frac{1}{16\pi G}\left[2\pi V_{D-2}N_{D-2} - \binom{D+1}{2}\left(\arccos\frac{1}{D}\right)V_{D-2}N_D\right]\,.$$

An Euclidean quantum gravity partition function (with no matter content) discretized on an equilateral triangulation [55, 4] therefore takes the form:

$$Z_{\text{grav}}(G,\Lambda) = \sum_{\text{topologies}}\int[\mathcal{D}g]\,e^{\;S_{EH}} \to \sum_{\text{Triangulations}} e^{\kappa_{D-2}N_{D}\;{}_{2}-\kappa_D N_D}\ ,$$

where the two coupling constants are:

$$\kappa_{D-2} = \frac{V_{D-2}}{8G}\ ,\qquad \kappa_D = \frac{\frac{D(D+1)}{4}\left(\arccos\frac{1}{D}\right)V_{D-2} + \Lambda V_D}{8\pi G}\,.$$

The link with tensor models comes from the following remark. Let us consider the logarithm of the partition function of tensor models in Eq. (8.3):

$$\frac{1}{N^D}\ln Z_{\text{TM}}(\lambda;\{t_{\mathcal{B}}\}) = \sum_{\substack{\mathcal{G}\text{ connected}\\ \mathcal{B}^{(2)}\not\subset\mathcal{G}}}\frac{\lambda^{k(\mathcal{G})}}{k(\mathcal{G})!k(\mathcal{G})!}\left(\prod_{\mathcal{B},\,\mathcal{B}\subset\mathcal{G}}(-t_{[\mathcal{B}]})\right)$$

$$\times N^{-\frac{2}{(D-1)!}\omega(\mathcal{G})+\sum_{\mathcal{B},\,\mathcal{B}\subset\mathcal{G}}\frac{2}{(D-2)!}\omega(\mathcal{B})},$$

and let us choose $t_{[\mathcal{B}]} = -N^{-\frac{2}{(D-2)!}\omega(\mathcal{B})}$. Recalling Eq. (3.1):

$$F(\mathcal{G}) = D + \frac{D(D-1)}{2}k(\mathcal{G}) - \frac{2}{(D-1)!}\omega(\mathcal{G}),$$

the logarithm of the partition function of this tensor model becomes:

$$\ln Z_{\mathrm{TM}}(\lambda, N) = \sum_{\substack{\mathcal{G}\text{ connected}\\ \mathcal{B}^{(2)}\not\subset\mathcal{G}}} \frac{\lambda^{k(\mathcal{G})}}{k(\mathcal{G})!k(\mathcal{G})!} N^{-\frac{D(D-1)}{2}k(\mathcal{G})+F(\mathcal{G})},$$

and taking into account that each graph is dual to a triangulation with $2k(\mathcal{G})$ simplices of dimension D and $F(\mathcal{G})$ simplices of dimension $(D-2)$, the gravity partition function is identified with the tensor model free energy if one sets:

$$\kappa_{D-2} \equiv \ln N, \qquad \kappa_D \equiv -\frac{1}{2}\ln\lambda + \frac{D(D-1)}{4}\ln N,$$

that is:

$$\boxed{\begin{aligned} &Z_{\mathrm{grav}}(G,\Lambda) \equiv \ln Z_{\mathrm{TM}}(\lambda, N), \\ &\qquad G = \frac{V_{D-2}}{8\ln N} \\ &\qquad \Lambda = \left(\frac{\pi}{2}\frac{V_{D-2}}{V_D}\right)\frac{\ln\frac{1}{\lambda}}{\ln N} + \frac{D}{4}\frac{V_{D-2}}{V_D}\left(\pi(D-1) - (D+1)\arccos\frac{1}{D}\right). \end{aligned}}$$

In order to track the dimensions of the cosmological and Newton's constant, it is convenient to reinstate the edge lengths a:

$$G = c_1\frac{1}{\ln N}a^{D-2}, \qquad \Lambda(G) = c_2\frac{1}{a^D}G\,\ln\left(\frac{1}{\lambda}\right) + \frac{c_3}{a^2},$$

with $c_1, c_2, c_3 > 0$ some constants of order 1 (and $c_3 = 0$ for $D = 2$). Let us denote $\alpha = \arctan\left[c_2\ln\left(\frac{1}{\lambda}\right)\right]$. The dimensionless constants, see Figure 12.1, are:

$$\tilde{G} = \frac{G}{a^{D-2}}, \quad \tilde{\Lambda} = \Lambda a^2,$$
$$\tilde{G} = c_1\frac{1}{\ln N}, \quad \tilde{\Lambda}(\tilde{G}) = \tan\alpha\,\tilde{G} + c_3.$$

We see that α is the *slope* of the line $\tilde{\Lambda}(\tilde{G})$ (represented as solid in Figure 12.1) with respect to the vertical axis. The coupling constant λ controls the slope α, while N controls the height of the dashed line in Figure 12.1. The physical constants $\tilde{G}$ and $\tilde{\Lambda}$ are the coordinates of the intersection point of the two lines. For $\lambda \in (0,\infty)$, $\alpha \in \left(\frac{\pi}{2}, -\frac{\pi}{2}\right)$

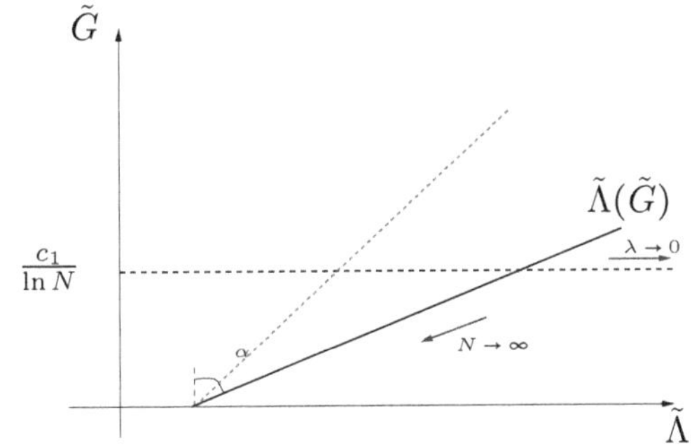

Figure 12.1: The $(\tilde{G}, \tilde{\Lambda})$ plane.

and the line $\tilde{\Lambda}(\tilde{G})$ sweeps the entire plane. As $N > 1$, the two parameters λ and N of tensor models allow one to explore the entire upper half plane of the physical constants. Observe that $\lambda \in (0, \infty)$ corresponds to the *unstable* side of the tensor models. If one considers the quartic tensor model of Chapter 9 for instance, one needs to replace the coupling constant λ above by $\imath\sqrt{\lambda_{\text{quart}}}$, where λ_{quart} is the coupling constant of the quartic models as defined in Chapter 9 (and to restrict the sum to only a subclass of triangulations).

As long as a is kept fixed, the dimensionfull constants are just a finite rescaling of the dimensionless ones. The correspondence between the various regimes of tensor models and the physical coupling constants can be readily understood from this figure, observing that $\lambda \to 0$ at fixed N moves the intersection point to the right $(\alpha \to \frac{\pi}{2})$ at constant height, while the $N \to \infty$ at fixed λ descends the intersection point along the line $\tilde{\Lambda}(\tilde{G})$ at fixed slope α.

The perturbative expansion. We have a and N fixed, $\lambda \to 0$. The perturbative expansion of tensor models is an expansion at *large cosmological constant* and *fixed Newton's constant*:

$$G \text{ constant (not too large)}, \quad \alpha \to \frac{\pi}{2}, \quad \tilde{\Lambda} \to \infty .$$

As expected, this is an expansion around the zero volume state corresponding to $\tilde{\Lambda} = \infty$.

The $1/N$ expansion. We have a and λ fixed, $N \to \infty$. For $D \geq 3$, the $1/N$ expansion is an expansion at *finite cosmological constant, small Newton's constant* and constant slope:

$$\tilde{G} \to 0, \quad \alpha \text{ constant}, \quad \tilde{\Lambda} \to c_3 .$$

It is convenient to rewrite:

$$e^{-S_{\text{EH}}} = e^{-a^{-D} \tan\alpha \text{Vol} + G^{-1}\left(\text{Ric} - c_3 a^{-2} \text{Vol}\right)} ,$$

with Ric and Vol the integrated Ricci scalar and total volume. It is transparent in this form that the $1/N$ expansion is an expansion around geometries for which Ric is maximal at fixed Vol, that is geometries of maximal possible positive curvature. These are the *melonic* geometries.

The situation is rather different in $D = 2$ for which the $1/N$ expansion is an expansion at *small cosmological constant* and *small Newton's constant* with fixed slope and, as the integral of the Ricci scalar is the Euler character, the $1/N$ expansion is a topological expansion.

The large N limit. We have a, λ fixed, $N = \infty$. In this limit one has a *finite cosmological constant* and *zero Newton's constant*, but this regime is approached along a line with fixed slope:

$$\tilde{G} = 0 \,, \quad \alpha \text{ constant} \,, \quad \tilde{\Lambda} = c_3 \,.$$

The large N limit projects onto melonic geometries with maximal positive curvature at given volume. For melonic geometries the numbers of $(D-2)$-simplices and D-simplices are proportional. That is, not only Ric is maximal at given Vol, but also Ric is proportional to Vol. Among the melonic geometries, one has a second expansion governed by the slope α around the zero volume state. That is, once projected onto melonic geometries, the slope α plays the role of an effective cosmological constant which is large and positive.

The critical regime. We have a fixed, $\lambda \to \lambda_c$, $N = \infty$. One still has a *finite cosmological constant* and *zero Newton's constant*, but the slope approaches a critical value:

$$\tilde{G} = 0 \,, \quad \alpha \to \alpha_c \,, \quad \tilde{\Lambda} = c_3 \,.$$

We represented the line with critical slope α_c in red in Figure 12.1. The critical value α_c corresponds to a divergent number of D simplices hence, as a is kept fixed, these geometries have infinite volume.

The continuum limit. We have $a \to 0$, $\lambda \to \lambda_c$, $a^D(\lambda_c - \lambda)^{-1}$ fixed, $N = \infty$. As we have seen, close to criticality, the number of D-simplices diverges, hence in order to keep the physical volume fixed one needs to send a to zero in such a way that $a^D(\lambda_c - \lambda)^{-1}$ is fixed. Due to the natural mass dimensions of the couplings, the dimensionless and dimensionfull constants have a different behavior:

$$\tilde{G} = 0 \,, \quad \alpha \to \alpha_c \,, \quad \tilde{\Lambda} = c_3 \,,$$

while:

$$G = 0 \,, \quad \Lambda = \infty \,.$$

The double scaling regime. We have a fixed, $\lambda \to \lambda_c$, $N \to \infty$, $N^{D-2}(\lambda_c - \lambda)$ fixed. This regime corresponds to a trajectory in the plane of physical coupling constants:

$$\tilde{G} \to 0 \,, \quad \alpha \to \alpha_c \,, \quad \tilde{\Lambda} \to c_3 \,,$$

along which:

$$e^{\frac{c_1(D-2)}{\tilde{G}}}(\alpha - \alpha_c) = \text{constant} \,.$$

The continuum double scaling limit. We have $a \to 0$, $\lambda \to \lambda_c$, $N \to \infty$, with:

$$N^{D-2}(\lambda_c - \lambda) = \text{constant}\,, \quad a^D(\lambda_c - \lambda)^{-1} = \text{constant}\,.$$

In this regime we have:

$$\tilde{G} \to 0\,, \quad \alpha \to \alpha_c\,, \quad \tilde{\Lambda} \to c_3\,,$$

while for the dimensionfull constants we obtain:

$$G \to 0\,, \quad \Lambda \to \infty\,.$$

In conclusion, the analytically accessible regimes of tensor models explore regions with *small positive (or zero)* Newton constant and *large (or very large) positive* cosmological constant. However, care should be taken when trying to relate these regimes of tensor models to infrared physics. As this theory is supposed to describe the ultraviolet behavior of gravity, the cosmological and Newton's constant we are discussing are the ultraviolet ones. In order to go from these values to the infrared coupling constants one should follow some sort of renormalization group flow, and the values of the constants are expected to vary substantially along such a trajectory, as neither the Newton constant nor the cosmological constant are marginal in $D \geq 3$.

The melonic geometries are rather simple, falling into the universality class of the continuous random tree. On the one hand this is good news, as one can treat analytically some statistical systems in random geometry [22, 32, 25, 29, 28]. On the other, one can only do so much with branched polymers and it is important to find regimes in which the emergent geometries are not just branched polymers.

The double scaling limit we have presented in detail in the main body of this book is one attempt to go beyond branched polymers. In this limit larger families of graphs are included and, more importantly, below $D = 6$ it seems likely that the double scaling regime can be followed by a triple scaling regime, revealing an even larger family of graphs. Observe that the strategy one employs in this context is to keep the metric interpretation of a graph as dual to an equilateral triangulation, and to extend (in a controlled manner) the family of graphs included in the statistical ensemble.

The alternative, namely to keep the melonic family but change the metric associated to the melonic triangulations can also be explored quite naturally by slightly enlarging the invariant tensor models.

Generalizations of tensor models

Several generalizations of invariant tensor models have been explored in the literature.

Matrix models extend naturally to matrix field theories, which are intimately related to noncommutative quantum field theory [83]. The same can be done for tensors, and tensor models extend naturally to tensor field theories (TFT) [143, 144, 146] by breaking the unitary invariance of the quadratic part. The covariance of such models possesses a nontrivial spectrum, which in turn is naturally divided into scales. The integration of the high scales leads to a genuine renormalization group flow [18, 19, 12] and the TFTs explore the *tensor theory space* [145] spanned by all the unitary invariant polynomial interactions. The TFTs provide examples of *renormalizable, nonlocal* field theories.

The melonic large N behavior of tensor models corresponds to the ultraviolet limit of TFTs. Typically the couplings grow in the infrared and such models exhibit a rich infrared behavior that remains largely to be explored either by analytic [41, 42] or by numeric [21] methods. We emphasize that the natural metric assignment for the triangulations dual to graphs in TFTs is not yet understood.

A second generalization of tensor models, going under the name of group field theory (GFT) [134, 9, 136, 135] has proven extremely fruitful. In GFT one considers tensors over some Lie group that are furthermore invariant under the diagonal action of the group. With minimal adaptations the $1/N$ expansion holds for GFTs [94, 96]. Due to the diagonal gauge invariance, the amplitude of a GFT graph is the discretized BF action on the dual triangulation. At least in three dimensions, this is exactly the gravity partition function on the dual triangulation (in higher dimensions the BF action must be supplemented by constraints).

The tantalizing fact is that, choosing for instance the group $SU(2)$ in three dimensions, the melonic triangulations (which still dominate) are endowed with a *flat connection*. This, most definitely, is *not* a branched polymer geometry. However, the reader should be aware of the following fact. While a metric on a (pseudo) manifold is uniquely encoded in the holonomies around all the closed loops in that (pseudo) manifold, when discretizing one loses a *local conformal factor*. While the fact that melonic geometries have a flat connection is very encouraging, before fixing this conformal factor issue one cannot discuss the emerging random geometry in GFTs.

Finally, putting TFT and GFT together one obtains tensor group field theories TGFTs [46, 45, 41]. In this case one uses both tensors with a diagonal gauge invariance and a nontrivial covariance. Such models are also perturbatively renormalizable (which is highly nontrivial in this case, and relies crucially on the ultraviolet dominance of the melonic graphs), and exhibit a very rich phase portrait. While much remains to be done, one could hope that this class of models can fix the local conformal factor problem of the usual GFTs.

Part III

Appendices

Appendix A

The Weingarten functions revisited

In this appendix we briefly discuss the Weingarten functions (introduced in [163]) appearing in the Eq. (2.2) and in particular we prove Eq. (2.4).

Integrals over the group of unitary $N \times N$ matrices arise in many instances in physics (see for instance [167] and references therein). As mentioned in the main text, they are computed explicitly [48, 49] as:

$$\int [dU] \prod_{j=1}^{k} U_{a_j b_j} \prod_{j=1}^{k} U^{\dagger}_{\bar{b}_j \bar{a}_j} = \sum_{\sigma,\tau \in \mathfrak{S}(k)} \mathrm{Wg}(N, \sigma^{-1}\tau) \prod_{j=1}^{k} \delta_{a_j \bar{a}_{\sigma(j)}} \delta_{b_j \bar{b}_{\tau(j)}} ,$$

where σ and τ run over the permutations of k elements, $U^{\dagger}$ is the Hermitian conjugate of U and the integral is zero if the numbers of Us and $U^{\dagger}$s are different.

The main properties of the Weingarten functions are:

- *Invariance* [48, 49, 40]. The function $\mathrm{Wg}(N, \pi)$ depends only on the class of π, that is it only depends on the numbers $C_q(\pi)$ (recall that $C_q(\pi)$ denotes the number of cycles of length q of π). In this appendix we will use the alternative notation:

  $$W_{1^{C_1(\pi)} 2^{C_2(\pi)} \ldots} ,$$

  for $\mathrm{Wg}(N, \pi)$.
- *Large N behavior* [48, 49, 40]. At large N we have:

  $$W_{1^{C_1} 2^{C_2} \ldots} = \prod_{q \geq 1} \left[N^{1-2q} (-1)^{q-1} \frac{1}{q} \binom{2q-2}{q-1} \right]^{C_q} + \\ + O(N^{-2+\sum_q (1-2q) C_q}) . \qquad (A.1)$$
- *Recursion* [40]. The Weingarten function obey the recursion relation:

  $$W_{1^0 2^0 \ldots q^0 \ldots} = 1 ,$$

$$NW_{1^{C_1}2^{C_2}\dots r^{C_r+1}\dots} = \delta_{r1}W_{1^{C_1}2^{C_2}\dots} - \sum_{s,t\geq 1}^{s+t=r} W_{1^{C_1}\dots s^{C_s+1}\dots t^{C_t+1}\dots} \\ - \sum_{t\geq 1} tC_t \, W_{1^{C_1}2^{C_2}\dots t^{C_t-1}\dots (r+t)^{C_{r+t}+1}\dots} , \tag{A.2}$$

where by convention W is set to zero if on of the Cs is negative.

The recursion relations uniquely fix the Weingarten functions. In order to find the Weingarten functions for permutations of k elements one notices that the sub system of equations with $r + \sum qC_q \leq k$ closes. Omitting the qs with $C_q = 0$, at first orders we find:

$$k=1:\begin{cases} r=1 & NW_{1^1}=1 \end{cases} \Rightarrow \begin{cases} W_{1^1}=\frac{1}{N} \end{cases} ,$$

$$k=2:\begin{cases} r=1, C_1=1 & NW_{1^2}=W_{1^1}-W_{2^1} \\ r=2 & NW_{2^1}=-W_{1^2} \end{cases} \Rightarrow \begin{cases} W_{1^2}=\frac{1}{(N^2-1)} \\ W_{2^1}=\frac{-1}{N(N^2-1)} \end{cases} ,$$

$$k=3:\begin{cases} r=1, C_1=2 & NW_{1^3}=W_{1^2}-2W_{1^12^1} \\ r=1, C_2=1 & NW_{1^12^1}=W_{2^1}-2W_{3^1} \\ r=2, C_1=1 & NW_{1^12^1}=-W_{1^3}-W_{3^1} \\ r=3 & NW_{3^1}=-W_{1^12^1}-W_{1^12^1} \end{cases} \Rightarrow \begin{cases} W_{1^3}=\frac{N^2-2}{N(N^2-1)(N^2-4)} \\ W_{1^12^1}=\frac{-N}{N(N^2-1)(N^2-4)} \\ W_{3^1}=\frac{2}{N(N^2-1)(N^2-4)} \end{cases} .$$

A.1 Generating function

Let us define the function of two $N \times N$ matrices A and B:

$$F(A,B) \equiv \int [dU] \, e^{\mathrm{Tr}[UA]+\mathrm{Tr}[U^\dagger B]} ,$$

where $[dU]$ denotes the Haar measure on the unitary group. This integral has been much studied in physics [131, 167], especially in the large N limit. Due to the invariance of the Haar measure, $F(A,B)$ is invariant under the action of the unitary group:

$$F(U_1AU_2, U_2^\dagger BU_1^\dagger) = F(A,B) .$$

By the fundamental theorem of classical invariant theory [132] (and also shown in Chapter 2), the symmetrized trace invariants:

$$P_{C_1C_2\dots C_q\dots} = \prod_{q\geq 1}^{\infty} \frac{1}{C_q!}\Big(\frac{1}{q}\,\mathrm{Tr}[(AB)^q]\Big)^{C_q} \\ = \frac{1}{(\sum qC_q)!} \sum_{\substack{\tau \\ C_q(\tau)=C_q}} \prod_q \Big(\mathrm{Tr}[(AB)^q]\Big)^{C_q(\tau)} ,$$

span the space of invariant functions, therefore the function $F(A,B)$ admits a decomposition:

$$F(A,B) = \sum_{\mathbf{C}} V_{1^{C_1}2^{C_2}\dots} P_{C_1C_2\dots} ,$$

where $\mathbf{C}$ runs over all the possible strings $C_1, C_2, \dots$ of nonnegative integers. Although at finite N these polynomials are not independent, they become so in the large N limit, hence we will treat them (formally) as a basis in the space of invariants. At finite N all the statements here hold only for the polynomials of degree less than N.

Proposition A.1. *$F(A,B)$ is the generating function for the Weingarten functions, that is:*

$$V_{1^{C_1}2^{C_2}\dots} = W_{1^{C_1}2^{C_2}\dots}\,.$$

Proof. Following [48], we first note that:

$$\begin{aligned}\mathrm{Wg}(N,\pi) &= \int [dU]\left(\prod_{j=1}^{\sum qC_q(\pi)} U_{jj}U^\dagger_{\pi(j)j}\right)\\ &= \left[\left(\prod_{j=1}^{\sum qC_q(\pi)} \frac{d}{dA_{jj}}\frac{d}{dB_{j\pi(j)}}\right)F(A,B)\right]_{A=B=0}.\end{aligned}$$

The function $F(A,B)$ can be expressed as:

$$\begin{aligned}F(A,B) &= \sum_{\mathbf{C}} V_{1^{C_1}\dots}P_{C_1\dots} = \sum_{\mathbf{C}} V_{1^{C_1}\dots}\frac{1}{(\sum qC_q)!}\sum_{\substack{\tau\\ C_q(\tau)=C_q}}\prod_q\Big(\mathrm{Tr}[(AB)^q]\Big)^{C_q(\tau)}\\ &= \sum_{\mathbf{C}} V_{1^{C_1}\dots}\frac{1}{(\sum qC_q)!}\sum_{\substack{\tau\\ C_q(\tau)=C_q}}\ \sum_{a_1,\dots,a_{\sum qC_q}=1}^{N}\ \prod_{j=1}^{\sum qC_q}[AB]_{a_j a_{\tau(j)}}\,.\end{aligned}$$

Combining the two equations we obtain:

$$\begin{aligned}\mathrm{Wg}(N,\pi) = &\sum_{\mathbf{C}} V_{1^{C_1}\dots}\frac{1}{(\sum qC_q)!}\sum_{\substack{\tau\\ C_q(\tau)=C_q}}\ \sum_{\substack{a_1,\dots,a_{\sum qC_q}=1\\ b_1,\dots,b_{\sum qC_q}=1}}^{N}\\ &\times\left[\left(\prod_{j=1}^{\sum qC_q(\pi)}\frac{d}{dA_{jj}}\frac{d}{dB_{j\pi(j)}}\right)\left(\prod_{j=1}^{\sum qC_q} A_{a_j b_j}B_{b_j a_{\tau(j)}}\right)\right]_{A=B=0}.\end{aligned}$$

The derivative is zero unless $\mathbf{C}$ is such that $\sum qC_q = \sum qC_q(\pi)$, and in this case it computes to:

$$\sum_{\sigma_1,\sigma_2\in\mathfrak{S}(\sum qC_q(\pi))}\ \prod_{j=1}^{\sum qC_q(\pi)} \delta_{a_j,\sigma_1(j)}\delta_{b_j,\sigma_1(j)}\delta_{b_j,\sigma_2(j)}\delta_{a_{\tau(j)},\sigma_2\pi(j)}\,,$$

hence summing over a and b we obtain:

$$\begin{aligned}\mathrm{Wg}(N,\pi) = &\sum_{\substack{\mathbf{C}\\ \sum qC_q=\sum qC_q(\pi)}} V_{1^{C_1}\dots}\frac{1}{(\sum qC_q)!}\sum_{\substack{\tau\\ C_q(\tau)=C_q}}\ \sum_{\sigma_1,\sigma_2\in\mathfrak{S}(\sum qC_q)}\\ &\times\prod_{j=1}^{\sum qC_q}\delta_{\sigma_1(j),\sigma_2(j)}\,\delta_{\sigma_1(j),\sigma_2\pi\tau^{-1}(j)}\end{aligned}$$

$$= \sum_{\substack{\mathbf{C} \\ \sum qC_q = \sum qC_q(\pi)}} V_{1^{C_1}\dots} \sum_{\substack{\tau \\ C_q(\tau)=C_q}} \prod_{j=1}^{\sum qC_q} \delta_{j,\pi\tau^{-1}(j)} \,.$$

The product over j is zero unless $\pi = \tau$, therefore only one term survives and:

$$\mathrm{Wg}(N,\pi) = V_{1^{C_1(\pi)}\dots} \,.$$

□

A.2 Schwinger–Dyson equations

Having established that the Weingarten functions are the coefficients of the expansion of $F(A,B)$ on the basis of symmetrized trace invariant polynomials, we will now recover the recursion relations they obey as a set of Schwinger–Dyson equations for $F(A,B)$.

Proposition A.2. *The function $F(A,B)$ respects the Schwinger–Dyson equation:*

$$\Big(\sum_a \frac{d^2}{dA_{na}dB_{am}}\Big) F(A,B) = \delta_{nm}\, F(A,B) \,. \tag{A.3}$$

Remark that $W_{1^0 2^0 \dots} = \int [dU] = 1$. Furthermore the Weingarten functions respect the recursion relation (A.2).

Proof. The Schwinger–Dyson equation is trivial: from the definition of $F(A,B)$ we immediately have:

$$\begin{aligned}
&\Big(\sum_a \frac{d^2}{dA_{na}dB_{am}}\Big) \int [dU]\, e^{\mathrm{Tr}[UA]+\mathrm{Tr}[U^\dagger B]} \\
&\quad = \int [dU] \Big(\sum_a U^\dagger_{ma} U_{an}\Big) e^{\mathrm{Tr}[UA]+\mathrm{Tr}[U^\dagger B]} \\
&\quad = \delta_{nm} \int [dU]\, e^{\mathrm{Tr}[UA]+\mathrm{Tr}[U^\dagger B]} \,.
\end{aligned}$$

In order to derive a relation for the Ws from this Schwinger–Dyson equation we will substitute the expansion of $F(A,B)$ in invariants into Eq. (A.3). The derivatives of the invariant polynomials $P_{C_1C_2\dots}$ are:

$$\frac{d}{dB_{am}} P_{C_1C_2,\dots} = \sum_{q\geq 1} [(AB)^{q-1}A]_{ma}\, P_{C_1C_2\dots C_q-1\dots} \,,$$

and $P_{\dots C\dots}$ is set to zero if $C<0$. We have:

$$\frac{d}{dB_{am}} F(A,B) = \sum_{\mathbf{C}} \sum_{q\geq 1} [(AB)^{q-1}A]_{ma}\, W_{1^{C_1}2^{C_2}\dots q^{C_q+1}\dots}\, P_{C_1C_2\dots C_q\dots} \,,$$

where we have shifted the dummy summation index C_q by 1.

It follows that the left hand side of Eq. (A.3) evaluates to:

$$\sum_{\mathbf{C}}\sum_{q\geq 1}\sum_{r=1}^{q}[(AB)^{r-1}]_{mn}\ \mathrm{Tr}[(AB)^{q-r}]\ W_{1^{C_1}\dots q^{C_q+1}\dots}\ P_{C_1\dots C_q\dots}$$
$$+\sum_{\mathbf{C}}\sum_{q\geq 1}\sum_{s\geq 1}[(AB)^{q+s-1}]_{mn}\ W_{1^{C_1}\dots q^{C_q+1}\dots s^{C_s+1}\dots}\ P_{C_1\dots C_q\dots C_s\dots}\ ,$$

where in the second term we have shifted C_s by one.

The first term requires more work. Exchanging the sums over q and r it rewrites as:

$$\sum_{\mathbf{C}}\sum_{r\geq 1}[(AB)^{r-1}]_{mn}$$
$$\times\Big[NW_{1^{C_1}\dots r^{C_r+1}\dots}\ P_{C_1\dots}+\sum_{q\geq r+1}\mathrm{Tr}[(AB)^{q-r}]\ W_{1^{C_1}\dots q^{C_q+1}\dots}\ P_{C_1\dots}\Big]\,.$$

Denoting $t=q-r$ in the second term and noting that:

$$\mathrm{Tr}[(AB)^{t}]\ P_{C_1\dots C_t\dots}=t(C_t+1)P_{C_1\dots C_t+1\dots}\ ,$$

the term becomes:

$$\sum_{\mathbf{C}}\sum_{r\geq 1}[(AB)^{r-1}]_{mn}\Big[NW_{1^{C_1}\dots r^{C_r+1}\dots}\ P_{C_1\dots}$$
$$+\sum_{t\geq 1}t(C_t+1)\ W_{1^{C_1}\dots (t+r)^{C_{t+r}+1}}\ P_{C_1\dots C_t+1\dots}\Big]$$
$$=\sum_{\mathbf{C}}\sum_{r\geq 1}[(AB)^{r-1}]_{mn}\Big[NW_{1^{C_1}\dots r^{C_r+1}\dots}\ P_{C_1\dots}$$
$$+\sum_{t\geq 1}tC_t\ W_{1^{C_1}\dots t^{C_t-1}\dots (t+r)^{C_{t+r}+1}\dots}\ P_{C_1\dots C_t\dots}\Big]\,.$$

Using again the fundamental theorem of classical invariant theory [132], the polynomials $[(AB)^{r-1}]_{mn}P_{C_1\dots}$ span the set of covariant polynomials, and become a basis in the large N limit, hence Eq. (A.3), yields in components:

$$\delta_{mn}P_{C_1\dots}:\qquad NW_{1^{C_1+1}\dots}+\sum_{t\geq 1}tC_tW_{1^{C_1}\dots t^{C_t-1}(t+1)^{C_{t+1}+1}\dots}=W_{1^{C_1}\dots}$$
$$[(AB)^{r-1}]_{mn}P_{C_1\dots}:\qquad NW_{1^{C_1}\dots r^{C_r+1}\dots}+\sum_{t\geq 1}tC_t\ W_{1^{C_1}\dots t^{C_t-1}\dots (t+r)^{C_{t+r}+1}\dots}$$
$$+\sum_{s,t\geq 1}^{s+t=r}W_{1^{C_1}\dots s^{C_s+1}\dots t^{C_t+1}\dots}=0\ ,$$

which can be encoded in the unique equation (A.2).

□

The recursion equation can be solved at leading order in $1/N$ by bootstrap. The functions W_{q^1} obey the system of equations obtained for $r=q, C_{q'}=0\forall q'$:

$$W_{1^1}=\frac{1}{N}\ ;\qquad 0=NW_{q^1}+\sum_{s,t\geq 1}^{s+t=q}W_{s^1t^1}\ ,\quad \forall q\geq 2\,.$$

Let us assume that at leading order in $1/N$ the functions W factor:

$$W_{1^{C_1}\dots q^{C_q}\dots} = \prod_{q\geq 1} \left(W_{q^1}\right)^{C_q} + \text{corrections}\,.$$

Then at leading order in $1/N$ the system of equations becomes:

$$W_{1^1} = \frac{1}{N}\,; \qquad 0 = NW_{q^1} + \sum_{s,t\geq 1}^{s+t=q} W_{s^1}W_{t^1}\,, \quad \forall q \geq 2\,,$$

and is solved in terms of the Catalan numbers $\text{Cat}(n) = \frac{1}{n+1}\binom{2n}{n}$, which obey the recursion $\text{Cat}(n+1) = \sum_{p,q\geq 0}^{p+q=n} \text{Cat}(p)\,\text{Cat}(q)$. We obtain:

$$W_{q^1} = N^{1-2q}(-1)^{q-1}\,\text{Cat}(q-1) + \text{corrections}\,,$$

and plugging this result in the full recursion, it is easy to check that the assumption of factorization at large N is consistent.

Appendix B

Probability measures

In this appendix we review some standard material on probability measures. We will first discuss in some detail the Gaussian measures and subsequently we will discuss the more interesting perturbed Gaussian measures and their so-called *perturbative expansion* in terms of Feynman graphs.

We deal with real or complex random variables. Let $\mathcal{N}$ be a nonempty set and denote its elements by $a \in \mathcal{N}$. The set $\mathcal{N}$ can be finite, countable (indexing for instance a basis in the Hilbert space) or uncountable. The last case is of no interest for this work. Let $\phi : \mathcal{N} \to \mathbb{R}$ (or $\mathbb{C}$) be a real (or complex)-valued random function defined on $\mathcal{N}$ (and if ϕ is a complex function, we denote $\bar{\phi}$ its complex conjugate). We denote the (vector) space of such functions Func($\mathcal{N}$). We denote by $\phi_a \equiv \phi(a)$ the value of ϕ at the point a. For complex functions we denote $\bar{\phi}_a$ the value of $\bar{\phi}$ at the point $\bar{a} \in \mathcal{N}$. The bar on an element a *does not* denote the complex conjugation, but merely signals that the complex conjugated function is evaluated at that element. From now on we will use the physics terminology and call the function ϕ the *field*.

A probability measure $\mu(\phi)$ (resp. $\mu(\phi, \bar{\phi})$) of the joint distribution of ϕ_a (resp. $\phi_a, \bar{\phi}_{\bar{a}}$) can be analyzed in terms of its *moments*:

$$\langle \phi_{a_1} \dots \phi_{a_q} \rangle_\mu = \int d\mu(\phi) \quad \phi_{a_1} \dots \phi_{a_q} ,$$
$$\langle \phi_{a_1} \dots \phi_{a_q} \bar{\phi}_{\bar{a}_1} \dots \bar{\phi}_{\bar{a}_k} \rangle_\mu = \int d\mu(\phi, \bar{\phi}) \quad \phi_{a_1} \dots \phi_{a_q} \bar{\phi}_{\bar{a}_1} \dots \bar{\phi}_{\bar{a}_k} ,$$

where μ is normalized $\langle 1 \rangle = 1$. For complex fields we often consider distributions such that the moments are nonzero only for $q = k$.

Let us denote $\mathbf{P}$ a partition of the set $\mathcal{V} = \{a_1, \dots a_q\}$ (respectively, for complex fields, of the set $\mathcal{V} = \{a_1, \dots a_q, \bar{a}_1, \dots \bar{a}_k\}$) into disjoint subsets. We denote by $|\mathbf{P}|$ the number of parts in $\mathbf{P}$, and we denote the parts of $\mathbf{P}$ by $\mathcal{V}(\alpha)$ for $\alpha = 1, 2, \dots |\mathbf{P}|$. For real fields we denote $|\mathcal{V}(\alpha)| = q(\alpha)$, while in the complex case we denote $|\mathcal{V}(\alpha) \cap \{a_1 \dots a_q\}| = q(\alpha)$ and $|\mathcal{V}(\alpha) \cap \{\bar{a}_1 \dots \bar{a}_k\}| = k(\alpha)$. Note that $\sum_{\alpha=1}^{|\mathbf{P}|} q(\alpha) = q$, (respectively $\sum_{\alpha=1}^{|\mathbf{P}|} q(\alpha) =$

$q, \sum_{\alpha=1}^{|\mathbf{P}|} k(\alpha) = k$ for the complex case). The cumulants of μ are defined implicitly by:

$$\langle \phi_{a_1} \dots \phi_{a_q} \rangle_\mu = \sum_{\mathbf{P}} \prod_{\alpha=1}^{|\mathbf{P}|} \kappa_{q(\alpha)} \Big[\phi_{a_i} | a_i \in \mathcal{V}(\alpha) \Big] ,$$

$$\langle \phi_{a_1} \dots \phi_{a_q} \bar{\phi}_{\bar{a}_1} \dots \bar{\phi}_{\bar{a}_k} \rangle_\mu = \sum_{\mathbf{P}} \prod_{\alpha=1}^{|\mathbf{P}|} \kappa_{q(\alpha)+k(\alpha)} \Big[\phi_{a_i}, \bar{\phi}_{\bar{a}_j} | a_i, \bar{a}_j \in \mathcal{V}(\alpha), \Big] .$$

Again in the complex case we often consider distributions such that only the cumulants with $q(\alpha) = k(\alpha)$ are nonzero.

The cumulants can be computed in terms of the moments using the Möbius inversion formula. The generating function of the moments of μ is defined as:

$$Z(J;\mu) = \int d\mu(\phi) \quad e^{\sum_a \phi_a J_a} ,$$

$$Z(J,\bar{J};\mu) = \int d\mu(\phi,\bar{\phi}) \quad e^{\sum_{\bar{a}} \bar{\phi}_{\bar{a}} \bar{J}_{\bar{a}} + \sum_a \phi_a J_a} ,$$

where $\sum_a$ designates the Lebesgue integral over $\mathcal{N}$. The cumulants of μ can be computed as the derivatives of the logarithm of the generating function of the moments:

$$\kappa_q(\phi_{a_1}, \dots \phi_{a_q}) = \left[\frac{\partial^{(q)}}{\partial J_{a_1} \dots \partial J_{a_q}} \ln Z(J;\mu) \right]_{J=0} ,$$

$$\kappa_{q+k}(\phi_{a_1}, \dots \phi_{a_q}, \bar{\phi}_{\bar{a}_1}, \dots \bar{\phi}_{\bar{a}_k}) =$$

$$\left[\frac{\partial^{(q+k)}}{\partial J_{a_1} \dots \partial J_{a_q} \partial \bar{J}_{\bar{a}_1} \dots \bar{J}_{\bar{a}_k}} \ln Z(J,\bar{J};\mu) \right]_{J=\bar{J}=0} .$$

B.1 Gaussian measures

While many subtleties arise when the cardinality of $\mathcal{N}$ is infinite, normalized Gaussian measures can always be defined. Consider a linear operator $C : \mathrm{Func}(\mathcal{N}) \to \mathrm{Func}(\mathcal{N})$ and denote its kernel $C_{a,a'}$:

$$(C\phi)_a = \sum_b C_{a,b} \, \phi_b ,$$

where $\sum_b$ denotes the Lebesgue integral on $\mathcal{N}$. If $\mathcal{N}$ is the set:

$$\mathcal{N} = \left\{ a^{\mathcal{D}} \;\middle|\; a^{\mathcal{D}} = (a^1, \dots a^D), \; a^c = 1 \dots N \right\} ,$$

the Lebesgue measure on $\mathcal{N}$ is a pure point measure (that is $\sum_{a^{\mathcal{D}}}$ is, as suggested by the notation, just the sum over all possible D-uples of integers), the random fields on $\mathcal{N}$ are random tensor with D indices $\mathbb{T}_{a^{\mathcal{D}}}$ and the kernels of operators C are $N^D \times N^D$ matrices $C_{a^{\mathcal{D}},b^{\mathcal{D}}}$.

Definition B.1. *The* normalized Gaussian probability measure of covariance C, *denoted* μ_C, *is the measure whose only nonzero moments are:*

- *for real fields $\phi : \mathcal{N} \to \mathbb{R}$,*

$$\left\langle \prod_{i=1}^{q} \phi_{a_i} \right\rangle_{\mu_C} = \sum_{\mathbf{P}_2} \prod_{\{a_i,a_j\}\in\mathbf{P}_2} C_{a_i,a_j} ,$$

where $\mathbf{P}_2$ runs over all the pairings *of $\{a_1,\dots a_{2q}\}$, that is the partitions of $\{a_1,\dots a_{2q}\}$ into two element subsets. If q is odd the set of pairings is empty and the right hand side is zero.*

- *for complex fields $\phi, \bar{\phi} : \mathcal{N} \to \mathbb{C}$,*

$$\left\langle \left(\prod_{i=1}^{q} \phi_{a_i}\right)\left(\prod_{j=1}^{k} \bar{\phi}_{\bar{a}_j}\right) \right\rangle_{\mu_C} = \delta_{qk} \sum_{\pi\in\mathfrak{S}(k)} \prod_{i=1}^{k} C_{a_i,\bar{a}_{\pi(i)}} ,$$

where π runs over the permutations of k elements.

The covariance C *needs not be an invertible operator.* While in the real case C must by symmetric, C does not need to be Hermitian in the complex case. The following proposition is very useful:

Proposition B.1. *The moments of the Gaussian measure are computed as:*

$$\left\langle \prod_{i=1}^{q} \phi_{a_i} \right\rangle_{\mu_C} = \left[e^{\frac{1}{2}\sum_{r,s} C_{r,s} \frac{\partial}{\partial\phi_r}\frac{\partial}{\partial\phi_s}} \prod_{i=1}^{q} \phi_{a_i} \right]_{\phi=0}$$

$$\left\langle \left(\prod_{i=1}^{q} \phi_{a_i}\right)\left(\prod_{j=1}^{k} \bar{\phi}_{\bar{a}_j}\right) \right\rangle_{\mu_C} = \left[e^{\sum_{r,\bar{s}} C_{r,\bar{s}} \frac{\partial}{\partial\phi_r}\frac{\partial}{\partial\bar{\phi}_{\bar{s}}}} \left(\prod_{i=1}^{q} \phi_{a_i}\right)\left(\prod_{j=1}^{k} \bar{\phi}_{\bar{a}_j}\right) \right]_{\substack{\phi=0\\ \bar{\phi}=0}} .$$

Proof. The proof is a direct computation. For complex fields for instance, expanding the exponential on the right hand side, we obtain:

$$\sum_{n\geq 0} \frac{1}{n!} \left[\left(\sum_{r,\bar{s}} C_{r,\bar{s}} \frac{\partial}{\partial\phi_r}\frac{\partial}{\partial\bar{\phi}_{\bar{s}}} \right)^n \left(\prod_{i=1}^{q} \phi_{a_i}\right)\left(\prod_{j=1}^{k} \bar{\phi}_{\bar{a}_j}\right) \right]_{\substack{\phi=0\\ \bar{\phi}=0}} ,$$

which is zero unless $q = k = n$. Evaluating the derivatives this becomes:

$$\frac{1}{k!} \sum_{r_1,\bar{s}_1} \cdots \sum_{r_k,\bar{s}_k} C_{r_1,\bar{s}_1} \dots C_{r_k,\bar{s}_k} \sum_{\sigma,\pi\in\mathfrak{S}(k)} \prod_{i=1}^{k} \delta_{a_i r_{\sigma(i)}} \delta_{\bar{a}_i \bar{s}_{\pi(i)}} = \sum_{\pi\in\mathfrak{S}(k)} \prod_{i=1}^{k} C_{a_i \bar{a}_{\pi(i)}} .$$

□

As, by the Stone–Weierstrass theorem, the polynomials are dense in the space of functions, it follows that for any functional $F(\phi)$ (resp. $F(\phi,\bar{\phi})$):

$$\langle F(\phi) \rangle_{\mu_C} = \left[e^{\frac{1}{2}\sum_{r,s} C_{r,s} \frac{\partial}{\partial\phi_r}\frac{\partial}{\partial\phi_s}} F(\phi) \right]_{\phi=0} ,$$

$$\langle F(\phi,\bar\phi)\rangle_{\mu_C} = \left[e^{\sum_{r,\bar s} C_{r,\bar s} \frac{\partial}{\partial \phi_r} \frac{\partial}{\partial \bar\phi_{\bar s}}} \; F(\phi,\bar\phi) \right]_{\substack{\phi=0\\ \bar\phi=0}} .$$

These formulae can be given a graphical representation in terms of *Feynman "graphs"*. We will sometimes be using the following shorthand notations:

$$\frac{\partial}{\partial\phi} C \frac{\partial}{\partial\bar\phi} \equiv \sum_{r,\bar s} C_{r,\bar s} \frac{\partial}{\partial\phi_r} \frac{\partial}{\partial\bar\phi_{\bar s}} , \qquad J C \bar J \equiv \sum_{r,\bar s} J_r C_{r,\bar s} \bar J_{\bar s} ,$$

$$\text{and} \quad \phi J + \bar\phi \bar J \equiv \sum_a \phi_a J_a + \sum_{\bar a} \bar\phi_{\bar a} \bar J_{\bar a} ,$$

and similarly for the real case.

B.1.1 Feynman graphs

The graphical representation we present here is model dependent. The class of combinatorial structures obtained varies slightly as a function of the details of the model, and in order to present this construction here we will choose a specific model. In the main text we will adapt this graphical representation to the various situations we encounter.

Let us consider a real field ϕ and a finite set $\mathcal{N}$. Any functional F expands in Taylor series as:

$$F(\phi) = \sum_q \sum_{a_1 \dots a_q} \left[\left(\prod_{i=1}^q \frac{\partial}{\partial \phi_{a_i}} \right) F(\phi) \right]_{\phi=0} \frac{\prod_{i=1}^q \phi_{a_i}}{q!} .$$

In order to compute the Gaussian expectation of F, we commute this Taylor sum with the Gaussian integral. This is a questionable step, as nothing ensures in principle that the series we obtain is not divergent. Of course, if F is a polynomial the series is just a finite sum, but for observables like $\frac{1}{1+\sum_a \phi_a^2}$ the series is not convergent. We continue here with this formal manipulation, keeping in mind that one needs in principle to study the convergence properties of the series obtained by this procedure.

The Gaussian expectation of F becomes:

$$\sum_{k\geq 0} \sum_{a_1 \dots a_{2k}} \left[\frac{1}{(2k)!} \partial_{\phi_{a_1}} \dots \partial \phi_{a_{2k}} F \right]_{\phi=0} \sum_{\mathbf{P}_2} \prod_{\{a_i,a_j\}\in \mathbf{P}_2} C_{a_i,a_j} , \tag{B.1}$$

where $\mathbf{P}_2$ runs over the partitions of $\{a_1, \dots a_{2k}\}$ into two element subsets.

The terms in the above sum can be given a graphical representation. We represent $F(\phi)$ as a vertex embedded in the plane, or a *fat vertex*. Each derivative is represented as a halfedge hooked to the vertex (hence the vertex has $2k$ halfedges), and the halfedge representing $\frac{\partial}{\partial\phi_{a_i}}$ is labeled by i. To the vertex we associate the weight:

$$\left[\frac{1}{(2k)!} \partial_{\phi_{a_1}} \dots \partial \phi_{a_{2k}} F \right]_{\phi=0} .$$

We order the halfedges in an arbitrary, but *fixed*, order around the fat vertex. For instance we can order them clockwise as $\frac{\partial}{\partial\phi_{a_1}}$, $\frac{\partial}{\partial\phi_{a_2}}$, and so on. The covariance $C_{a_i a_j}$

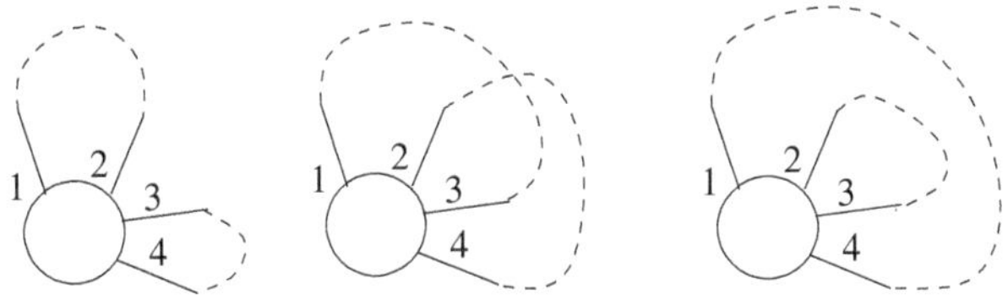

Figure B.1: Feynman maps.

is represented as a dashed edge connecting the halfedges i and j. For the case $k = 2$, the three resulting drawings are presented in Figure B.1.

These drawings are known as Feynman "graphs". However, this is a misnomer: as graphs, the three terms above are identical and consist in a vertex with two self loops (i.e. two edges starting and ending at the vertex). However, they are distinguished by the order of the halfedges at the vertex and therefore they represent distinct combinatorial maps (see Definition 3.9) with one vertex. Note that the labeling of the halfedges is completely arbitrary, hence a map, which is associated to a triple $(\mathcal{S}, \sigma, \alpha)$ according to Definition 3.9, is counted up to relabelings of the elements of $\mathcal{S}$.

B.1.2 Properties of the Gaussian measure

The normalized Gaussian measures have many very interesting properties, some of which are listed here.

Integral representation. If $\mathcal{N}$ is a finite set, then the covariance is a matrix. If the covariance is invertible the normalized Gaussian measure of covariance C can be written alternatively as:

- for a real field $\phi : \mathcal{N} \to \mathbb{R}$,

$$d\mu_C(\phi) = \frac{1}{\sqrt{\det(C)}} \left(\prod_{a \in \mathcal{N}} \frac{d\phi_a}{\sqrt{2\pi}} \right) \quad e^{-\frac{1}{2} \sum_{a,b} \phi_b (C^{-1})_{b,a} \phi_a} .$$

- for a complex field $\phi : \mathcal{N} \to \mathbb{C}$,

$$d\mu_C(\phi, \bar{\phi}) = \frac{1}{\det(C)} \left(\prod_{a \in \mathcal{N}} \frac{d\bar{\phi}_{\bar{a}} d\phi_a}{2\imath\pi} \right) \quad e^{-\sum_{a,\bar{b}} \bar{\phi}_{\bar{b}} (C^{-1})_{\bar{b},a} \phi_a} . \tag{B.2}$$

Bounds. If the covariance is a Hermitian operator, then the Gaussian expectation of any functional is bounded by the sup norm of that functional:

$$\|F(\phi, \bar{\phi})\|_\infty \equiv \sup_{\phi, \bar{\phi}} |F(\phi, \bar{\phi})| , \qquad \left[e^{\frac{\partial}{\partial \phi} C \frac{\partial}{\partial \bar{\phi}}} \; F(\phi, \bar{\phi}) \right]_{\substack{\phi = 0 \\ \bar{\phi} = 0}} \leq \|F(\phi, \bar{\phi})\|_\infty ,$$

as can be seen from the integral expression in Eq. (B.2). If C is not invertible this bound still holds, as it holds for all the measures with covariance:

$$C^\epsilon = C + \epsilon 1^{\text{Ker}} ,$$

where 1^{Ker} is the identity operator on the kernel of C, which are invertible.

The generating function of the moments. The generating function of the moments of a normalized Gaussian measure is evaluated explicitly as:

$$
\begin{aligned}
Z(J;\mu_C) = \left\langle e^{J\phi}\right\rangle_{\mu_C} &= \sum_{q\geq 0}\frac{1}{(2q)!}\sum_{a_1,\dots a_{2q}}\left(\prod_{i=1}^{2q}J_{a_i}\right)\sum_{\mathbf{P}_2}\prod_{\{a_i,a_j\}\in\mathbf{P}_2}C_{a_i,a_j}\\
&= \sum_{q\geq 0}\frac{1}{(2q)!}\frac{(2q)!}{2^q q!}\left(\sum_{a,b}J_aC_{a,b}J_b\right)^q = e^{\frac{1}{2}\sum_{a,b}J_aC_{a,b}J_b}\\
Z(J,\bar J;\mu_C) = \left\langle e^{\phi J+\bar\phi\bar J}\right\rangle_{\mu_C} &\\
&= \sum_{k\geq 0}\frac{1}{k!k!}\sum_{\substack{a_1,\dots a_k\\ \bar b_1,\dots \bar b_k}}\left(\prod_{i=1}^{k}J_{a_i}\bar J_{\bar a_i}\right)\sum_{\pi\in\mathfrak{S}(k)}\prod_{i=1}^{k}C_{a_i,\bar a_{\pi(i)}}\\
&= e^{\sum_{a,\bar a}J_aC_{a,\bar a}\bar J_{\bar a}}\ ,
\end{aligned}
$$

that is the only nonzero cumulant is κ_2, and it equals the covariance.

Covariance split. An important property of normalized Gaussian measures is the so-called *covariance split* property. Namely, if one splits the covariance of a Gaussian measure into the sum of two covariances, $C = C^1 + C^2$, then there exists an associated split of the field ϕ into the sum of two *independent* fields $\phi = \phi^1 + \phi^2$ (and $\bar\phi = \bar\phi^1 + \bar\phi^2$), such that ϕ^1 is distributed on a normalized Gaussian of covariance C^1 and ϕ^2 is distributed on a normalized Gaussian of covariance C^2. Taking into account that for any polynomials P and Q:

$$
\begin{aligned}
&\left[P(\partial_\phi,\partial_{\bar\phi})Q(\phi,\bar\phi)\right]_{\substack{\phi=0\\ \bar\phi=0}}\\
&= \sum_{p,\bar p,q,\bar q}\sum_{\substack{a_1,\dots a_p\\ \bar a_1,\dots \bar a_{\bar p}}}\sum_{\substack{b_1,\dots b_q\\ \bar b_1,\dots \bar b_{\bar q}}} P_{a_1\dots a_p;\bar a_1\dots\bar a_{\bar p}}Q_{b_1\dots b_q;\bar b_1\dots \bar b_{\bar q}}\\
&\qquad\times\left[\left(\prod_{i=1}^{p}\partial_{\phi_{a_i}}\right)\left(\prod_{\bar i=1}^{\bar p}\partial_{\bar\phi_{\bar a_{\bar i}}}\right)\left(\prod_{j=1}^{q}\phi_{b_j}\right)\left(\prod_{\bar j=1}^{\bar q}\phi_{\bar b_{\bar j}}\right)\right]_{\substack{\phi=0\\ \bar\phi=0}}\\
&= \sum_{p,\bar p}\sum_{\substack{a_1,\dots a_p\\ \bar a_1,\dots \bar a_{\bar p}}}\sum_{\substack{b_1,\dots b_p\\ \bar b_1,\dots \bar b_{\bar p}}} P_{a_1\dots a_p;\bar a_1\dots\bar a_{\bar p}}Q_{b_1\dots b_p;\bar b_1\dots \bar b_{\bar p}}\\
&\qquad\times\sum_{\sigma\in\mathfrak{S}(p),\tau\in\mathfrak{S}(\bar p)}\prod_{i=1}^{p}\delta_{b_ia_{\sigma(i)}}\prod_{\bar i=1}^{\bar p}\delta_{\bar b_{\bar i}\bar a_{\tau(\bar i)}}\\
&= \sum_{p,\bar p}\sum_{\sigma,\tau}\sum_{\substack{a_1,\dots a_p\\ \bar a_1,\dots \bar a_{\bar p}}} P_{a_1\dots a_p;\bar a_1\dots\bar a_{\bar p}}Q_{a_{\sigma(1)}\dots a_{\sigma(p)};\bar a_{\tau(1)}\dots\bar a_{\tau(\bar p)}}\\
&= \left[Q(\partial_\phi,\partial_{\bar\phi})P(\phi,\bar\phi)\right]_{\substack{\phi=0\\ \bar\phi=0}}\ ,
\end{aligned}
$$

the split of the normalized Gaussian follows from:

$$\left[e^{\frac{\partial}{\partial\phi}C\frac{\partial}{\partial\bar{\phi}}}\ F(\phi,\bar{\phi})\right]_{\substack{\phi=0\\ \bar{\phi}=0}} = \left[F\left(\frac{\partial}{\partial J},\frac{\partial}{\partial\bar{J}}\right)e^{JC\bar{J}}\right]_{\substack{J=0\\ \bar{J}=0}}$$
$$= \left[F\left(\frac{\partial}{\partial J},\frac{\partial}{\partial\bar{J}}\right)e^{JC^1\bar{J}}e^{JC^2\bar{J}}\right]_{\substack{J=0\\ \bar{J}=0}}$$
$$= \left[F\left(\frac{\partial}{\partial J},\frac{\partial}{\partial\bar{J}}\right)\left[e^{\frac{\partial}{\partial\phi^1}C^1\frac{\partial}{\partial\bar{\phi}^1}}\ e^{\phi^1 J+\bar{\phi}^1\bar{J}}\right]_{\substack{\phi^1=0\\ \bar{\phi}^1=0}}\left[e^{\frac{\partial}{\partial\phi^2}C^2\frac{\partial}{\partial\bar{\phi}^2}}\ e^{\phi^2 J+\bar{\phi}^2\bar{J}}\right]_{\substack{\phi^2=0\\ \bar{\phi}^2=0}}\right]_{\substack{J=0\\ \bar{J}=0}}$$
$$= \left[e^{\frac{\partial}{\partial\phi^1}C^1\frac{\partial}{\partial\bar{\phi}^1}}e^{\frac{\partial}{\partial\phi^2}C^2\frac{\partial}{\partial\bar{\phi}^2}}\left[F\left(\frac{\partial}{\partial J},\frac{\partial}{\partial\bar{J}}\right)e^{\phi^1 J+\bar{\phi}^1\bar{J}}e^{\phi^2 J+\bar{\phi}^2\bar{J}}\right]_{\substack{J=0\\ \bar{J}=0}}\right]_{\substack{\phi^1=\phi^2=0\\ \bar{\phi}^1=\bar{\phi}^2=0}}$$
$$= \left[e^{\frac{\partial}{\partial\phi^1}C^1\frac{\partial}{\partial\bar{\phi}^1}}e^{\frac{\partial}{\partial\phi^2}C^2\frac{\partial}{\partial\bar{\phi}^2}}\ F\left(\phi^1+\phi^2,\bar{\phi}^1+\bar{\phi}^2\right)\right]_{\substack{\phi^1=\phi^2=0\\ \bar{\phi}^1=\bar{\phi}^2=0}} .$$

The replica trick. We will often use in this book the so-called *replica trick*. Suppose that we are interested in evaluating the expectation of the nth power of some functional $V(\phi,\bar{\phi})$:

$$\left[e^{\frac{\partial}{\partial\phi}C\frac{\partial}{\partial\bar{\phi}}}\ \left[V(\phi,\bar{\phi})\right]^n\right]_{\substack{\phi=0\\ \bar{\phi}=0}} = \left[e^{\frac{\partial}{\partial\phi}C\frac{\partial}{\partial\bar{\phi}}}\ \underbrace{V(\phi,\bar{\phi})\dots V(\phi,\bar{\phi})}_{n\text{ times}}\right]_{\substack{\phi=0\\ \bar{\phi}=0}} .$$

When expanding the exponential, a product of operators $\frac{\partial}{\partial\phi}C\frac{\partial}{\partial\bar{\phi}}$ will act on a product of $V(\phi,\bar{\phi})$s. The two derivatives in an operator can either both act on the same $V(\phi,\bar{\phi})$ or they can act on two different $V(\phi,\bar{\phi})$s. In all cases, the weight associated with a couple of derivatives is C. It follows that we can rewrite this expectation by introducing a distinct replica of the field $\phi^{(i)}$ for each term $V(\phi,\bar{\phi})$ in the product, provided that the covariance is degenerated between the replicas (i.e. the covariance does not tell the difference between the replicas). That is, the expectation above is equal to:

$$\left[e^{\sum_{i,j=1}^n \frac{\partial}{\partial\phi^{(i)}}C\frac{\partial}{\partial\bar{\phi}^{(j)}}}\ \prod_{i=1}^n V(\phi^{(i)},\bar{\phi}^{(i)})\right]_{\substack{\phi^{(1)}=\dots\phi^{(n)}=0\\ \bar{\phi}^{(1)}=\dots=\bar{\phi}^{(n)}=0}} .$$

B.2 Perturbed Gaussian measures

We now consider the more interesting case of perturbed Gaussian measures. We start by the simple case of a quadratic perturbation, and subsequently discuss the general case.

B.2.1 Quadratic perturbation

Let M be a linear operator on Func($\mathcal{N}$). We are interested in studying the measure with expectations defined by:

$$\big\langle F(\phi,\bar\phi)\big\rangle_{C;-\bar\phi M\phi} = \frac{\left[e^{\frac{\partial}{\partial\phi}C\frac{\partial}{\partial\bar\phi}}\, e^{-\bar\phi M\phi}\, F(\phi,\bar\phi)\right]_{\substack{\phi=0\\ \bar\phi=0}}}{\left[e^{\frac{\partial}{\partial\phi}C\frac{\partial}{\partial\bar\phi}}\, e^{-\bar\phi M\phi}\right]_{\substack{\phi=0\\ \bar\phi=0}}}\,,$$

where the denominator ensures that the measure is normalized:

$$\langle 1\rangle_{C;-\bar\phi M\phi} = 1\,.$$

Normalization. Before computing the moments of this measure, let us first compute the normalization factor. Only the terms with equal numbers of derivatives and powers of the fields survive, therefore we have:

$$\begin{aligned}\left[e^{\frac{\partial}{\partial\phi}C\frac{\partial}{\partial\bar\phi}}\, e^{-\bar\phi M\phi}\right]_{\substack{\phi=0\\ \bar\phi=0}} &= \sum_{n\geq 0}\frac{1}{n!n!}\left[\left(\frac{\partial}{\partial\phi}C\frac{\partial}{\partial\bar\phi}\right)^n\left(-\bar\phi M\phi\right)^n\right]_{\substack{\phi=0\\ \bar\phi=0}} \\ &= \sum_{n\geq 0}\frac{(-1)^n}{n!n!}\sum_{\sigma,\tau\in\mathfrak{S}(n)}\sum_{\substack{a_1\dots a_n\\ \bar a_1\dots\bar a_n}}\prod_{i=1}^{n} C_{a_{\tau(i)},\bar a_i} M_{\bar a_{\sigma(i)},a_i}\,.\end{aligned}$$

The product of Cs and Ms traces the cycles of the permutation $\pi=\sigma\tau$, and (using the same notations as in Appendix A) we obtain:

$$\sum_{n\geq 0}\frac{(-1)^n}{n!}\sum_{\substack{\pi\in\mathfrak{S}(n)\\ C_q(\pi)=C_q,\forall q}}\prod_{q\geq 1}\Big(\mathrm{Tr}[(CM)^q]\Big)^{C_q}\,.$$

Taking into account that there are $\frac{n!}{\prod_q C_q! q^{C_q}}$ permutations of n elements in the class $1^{C_1}\dots q^{C_q}\dots$ we obtain:

$$\sum_{\mathbf{C}}\prod_{q\geq 1}\frac{1}{C_q!}\left[\frac{(-1)^q}{q}\,\mathrm{Tr}[(CM)^q]\right]^{C_q} = e^{-\sum_q \frac{(-1)^{q+1}}{q}\mathrm{Tr}[(CM)^q]} = e^{-\mathrm{Tr}[\ln(1+CM)]}\,,$$

where 1 is the identity operator on Func($\mathcal{N}$).

Moments. Let us now compute the moments of this distribution. A moment:

$$\frac{\left[e^{\frac{\partial}{\partial\phi}C\frac{\partial}{\partial\bar\phi}}\, e^{-\bar\phi M\phi}\,\left(\prod_{i=1}^q\phi_{a_i}\right)\left(\prod_{j=1}^k\bar\phi_{\bar a_j}\right)\right]_{\substack{\phi=0\\ \bar\phi=0}}}{\left[e^{\frac{\partial}{\partial\phi}C\frac{\partial}{\partial\bar\phi}}\, e^{-\bar\phi M\phi}\right]_{\substack{\phi=0\\ \bar\phi=0}}}\,,$$

is nonzero only if $q=k$, and in this case it can be written as a derivative of the normalization factor:

$$\left\langle\left(\prod_{i=1}^k\phi_{a_i}\right)\left(\prod_{i=1}^k\bar\phi_{\bar a_i}\right)\right\rangle_{C,-\bar\phi M\phi} = \frac{\left[\prod_{i=1}^k\left(-\frac{\partial}{\partial M_{a_j,\bar a_j}}\right)\right]\left[e^{\frac{\partial}{\partial\phi}C\frac{\partial}{\partial\bar\phi}}\, e^{-\bar\phi M\phi}\right]_{\substack{\phi=0\\ \bar\phi=0}}}{\left[e^{\frac{\partial}{\partial\phi}C\frac{\partial}{\partial\bar\phi}}\, e^{-\bar\phi M\phi}\right]_{\substack{\phi=0\\ \bar\phi=0}}}$$

$$= e^{\mathrm{Tr}[\ln(1+CM)]} \left[\prod_{i=1}^{k} \left(-\frac{\partial}{\partial M_{\bar{a}_i,a_i}} \right) \right] e^{-\mathrm{Tr}[\ln(1+CM)]} .$$

Note that, while we choose to pair the index a_i with the index $\bar{a}_i$ any other pairing of the indices a and $\bar{a}$ is just as good. This measure is in fact *a normalized Gaussian measure itself*, with covariance $C\frac{1}{1+MC}$ due to the following proposition.

Proposition B.2. *We have:*

$$e^{\mathrm{Tr}[\ln(1+CM)]} \left[\prod_{i=1}^{k} \left(-\frac{\partial}{\partial M_{\bar{a}_i,a_i}} \right) \right] e^{-\mathrm{Tr}[\ln(1+CM)]}$$
$$= \sum_{\sigma\in\mathfrak{S}(k)} \prod_{i=1}^{k} \left(C\frac{1}{1+MC} \right)_{a_i,\bar{a}_{\sigma(i)}} .$$

Proof. We first evaluate the derivatives:

$$-\frac{\partial}{\partial M_{\bar{a}_i,a_i}} \Big(-\mathrm{Tr}[\ln(1+CM)] \Big) = \left[C\frac{1}{1+MC} \right]_{a_i,\bar{a}_i} ,$$
$$-\frac{\partial}{\partial M_{\bar{a}_i,a_i}} \left[C\frac{1}{1+MC} \right]_{a_j,\bar{a}_j} = \left[C\frac{1}{1+MC} \right]_{a_j,\bar{a}_i} \left[C\frac{1}{1+MC} \right]_{a_i,\bar{a}_j} .$$

The proof proceeds by induction. For $k=1$ the statement is trivial. Using the lemma for k, at $k+1$ the left hand side writes:

$$e^{\mathrm{Tr}[\ln(1+CM)]} \frac{\partial}{\partial M_{\bar{a}_{k+1},a_{k+1}}} \left[e^{-\mathrm{Tr}[\ln(1+CM)]} \sum_{\sigma\in\mathfrak{S}(k)} \prod_{i=1}^{k} \left(C\frac{1}{1+MC} \right)_{a_i,a_{\sigma(i)}} \right]$$
$$= \left(C\frac{1}{1+MC} \right)_{a_{k+1},\bar{a}_{k+1}} \sum_{\sigma\in\mathfrak{S}(k)} \prod_{i=1}^{k} \left(C\frac{1}{1+MC} \right)_{a_i,\bar{a}_{\sigma(i)}}$$
$$+ \sum_{\sigma\in\mathfrak{S}(k)} \sum_{r=1}^{k} \left[\prod_{i=1}^{r} \left(C\frac{1}{1+MC} \right)_{a_i,\bar{a}_{\sigma(i)}} \right] \left(C\frac{1}{1+MC} \right)_{a_r,\bar{a}_{k+1}}$$
$$\times \left(C\frac{1}{1+MC} \right)_{a_{k+1},\bar{a}_{\sigma(r)}} \left[\prod_{i=r+1}^{k} \left(C\frac{1}{1+MC} \right)_{a_i,\bar{a}_{\sigma(i)}} \right] ,$$

and the first term sums over the permutations of $k+1$ elements having an extra cycle $(k+1)$ with respect to σ, while the second one sums over all the possible ways to insert $k+1$ in one of the cycles of σ.

□

B.2.2 Generic perturbation

We present in this section the case of a real field, as the case of a complex field follows with minimal modifications. Let $V(\phi)$ be an arbitrary functional of the field. The expectations of the Gaussian measure perturbed with $V(\phi)$ are:

$$\langle F(\phi)\rangle_{C;V(\phi)} = \frac{\left[e^{\frac{1}{2}\frac{\partial}{\partial\phi}C\frac{\partial}{\partial\phi}}\, e^{V(\phi)}\, F(\phi)\right]_{\phi=0}}{\left[e^{\frac{1}{2}\frac{\partial}{\partial\phi}C\frac{\partial}{\partial\phi}}\, e^{V(\phi)}\right]_{\phi=0}} ,$$

where the denominator is again chosen so as to ensure that the measure is normalized.

Let us first discuss the normalization factor. We perform a Taylor expansion in the perturbation $V(\phi)$ and we commute the sum over the expansion order n with the Gaussian integral:

$$\begin{aligned} Z(V) &\equiv \left[e^{\frac{1}{2}\frac{\partial}{\partial\phi}C\frac{\partial}{\partial\phi}}\, e^{V(\phi)}\right]_{\phi=0} = \left[e^{\frac{1}{2}\frac{\partial}{\partial\phi}C\frac{\partial}{\partial\phi}} \sum_{n\geq 0}\frac{1}{n!}\big[V(\phi)\big]^n\right]_{\phi=0} \\ &= \sum_{n\geq 0}\frac{1}{n!}\left[e^{\frac{1}{2}\frac{\partial}{\partial\phi}C\frac{\partial}{\partial\phi}}\big[V(\phi)\big]^n\right]_{\phi=0} . \end{aligned} \tag{B.3}$$

Again, the commutation of the sum with the Gaussian integral is problematic. In the case we dealt with before, leading to Eq. (B.1), the sum was generated by the Taylor expansion of the observable, and in particular was finite when considering a moment of the distribution. This is not the case now, as even in order to compute the normalization factor one needs to deal with an infinite series. This series is called the *perturbative series*. Concerning the convergence of this series, there are several cases:

- The perturbative series is absolutely convergent in some domain. This is the case for instance if we consider a complex field and a perturbation $V(\phi,\bar\phi) = \lambda\bar\phi\phi^q$ for some fixed q.
- The perturbative series is not convergent but it is Borel summable (see Appendix C). This is the most common case and a typical example is $V(\phi) = \lambda\phi^4$.
- The perturbative series is not Borel summable. While in many cases Borel summability has not been proven, we know of no example of interest in which some perturbative series has been *proven to not be* Borel summable (to be more precise, Borel-Leroy of a certain order).

We extend in a straightforward manner the graphical representation of Section B.1.1 to this case. Taylor expanding the perturbation $V(\phi)$ we obtain:

$$\begin{aligned} Z(V) = \sum_{n\geq 0}\frac{1}{n!}\sum_{d_1,\dots d_n\geq 0}&\left[\prod_{i=1}^{n}\frac{1}{d_i!}\left[\partial_{\phi_{a_1^{(i)}}}\cdots\partial_{\phi_{a_{d_i}^{(i)}}}V(\phi)\right]_{\phi=0}\right] \\ \times\sum_{\mathbf{P}_2}&\prod_{\{a_{l_i}^{(i)}a_{l_j}^{(j)}\}\in\mathbf{P}_2} C_{a_{l_i}^{(i)},a_{l_j}^{(j)}} , \end{aligned}$$

where $\mathbf{P}_2$ runs over the partitions of $\cup_{i=1}^{n}\{a_1^{(i)},\dots a_{d_i}^{(i)}\}$ into two element subsets. We now represent every factor:

$$\frac{1}{d_i!}\left[\partial_{\phi_{a_1^{(i)}}}\cdots\partial_{\phi_{a_{d_i}^{(i)}}}V(\phi)\right]_{\phi=0} ,$$

by a black embedded vertex with d_i ordered halfedges, and every covariance by a dashed edge, as in Figure B.2.

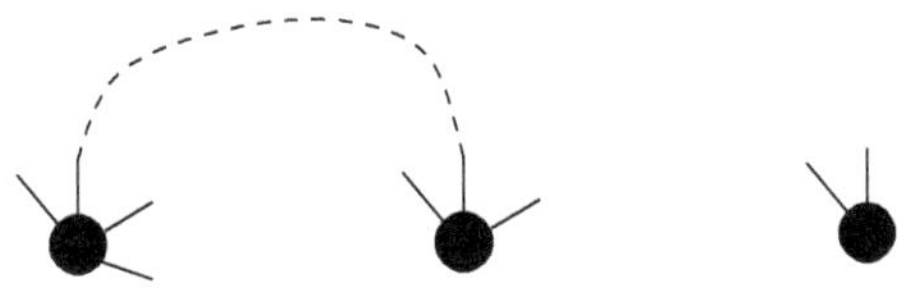

Figure B.2: Feynman vertices and propagators.

The vertices of the map are labeled $1, \dots n$. We again order the halfedges on a vertex in some arbitrary order, say $\partial_{\phi_{a_1^{(i)}}}, \dots \partial_{\phi_{a_{d_i}^{(i)}}}$. The sum over pairings generates again a sum over combinatorial maps, up to relabelings of the halfedges belonging to a vertex. If one groups together the maps that differ only by a relabeling of the vertices, one obtains a sum over unlabeled combinatorial maps. For the numerator in Eq. (B.3) we obtain the same representation, but with an extra vertex representing $F(\phi)$. Note that the maps we obtain can be disconnected. The contribution of a disconnected map factors over its connected components.

Appendix C

Borel summability

Many of the so-called "perturbative series" in tensor models and quantum field theory in general are not summable but only Borel summable. Before presenting the formal definition of Borel summability, let us start with an informal discussion about the meaning of this notion and its uses in quantum field theory. Let us consider the example of the one dimensional integral ("0 dimensional field theory"):

$$Z(\lambda) = \int \frac{d\phi}{\sqrt{2\pi}} e^{-\frac{\phi^2}{2}-\lambda\phi^4} . \tag{C.1}$$

The perturbative series is obtained by Taylor expanding around $\lambda = 0$. It is clear that this expansion is bound to run into troubles. The reason for this is that the "interaction term" $(-\lambda\phi^4)$ is stable for $\lambda > 0$ but unstable for $\lambda < 0$: the integral *diverges* for λ real and negative.

The partition function is well defined for λ real and positive. As such $Z(\lambda)$ might admit an analytic continuation in some domain in the complex plane, but the maximal domain of analyticity of $Z(\lambda)$ cannot intersect the negative real axis. It follows that $\lambda = 0$ belongs to the *boundary* of the analyticity domain of $Z(\lambda)$. By definition, the Taylor expansion of a function around a point belonging to the boundary of its maximal analyticity domain must diverge (otherwise the function would admit an analytic continuation in a neighborhood of the point and the point would not belong to the boundary of the maximal domain of analyticity).

However, all is not lost. In some cases (and $Z(\lambda)$ in Eq. (C.1) is one of them), although divergent, the perturbative expansions turn out to be Borel summable. Like the notion of convergence, Borel summability provides a uniqueness criterion: the sum of a summable series is unique, and the Borel sum of a Borel summable series is also unique. More importantly (and it is in this way that Borel summability is used in quantum field theory): if a series is the Taylor expansion of a Borel summable function $Z(\lambda)$ at $\lambda = 0$, then $Z(\lambda)$ is the *unique* Borel summable function whose Taylor series is the original series. Moreover, the function $Z(\lambda)$ can be reconstructed starting from its (divergent) Taylor series. The point is that the problem has now shifted from trying to show that the perturbative series is Borel summable to:

- prove that *a priori* (that is not by using the perturbative series) the function $Z(\lambda)$ is Borel summable.
- try to Borel sum the perturbative series in order to reconstruct the function $Z(\lambda)$. It should be stressed that this second step is quite hard [39, 166]. However, proving just the first point is reassuring, as it guarantees that the resummation procedure must in principle work.

If a function is Borel summable then all the information about this function can be reconstructed from its perturbative series. Of course this cannot be done directly: one needs to reconstruct first $Z(\lambda)$ in some domain and then extend this domain to the maximal domain of analyticity. Typically the cut on the negative real axis is a branch cut, $\lambda = 0$ is a branch point, and $Z(\lambda)$ can be extended to a Riemann surface. This nontrivial procedure leads to some surprising effects: the discontinuity of $Z(\lambda)$ at the cut must capture *instanton contributions.*

Instanton effects are generated by the nontrivial solutions of the classical equations of motion and are (naively) invisible in the perturbative expansion. Indeed, the classical equations of motion $-\phi - 4\lambda\phi^3 = 0$ admit two solutions $\phi = 0$ and $\phi^2 = -\frac{1}{4\lambda}$ that act as saddle points for the integral. The nontrivial solution becomes physical for λ real negative (the contour of integration passes through the saddle point) and its leading contribution to $Z(\lambda)$ is:

$$e^{-\frac{1}{2}(-\frac{1}{4\lambda})-\lambda(-\frac{1}{4\lambda})^2} = e^{\frac{1}{16\lambda}} = e^{-\frac{1}{16|\lambda|}} .$$

Such an exponential effect is invisible in perturbation series: the Taylor expansion of $e^{-\frac{1}{16|\lambda|}}$ around $\lambda = 0$ is identically zero.

The strength of the Borel resummation is that even such effects, which are naively missed by the perturbative series, are in fact recovered once one applies the correct procedure of Borel resuming the perturbative series, finding the analytic continuation of the result on a Riemann surface and then computing the discontinuity at the branch cut.

In the case of tensor and matrix models a further twist comes from the fact that, as we are interested in $N \to \infty$ limit, the Borel summability should be uniform in N. This ensures that the domains in the complex plane do not shrink when we go to the limit (hence our limit objects are well defined functions) and that, after taking a Borel transform of the perturbative series, we can take the large N limit term by term.

The main theorem on Borel summability we will be using in this book is the Nevanlinna-Sokal theorem.

Theorem C.1 (Nevanlinna-Sokal, [157]). *A function $f(\lambda, N)$ with $\lambda \in \mathbb{C}$ and $N \in \mathbb{R}_+$ is said to be Borel summable in λ uniformly in N if:*

- *$f(\lambda, N)$ is analytic in a disk $\Re(\lambda^{-1}) > R^{-1}$ with $R \in \mathbb{R}_+$ independent of N.*
- *$f(\lambda, N)$ admits a Taylor expansion at the origin with uniform bound on the Taylor remainder:*

$$f(\lambda, N) = \sum_{k=0}^{r-1} f_{N,k}\lambda^k + R_{N,r}(\lambda) , \qquad |R_{N,r}(\lambda)| \le K\, \sigma^r\, r!|\lambda|^r ,$$

for some constants K and σ independent of N.

If $f(\lambda, N)$ is Borel summable in λ uniformly in N then:

$$B(t,N) = \sum_{k=0}^{\infty} \frac{1}{k!} f_{N,k} t^k ,$$

is an analytic function for $|t| < \sigma^{-1}$ that admits an analytic continuation in the strip $\{z| \, |\Im z| < \sigma^{-1}\}$ such that $|B(t,N)| \leq Be^{t/R}$ for some constant B independent of N and $f(\lambda, N)$ is given by the absolutely convergent integral:

$$f(\lambda, N) = \frac{1}{\lambda} \int_0^{\infty} dt \, B(t,N) \, e^{-\frac{t}{\lambda}} .$$

In other words, the Taylor expansion of $f(\lambda, N)$ at the origin is Borel summable, and $f(\lambda, N)$ is its Borel sum.

The set $\{\lambda | \Re(\lambda^{-1}) > R^{-1}, R \in \mathbb{R}_+\}$ is a disk (which we call a Borel disk) in the complex plane centered at $\frac{R}{2}$ and of radius $\frac{R}{2}$ (hence tangent to the imaginary axis, see Figure C.1) as, denoting $\lambda = \frac{R}{2} + ae^{\imath\gamma}$,

$$\Re(\lambda^{-1}) > R^{-1} \Leftrightarrow \frac{R^2}{4} > a .$$

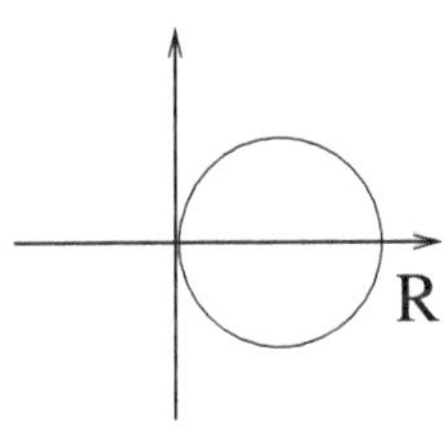

Figure C.1: A Borel disk.

Example C.1. Let us show that $Z(\lambda)$ in Eq. (C.1) is Borel summable in λ. The integral is absolutely convergent for $\Re\lambda \geq 0$, hence it is convergent in some Borel disk. The Taylor expansion with integral rest of $Z(\lambda)$ is:

$$\begin{aligned} Z(\lambda) &= \int \frac{d\phi}{\sqrt{2\pi}} e^{-\frac{\phi^2}{2}} \left[\sum_{q=0}^{r-1} \frac{1}{q!} (-\lambda)^q \phi^{4q} + \int_0^1 du \, \frac{(1-u)^{r-1}}{(r-1)!} (-\lambda)^r \phi^{4r} e^{-u\lambda\phi^4} \right] \\ &= \sum_{q=0}^{r-1} \frac{1}{q!} (-\lambda)^q \frac{(4q)!}{2^{2q}(2q)!} + R_r(\lambda) , \\ R_r(\lambda) &= \int \frac{d\phi}{\sqrt{2\pi}} e^{-\frac{\phi^2}{2}} \int_0^1 du \, \frac{(1-u)^{r-1}}{(r-1)!} (-\lambda)^r \phi^{4r} e^{-u\lambda\phi^4} . \end{aligned}$$

For all λ with $\Re\lambda \geq 0$ the rest term is bounded as:

$$|R_r(\lambda)| \leq \frac{1}{(r-1)!}|\lambda|^r \frac{(4r)!}{2^{2r}(2r)!} \leq \frac{3^{4r}}{2^{2r}} r\ |\lambda|^r r! \leq \left(\frac{3^4}{2^2}e\right)^r |\lambda|^r r!\ ,$$

and $Z(\lambda)$ is Borel summable. In particular, $Z(\lambda)$ admits the integral representation:

$$Z(\lambda) = \frac{1}{\lambda}\int_0^\infty dt\ e^{-\frac{t}{\lambda}} B(t)\ ,$$

where $B(t)$ is the analytic continuation of:

$$B(t) = \sum_{q=0}^{\infty} \frac{1}{q!q!}(-t)^q \frac{(4q)!}{2^{2q}(2q)!}\ ,$$

on the positive real axis.

Appendix D

The BKAR formula

The Brydges-Kennedy-Abdesselam-Rivasseau (BKAR) forest formula is a Taylor formula for functions of several variables. Due to its symmetry and positivity properties it is very well adapted for non perturbative quantum field theory.

Let us consider a set of n points labeled $i, i = 1 \dots n$, which we identify with the set $\mathcal{V}_n$ of vertices of the complete graph $\mathcal{K}_n$. The set of unordered pairs of such points has $n(n-1)/2$ elements $e = (i,j)$ for $1 \le i,j \le n$, $i \ne j$ and can be identified with the set $\mathcal{E}_n$ of edges of $\mathcal{K}_n$. Let us consider a smooth (and arbitrarily derivable) function $f : [0,1]^{n(n-1)/2} \to \mathbb{R}$ depending on the edge variables $x_e \equiv x_{ij}$, $e = (i,j)$.

Theorem D.1. *[The Forest Formula, [1]] We have (with the convention that empty products are* 1*):*

$$f(1,\dots 1) = \sum_{\mathfrak{F}} \underbrace{\int_0^1 \cdots \int_0^1}_{|\mathfrak{F}| \text{ times}} \left(\prod_{e \in \mathfrak{F}} du_e \right) \left[\left(\prod_{e \in \mathfrak{F}} \frac{\partial}{\partial x_e} \right) f \right] \left(w_{ij}^{\mathfrak{F}}(u_{\mathfrak{F}}) \right),$$

where:

- *the sum runs over the forests*[1] $\mathfrak{F}$ *drawn over the* n *labeled vertices* i*, including the empty forest (having no edge). To each edge* $e \in \mathfrak{F}$ *we attribute a variable* u_e *that is integrated from* 0 *to* 1 *and* $u_{\mathfrak{F}} = \{u_e | e \in \mathfrak{F}\}$.
- *the derivative* $\left[\left(\prod_{e \in \mathfrak{F}} \frac{\partial}{\partial x_e} \right) f \right]$ *is evaluated at the point:*

$$w_{ij}^{\mathfrak{F}}(u_{\mathfrak{F}}) = \inf_{e' \in P_{i \to j}^{\mathfrak{F}}} (u_{e'}),$$

where $P_{i \to j}^{\mathfrak{F}}$ *denotes the unique path in the forest* $\mathfrak{F}$ *joining the vertices* i *and* j*, and the infimum is set to zero if such a path does not exist.*

[1] Acyclic edge-subgraphs of the complete graph $\mathcal{K}_n$.

Setting by convention $w^{\mathfrak{F}}_{ii}(u_{\mathfrak{F}}) \equiv 1$, the symmetric $n \times n$ matrix:

$$\left(w^{\mathfrak{F}}_{ij}(u_{\mathfrak{F}}) \right)_{1 \le i,j \le n} ,$$

is positive *for all $u_{\mathfrak{F}}$.*

The case $n = 2$ is particularly simple. The function f is a function of one variable x_{12}, and there are two forests over two labeled vertices: the empty forest and the forest with one edge $\{(1,2)\}$:

$$f(1) = f(0) + \int_0^1 du_{12} \left(\frac{\partial f}{\partial x_{12}} \right)(u_{12}) .$$

For $n = 3$ we have the seven forests:

$$\begin{aligned} &\varnothing, \{(1,2)\}, \{(1,3)\}, \{(2,3)\}, \\ &\{(1,2),(1,3)\}, \{(1,2),(2,3)\}, \{(1,3),(2,3)\} , \end{aligned}$$

and we get:

$$\begin{aligned} f(1,1,1) =& f(0,0,0) \\ &+ \int_0^1 du_{12} \left(\frac{\partial f}{\partial x_{12}} \right)(u_{12},0,0) + \int_0^1 du_{13} \left(\frac{\partial f}{\partial x_{13}} \right)(0,u_{13},0) \\ &+ \int_0^1 du_{23} \left(\frac{\partial f}{\partial x_{23}} \right)(0,0,u_{23}) \\ &+ \int_0^1 \int_0^1 \int_0^1 du_{12} du_{13} \left(\frac{\partial^2 f}{\partial x_{12} \partial x_{13}} \right) \big(u_{12}, u_{13}, \inf(u_{12},u_{13})\big) \\ &+ \int_0^1 \int_0^1 \int_0^1 du_{12} du_{23} \left(\frac{\partial^2 f}{\partial x_{12} \partial x_{23}} \right) \big(u_{12}, \inf(u_{12},u_{23}), u_{23},\big) \\ &+ \int_0^1 \int_0^1 \int_0^1 du_{13} du_{23} \left(\frac{\partial^2 f}{\partial x_{13} \partial x_{23}} \right) \big(\inf(u_{13},u_{23}), u_{13}, u_{23}\big) . \end{aligned}$$

The proof of Theorem D.1 goes as follows.

Proof. Let $\mathfrak{F}$ be a forest and associate to any edge $e \in \mathfrak{F}$ a variable u_e. We furthermore associate to $\mathfrak{F}$ the $n \times n$ symmetric matrix $W^{\mathfrak{F}}(u_{\mathfrak{F}};t)$ defined as:

$$W^{\mathfrak{F}}_{ij}(u_{\mathfrak{F}};t) = \begin{cases} \inf_{e' \in P^{\mathfrak{F}}_{i \to j}} (u_{e'}) , & i \ne j \text{ and } P^{\mathfrak{F}}_{i \to j} \text{ exists} \\ t , & i \ne j \text{ and } i \text{ and } j \text{ are not} \\ & \quad \text{connected by a path in } \mathfrak{F} \\ 1 , & i = j \end{cases} .$$

Note that $W^{\mathfrak{F}}_{ij}(u_{\mathfrak{F}};0) = w^{\mathfrak{F}}_{ij}(u_{\mathfrak{F}})$. Also, if $|\mathfrak{F}| = n-1$ then all the vertices are connected by a path in $\mathfrak{F}$ and $W^{\mathfrak{F}}_{ij}(u_{\mathfrak{F}};t) = w^{\mathfrak{F}}_{ij}(u_{\mathfrak{F}})$. Furthermore, as the empty forest has no edges, $u_{\varnothing}$ is empty and:

$$W^{\varnothing}_{ij}(u_{\varnothing},t)|_{t=1} = 1 , \quad W^{\varnothing}_{ij}(u_{\varnothing},t)|_{t=0} = w^{\varnothing}_{ij}(u_{\varnothing}) = \delta_{ij} .$$

A forest $\mathfrak{F}$ divides the complete graph $\mathcal{K}_n$ into several connected components (or blocks). For instance, if $\mathfrak{F}$ is the empty forest, the blocks are all singletons consisting in a unique vertex per block. The entries in W corresponding to vertices belonging to the same block of $\mathfrak{F}$ are frozen, and all the trans block entries are equal to t. In particular, the matrix:

$$B^{\mathfrak{F}}_{ij} = \begin{cases} 1 , & \text{if } i,j \text{ belong to the same block of } \mathfrak{F} \\ 0 , & \text{otherwise} \end{cases} ,$$

can be written as $B^{\mathfrak{F}} = W^{\mathfrak{F}}(u_{\mathfrak{F}};t)\big|_{\substack{t=0 \\ u_e=1,\ \forall e\in\mathfrak{F}}}$. Remark that $B^{\varnothing}$ is the identity matrix. For any forest $\mathfrak{F}$, the matrix $B^{\mathfrak{F}}$ is positive. Indeed, denoting $b \in \mathfrak{F}$ the blocks of $\mathfrak{F}$, and $i \in b$ the vertices in the block b,

$$\sum_{i,j} B^{\mathfrak{F}}_{ij} a_i a_j = \sum_{b\in\mathfrak{F}} \Big(\sum_{i\in b} a_i \Big)^2 .$$

Proposition D.1. *For all t such that $\forall e \in \mathfrak{F}$, $t \le u_e$, the matrix $W^{\mathfrak{F}}(u_{\mathfrak{F}};t)$ is positive. In particular, $\left(w^{\mathfrak{F}}_{ij}(u_{\mathfrak{F}}) \right)_{1\le i,j\le n}$ is a positive matrix.*

Proof. Let us denote the number of edges in $\mathfrak{F}$ by $q \equiv |\mathfrak{F}|$. We order the edges of $\mathfrak{F}$ in decreasing order of their parameters u:

$$1 \ge u_{e_1} \ge u_{e_2} \ge \dots u_{e_q} \ge t .$$

Adding edges one by one starting from the highest edge we obtain a family of subforests of $\mathfrak{F}$:

$$\mathfrak{F}^0 = \varnothing , \ \mathfrak{F}^1 = \{e_1\} , \ \mathfrak{F}^2 = \{e_1, e_2\} , \dots , \ \mathfrak{F}^q = \{e_1, \dots e_q\} = \mathfrak{F} .$$

We denote $\mathit{1}$ the matrix whose entries are all 1, $\mathit{1}_{ij} = 1$, $\forall i,j$. We have:

$$W^{\mathfrak{F}}(u_{\mathfrak{F}};t) = (1-u_{e_1})B^{\mathfrak{F}^0} + (u_{e_1} - u_{e_2})B^{\mathfrak{F}^1} + \dots + (u_{e_q} - t)B^{\mathfrak{F}^q} + t\mathit{1} .$$

Indeed, if i and j do not belong to the same block of $\mathfrak{F}^q = \mathfrak{F}$, then they do not belong to the same block in any of the $\mathfrak{F}^s$, $s \le q$, and only the last term above contributes to $W^{\mathfrak{F}}_{ij}(u_{\mathfrak{F}};t)$.

If, on the other hand, i and j belong to the same block of $\mathfrak{F}$, then:

$$\Big[(1-u_{e_1})B^{\mathfrak{F}^0} + (u_{e_1} - u_{e_2})B^{\mathfrak{F}^1} + \dots + (u_{e_q} - t)B^{\mathfrak{F}^q} + t\mathit{1} \Big]_{ij} = u_{e_s} ,$$

where s is such that i and j belong to the same block of $\mathfrak{F}^s$, but belong to two different blocks of $\mathfrak{F}^{s-1}$. As:

$$u_{e_s} \le u_{e_{s-1}} \le u_{e_{s-2}} \le \dots ,$$

it follows that u_{e_s} is the infimum of the us in the unique path in $\mathfrak{F}^s$ joining i and j, hence it is also the infimum of the us in the unique path in $\mathfrak{F}$ joining i and j.

The matrix $W^{\mathfrak{F}}(u_{\mathfrak{F}};t)$ is a convex combination of positive matrices, hence it is itself a positive matrix.

□

Consider now a function of the edge variables $x_e \equiv x_{ij}$, $e = (i,j)$. Then:

$$\frac{d}{dt}\Big[f\Big(W_{ij}^{\mathfrak{F}}(u_{\mathfrak{F}};t)\Big)\Big] = \sum_{\substack{e \\ e\cup\mathfrak{F} \text{ is a forest}}} \Big(\frac{\partial}{\partial x_e} f\Big)\Big(W_{ij}^{\mathfrak{F}}(u_{\mathfrak{F}};t)\Big),$$

because t is the variable associated in $W_{ij}^{\mathfrak{F}}(u_{\mathfrak{F}};t)$ to pairs of vertices k and l that are *not* connected by $\mathfrak{F}$, hence $\mathfrak{F}\cup(k,l)$ is still a forest.

The proof of Theorem D.1 consists in iterated Taylor expansions with respect to the variable t. We first have:

$$\begin{aligned} f(1,\dots 1) &= f\Big(W_{ij}^{\varnothing}(u_{\varnothing},t)\Big)\Big|_{t=1} = f\Big(W_{ij}^{\varnothing}(u_{\varnothing},0)\Big) + \int_0^1 dt\, \frac{d}{dt}\Big[f\Big(W_{ij}^{\varnothing}(u_{\varnothing},t)\Big)\Big] \\ &= f\Big(w_{ij}^{\varnothing}(u_{\varnothing})\Big) + \sum_e \int dt\, \Big(\frac{\partial}{\partial x_e} f\Big)\Big(W_{ij}^{\varnothing}(u_{\varnothing},t)\Big). \end{aligned}$$

The sum in the second term is a sum over forests with one edge, $\mathfrak{F} = \{e\}$. Relabeling for each of the forests the dummy variable t by u_e, we get:

$$f(1,\dots 1) = f\Big(w_{ij}^{\varnothing}(u_{\varnothing})\Big) + \sum_{\mathfrak{F}=\{e\}} \int du_e\, \Big(\frac{\partial}{\partial x_e} f\Big)\Big(W_{ij}^{\varnothing}(u_{\varnothing},t=u_e)\Big).$$

In order to iterate the Taylor formula we must choose for each forest $\mathfrak{F} = \{e\}$ an interpolation of $W_{ij}^{\varnothing}(u_{\varnothing},t=u_e)$. This is done independently for each term. One should be aware that this choice is not canonical. For the term corresponding to $\mathfrak{F} = \{e\}$ we use the interpolation:

$$W_{ij}^{\varnothing}(u_{\varnothing},t=u_e) = W_{ij}^{\{e\}}(\{u_e\};t)|_{t=u_e},$$

and perform a Taylor expansion at first order on t to obtain:

$$\begin{aligned} f(1,\dots 1) &= f\Big(w_{ij}^{\varnothing}(u_{\varnothing})\Big) + \sum_{\mathfrak{F}=\{e\}} \int du_e\, \Big(\frac{\partial}{\partial x_e} f\Big)\Big(w_{ij}^{\{e\}}(u_{\{e\}})\Big) \\ &+ \sum_{\{e_1,e_2\}\text{ forest}} \int_0^1 du_{e_1} \int_0^{u_{e_1}} du_{e_2} \Big[\Big(\frac{\partial^2}{\partial x_{e_2}\partial x_{e_1}}\Big) f\Big]\Big(W^{\{e_1\}}(\{u_{e_1}\};t=u_{e_2})\Big). \end{aligned}$$

We are now in the position to iterate. At each step we interpolate the W in the term corresponding to a forest $\{e_1,\dots e_q\}$ as:

$$\begin{aligned} &W^{\{e_1,\dots e_{q-1}\}}(\{u_{e_1},\dots u_{e_{q-1}}\};t=u_{e_q}) \\ &\quad = W^{\{e_1,\dots e_{q-1},e_q\}}(\{u_{e_1},\dots u_{e_{q-1}},u_{e_q}\};t)|_{t=u_{e_q}}. \end{aligned}$$

The iteration stops when reaching maximal forests with $n-1$ edges, and:

$$f(1,\dots 1) = \sum_{q=0}^{n-1} \sum_{\{e_1,\dots e_q\}\text{ forest}} \int_0^1 du_{e_1} \int_0^{u_1} du_{e_2} \dots \int_0^{u_{e_{q-1}}} du_{e_q}$$

$$\times \left[\prod_{e \in \{e_1, \dots e_q\}} \left(\frac{\partial}{\partial x_e} \right) f \right] \left(w_{ij}^{\{e_1, \dots e_q\}} (u_{\{e_1, \dots e_q\}}) \right) .$$

The same forest $\mathfrak{F}$ appears $|\mathfrak{F}|!$ times in this sum, corresponding to all the orderings of its edges. Grouping these contributions together proves Theorem D.1.

□

D.1 The forest formula and the connected moments

The BKAR forest formula is used in quantum field theory in order to compute the generating function of the connected moments of perturbed Gaussian measures (see Appendix B.2). One first starts as in the Feynman expansion by expanding the perturbation, but, instead of directly computing the Gaussian integral term by term, one uses a replica trick and the forest formula to obtain a convergent representation.

Lemma D.1. *Let $\{\phi_a | a = 1, \dots N\}$ be a complex vector and let Z be the generating function of the moments of the perturbed Gaussian measure of covariance C defined as:*

$$Z = \left[e^{\frac{\partial}{\partial \phi} C \frac{\partial}{\partial \bar\phi}} e^{S(\bar\phi, \phi)} \right]_{\phi = \bar\phi = 0} , \qquad S(\bar\phi, \phi) = V(\bar\phi, \phi) + \phi J + \bar\phi \bar J ,$$

where $\frac{\partial}{\partial \phi} C \frac{\partial}{\partial \bar\phi} = \sum_{a, \bar b} \phi_a C_{a \bar b} \bar\phi_{\bar b}$, $\bar\phi \bar J = \sum_{\bar a} \bar\phi_{\bar a} \bar J_{\bar a}$ and $\phi J = \sum_a \phi_a J_a$. Then the generating function of the connected moments is:

$$\ln Z = \sum_{n \ge 1} \frac{1}{n!} \sum_{\mathfrak{T}_n} \int_0^1 \Big(\prod_{(i,j) \in \mathfrak{T}_n} du_{ij} \Big) \Big\{ e^{\sum_{i,j} w_{ij}^{\mathfrak{T}_n} (u_{\mathfrak{T}_n}) \frac{\partial}{\partial \phi^{(i)}} C \frac{\partial}{\partial \bar\phi^{(j)}}}$$
$$\times \left[\prod_{(i,j) \in \mathfrak{T}_n} \left(\frac{\partial}{\partial \phi^{(i)}} C \frac{\partial}{\partial \bar\phi^{(j)}} + \frac{\partial}{\partial \phi^{(j)}} C \frac{\partial}{\partial \bar\phi^{(i)}} \right) \right] \prod_{i=1}^n S(\bar\phi^{(i)}, \phi^{(i)}) \Big\}_{\phi^{(i)} = \bar\phi^{(i)} = 0} ,$$

where $\mathfrak{T}_n$ runs over all the combinatorial trees with n vertices labeled 1, 2 up to n, (i,j) denotes the tree edge connecting the vertices i and j, and the parameters $w_{ij}^{\mathfrak{T}_n}(u_{\mathfrak{T}_n})$ are defined as:

$$w_{ii}^{\mathfrak{T}_n}(u_{\mathfrak{T}_n}) = 1 , \qquad w_{ij}^{\mathfrak{T}_n}(u_{\mathfrak{T}_n}) = \inf_{(k,l) \in P_{i \to j}^{\mathfrak{T}_n}} u_{kl} ,$$

where $P_{i \to j}^{\mathfrak{T}_n}$ denotes the unique path in the tree $\mathfrak{T}_n$ joining i and j.

Proof. We Taylor expand in $S(\bar\phi, \phi)$ to get:

$$Z = \sum_{n \ge 0} \frac{1}{n!} \left[e^{\frac{\partial}{\partial \phi} C \frac{\partial}{\partial \bar\phi}} [S(\bar\phi, \phi)]^n \right]_{\phi = \bar\phi = 0} ,$$

and use the replica trick (see Appendix B.2) to express the generating function as:

$$Z = \sum_{n \ge 0} \frac{1}{n!} \left[e^{\sum_{i,j=1}^n \frac{\partial}{\partial \phi^{(i)}} C \frac{\partial}{\partial \bar\phi^{(j)}}} \prod_{i=1}^n S(\bar\phi^{(i)}, \phi^{(i)}) \right]_{\phi^{(i)} = \bar\phi^{(i)} = 0} .$$

For each n we introduce the weakening parameters x_{ii} and $x_{ij} = x_{ij}$ and write:

$$Z = \sum_{n\geq 0} \frac{1}{n!} \left[e^{\sum_{i,j=1}^{n} x_{ij} \frac{\partial}{\partial \phi^{(i)}} C \frac{\partial}{\partial \bar{\phi}^{(j)}}} \prod_{i=1}^{n} S(\bar{\phi}^{(i)}, \phi^{(i)}) \right]_{\substack{\phi^{(i)} = \bar{\phi}^{(i)} = 0 \\ x_{ij} = 1, \forall i,j}} .$$

Taking into account that for all $k \neq l$:

$$\frac{\partial}{\partial x_{kl}} \left[e^{\sum_{i,j=1}^{n} x_{ij} \frac{\partial}{\partial \phi^{(i)}} C \frac{\partial}{\partial \bar{\phi}^{(j)}}} \right] =$$
$$= e^{\sum_{i,j=1}^{n} x_{ij} \frac{\partial}{\partial \phi^{(i)}} C \frac{\partial}{\partial \bar{\phi}^{(j)}}} \left(\frac{\partial}{\partial \phi^{(k)}} C \frac{\partial}{\partial \bar{\phi}^{(l)}} + \frac{\partial}{\partial \phi^{(l)}} C \frac{\partial}{\partial \bar{\phi}^{(k)}} \right) ,$$

we get by applying Theorem D.1:

$$Z = \sum_{n\geq 0} \frac{1}{n!} \sum_{\mathfrak{F}_n} \int_0^1 \Big(\prod_{(i,j)\in \mathfrak{F}_n} du_{ij} \Big) \Big\{ e^{\sum_{i,j} w_{ij}^{\mathfrak{F}_n}(u_{\mathfrak{F}_n}) \frac{\partial}{\partial \phi^{(i)}} C \frac{\partial}{\partial \bar{\phi}^{(j)}}}$$
$$\times \left[\prod_{(i,j)\in \mathfrak{F}_n} \left(\frac{\partial}{\partial \phi^{(i)}} C \frac{\partial}{\partial \bar{\phi}^{(j)}} + \frac{\partial}{\partial \phi^{(j)}} C \frac{\partial}{\partial \bar{\phi}^{(i)}} \right) \right] \prod_{i=1}^{n} S(\bar{\phi}^{(i)}, \phi^{(i)}) \Big\}_{\phi^{(i)} = \bar{\phi}^{(i)} = 0} ,$$

where $\mathfrak{F}_n$ runs over the forests over n labeled vertices. The lemma follows by noticing that the contribution of a forest factors over the blocks in the forest, which correspond to the trees, and by using Lemma 2.1.

□

D.2 Hepp sectors

To any graph $\mathcal{G}$ with vertices labeled $i, i = 1, \dots n$ one can associate a *characteristic* function depending on edge variables x_{ij}:

$$\prod_{(i,j)\in \mathcal{E}(\mathcal{G})} x_{ij} .$$

The forest formula applied to this function yields:

$$1 = \sum_{\mathfrak{F}\subset \mathcal{G}} \underbrace{\int_0^1 \cdots \int_0^1}_{|\mathfrak{F}| \text{ times}} \Big(\prod_{(i,j)\in \mathfrak{F}} du_{ij} \Big) \Big[\prod_{(k,l)\notin \mathfrak{F}} w_{kl}^{\mathfrak{F}}(u_{\mathfrak{F}}) \Big] .$$

Remark that the derivative of the characteristic function is nonzero only if the forest $\mathfrak{F}$ is made of edges of $\mathcal{G}$ (which we signify by $\mathfrak{F} \subset \mathcal{G}$). Furthermore, if $\mathfrak{F}$ has more than one block, then one of the ws in the product is set to zero. It follows that only trees contribute:

$$1 = \sum_{\mathfrak{T}\subset \mathcal{G}} \underbrace{\int_0^1 \cdots \int_0^1}_{|\mathfrak{T}| \text{ times}} \Big(\prod_{(i,j)\in \mathfrak{T}} du_{ij} \Big) \Big[\prod_{(k,l)\notin \mathfrak{T}} w_{kl}^{\mathfrak{T}}(u_{\mathfrak{T}}) \Big] .$$

This formula defines normalized weights associated with the graph $\mathcal{G}$ and the spanning trees $\mathfrak{T} \subset \mathcal{G}$:

$$w(\mathcal{G}, \mathfrak{T}) = \underbrace{\int_0^1 \cdots \int_0^1}_{|\mathfrak{T}| \text{ times}} \left(\prod_{(i,j)\in\mathfrak{T}} du_{ij} \right) \left[\prod_{(k,l)\notin\mathfrak{T}} w_{kl}^{\mathfrak{T}}(u_{\mathfrak{T}}) \right], \qquad \sum_{\mathfrak{T}\subset\mathcal{G}} w(\mathcal{G}, \mathfrak{T}) = 1 ,$$

which admit a striking combinatorial interpretation [147].

We define a *Hepp sector* as a total ordering σ of the edges of $\mathcal{G}$:

$$\sigma : \mathcal{E}(\mathcal{G}) \to \{1, \ldots |\mathcal{E}(\mathcal{G})|\} , \qquad \sigma \text{ one to one.}$$

For any σ *the leading spanning tree* of σ, $\mathfrak{T}(\sigma)$ is the tree such that:

$$\sum_{e\in\mathfrak{T}(\sigma)} \sigma(e)$$

is minimal. The tree $\mathfrak{T}(\sigma)$ is obtained by Kruskal's greedy algorithm [119]: at each step one adds the edge $e \in \mathcal{G}$ with minimal $\sigma(e)$ that does not form a loop.

Lemma D.2. *We have:*

$$w(\mathcal{G}, \mathfrak{T}) = \frac{N(\mathcal{G}, \mathfrak{T})}{|\mathcal{E}(\mathcal{G})|!}$$

where $N(\mathcal{G}, \mathfrak{T})$ is the number of sectors σ such that $\mathfrak{T}(\sigma) = \mathfrak{T}$.

Proof. Let us define the function:

$$\chi(u_{\mathcal{E}(\mathcal{G})\setminus\mathfrak{T}} \le u_{\mathfrak{T}}) \begin{cases} 1 , & \text{if } \forall (i,j) \in \mathcal{E}(\mathcal{G}) \setminus \mathfrak{T}, \quad u_{ij} \le \inf_{(k,l)\in P^{\mathfrak{T}}_{i\to j}} u_{kl} \\ 0 , & \text{otherwise} \end{cases} .$$

We observe that:

$$\begin{aligned} w(\mathcal{G}, \mathfrak{T}) &= \int_0^1 \left(\prod_{e\in\mathcal{E}(\mathcal{G})} du_e \right) \chi(u_{\mathcal{G}\setminus\mathfrak{T}} \le u_{\mathfrak{T}}) \\ &= \sum_{\sigma, \mathfrak{T}(\sigma)=\mathfrak{T}} \int_0^1 du_1 \int_0^{u_1} du_2 \ldots \int_0^{u_{|\mathcal{E}(\mathcal{G})|-1}} du_{|\mathcal{E}(\mathcal{G})|} . \end{aligned}$$

□

Bibliography

[1] A. Abdesselam and V. Rivasseau, *Trees, forests and jungles: a botanical garden for cluster expansions*. New York: Springer Verlag, 1995.

[2] D. Aldous, *The continuum random tree II: an overview*, Stochastic analysis, vol. 167, pp. 23–70, 1991.

[3] J. Ambjorn, B. Durhuus and T. Jonsson, *Three-dimensional simplicial quantum gravity and generalized matrix models*, Mod. Phys. Lett., vol. A6, pp. 1133–1146, 1991.

[4] ——, *Quantum geometry. A statistical field theory approach*. Cambridge: Cambridge University Press, 1997.

[5] J. Ambjorn, A. Goerlich, J. Jurkiewicz and R. Loll, *Nonperturbative Quantum Gravity*, Phys. Rept., vol. 519, pp. 127–210, 2012. arXiv: 1203.3591 [hep-th].

[6] J. Ambjorn, J. Jurkiewicz and Y. M. Makeenko, *Multiloop correlators for two-dimensional quantum gravity*, Phys. Lett., vol. B251, pp. 517–524, 1990.

[7] A. Baratin, S. Carrozza, D. Oriti, J. Ryan and M. Smerlak, *Melonic phase transition in group field theory*, Lett. Math. Phys., vol. 104, pp. 1003–1017, 2014. arXiv: 1307.5026 [hep-th].

[8] A. Baratin and D. Oriti, *Group field theory with non-commutative metric variables*, Phys. Rev. Lett., vol. 105, p. 221 302, 2010. arXiv: 1002.4723 [hep-th].

[9] ——, *Ten questions on group field theory (and their tentative answers)*, J. Phys. Conf. Ser., vol. 360, p. 012 002, 2012. arXiv: 1112.3270 [gr-qc].

[10] G. Ben Arous and A. Guionnet, *Large deviations for Wigner's law and Voiculescu's non-commutative entropy*, Probability theory and related fields, vol. 108, no. 4, pp. 517–542, 1997.

[11] J. Ben Geloun, *Asymptotic Freedom of Rank 4 Tensor Group Field Theory*, in 29th International Colloquium on Group-Theoretical Methods in Physics (GROUP 29) Tianjin, China, August 20-26, 2012, 2012. arXiv: 1210.5490 [hep-th].

[12] ——, *Two and four-loop β-functions of rank 4 renormalizable tensor field theories*, Class. Quant. Grav., vol. 29, p. 235 011, 2012. arXiv: 1205.5513 [hep-th].

[13] ——, *On the finite amplitudes for open graphs in Abelian dynamical colored BoulatovOoguri models*, J. Phys., vol. A46, p. 402 002, 2013. arXiv: 1307.8299 [hep-th].

[14] ——, *Renormalizable Models in Rank $d \geq 2$ Tensorial Group Field Theory*, Commun. Math. Phys., vol. 332, pp. 117–188, 2014. arXiv: 1306.1201 [hep-th].

[15] J. Ben Geloun and E. R. Livine, *Some classes of renormalizable tensor models*, J. Math. Phys., vol. 54, p. 082 303, 2013. arXiv: 1207.0416 [hep-th].

[16] J. Ben Geloun, J. Magnen and V. Rivasseau, *Bosonic colored group field theory*, Eur. Phys. J., vol. C70, pp. 1119–1130, 2010. arXiv: 0911.1719 [hep-th].

[17] J. Ben Geloun and S. Ramgoolam, *Counting tensor model observables and branched covers of the 2-Sphere*, 2013. arXiv: 1307.6490 [hep-th].

[18] J. Ben Geloun and V. Rivasseau, *A renormalizable 4-Dimensional tensor field theory*, Commun. Math. Phys., vol. 318, pp. 69–109, 2013. arXiv: 1111.4997 [hep-th].

[19] J. Ben Geloun and D. O. Samary, *3D tensor field theory: renormalization and one-loop β-functions*, Ann. H. Poincaré, vol. 14, pp. 1599–1642, 2013. arXiv: 1201.0176 [hep-th].

[20] J. Ben Geloun and R. Toriumi, *Parametric representation of rank d tensorial group field theory: Abelian models with kinetic term $\sum_s |p_s| + \mu$*, J. Math. Phys., vol. 56, no. 9, p. 093 503, 2015. arXiv: 1409.0398 [hep-th].

[21] D. Benedetti, J. Ben Geloun and D. Oriti, *Functional Renormalisation Group Approach for Tensorial Group Field Theory: a Rank-3 Model*, JHEP, vol. 03, p. 084, 2015. arXiv: 1411.3180 [hep-th].

[22] D. Benedetti and R. Gurau, *Phase transition in dually weighted colored tensor models*, Nucl. Phys., vol. B855, pp. 420–437, 2012. arXiv: 1108.5389 [hep-th].

[23] ——, *Symmetry breaking in tensor models*, Phys. Rev., vol. D92, no. 10, p. 104 041, 2015. arXiv: 1506.08542 [hep-th].

[24] F. Bergeron, *Combinatorial species and tree-like structures*. Cambridge: Cambridge University Press, 1998, vol. 67.

[25] V. Bonzom, *Multicritical tensor models and hard dimers on spherical random lattices*, Phys. Lett., vol. A377, pp. 501–506, 2013. arXiv: 1201.1931 [hep-th].

[26] ——, *New 1/N expansions in random tensor models*, JHEP, vol. 1306, p. 062, 2013. arXiv: 1211.1657 [hep-th].

[27] ——, *Revisiting random tensor models at large N via the Schwinger-Dyson equations*, JHEP, vol. 1303, p. 160, 2013. arXiv: 1208.6216 [hep-th].

[28] V. Bonzom and F. Combes, *Tensor models from the viewpoint of matrix models: the case of loop models on random surfaces*, 2013. arXiv: 1304.4152 [hep-th].

[29] V. Bonzom and H. Erbin, *Coupling of hard dimers to dynamical lattices via random tensors*, J. Stat. Mech., vol. 1209, P09009, 2012. arXiv: 1204.3798 [cond-mat.stat-mech].

[30] V. Bonzom, R. Gurau, A. Riello and V. Rivasseau, *Critical behavior of colored tensor models in the large N limit*, Nucl. Phys., vol. B853, pp. 174–195, 2011. arXiv: `1105.3122 [hep-th]`.

[31] V. Bonzom, R. Gurau and V. Rivasseau, *Random tensor models in the large N limit: Uncoloring the colored tensor models*, Phys. Rev., vol. D85, p. 084 037, 2012. arXiv: `1202.3637 [hep-th]`.

[32] ——, *The Ising model on random lattices in arbitrary dimensions*, Phys. Lett., vol. B711, pp. 88–96, 2012. arXiv: `1108.6269 [hep-th]`.

[33] V. Bonzom, R. Gurau, J. P. Ryan and A. Tanasa, *The double scaling limit of random tensor models*, JHEP, vol. 1409, p. 051, 2014. arXiv: `1404.7517 [hep-th]`.

[34] D. V. Boulatov, *A model of three-dimensional lattice gravity*, Mod. Phys. Lett., vol. A7, pp. 1629–1646, 1992. arXiv: `hep-th/9202074 [hep-th]`.

[35] D. V. Boulatov and V. A. Kazakov, *The Ising model on random planar lattice: the structure of phase transition and the exact critical exponents*, Phys. Lett., vol. B186, p. 379, 1987.

[36] E. Brezin, M. R. Douglas, V. Kazakov and S. H. Shenker, *The Ising model coupled to 2-D gravity: a nonperturbative analysis*, Phys. Lett., vol. B237, p. 43, 1990.

[37] E. Brezin, C. Itzykson, G. Parisi and J. B. Zuber, *Planar diagrams*, Commun. Math. Phys., vol. 59, p. 35, 1978.

[38] E. Brezin and V. A. Kazakov, *Exactly solvable field theories of closed strings*, Phys. Lett., vol. B236, pp. 144–150, 1990.

[39] E. Brezin, J. C. Le Guillou and J. Zinn-Justin, *Perturbation theory at large order. 1. The phi**2N interaction*, Phys. Rev., vol. D15, pp. 1544–1557, 1977.

[40] P. W. Brouwer and C. W. J. Beenakker, *Diagrammatic method of integration over the unitary group, with applications to quantum transport in mesoscopic systems*, J. Math. Phys., vol. 37, p. 4904, 1996. arXiv: `9604059 [cond-mat]`.

[41] S. Carrozza, *Discrete Renormalization Group for SU(2) Tensorial Group Field Theory*, Ann. Inst. Henri Poincaré Comb. Phys. Interact., vol. 2, pp. 49–112, 2015. arXiv: `1407.4615 [hep-th]`.

[42] ——, *Group field theory in dimension* $4-\epsilon$, Phys. Rev., vol. D91, no. 6, p. 065 023, 2015. arXiv: `1411.5385 [hep-th]`.

[43] S. Carrozza and D. Oriti, *Bounding bubbles: the vertex representation of 3d group field theory and the suppression of pseudo-manifolds*, Phys. Rev., vol. D85, p. 044 004, 2012. arXiv: `1104.5158 [hep-th]`.

[44] ——, *Bubbles and jackets: new scaling bounds in topological group field theories*, JHEP, vol. 1206, p. 092, 2012. arXiv: `1203.5082 [hep-th]`.

[45] S. Carrozza, D. Oriti and V. Rivasseau, *Renormalization of a SU(2) Tensorial Group Field Theory in Three Dimensions*, Commun. Math. Phys., vol. 330, pp. 581–637, 2014. arXiv: `1303.6772 [hep-th]`.

[46] ——, *Renormalization of Tensorial Group Field Theories: Abelian U(1) Models in Four Dimensions*, Commun. Math. Phys., vol. 327, pp. 603–641, 2014. arXiv: 1207.6734 [hep-th].

[47] G. Chapuy, M. Marcus and G. Schaeffer, *A bijection for rooted maps on orientable surfaces*, SIAM J. Discrete Math., vol. 23, no. 3, pp. 1587–1611, 2009. arXiv: 0712.3649 [math.CO].

[48] B. Collins, *Moments and cumulants of polynomial random variables on unitary groups, the Itzykson-Zuber integral, and free probability*, Int. Math. Res. Not., vol. 17, p. 953, 2003. arXiv: 0205010 [math-ph].

[49] B. Collins and P. Sniady, *Integration with respect to the Haar measure on unitary, orthogonal and symplectic group*, Commun. Math. Phys., vol. 264, p. 773, 2006. arXiv: 0402073 [math-ph].

[50] R. Cori and G. Schaeffer, *Description trees and Tutte formulas*, Theor. Comput. Sci., vol. 292, no. 1, pp. 165–183, 2003.

[51] S. Dartois, R. Gurau and V. Rivasseau, *Double scaling in tensor models with a quartic interaction*, JHEP, vol. 1309, p. 088, 2013. arXiv: 1307.5281 [hep-th].

[52] S. Dartois, V. Rivasseau and A. Tanasa, *The $1/N$ expansion of multi-orientable random tensor models*, Ann. H. Poincaré, vol. 15, pp. 965–984, 2014. arXiv: 1301.1535 [hep-th].

[53] F. David, *Planar diagrams, two-dimensional lattice gravity and surface models*, Nucl. Phys., vol. B257, p. 45, 1985.

[54] ——, *Conformal field theories coupled to 2D gravity in the conformal gauge*, Mod. Phys. Lett., vol. A3, p. 1651, 1988.

[55] ——, *Simplicial quantum gravity and random lattices*, in Gravitation and quantizations. Proceedings, 57th Session of the Les Houches Summer School in Theoretical Physics, NATO Advanced Study Institute, Les Houches, France, July 5 - August 1, 1992, 1992. arXiv: hep-th/9303127 [hep-th].

[56] R. De Pietri and C. Petronio, *Feynman diagrams of generalized matrix models and the associated manifolds in dimension 4*, J. Math. Phys., vol. 41, pp. 6671–6688, 2000. arXiv: gr-qc/0004045 [gr-qc].

[57] T. Delepouve and R. Gurau, *Phase Transition in Tensor Models*, JHEP, vol. 06, p. 178, 2015. arXiv: 1504.05745 [hep-th].

[58] T. Delepouve, R. Gurau and V. Rivasseau, *Universality and Borel Summability of Arbitrary Quartic Tensor Models*, Annales de l'Institut Henri Poincaré - Probabilités et Statistiques, vol. 52, no. 2, pp. 821–848, 2016. arXiv: 1403.0170 [hep-th].

[59] P. Di Francesco, P. H. Ginsparg and J. Zinn-Justin, *2-D Gravity and random matrices*, Phys. Rept., vol. 254, pp. 1–133, 1995. arXiv: hep-th/9306153 [hep-th].

[60] R. Dijkgraaf, H. L. Verlinde and E. P. Verlinde, *Loop equations and Virasoro constraints in nonperturbative 2-D quantum gravity*, Nucl. Phys., vol. B348, pp. 435–456, 1991.

[61] M. Disertori, R. Gurau, J. Magnen and V. Rivasseau, *Vanishing of beta function of non commutative Phi**4(4) theory to all orders*, Phys. Lett., vol. B649, pp. 95–102, 2007. arXiv: hep-th/0612251 [hep-th].

[62] J. Distler and H. Kawai, *Conformal field theory and 2D quantum gravity or who's afraid of Joseph Liouville?* Nucl. Phys., vol. B321, p. 509, 1989.

[63] M. R. Douglas and S. H. Shenker, *Strings in less than one-dimension*, Nucl. Phys., vol. B335, p. 635, 1990.

[64] B. Duplantier, *Conformal random geometry*. Amsterdam: Elsevier Academic Press, 2006, pp. 101–217, Les Houches, Session LXXXIII, Mathematical Statistical Physics.

[65] L. Erdős, *Universality of wigner random matrices: a survey of recent results*, Russian Mathematical Surveys, vol. 66, no. 3, p. 507, 2011. arXiv: 1004.0861 [math-ph].

[66] B. Eynard, *Topological expansion for the 1-Hermitian matrix model correlation functions*, JHEP, vol. 0411, p. 031, 2004. arXiv: hep-th/0407261 [hep-th].

[67] ——, *Another algebraic variational principle for the spectral curve of matrix models*, 2014. arXiv: 1407.8324 [hep-th].

[68] B. Eynard and N. Orantin, *Invariants of algebraic curves and topological expansion*, Commun. Num. Theor. Phys., vol. 1, pp. 347–452, 2007. arXiv: math-ph/0702045 [math-ph].

[69] M. Ferri and C. Gagliardi, *Crystallisation moves*, Pacific J. of Math., vol. 100, p. 1, 1982.

[70] M. Ferri, C. Gagliardi and L. Grasselli, *A graph-theoretical representation of PL-manifolds – A survey on crystallizations*, Aequationes Mathematicae, vol. 31, p. 121, 1986.

[71] P. Flajolet and R. Sedgewick, *Analytic combinatorics*. Cambridge: Cambridge University Press, 2009.

[72] M. Fukuma, H. Kawai and R. Nakayama, *Continuum Schwinger-Dyson equations and universal structures in two-dimensional quantum gravity*, Int. J. Mod. Phys., vol. A6, pp. 1385–1406, 1991.

[73] E. Fusy and A. Tanasa, *Asymptotic expansion of the multi-orientable random tensor model*, The electronic journal of combinatorics, vol. 22, pp. 1–52, 1 2015. arXiv: 1408.5725 [math.CO].

[74] I. M. Gelfand, M. M. Kapranov and A. V. Zelevinsky, *Discriminants, resultants and multidimensional determinants*. Boston: Birkhauser, 1994.

[75] S. L. Glashow, *Partial symmetries of weak interactions*, Nucl. Phys., vol. 22, pp. 579–588, 1961.

[76] J. Glimm and A. Jaffe, *Quantum physics. A functional integral point of view*. New York: Springer, 1987.

[77] M. H. Goroff and A. Sagnotti, *The ultraviolet behavior of Einstein gravity*, Nucl. Phys., vol. B266, p. 709, 1986.

[78] D. J. Gross and A. A. Migdal, *Nonperturbative two-dimensional quantum gravity*, Phys. Rev. Lett., vol. 64, p. 127, 1990.

[79] D. J. Gross and F. Wilczek, *Asymptotically Free Gauge Theories. 1*, Phys. Rev., vol. D8, pp. 3633–3652, 1973.

[80] ——, *Ultraviolet behavior of nonabelian gauge theories*, Phys. Rev. Lett., vol. 30, pp. 1343–1346, 1973.

[81] ——, *Asymptotically free gauge theories. 2*, Phys. Rev., vol. D9, pp. 980–993, 1974.

[82] M. Gross, *Tensor models and simplicial quantum gravity in > 2-D*, Nucl. Phys. Proc. Suppl., vol. 25A, pp. 144–149, 1992.

[83] H. Grosse and R. Wulkenhaar, *Renormalization of phi**4 theory on noncommutative R**4 in the matrix base*, Commun. Math. Phys., vol. 256, pp. 305–374, 2005. arXiv: `hep-th/0401128 [hep-th]`.

[84] ——, *Progress in solving a noncommutative quantum field theory in four dimensions*, 2009. arXiv: `0909.1389 [hep-th]`.

[85] ——, *Construction of the Φ^4_4-quantum field theory on noncommutative Moyal space*, RIMS Kokyuroku, vol. 1904, pp. 67–105, 2013. arXiv: `1402.1041 [math-ph]`.

[86] ——, *Solvable limits of a 4D noncommutative QFT*, 2013. arXiv: `1306.2816 [math-ph]`.

[87] ——, *Solvable 4D noncommutative QFT: phase transitions and quest for reflection positivity*, 2014. arXiv: `1406.7755 [hep-th]`.

[88] A. Guionnet, G. W. Anderson and O. Zeitouni, *An Introduction to random matrices.* Cambridge: Cambridge University Press, 2009.

[89] A. Guionnet and O. Zeitouni, *Concentration of the spectral measure for large matrices*, Electron. Comm. Probab, vol. 5, pp. 119–136, 2000.

[90] R. Gurau, *Lost in translation: topological singularities in group field theory*, Class. Quant. Grav., vol. 27, p. 235 023, 2010. arXiv: `1006.0714 [hep-th]`.

[91] ——, *Topological graph polynomials in colored group field theory*, Ann. H. Poincaré, vol. 11, pp. 565–584, 2010. arXiv: `0911.1945 [hep-th]`.

[92] ——, *A generalization of the Virasoro algebra to arbitrary dimensions*, Nucl. Phys., vol. B852, pp. 592–614, 2011. arXiv: `1105.6072 [hep-th]`.

[93] ——, *Colored group field theory*, Commun. Math. Phys., vol. 304, pp. 69–93, 2011. arXiv: `0907.2582 [hep-th]`.

[94] ——, *The 1/N expansion of colored tensor models*, Ann. H. Poincaré, vol. 12, pp. 829–847, 2011. arXiv: `1011.2726 [gr-qc]`.

[95] ——, *The double scaling limit in arbitrary dimensions: a toy model*, Phys. Rev., vol. D84, p. 124 051, 2011. arXiv: `1110.2460 [hep-th]`.

[96] ——, *The complete 1/N expansion of colored tensor models in arbitrary dimension*, Ann. H. Poincaré, vol. 13, pp. 399–423, 2012. arXiv: `1102.5759 [gr-qc]`.

[97] ——, *The Schwinger Dyson equations and the algebra of constraints of random tensor models at all orders*, Nucl. Phys., vol. B865, pp. 133–147, 2012. arXiv: 1203.4965 [hep-th].

[98] ——, *The 1/N expansion of tensor models beyond perturbation theory*, Commun. Math. Phys., vol. 330, pp. 973–1019, 2014. arXiv: 1304.2666 [math-ph].

[99] ——, *Universality for Random Tensors*, Ann. Inst. H. Poincare Probab. Statist., vol. 50, no. 4, pp. 1474–1525, 2014. arXiv: 1111.0519 [math.PR].

[100] R. Gurau and T. Krajewski, *Analyticity results for the cumulants in a random matrix model*, 2014. arXiv: 1409.1705 [math-ph].

[101] R. Gurau, J. Magnen, V. Rivasseau and F. Vignes-Tourneret, *Renormalization of non-commutative phi(4)**4 field theory in x space*, Commun. Math. Phys., vol. 267, pp. 515–542, 2006. arXiv: hep-th/0512271 [hep-th].

[102] R. Gurau and V. Rivasseau, *The 1/N expansion of colored tensor models in arbitrary dimension*, Europhys. Lett., vol. 95, p. 50 004, 2011. arXiv: 1101.4182 [gr-qc].

[103] ——, *The Multiscale Loop Vertex Expansion*, Ann. H. Poincaré, vol. 16, no. 8, pp. 1869–1897, 2015. arXiv: 1312.7226 [math-ph].

[104] R. Gurau, V. Rivasseau and A. Sfondrini, *Renormalization: an advanced overview*, 2014. arXiv: 1401.5003 [hep-th].

[105] R. Gurau and J. P. Ryan, *Colored tensor models - a review*, SIGMA, vol. 8, p. 020, 2012. arXiv: 1109.4812 [hep-th].

[106] ——, *Melons are branched polymers*, Ann. H. Poincaré, vol. 15, no. 11, pp. 2085–2131, 2014. arXiv: 1302.4386 [math-ph].

[107] R. Gurau and G. Schaeffer, *Regular colored graphs of positive degree*, 2013. arXiv: 1307.5279 [math.CO].

[108] G. 't Hooft, *A planar diagram theory for strong interactions*, Nucl. Phys., vol. B72, p. 461, 1974.

[109] G. 't Hooft and M. J. G. Veltman, *Regularization and Renormalization of Gauge Fields*, Nucl. Phys., vol. B44, pp. 189–213, 1972.

[110] ——, *One loop divergencies in the theory of gravitation*, Ann. Poincaré Phys. Theor., vol. A20, pp. 69–94, 1974.

[111] T. Jonsson and J. F. Wheater, *The spectral dimension of the branched polymer phase of two-dimensional quantum gravity*, Nucl. Phys., vol. B515, pp. 549–574, 1998. arXiv: hep-lat/9710024 [hep-lat].

[112] W. Kaminski, D. Oriti and J. P. Ryan, *Towards a double-scaling limit for tensor models: probing sub-dominant orders*, New J. Phys., vol. 16, p. 063 048, 2014. arXiv: 1304.6934 [hep-th].

[113] V. A. Kazakov, *Bilocal regularization of models of random surfaces*, Phys. Lett., vol. B150, pp. 282–284, 1985.

[114] ——, *Ising model on a dynamical planar random lattice: Exact solution*, Phys. Lett., vol. A119, pp. 140–144, 1986.

[115] ——, *The appearance of matter fields from quantum fluctuations of 2D gravity*, Mod. Phys. Lett., vol. A4, p. 2125, 1989.

[116] V. G. Knizhnik, A. M. Polyakov and A. B. Zamolodchikov, *Fractal structure of 2D quantum gravity*, Mod. Phys. Lett., vol. A3, p. 819, 1988.

[117] D. Kozlov, *Combinatorial algebraic topology*, ser. Algorithms and Computation in Mathematics. New York: Springer, 2008.

[118] T. Krajewski, *Schwinger-Dyson Equations in Group Field Theories of Quantum Gravity*, in 29th International Colloquium on Group-Theoretical Methods in Physics (GROUP 29) Tianjin, China, August 20-26, 2012, 2012. arXiv: 1211.1244 [math-ph].

[119] J. B. Kruskal, *On the shortest spanning subtree of a graph and the traveling salesman problem*, in Proceedings of the American Mathematical Society, 7, 1956.

[120] J. F. Le Gall, *The topological structure of scaling limits of large planar maps*, Inventiones Mathematicae, vol. 169, no. 3, pp. 621–670, 2007. arXiv: 0607567 [math.PR].

[121] ——, *Geodesics in large planar maps and in the Brownian map*, Acta Mathematica, vol. 205, no. 2, pp. 287–360, 2010. arXiv: 0804.3012 [math.PR].

[122] ——, *Uniqueness and universality of the Brownian map*, The Ann. of Probab., vol. 41, no. 4, pp. 2880–2960, 2013. arXiv: 1105.4842 [math.PR].

[123] J. Magnen, K. Noui, V. Rivasseau and M. Smerlak, *Scaling behaviour of three-dimensional group field theory*, Class. Quant. Grav., vol. 26, p. 185 012, 2009. arXiv: 0906.5477 [hep-th].

[124] J. Magnen and V. Rivasseau, *Constructive phi**4 field theory without tears*, Ann. H. Poincaré, vol. 9, pp. 403–424, 2008. arXiv: 0706.2457 [math-ph].

[125] Y. Makeenko, *Loop equations and Virasoro constraints in matrix models*, in 25th International Ahrenshoop Symposium on the Theory of Elementary Particles Gosen, Germany, September 23-26, 1991, 1991. arXiv: hep-th/9112058 [hep-th].

[126] O. Marchal, B. Eynard and M. Bergere, *The sine-law gap probability, Painlevé 5, and asymptotic expansion by the topological recursion*, 2013. arXiv: 1311.3217 [math-ph].

[127] M. Marino, *Lectures on non-perturbative effects in large N gauge theories, matrix models and strings*, Fortsch. Phys., vol. 62, pp. 455–540, 2014. arXiv: 1206.6272 [hep-th].

[128] M. L. Mehta, *Random matrices*. Amsterdam: Elsevier Academic Press, 2004.

[129] L. Nachbin and L. Bechtolsheim, *The Haar integral*, ser. University series in higher mathematics. Van Nostrand, 1965.

[130] A. Nica and R. Speicher, *Lectures on the combinatorics of free probability (London Mathematical Society Lecture Note Series)*. Cambridge: Cambridge University Press, 2006.

[131] K. H. O'Brien and J. B. Zuber, *A note on U(N) integrals in the large N limit*, Phys. Lett., vol. B144, p. 407, 1984.

[132] P. J. Olver, *Classical invariant theory*. London: London Mathematical Society, 1999.

[133] H. Ooguri, *Topological lattice models in four-dimensions*, Mod. Phys. Lett., vol. A7, pp. 2799–2810, 1992. arXiv: `hep-th/9205090 [hep-th]`.

[134] D. Oriti, *The microscopic dynamics of quantum space as a group field theory*, in Proceedings, Foundations of Space and Time: Reflections on Quantum Gravity, 2011, pp. 257–320. arXiv: `1110.5606 [hep-th]`.

[135] ——, *Group field theory and loop quantum gravity*, 2014. arXiv: `1408.7112 [gr-qc]`.

[136] D. Oriti, J. P. Ryan and J. Thurigen, *Group field theories for all loop quantum gravity*, New J. Phys., vol. 17, no. 2, p. 023042, 2015. arXiv: `1409.3150 [gr-qc]`.

[137] M. Petro, *Moduli Spaces of Riemann Surfaces*. Wisconsin: University of Wisconsin–Madison, 2008.

[138] M. Pezzana, *Sulla struttura topologica delle varietà compatte*, Atti Sem. Mat. Fis. Univ. Modena, vol. 23, pp. 269–277, 1974.

[139] H. D. Politzer, *Reliable Perturbative Results for Strong Interactions?* Phys. Rev. Lett., vol. 30, pp. 1346–1349, 1973.

[140] T. Regge, *General relativity without coordinates*, Il Nuovo Cimento Series 10, vol. 19, no. 3, pp. 558–571, 1961.

[141] V. Rivasseau, *Constructive matrix theory*, JHEP, vol. 0709, p. 008, 2007. arXiv: `0706.1224 [hep-th]`.

[142] ——, *Constructive field theory in zero dimension*, Adv.Math.Phys., vol. 2010, p. 180159, 2010. arXiv: `0906.3524 [math-ph]`.

[143] ——, *Quantum gravity and renormalization: the tensor track*, AIP Conf. Proc., vol. 1444, pp. 18–29, 2011. arXiv: `1112.5104 [hep-th]`.

[144] ——, *The Tensor Track: an Update*, in 29th International Colloquium on Group-Theoretical Methods in Physics (GROUP 29) Tianjin, China, August 20-26, 2012, 2012. arXiv: `1209.5284 [hep-th]`.

[145] ——, *The Tensor Theory Space*, Fortsch. Phys., vol. 62, pp. 835–840, 2014. arXiv: `1407.0284 [hep-th]`.

[146] ——, *The Tensor Track, III*, Fortsch. Phys., vol. 62, pp. 81–107, 2014. arXiv: `1311.1461 [hep-th]`.

[147] V. Rivasseau and Z. Wang, *How to Resum Feynman Graphs*, Ann. H. Poincaré, vol. 15, no. 11, pp. 2069–2083, 2014. arXiv: `1304.5913 [math-ph]`.

[148] J. P. Ryan, *Tensor models and embedded Riemann surfaces*, Phys. Rev., vol. D85, p. 024010, 2012. arXiv: `1104.5471 [gr-qc]`.

[149] A. Salam, *Weak and Electromagnetic Interactions*, Conf. Proc., vol. C680519, pp. 367–377, 1968.

[150] D. O. Samary, *Beta functions of $U(1)^d$ gauge invariant just renormalizable tensor models*, Phys. Rev., vol. D88, p. 105003, 2013. arXiv: `1303.7256 [hep-th]`.

[151] ——, *Closed equations of the two-point functions for tensorial group field theory*, Class. Quant. Grav., vol. 31, p. 185005, 2014. arXiv: `1401.2096 [hep-th]`.

[152] D. O. Samary and F. Vignes-Tourneret, *Just Renormalizable TGFT's on $U(1)^d$ with Gauge Invariance*, Commun. Math. Phys., vol. 329, pp. 545–578, 2014. arXiv: `1211.2618 [hep-th]`.

[153] N. Sasakura, *Tensor model for gravity and orientability of manifold*, Mod. Phys. Lett., vol. A6, pp. 2613–2624, 1991.

[154] ——, *Super tensor models, super fuzzy spaces and super n-ary transformations*, Int. J. Mod. Phys., vol. A26, pp. 4203–4216, 2011. arXiv: `1106.0379 [hep-th]`.

[155] ——, *Tensor models and hierarchy of n-ary algebras*, Int. J. Mod. Phys., vol. A26, pp. 3249–3258, 2011. arXiv: `1104.5312 [hep-th]`.

[156] G. Schaeffer, *Bijective census and random generation of Eulerian planar maps with prescribed degrees*, Electr. J. Comb., vol. 4, no. 1, 1997.

[157] A. D. Sokal, *An improvement of Watson's theorem on Borel summability*, J. Math. Phys., vol. 21, pp. 261–263, 1980.

[158] T. Tao and V. Vu, *Random matrices: universality of local eigenvalue statistics*, Acta Mathematica, vol. 206, p. 127, 2009. arXiv: `0906.0510 [math.PR]`.

[159] D. Voiculescu, *Symmetries of some reduced free product C^*-algebras*. New York: Springer-Verlag, 1983.

[160] ——, *Limit laws for random matrices and free products*, Invent. Math., vol. 104, p. 201, 1991.

[161] D. Voiculescu, K. Dykema and A. Nica, *Free random variables*. AMS: CRM Monograph Series, 1992.

[162] S. Weinberg, *A model of leptons*, Phys. Rev. Lett., vol. 19, pp. 1264–1266, 1967.

[163] D. Weingarten, *Asymptotic behavior of group integrals in the limit of infinite rank*, J. Math. Phys., vol. 19, p. 999, 1978.

[164] E. Wigner, *Characteristic vectors of bordered matrices with infinite dimensions*, Ann. Of Math., vol. 62, p. 548, 1955.

[165] J. Wishart, *The generalised product moment distribution in samples from a normal multivariate population*, Biometrika, vol. 20, p. 32, 1928.

[166] J. Zinn-Justin, *Perturbation series at large orders in quantum mechanics and field theories: Application to the problem of resummation*, Phys. Rept., vol. 70, p. 109, 1981.

[167] P. Zinn-Justin and J. B. Zuber, *On some integrals over the U(N) unitary group and their large N limit*, J. Phys., vol. A36, pp. 3173–3194, 2003. arXiv: `math-ph/0209019 [math-ph]`.

Index

The manufacturer's authorised representative in the EU for product safety is Oxford University Press España S.A. of el Parque Empresarial San Fernando de Henares, Avenida de Castilla, 2 – 28830 Madrid (www.oup.es/en or product.safety@oup.com). OUP España S.A. also acts as importer into Spain of products made by the manufacturer.

www.ingramcontent.com/pod-product-compliance
Lightning Source LLC
LaVergne TN
LVHW062206200926
848484LV00006B/1336

* 9 7 8 0 1 9 8 7 8 7 9 3 8 *